Recent Advances in Oxidative Stress Associated Chronic Diseases Volume 1

Kaviyarasu Kasinathan ·
Azhaguchamy Muthukumaran
Editors

Recent Advances in Oxidative Stress Associated Chronic Diseases Volume 1

A Review of The Health Benefits and Risks of The Substance

Springer

Editors
Kaviyarasu Kasinathan
The University of South Africa
Roodepoort, Gauteng, South Africa

Azhaguchamy Muthukumaran
Department of Biotechnology
School of Biochemical and Processing Engineering
Kalasalingam Academy of Research and Education
Krishnankoil, Tamil Nadu, India

ISBN 978-981-96-9669-7 ISBN 978-981-96-9670-3 (eBook)
https://doi.org/10.1007/978-981-96-9670-3

This Springer imprint is published by the registered company Springer Nature Singapore Pte Ltd.
The registered company address is: 152 Beach Road, #21-01/04 Gateway East, Singapore 189721, Singapore

Contents

Contributors

D. Abilash Gene Cloning and Technology Lab, Department of Biotechnology, SBST, VIT Vellore, Vellore, Tamil Nadu, India

L. Anandaraj Department of Physics, Sacred Heart College (Autonomous), (Affiliated to Thiruvalluvar University), Tirupattur, Tamil Nadu, India

Vijaya Chitra Arumugam Department of Microbiology, Sri Ramakrishnma College of Arts & Science for Women, Coimbatore, Tamil Nadu, India

R. Babujanarthanam Department of Biotechnology, Thiruvalluvar University, Serkkadu, Vellore, Tamil Nadu, India

Ayesha Baje Babu Regeneration and Stem Cell Biology Lab, Centre for Molecular and Nanomedical Sciences, International Research Centre, Sathyabama Institute of Science and Technology, Chennai, Tamil Nadu, India

Sathyabama Balaji Department of Physics, SSN Research Centre, Sri Sivasubramaniya Nadar College of Engineering, Chennai, Tamil Nadu, India

Nivethitha Chandran Regeneration and Stem Cell Biology Lab, Centre for Molecular and Nanomedical Sciences, International Research Centre, Sathyabama Institute of Science and Technology, Chennai, Tamil Nadu, India

Subashini Chandrasekaran Regeneration and Stem Cell Biology Lab, Centre for Molecular and Nanomedical Sciences, International Research Centre, Sathyabama Institute of Science and Technology, Chennai, Tamil Nadu, India

Maria Cynthiya Charles Raj Regeneration and Stem Cell Biology Lab, Centre for Molecular and Nanomedical Sciences, International Research Centre, Sathyabama Institute of Science and Technology, Chennai, Tamil Nadu, India

Pandian Chithamparam Department of Biotechnology, Kalasalingam Academy of Research and Education, Krishnankoil, Tamil Nadu, India

S. Diviya Priya Department of Physics, Sacred Heart College (Autonomous), (Affiliated to Thiruvalluvar University), Tirupattur, Tamil Nadu, India

R. Ganesamoorthy Department of Chemistry, Vinayaka Mission's Kirupananda Variyar Arts and Science College, Vinayaka Mission's Research Foundation (DU), Salem, Tamil Nadu, India

Mithra Jayan Department of Biotechnology, Karpagam Academy of Higher Education, Coimbatore, Tamil Nadu, India

Meenakshi Kaniyur Chandrasekaran Department of Biochemistry, Karpagam Academy of Higher Education, Coimbatore, Tamil Nadu, India

Mageshwari Kannan Regeneration and Stem Cell Biology Lab, Centre for Molecular and Nanomedical Sciences, International Research Centre, Sathyabama Institute of Science and Technology, Chennai, Tamil Nadu, India

Ramanathan Karuppasamy Department of Biotechnology, School of Bio Sciences and Technology, Vellore Institute of Technology, Vellore, India

K. Kasthuri Department of Biochemistry, Meenakshi Medical College Hospital and Research Institute, Meenakshi Academy of Higher Education and Research, Kanchipuram, Tamil Nadu, India

Harsath Ram Govindan Radha Krishnan Department of Biotechnology, Kalasalingam Academy of Research and Education, Krishnankoil, Tamil Nadu, India

P. Madhan Kumar Department of Biotechnology, Thiruvalluvar University, Serkkadu, Vellore, Tamil Nadu, India

Sadananda Mal Department of Biotechnology, School of Bio Sciences and Technology, Vellore Institute of Technology, Vellore, India

Asma Shereen Mohammed Yaseen Regeneration and Stem Cell Biology Lab, Centre for Molecular and Nanomedical Sciences, International Research Centre, Sathyabama Institute of Science and Technology, Chennai, Tamil Nadu, India

Kamalaharshini Mohanarangam Regeneration and Stem Cell Biology Lab, Centre for Molecular and Nanomedical Sciences, International Research Centre, Sathyabama Institute of Science and Technology, Chennai, Tamil Nadu, India

Rathi Muthaiyan Ahalliya Department of Biochemistry, Karpagam Academy of Higher Education, Coimbatore, Tamil Nadu, India;
Department of Biotechnology, Karpagam Academy of Higher Education, Coimbatore, Tamil Nadu, India

Azhaguchamy Muthukumaran Department of Biotechnology, Kalasalingam Academy of Research and Education, Krishnankoil, Tamil Nadu, India

Smriti Narendra Kumar Regeneration and Stem Cell Biology Lab, Centre for Molecular and Nanomedical Sciences, International Research Centre, Sathyabama Institute of Science and Technology, Chennai, Tamil Nadu, India

Shweta Panchal Department of Biotechnology, School of Bio Sciences and Technology, Vellore Institute of Technology, Vellore, India

Muthu Senthil Pandian Department of Physics, SSN Research Centre, Sri Sivasubramaniya Nadar College of Engineering, Chennai, Tamil Nadu, India

Nirmaladevi Panneerselvam Department of Biotechnology, Karpagam Academy of Higher Education, Coimbatore, Tamil Nadu, India;
Department of Biochemistry, Karpagam Academy of Higher Education, Coimbatore, Tamil Nadu, India

R. Parameswari Department of Biotechnology, Thiruvalluvar University, Serkkadu, Vellore, Tamil Nadu, India

Varshni Premnath Department of Biotechnology, School of Bio Sciences and Technology, Vellore Institute of Technology, Vellore, Tamil Nadu, India

Harsath Ram Govindan Radhakrishnan Department of Biotechnology, Kalasalingam Academy of Research and Education, Krishnankoil, Tamilnadu, India

R. Ramesh Department of Physics, Sacred Heart College (Autonomous), (Affiliated to Thiruvalluvar University), Tirupattur, Tamil Nadu, India

Johnson Retnaraj Samuel Selvan Christyraj Regeneration and Stem Cell Biology Lab, Centre for Molecular and Nanomedical Sciences, International Research Centre, Sathyabama Institute of Science and Technology, Chennai, Tamil Nadu, India

S. Subhashini School of Physiotherapy, Dhanalakshmi Srinivasan University, Tiruchirappalli, Tamil Nadu, India

Renuka Suresh Department of Biotechnology, School of Bio Sciences and Technology, Vellore Institute of Technology, Vellore, India

Manikandan Vani Raju Department of Biochemistry, Karpagam Academy of Higher Education, Coimbatore, Tamil Nadu, India

Shanthi Veerappapillai Department of Biotechnology, School of Bio Sciences and Technology, Vellore Institute of Technology, Vellore, Tamil Nadu, India

Gopalakrishnan Velliyur Kanniappan Department of Physiology, Saveetha Medical College and Hospital, Saveetha Institute of Medical and Technical Sciences (SIMATS), Chennai, Tamil Nadu, India

Abbreviations

$(CH_3)_2S$	Dimethyl sulfide
$_1O^2$	Singlet oxygen
$^1\Delta g$	Singlet oxygen
3-MST	3-mercaptopyruvate sulfur transferase
$3O_2$	Reactive triplet ground state
4HB	4-hydroxybenzoate
4-HNE	4-hydroxy-2-nonenal
4HNE	4-hydroxy-2-nonenal
8-OHdG	8-hydroxy-2′-deoxyguanosine
8-oxo-dG	8-oxo-2'-deoxyguanosine
AAPH	2,2'-azobis(2-amidinopropane) dihydrochloride
AAT	Alpha-1 antitrypsin
AAT	Aspartate aminotransferase
AAV	Adeno-associated virus
ABA	Abscisic acid
ABC1	ABC transporter
ABTS	2,2'-azino-bis(3-ethylbenzothiazoline-6-sulfonic acid)
ABTS	3-ethylbenzothiazoline-6-sulfonic acid
ACOG	American college of obstetricians and gynecologists
ACPA	Anti-citrullinated protein antibodies
aCSF	Artificial cerebrospinal fluid
AD	Alzheimer's disease
ADMA	Asymmetric dimethyl arginine
ADME	Absorption, distribution, metabolism, and excretion
ADP	Adenosine diphosphate
AF	Atrial fibrillation

AGE	Advanced glycation end-product
AgNPs	Silver nanoparticles
ALA	Alpha-linolenic acid
ALD	Alcoholic liver disease
ALEs	Advanced lipoxidation end products
ALS	Amyotrophic lateral sclerosis
AMD	Age-related macular degeneration
AMR	Antimicrobial resistance
ANI	Average nucleotide identity
ANNs	Artificial neural networks
AOPPs	Advanced oxidation protein products
AO_X	Alternative oxidase
AP	Apurinic/apyrimidinic
APX	Ascorbate peroxidase
ARE	Antioxidant response element
AsA	Ascorbic acid
AT1	Angiotensin II receptor type 1
ATF6	Activating transcription factor 6
ATM	Ataxia–telangiectasia mutated
ATP	Adenosine triphosphate
AUC	Area under curve
AuNMs	Gold-based nanomaterials
Aβ	Amyloid-beta
BAC	Blood alcohol concentration
BBB	Blood–brain barrier
BER and NER	Excisions and nucleotides
BER	Base excision repair
bFGF	Basic fibroblast growth factor
BH4	Tetrahydrobiopterin
BHA	Butylated 4-hydroxyanisole
BHA	Butylated hydroxyl anisole
BHT	Butylated 4-hydroxytoluene
BMD	Bone mineral density
BOD	Biochemical oxygen demand
BP	Blood pressure
CAT	Catalase
CAT	Enzyme catalase
CBD	Cannabidiol
CBS	Cystathionine β-synthase
CCl_4	Carbon tetrachloride
CDC	Centres for disease control and prevention
CDK2	Cyclin-dependent kinase 2
CDKN1A, GADD45B, MAFF	Mitochondrial oxidative stress-related gene signature
CDO	Cysteine dioxygenase

CF	Cystic fibrosis
cGMP	3′,5′-cyclic guanosine monophosphate
CH_3SH	Methyl mercaptan
CHD	Coronary heart disease
CHIPS	Control of hypertension in pregnancy study
CHMs	Chinese herbal medicines
CIM	Critical illness myopathy
CKD	Chronic kidney disease
CKD/AKI	Chronic or acute
Cl^-	Chloride ions
CMD	Coronary microvascular dysfunction
CMs	Compromised cardiomyocytes
CNC	Cap "n" collar
−COOH	Carboxyl
COPD	Chronic obstructive pulmonary disease
COS	Carbonyl sulfide
COX	Cyclooxygenase
COX-2	Cyclooxygenase-2
CPDs	Cyclobutane pyrimidine dimers
CRC	Colorectal cancer
CRP and TNF-α	Inflammatory biomarkers
CRP	C-reactive protein
crtI	Phytoene desaturase
CSE	Cystathionine γ-lyase
CT	Computed tomography
CTGF	Connective tissue growth factor
CUD	Cannabis use disorder
CUPRAC	Cupric reducing antioxidant capacity
CVD	Cardiovascular disease
CYP450s	Cytochrome P450 monooxygenases
CYPs	Cytochrome P450s
CYS 34	Free thiol groups
CysSSH	Cysteine persulfide
cytochrome bc1 complex	Complex III
D_2	Ergocalciferol
D_3	Cholecalciferol
DAG	Diacylglycerol
DAMPs	Danger-related molecular patterns
DAS28	Disease activity score
DbCM	Diabetic cardiomyopathy
$DCFH_2$-DA	2′7′-Dichlorodihydrofluorescein Diacetate
DCW	Dry cell weight
ddsA gene	Decaprenyl diphosphate synthase gene
DHA	Dehydroascorbic acid
DHA	Docosahexaenoic acid

DHAR	Dehydroascorbate reductase
DHE	Dihydroethidium
DISC	Death-inducing signaling complex
DM	Diabetes mellitus
DMAPP	Dimethylallyl pyrophosphate
DMARDs	Disease-modifying anti-rheumatic drugs
DMD	Mainly Duchenne
DMF	Dimethyl fumarate
DMS	Dimethyl sulfide
DNA	Deoxyribonucleic acid
DOX	Doxorubicin
DPCs	DNA–protein cross-links
DPP	Decaprenyl diphosphate
DPPH	2,2-diphenyl-1-picrylhydrazyl
dRP	5'-deoxyribose phosphate residue
DSB	Double-strand break
DSC	Differential scanning calorimetry
DTNB	5,5'-dithiobis-2-nitrobenzoic acid
Dxs	1-deoxy-D-xylulose-5-phosphate synthase
EA	Ellagic acid
EAMs	Extra-articular manifestations
EAs	Ergot alkaloids
ECG	Electrocardiogram
ECSOD	Extracellular fluids
EDHFs	Endothelium-derived hyperpolarizing
E-DNA	Electrochemical DNA
EDRFs	Endothelium-derived relaxing
EDTA	Ethylenediaminetetraacetic acid
EGCG	Epigallocatechin-3-gallate
EGF	Epidermal growth factor
EGFR	Epidermal growth factor receptor
ELEM	Leukoencephalomalacia
ELISA	Enzyme-linked immunosorbent assay
eNOS	Endothelial
eNOS	Endothelial nitric oxide synthase
EPA	Eicosapentaenoic acid
EPR	Extended producer responsibility
EPS	Exopolysaccharides
EPS	Microbial exopolysaccharides
EPSPS	5-enolpyruvylshikimate-3-phosphate synthase
EPSs	Extracellular polymeric substances
ER	Endoplasmic reticulum
ERK1/2	Extracellular signal-regulated protein kinases 1 and 2

ERKs	Transmit signals from the membrane to the nucleus, promoting either survival
ERT	Enzyme replacement therapy
ERTs	Enzyme replacement therapies
ES	Embryonic stem
ESKD	End-stage kidney disease
ESR	Erythrocyte sedimentation rate
ET	Electron transfer method
ETC	Electron transport chain
ETC	Mitochondrial electron transport chain
FA	Ferulic acid
FAD	Flavin adenine dinucleotide
Fdx	Ferredoxin
FDX–FTR–TRX	Ferredoxin–thioredoxin
Fe^{2+}-TPTZ	Ferrous form
Fe^{3+}-TPTZ	Ferric-tripyridyl triazine
FET	ROS form via forward
FEV1	Forced expiratory volume in one second
FFA	Free fatty acids
FGF	Fibroblast growth factor
FLS	Fibroblast-like synoviocytes
FMN	Flavin mononucleotide
FRAP	Ferric reducing antioxidant power
FRASC	Ferric reducing ability of plasma with stabilized ascorbate
FRs	Free radicals
FUM	Fumarase
FUM	Fumonisins
FVC	Forced vital capacity
GA	Gallic acid
GFP	Green fluorescent protein
GGPP	Geranylgeranyl diphosphate synthase
GI	Gastrointestinal
GOX	Glycolate oxidase
GPP	Geranyl pyrophosphate
GPx	Glutathione peroxidase
GR	Glutathione reductase
GS	Glutamine synthetase
GS^-	Thiolate anion
GSH	Glutathione
GSH	Glycine, cysteine, and glutamate assay
GSNO	Nitrosoglutathione
GSNOR	S-nitrosoglutathione reductase
GSSG	Glutathione disulfide
GSSG	Reduced and oxidized GSH

GSSH	Glutathione persulfide
GST	Glutathione S-transferase
H/R	Hypoxia and hypoxia-reoxygenation
H_2O_2	Hydrogen peroxide
H_2S	Hydrogen sulfide
HAS	Human serum albumin
HAT	Hydrogen atom transfer
HBV	Hepatitis B
HCC	Hepatocellular carcinoma
HCL	Hydrachloric acid
HCO_3^-	Bicarbonate
HCV	Hepatitis C
HD	Huntington's disease
HDAC	Histone deacetylase
HDACs	Inhibiting histone deacetylases
HDL	High-density lipoprotein
HDP	Hypertensive conditions of pregnancy
HFrEF	Heart failure with reduced ejection fraction
HGF or VEGF	Bioactive molecules such as growth factors
HHP	High hydrostatic pressure
HIF-1	Hypoxia-inducible factor-1
HIF-1α	Hypoxia-inducible factor 1α
HNE	4-hydroxy-2-nonenal
HNE	Hydro-mechanical neonatal encephalopathy
HO-1	Haem oxygenase-1
HO-1	Heme oxygenase-1
HOCl	Hydrochlorous acid
HOCl	Hypochlorous acid
HPLC	High-performance liquid chromatography
HSC	Hepatic stellate cell
HSF1	Heat shock factor 1
HSPs	Heat shock proteins
HTT	Genetic mutation in the huntingtin
HεdC	Heptanone-etheno-2′-deoxycytidine
IBD	Inflammatory bowel disease
IBS	Irritable bowel syndrome
IC50	Half-maximal inhibitory concentration
ICH	Intracerebral hemorrhage
IGF-I	Insulin-like growth factor-1
IHD	Ischemic heart disease
ILs	Iliotibialis
iNOS	Inducible
iNOS	Inducible nitric oxide synthase
IONs	Iron oxide nanoparticles

IPP	Isopentenyl pyrophosphate
IRE1	Inositol requiring enzyme-1
Itga-8	Integrin subunit alpha 8
IVC	Integrated virtual care
IVF	In vitro fertilization
JNK	Jun N-terminal kinase
K_1	Phylloquinone
K_2	Menaquinone
Kelch-like ECH-associated protein 1	Keap1
L-AA	L-ascorbic acid
LCAs	Life cycle assessments
LC–MS	Liquid chromatography–mass spectrometry
LCY-β	Lycopene β-cyclase
LDL	Low-density lipoprotein
LDPE	Low-density polyethylene
LOH	Harmful alcohols
LOOH	Lipid hydroperoxide
LOOH	Lowering lipid hydroperoxides
LOONO	Forming alkyl peroxynitrate
LOXs	Lipoxygenases
LPO	Lipid peroxidation
LPO	Lipoxygenase
LPS	Lipopolysaccharide
MAGE	Multiplex automated genome engineering
MAMPs/PAMPs	Microbes or pathogens' molecular patterns
MAO	Monoamine oxidase
MAPK/PI3K/AKT	Mitogen-activated protein kinase/phosphoinositide 3-kinase/protein kinase B
MAPKs	Mitogen-activated protein kinases
MCP-1	Monocyte chemotactic protein-1
MCPs	Monocyte chemoattractant proteins
MDA	Malondialdehyde
MDA	Malondialdehyde assay
MDA	The malondialdehyde
MD–CE/ME	Microdialysis–capillary electrophoresis/microchip electrophoresis
MDD	Major depressive disorder
MDHA	Mono-dehydroascorbate
MDHAR	Monodehydroascorbate reductase
MD–LC	Microdialysis–liquid chromatography
MDs	Muscular dystrophies
MeJA	Methyl jasmonate
MeOH	Methanol
MEP	2-C-methyl-D-erythritol 4-phosphate
METLIN	Metabolite and chemical entity database

MFC-Bs	Microbial fuel cell-based biosensors
mHtt	Misfolded mutant huntingtin protein
mHTT	Mutant huntingtin
MI	Maximum impulse
MI	Myocardial infarction
MIP	Macrophage inflammatory proteins
miRNAs	MicroRNAs
MLR	Monocyte-to-lymphocyte ratio
mM	Millimole
MMIF	Macrophage migration inhibitory factor
MM-LDL	Minimally oxidized LDL
MMPs	Matrix metalloproteinases
MnSOD	Mitochondrial matrix
MnSODs	Manganese superoxide dismutases
MOFs	Metal-organic frameworks
MPO	Myeloperoxidase
MPST	Mercaptopyruvate sulfurtransferase
mPTP	Mitochondrial permeability transition pore
MRI	Magnetic resonance imaging
mRNA	Messenger RNA
MRSA	Methicillin-resistant *Staphylococcus aureus*
MS	Multiple sclerosis
MSH	Mycothiol
Msr	Methionine sulfoxide reductase
MT	Metallothionein
mtDNA	Mitochondrial DNA
mtETC	Mitochondrial electron transport chain
mTOR	Mammalian target of rapamycin
MVA	Mevalonate
N_2O_3	Dinitrogen trioxide
NAC	N-acetylcysteine
N-acetyl-5-methoxytryptamine	Melatonin
NAC-N-	Acetylcysteine
NADH dehydrogenase	Complex I
NADPH	Nicotinamide adenine dinucleotide phosphate
NAFL	Non-alcoholic fatty liver
NAFLD	Non-alcoholic fatty liver disease
$NaHCO_3$	Sodium bicarbonate
NASH	Non-alcoholic steatohepatitis
NBT	Nitro blue tetrazolium
Nc	Neocuproine
N-DCs	Neutrophils transdifferentiate into dendritic cell-like phenotypes

NDG	Neurodegenerative
NDs	NADH dehydrogenases
NEG	Non-enzymatic glycosylation
NER	Nucleotide excision repair
NF-kB	Nuclear factor kappa B
NF-κB	Nuclear factor kappa-light-chain-enhancer of activated B cells
NF-κB	Nuclear factor-κB
NLR	Neutrophil-to-lymphocyte ratio
NNAT	Neuronatin
nNOS	Neuronal
nNOS	Neuronal NOS
NO–	Nitric oxide
NO^+	Nitroxyl (HNO), nitrosyl cations
NO_2^-	Nitrite
NO^{2+}	Nitronium ions
$NO_2^{\bullet}$	Nitrogen dioxide
NO_3^-	Nitrate
NOFNiR	Forming nitrite reductase
NOS	Nitric oxide synthase
NOS2	Nitrosative synthase-2
NOS^{-1}	NO synthase-like
NOX	NADPH oxidases
NOX	Oxidase
NOX2	Gp91phox
NOX2, NOX4	NADPH oxidases
NP-SH	Non-protein thiols
NR	Nitrate reductase
Nrf2	Nuclear factor erythroid 2-related factor 2
NROs	Non-radical oxidants
NSCLC	Non-small cell lung cancer
O_2	Oxygen
$O_2^{\bullet -}$	Superoxide
$O_2^{\bullet -}$	Superoxide anions
OBS	Oxidative balance score
OFR	Oxygen-free radical
OGG1	8-oxoguanine DNA glycosylase 1
OGTT	Oral glucose tolerance test
−OH	Hydroxyl
OH	Hydroxyl radical
$ONOO^-$	Peroxynitrite
$ONOOCO_2$	Nitroso peroxocarbonate
OPP	Pentose phosphate
ORAC	Oxygen radical absorbance capacity
ORMs	Oxygen-reactive molecules

OTA	Ochratoxin A
OVX	Ovariectomy
ox-LDL	Distinctive features of oxidatively modified LDL
ox-LDL	Oxidatively modified LDL
OXPHOS	Oxidative phosphorylation
oxPTM	Oxidative post-translational modification
OxS	Oxidative stress
PAHS	Polycyclic aromatic hydrocarbons
PAI	Plasminogen activator inhibitor
PAL	Phenylalanine ammonium lyase
PAMPs	Pathogen-related molecular patterns
PBM	Photobiomodulation
PBMCs	Peripheral blood mononuclear cells
PBS	Phosphate-buffered saline
pCA	P-hydroxybenzoic (pHBA), and p-coumaric
PCa	Prostate cancer
PCS	Phytochelatin synthase
PD	Parkinson's disease
PDGF	Platelet-derived growth factor
PDIs	Protein disulfide isomerases
PDMPs	Prescription drug monitoring programs
PDS	Phytoene desaturase
PEF	Pulse electric field
PEOz	Polymeric scaffolds made from poly (2-ethyl-2-oxazoline)
PERK	Protein kinase R-like endoplasmic reticulum kinase
PERP polyphenol-enriched plants	Polyphenols
PG	Propyl gallate
PGPB	Plant growth-promoting bacteria
PGSA-g-EG	Poly (glycerol-co-sebacate) acrylate ethylene glycol
Pi	Inorganic phosphate
PI3K	Phosphoinositide 3-kinase
PIH	Pregnancy-induced hypertension
PKC	Protein kinase C
PKs	Protein kinases
PLA	Polylactic acid
PLCL	Poly(L-lactide-co-caprolactone)
PLGA	Poly(lactic-co-glycolic acid)
PM	Pectoralis major
POD	Peroxidases
PON 1	Paraoxonase 1

PON-1	Paroxonase-1
PP	Polypropylene
PPP	Pentose phosphate pathway
Prdx6	Peroxiredoxin-6
pre-COPD	Symptoms of the disease developing the condition
PROs	Patient-reported outcomes
PR_X	Peroxiredoxin
PSH	Protein-SH
PSH	Protoporphyrinogen oxidase
PSII and PSI	Photosynthesis through light-dependent photosystems
PTEN	Phosphatase and tensin homolog
PTEN	Phosphatase and tensin homolog on chromosome 10
PTM	Post-translational modification
PTPs	Tyrosine phosphatases
PUFAs	Polyunsaturated fatty acids
PUFAs	Proteins and polyunsaturated fats
RA	Rheumatoid arthritis
rAAT	Recombinant human alpha-1 antitrypsin
rADI	Recombinant arginine deiminase
RAGEs	Receptor for advanced glycation end products
RANTES	Regulated activation, normal T cell expressed and secreted
RANTES	Regulated upon activation, normal T cell expressed and presumably secreted
RAS	Renin–angiotensin system
RBOHD	Mitochondrial burst oxidase homolog D
RBOHD	Mitochondrial calcium uniporter
RCS	Reactive carbonyl species
RDI	Dietary reference intake
REACH	European Union regulation concerning the registration, evaluation, authorization, and restriction of chemicals
RET	Reverse electron transfer
RET	Reverse electron transport
rhASM	Recombinant acid sphingomyelinase
rhDNase	Recombinant human DNase
rhGAA	Recombinant human alpha-glucosidase
r-HuEPO	Recombinant human erythropoietin
Riboflavin	Vitamin B_2
RNS	Reactive nitrogen species
RO	Alkoxyl radical

RONs	Recepteur d'Origine nantais
RONSS	Reactive oxygen, nitrogen, and sulfur species
ROO•	Peroxyl radical
ROS	Reactive oxygen species
ROS/RNS	Oxygen and nitrogen free radicals
$R\text{-}S(O)NH_2$	Sulfenic amide
$R\text{-}S(O_2)\ NH_2$	Sulfonamide
R-S•	Sulfur-centered radicals
RS•	Thiol radicals
RSI	Reactive species interactome
RSM	Response surface methodology
RSNOs	S-nitroso thiols
R-SOH	Sulfenic acid
R-SOH	Tosylsulfenic acid
RSOO•	Peroxysulfur radicals
RSS	Peroxysulfur radicals
RSS	Reactive sulfur species
RSSH	Persulfides
RTM	Resin transfer molding
RuBP	Produces urea ribulose bisphosphate
S	Sulfur
S_2^-	Sulfide
$S_2O_3^{2-}$	Thiosulfates
SAPK	Stress-activated protein kinase
SASP	Senescence-associated secretory phenotype
SCFAs	Short-chain fatty acids
SCN5A	Splicing of the Na^+ voltage-gated channel alpha subunit 5
SDF-1	Stromal cell-derived factor-1
SeNPs	Selenium nanoparticles
SET	Single electron transfer
sGC	Guanylate cyclase
−SH	Cysteine thiol
−SH	Sulfhydryl group
−SH	Thiol
SHS	Sodium hydrosulfide
SLE	Systemic lupus erythematosus
SNOs	S-nitrosothiols
SNP	Single-nucleotide polymorphism
SNP	Sodium nitroprusside
SO_2	Sulfur dioxide
SO_3	Sulfur trioxide
SO_4^{2-}	Sulfur or sulfate
SOD	Peroxisomal superoxide dismutase

SOD	Superoxide dismutase
SOD, CAT, GPx	Erythrocyte antioxidant enzymes
SOD1	SOD isoforms are localized in one-of-a-kind cellular cubicles: cytosolic
SOD2	SOD isoforms are localized in one-of-a-kind cellular cubicles: mitochondrial
SOD3	SOD isoforms are localized in one-of-a-kind cellular cubicles: extracellular
−SOH	Sulfenic acids
−SOH	Sulfinylation
SPAs	Synthetic phenolic antioxidants
SPIONs	Superparamagnetic ions
SQR	Sulfide quinone oxidoreductase
SR	Sarcoplasmic reticulum
SSB	Single-strand break
SSBs	Small single base lesions
STC	Sterigmatocystin
SUD	Substance use disorder
SUDs	Substance use disorders
S_x^{2-}	Polysulfides
Sypl2	Synaptophysin-like protein 2
T1D	Type 1 diabetes
T2DM	Type 2 diabetes mellitus
T3	Triiodothyronine
T4	Hormone thyroxine
TAC	Total antioxidant capacity
TBA	Thiobarbituric acid
TBARS	Reduced TBA-reactive chemicals
TBARS	Thiobarbituric acid reactive substances
TBHQ	Tert-butylhydroquinone
TBI	Traumatic brain injury
TCA	Tricarboxylic acid
TCA	Trichloroacetic acid
TEAC	Trolox equivalent antioxidant capacity
TEs	Trolox equivalents
TGA	Thermogravimetric analysis
TLC	Thin-layer chromatography
TME	The tumor microenvironment
TME	Tumor microenvironment
TMJ	Temporomandibular joint
TNB	2-nitro-5-thiobenzoate
TNF	Tumor necrosis factor
TNF-α, IL-1β	Astrocytes release cytokines

tPA	Tissue plasminogen activator
TPTZ	2,4,6-tripyridyl-s-triazine
TRC	Tumor-repopulating cell
tRNA	Transfer RNA
Trx	Thioredoxin
TrxR	Thioredoxin reductase
UC	Ulcerative colitis
UCP	Uncoupling protein
UPR	Unfolding protein response
UVA	Ultraviolet A radiation
UVB	Ultraviolet B radiation
UVC	Ultraviolet C radiation
VEGF	Vascular endothelial growth factor
VLDL	Very low-density lipoprotein
WCBs	Whole-cell bacterial biosensors
WHO	The World Health Organization
WMS	Whole metagenome sequencing
WUE	Crop water use efficiency
Xo	Xanthine oxidase
YAP	Yes-associated protein
ZEA	Zaralenone
$\Delta\Psi m$	Mitochondrial membrane potential

Reactive Species Such as ROS, RNS, and RSS

P. Madhan Kumar, D. Abilash, R. Ganesamoorthy, R. Babujanarthanam, and R. Parameswari

Abstract Aerobic oxidative metabolism produces reactive oxygen species (ROS) as a by-product. The hereditary chemical characteristics of hydrogen peroxide (H_2O_2), superoxide anion (O_2^-), and hydroxy radicals (OH) can cause harm or modify the pathophysiology of target molecules. ROS constantly admits that its primary mechanism is oxidative stress caused by ROS by-products, which exacerbates the destructive pathologies of macromolecules such as DNA, protein, and lipids. However, during the last two decades, it has been shown that ROS appears to regulate biological and physiological pathways through cellular signaling systems. At the early stages of evolution, ROS is chosen by nature for signal transmission and to allow by-products of physiological signaling release molecules to adhere to ambient nutrients in an oxidative environment. According to research, ROS was directly implicated and triggered transcription factors that adapted to rising stress molecules, with better-documented molecular pathways in prokaryotes than in eukaryotes. RNS and RSS are principally involved in plant metabolism, which includes redox-active compounds. The term "*reactive species interactome*" (RSI) refers to the reactive oxygen, nitrogen, and sulfur species (RONSS) that occur more often in prokaryotes than in eukaryotes. Furthermore, it contributes to disorders like cancer, diabetes, and chronic inflammation, producing infectious diseases while maintaining homeostatic systems.

P. Madhan Kumar · R. Babujanarthanam · R. Parameswari (✉)
Department of Biotechnology, Thiruvalluvar University, Vellore, Tamil Nadu, India
e-mail: eashvit@gmail.com

D. Abilash
Gene Cloning and Technology Lab, Department of Biotechnology, SBST, VIT, Vellore, Tamil Nadu, India

R. Ganesamoorthy
Department of Chemistry, Vinayaka Mission's Kirupananda Variyar Arts and Science College, Vinayaka Mission's Research Foundation (DU), Salem, Tamil Nadu, India

K. Kasinathan et al. (eds.), *Recent Advances in Oxidative Stress Associated Chronic Diseases Volume 1*, https://doi.org/10.1007/978-981-96-9670-3_1

1 Introduction

(i) *Reactive Species Interactome (RSI)*

The redox system's RSI undermines chemical interactions achieved by the RSS, ROS, and RNS, as well as the biological intended function. RSI is distinguished by the biological molecule's adaptability, flexibility, and resilience, as well as its fast attention to sensing surroundings, which is followed by commands to targets via numerous channels of molecular transduction. Thus, the chemical reactions of RSI are carried out by the specific chemical supply of redundancy via enzymatic processes. Meanwhile, RSI's enzymatic chemical interactions include one or more electrons in oxidation, sulfation, polysulfidation, and nitration reactions [1]. The reactions described rely on enzyme pH, temperature, and O_2, but the distance to the maximum travels to radii. Enzymatic regulatory activities are also performed by nutrients such as amino acids, sulfate, and inorganic substrates, which provide additional kinetic parameters of pH, concentration, and temperature for posttranslational gene expression. As a result, it stimulates downstream processes that improve the balance in altering cell membranes. This chapter deals with reactive oxygen, nitrogen, and sulfur species of redox substances and their importance in pathophysiological molecular mechanisms [2].

(ii) *Reactive Oxygen, Nitrogen, and Sulfur Species (RONSS)*

The mechanisms of cell metabolism are derived from prehistoric abiotic processes that distributed free energy from chemical and physical sources before the formation of regulated cell life. Abiotic processes like these are thought to have developed independently as one of the many physicochemical processes the early Earth system used to release energy free via the Sun and the material resources in the crust. Metabolism is an accumulation of cellular processes that enable the extraction, transformation, and distribution of free energy from the environment, along with the breakdown of matter [3]. A collection of supramolecular aggregates or clusters that cooperate to produce the various phenomena that enable cells to function constitute metabolism. Reactive species originating from RONSS, including reactive carbonyl species (RCS), malondialdehyde (MDA), and lipid hydroperoxides (LOOH), share comparable signaling roles. They function at lower concentrations than reactive species S, N, and O as shown in Fig. 1. The effects and biological roles of reactive species on plants have been examined separately in scientific research for ROS, RNS, and RSS [4]. The oxidizing ability of H_2S might be immediately offset by its reducing ability. H_2S's direct antioxidant activity is not highly significant during physiological applications compared to NO and O_2. This stems from H_2S's volatility and reactivity, which make it an infrequent biochemical species in cells where HS and RSS are more prevalent. Thus, through the interactions between RSS, RNS, and ROS, H_2S has an indirect antioxidant effect. Signaling molecules have been the primary focus of research on RSS, RNS, and ROS interactions of H_2S, NO, and O_2. For instance, 10 mmol (mM) of H_2O_2 promotes nitric oxide (NO) synthesis in *Phaseolus aureus leaf* epithelial arrangements [5]. It was also demonstrated that H_2O_2 increases NO synthesis, which in effect causes H_2S when applied to maize seedlings to induce thermotolerance. Heat shock (450 °C for 30 min) increased NO in *A. thaliana* plants sprayed with H_2O_2 (20–200 μM); NO consequently boosted the activity of glutathione reductase, catalase, along with ascorbate peroxidase, and these removed excess H_2O_2, lowering the peril during oxidative damage; NO promotes proliferation of the mitochondrial oxidase with salt exposure [6]. The reverse has also been discovered since administering gibberellic acid causes the body to produce H_2S naturally, hence lowering oxidative damage caused by boron poisoning. NO is additionally related to how plants react to nanomaterials (NMs), either in the way that NMs induce stress tolerance or in the way that NMs produce stress. RSS, RNS, and ROS subsequently affect growing regulators and signaling molecules. Selenium serves various purposes and is a necessary component of macroalgae and mammals. One of the most researched roles is that of antioxidant selenium proteins, which neutralize ROS and RNS and guard against oxidative stress. Ascorbate peroxidase, glutathione peroxidase, dehydroascorbate reductase, superoxide dismutase, and monodehydroascorbate reductase all exhibit elevated activity [7]. Because of the information that has been released to the public on the interaction between reactive species, these antioxidant enzymes have an indirect effect on RNS and RSS, whereas they have a direct impact on ROS. Other noncatalytic proteins like protein "P" and thioredoxin (TrxR) have also been linked to increased activity when Se is prevalent. Due to their shared absorption and assimilation mechanisms, the effects of selenium on antioxidant metabolites are linked to sulfur metabolism; *Allium* and *Prunus domestica* are known to exhibit these effects on GSH and GSSH concentrations. Glucosinolates accumulate more in plants that may contain high concentrations, including broccoli. Selenium activation of phenylalanine ammonium lyase (PAL) and accumulation of phenolic compounds [8]. Because of their reduced ability, these chemicals can alter the RONSS balance. Additionally, the activation of antioxidant action through the impulsive decrease of O_2-by GPx or the promotion of ascorbic acid production has been hypothesized as ways in which Se directly affects redox homeostasis. As a pro-oxidant, selenium also directly contributes to moderate oxidative stress and the generation of reactive oxygen species (ROS), which triggers the synthesis of both enzymatic and nonenzymatic antioxidants [9].

2 Role of RONSS in Plant Metabolism

Plant RONSS is primarily caused by $O_2^{\bullet -}$ from atmospheric O_2; •NO generated primarily from nitrogenous compounds (NO_3^-, NO_2^-, along with protein building blocks) plant utilizes minerals as well as to a far smaller level from traces of •NO existing throughout the environment; along with H_2S formed as the assimilating of sulfur-based compounds that plant consumes as good nutrients and to a much less extent via traces of H_2S and other compounds like dimethyl sulfide (DMS) as well as carbonyl sulfide (COS) existing in the atmosphere [6].

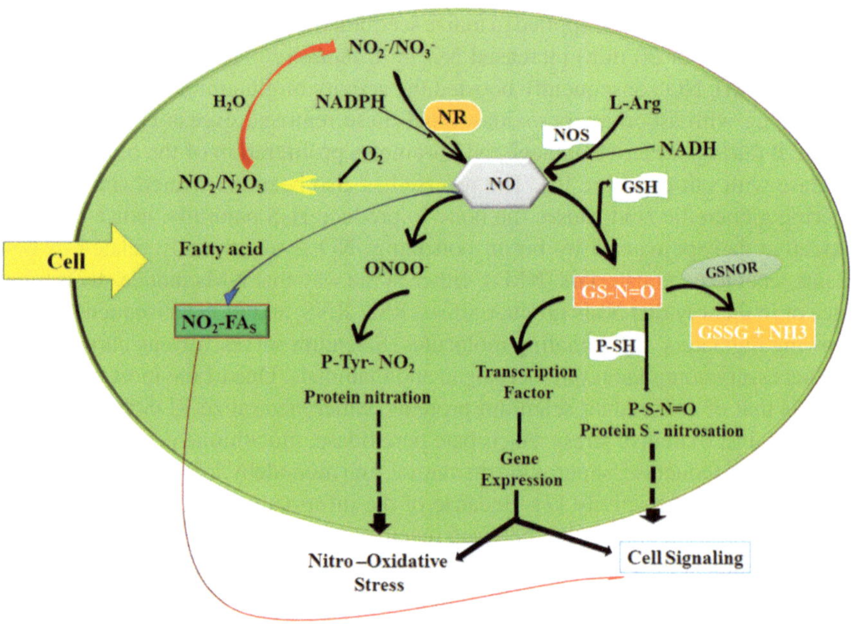

Fig. 1 Reactive species interactome (RSI) of ROS, RNS, and RSS signaling pathway—an overview

2.1 *ROS in Plant Metabolism*

Excess decreasing potential is dissipated, for instance, in chloroplasts under higher irradiance, at a high or low temperature, and in mitochondria during lower temperature, normally at any condition resulting in an imbalance among the reducing potential's production as well as metabolic use. O_2's strong electronegativity, which is only slightly lower than that of "F," its existence of unpaired electrons, and its several oxidations states It highlights that various ROS can be formed by successively accepting electrons. Additionally, "S" and "N" are strongly electronegative elements (S > N), which helps to explain how they can generate reactive species [10]. ROS are signaling molecules in almost every metabolic and plant growth mechanism due to their energy dissipation function. Membranes, polypeptide cysteine, including gene transcription factors, amino acids or histones, and low molecular size metabolites exhibit peroxidative alterations when interacting with ROS. By influencing how biomolecules work, the modifications enable cellular behavior and their energy dissipation function. Reactive oxygen species function as signaling agents in nearly every metabolic and cellular growth process. The modifications alter how biomolecules work, enabling organisms to modify their behavior to alterations in their redox balance [11]. Tosulfenic acid (RSOH) is produced when protein cysteines undergo ROS peroxidation. Proteins that undergo this type of oxidative posttranslational modification (oxPTM) alter their redox characteristics and ability to interact with their biological environment. The altered protein's function reacts to alterations in the redox balance. One example is protein cysteine peroxidation by ROS to sulfenic acid (RSOH). Proteins that undergo the process of oxidative post-translational modification (oxPTM) have their redox characteristics and ability to function in the cellular environment modified. This can impair an enzyme's consistency or ability to bind to its base material and damage the ability of genetic genes, such as histones, to communicate with DNA [12]. Examples of the ROS (oxPTM) on proteins include enzymes from the Calvin cycle, the cellular breakdown of sulfur and starch, and the hormone-responsive proteins associated with stress adaptation. Plants undergo significant changes in their chemical composition, morphological evolution, and overall ability to adapt to environmental shifts due to epigenomic and proteomic changes, including metabolomic alterations that alter their phenotypic or growth processes. The degradation of the cell membrane's lipids, proteins' thiol groups, and several metabolites, on the other hand, creates a signaling mechanism by detecting reduction–oxidation balances, such as NADPH: $NADP^+$, ascorbate: dehydroascorbate, and GSH: GSSH, which serve as an indicator for detecting the system's inbuilt the energy asserts [13]. The main RNS is nitric oxide (NO). Prokaryotes and eukaryotes, including animals and plants, do have this primitive signaling molecule. The chemical theme of NO is prevalent in the signaling. The mechanism by which the signal is used to stimulate cellular responses is also known as the signaling enzymes. It appears aged among the many eukaryotic lineages and differs among plants. These contribute to clarifying whether animals and plants respond significantly to RNS exposure. Animals are poisoned and allergic to nitric oxide, yet plants use it as a signaling chemical [14]. In animals and humans, S-nitrosothiols function as signaling molecules in plants. Conversely, peroxynitrites are classified as signaling agents that can cause animals to become more toxic than plants. It can result in guanine nitration in animals, which can produce guanine mispairs that cause mutations and cancer. On the other hand, peroxynitrites in plants can be deactivated by CO_2, producing CO_3^- and NO_2. Because NO-derived oxidants are more reactive than NO, they primarily cause NO's toxicity. At the same time, nitric oxide synthase is the primary synthesizer. In animals, many plant enzyme systems endogenously create NO [15]. Among these are the L–Arg NO synthase analogues' oxidative method, which is accomplished by reducing mechanisms like either the plasma bound to the membrane NO-forming nitrite reductase (NOFNiR) or nitrate reductase (NR), which generates NO_2, which is subsequently converted to NO by the NR itself. Alternative dehydrogenase, cytochrome c, complexes I and II, and the mitochondrial complexes III and IV produce NO reductively from NO_2. In anoxic or hypoxic environments, alternative oxidase (AO_X) in the mitochondria is also generated; however, in normoxic settings, AO_X eliminates excess NO [16]. A Phytogb1-NO cycle is produced in N_2 fixation nodules and anoxic settings by nonsymbiotic hemoglobin's interacting with mitochondrial NO $\rightarrow$ NO_3^- $\rightarrow$ –NO, which permits NADPH levels to be controlled and produces anoxic ATP. In additionally to NR & NOFNiR, several molybdoenzymes, like xanthine oxidases, including sulfite, aldehyde, and oxidases, do have the ability to reduce NO_2. The existence of RNS is linked to free energy/reducing potential dissipation mechanisms in the case of ROS. The primary methods by which RNS alters how cells behave are metal nitrosylation, nitration, and S-nitrosation. S-nitrosation is the process by which NO and the thiol (–SH) of cysteine (Cys) come together to generate S–N = O [6]. Proteins undergo reversible posttranslational modification (PTM), one of the key processes for NO signaling. Additionally, the S–N = O group serves as a reservoir and donor of NO. The function of proteins is altered by S-nitrosation, resulting in quick and reversible modifications to the cellular proteome. Nitration is when a nitro group ($-NO_2$) is added to protein molecules, fatty acids, and nucleic acids, mediated by ONOO–. Nitro-tyrosine is the product of the most researched nitration type in proteins. However, it also appears in additional amino compounds, including methionine, cysteine, and tryptophan. Gaining or losing protein function or possibly no effect can result from the nitration of amino acids, loss of function is the most typical outcome of nitrosylation of metals. Transition metals found in peptides interact with NO, and the plants involved in this process are not widely recognized [6, 17].

3 Reactive Sulfur Species in Plant Metabolism

Reactive sulfur species (RSS), crucial for cell functions, including communication, play a major role in plant metabolism, redox control, and stress response. A crucial component of plant biochemistry, sulfur (S) metabolism, is strongly linked to the generation of hydrogen sulfide (H_2S) along with RSS. Sulfur is an essential element that plants primarily absorb from the soil as oxidized sulfur or sulfate (SO_4^{2-}). A vital component for forming sulfur-containing biomolecules, including cysteine and methionine, sulfate is reduced in the cytoplasm and plastids after being taken in, producing sulfide (S^{2-}) [18]. NADPH and other reducing equivalents are consumed during this energetically expensive reduction process. Sulfur metabolism produces a delicate equilibrium between reduced and oxidized sulfur species, eventually forming reduced sulfur intermediates such as H_2S, which are crucial signaling molecules and are involved in several cellular developments. Recently, H_2S has gained more attention in research because of its dual roles as neither a therapeutic signaling ligand at low concentrations nor a fatal by-product at high quantities. The great reducing potential of H_2S, which may reduce a range of cell components, especially disulfides, is the mechanism that allows plants to generate and utilize it [19]. An important antioxidant and redox buffer in plant cells, oxidized glutathione (GSSG) is converted to its reduced form (GSH) by H_2S. Reducing oxidative stress aids in preserving the cell's redox balance. H_2S can also create sulfur-modified thiols called persulfides (R–S–SH) by reducing protein-based disulfides, like those in different enzymes. Protein persulfuration is an essential mechanism that controls the activity of specific enzymes and receptors, enabling the modification of signaling pathways. RSS's persulfuration and polysulfuration of proteins also impact cellular signaling and gene regulation. Protein thiols can form persulfides and polysulfides when one or more sulfur atoms connect to them. These compounds can alter the function of transcription factors and signaling molecules [20]. These sulfur alterations act as molecular switches that allow proteins implicated in apoptosis, growth control, and stress responses to be activated or deactivated. These mechanisms are closely controlled in plants to maintain a balanced redox indicator within the cell, limiting the buildup of harmful oxidized sulfur species and ensuring a sufficient supply of reduction sulfur intermediate for cellular operations [21].

RSS is involved in more than only protein modification within plant metabolism. H_2S_x, S_2^{2-}, S_3^{2-}, and S_5^{2-} are polysulfides that can be formed when H_2S is oxidized. These molecules are crucial for cell signaling. These polysulfides may interact with a wide range of biomolecules, influencing how they function and how cells react to abiotic and biotic stressors. For instance, synthesizing H_2S and its subsequent oxidization to polysulfides may trigger mechanisms of defense in plants, including reacting to oxidative stress or pathogen infection by starting antioxidant enzymes or controlling ion channels [22]. Moreover, protein persulfuration brought on by RSS may behave as a modification after translation, altering proteins reaerates signals. RSS's function in the metabolic process is made more complex by their interactions with additional reactive species, which include reactive oxygen species (ROS). Despite being commonly associated with oxidative stress, ROS and RSS can cooperate to modulate stress responses and cell signaling. Numerous physiological functions, including cell development, differentiation, and environmental adaption, can be regulated by the interaction between ROS and RSS, which includes the production of certain sulfur and oxygen species [23]. Reactive oxygen species and hydrogen sulfide, for instance, have been demonstrated to cooperate in controlling stomatal closure in response to irritation, a crucial mechanism for preserving plant transpiration balance. The complex interactions among RSS, H_2S, and various other sulfur-containing organisms, including sulfur dioxide (SO_2) and thiosulfates ($S_2O_3^{2-}$), highlight the significance of sulfur within plant metabolism. These interactions help the plant notice and respond to fluctuations in its surroundings and preserve cellular homeostasis. Because RSS plays a part in regulating plant resistance mechanisms and the response to pathogenic attack, the role of sulfur within plant immunization has received an abundance of interest. Although they affect the synthesis of defense-related molecules, including phytoalexins and pathogenesis-related proteins, the signaling roles of H_2S, persulfides, and polysulfides have become progressively recognized as significant regulators for plant immune responses [24]. Plants have extremely difficult and context-dependent control of H_2S along with other RSS, with various tissues and developmental stages displaying distinctive RSS production and usage variations. Existing research focuses on the structure and function of RSS synthesis and their interactions with other molecules that signal, including their impact on plant physiology.

Additionally, RSS may be used in agricultural applications to boost growth or increase crop resistance to stress. These species depend on redox equilibrium, protein changes, and cellular signaling pathways that control plant growth, stress reactions, and immunology (Table 1). Sulfur metabolism is crucial for cellular function, as demonstrated by plants' complex equilibrium between reduction and oxidized sulfur types [25].

Table 1 Various plant species and its associated reactive species impact in agriculture

Plant species	Impact	Reactive species	Reference
M. sativa, S. indicum *T. aestivum*	Reduced heavy metals toxicity along with absorption. The concentration of vital components has increased	H_2S, NO	[34–36]
Ocimum basilicum *Oriza sativa,* *Cynodon dactylon,* *Medicago sativa*	Increase in biomass along with relative growth rate (RGR)	H_2O_2, NO H_2S, NO	[37]
Triticum aestivum *Fragaria × ananassa*	Rise in the amount of relative water (RWC)	H_2O_2, NO H_2S, NO	[36]
Citrus aurantium *Fragaria × ananassa*	PSII's quantum efficiency (Fv/Fm) increased	H_2O_2, NO	[38]
Citrus aurantium	Gene expression alterations with reversible or irreversible posttranslational changes (PTM) can reprogram the proteome	H_2S, NO H_2O_2, NO	[39]
Medicago sativa *Triticum aestivum* *Solanum lycopersicum*	Enhanced accumulation of proline and additional osmolytes. Reduction of lipid peroxidation during acute conditions	H_2S, NO H_2O_2, NO	[34, 40]

3.1 RONSS as Biostimulants

RONSS biostimulation or priming, ROS, RNS, or RSS application, or the usage of ROS–RNS, ROS–RSS, or RNS–RSS in pairs are all pertinent and active topics associated with plant research. A range of biostimulants, including silicon, melatonin, and salicylic acid, as well as seed magneto priming, work by acting as signaling agents through RONSS [6]. Although RONSS application results in positive reactions to stress and increased yield or productivity in the nutritive makeup of florae, there are still numerous remaining inquiries concerning the molecular mechanisms influencing cellular reactions. The mixture of ROS–RNS and RSS–RNS can reduce lipid peroxidation. Cells still use the biochemical offspring of such ancient processes. Natural selection modified and optimized them over billions of decades of biological evolution to enable all cells and animals with multiple cells to utilize free energy and convert it into entropy to the maximum extent potential [26]. To achieve maximum entropy, organisms must have a method for gathering information that enables them to adapt to changes in their surroundings. This is accomplished by assessing the system's redox status to determine the energy situation. Redox status results in proteomic and metabolomic reactions, phenotypic modifications, and modifications in gene expression that modify metabolism to the organism's wishes in its specific atmosphere. The metabolic changes mentioned above are specifically expressed by the

RONSS. The number of chemical agents associated with cell signaling and biostimulation is expected to increase [27]. RONSS should be used with biostimulants like silicon, selenium, or iodine, plant and seaweed extracts, chitosan, additional biopolymers, humic compounds, and metabolites such as salicylic acid and melatonin. The table states that using RONSS as a biostimulant and for signaling has been tested in several economically valuable plants, including *Citrus aurantium*, with positive findings [28]. Early studies on Cu-stressed *T. aestivum* seedlings employing exogenous sodium hydrosulphide (SHS) as an H_2S donor indicated a decrease in the initiation of gamma-glutamylcysteine synthetase. Further reducing the damage caused by Cu were increases in ascorbic acid, glutathione, and total ascorbate levels. Additionally, it was shown that *T. aestivum seeds* treated with Cu suffered less damage to their plasma membrane integrity, that amylase and esterase activities were enhanced, and that germinating seeds that received H_2S donors had lower levels of malondialdehyde and H_2O_2. Tolerance to Cd stress in *T. aestivum* by applying NO and H_2S with donors such as sodium nitroprusside (SNP) and SHS [29]. Growing *S. lycopersicum* under high CO_2 concentrations has been linked to NO–induced root development, particularly inside roots, or increased nitric oxide synthesis action. *S. lycopersicum* showed a strong ability to stop as in the root while simultaneously facilitating its movement in the shoots through the overexpression of the ABC transporter (ABC1), metallothionein (MT), phytochelatin synthase (PCS), GSH1 and GSH2. According to the study, proteomics analysis shows that plants exposed to salinity have quantitative differences in 85 leaf proteins [30]. Plants treated with H_2O_2 or SNP lacked several of them. RONSS has also been assessed in medicinal plants for its potential as a biostimulant or signaling molecule. Medicinal plants have also been utilized as models in certain tests despite their lack of study. One significant vegetative source of protection against malaria is Artemisia annua. By applying H_2S (200 M) exogenously, the negative effects of Cu^{+2} (20 mg to 40 kg^{-1}) on *A. annua* can be mitigated, restoring physiological and biochemical parameters, decreasing lipid peroxidation and increasing the antioxidant potential of medicinal plants. The administration of H_2S also enhanced the trichome density and photosynthetic efficiency, as well as the generation of artemisinin. This well–known substance has antibacterial, anti-inflammatory, and antioxidant properties and is used to treat malaria [31]. Another Asteraceae with vital medicinal qualities and an important form of food—grade color for the food industry, Carthamus tinctorius, has also benefited from the efficient utilization of H_2S. Exogenous H_2S (1 mM) treatment of drought—stricken *C. tinctorius plants* showed that the negative impacts of the water shortage were mitigated, boosting the antioxidant capacity and secondary metabolite development. The biological effectiveness of therapeutic plant extracts in halting oxidative damage is finally worth mentioning [32]. However, their ability to scavenge free radicals or act as bio stimulant agents that promote the development of RONSS, or the production of more antioxidant enzymes has primarily been employed to treat wounds or inflammation in humans. However, we cannot dispute that plant extracts can function as biostimulants in sustainable crop production. When plant-based products are applied systematically, they have been demonstrated to reduce environmental stress-induced damage and encourage plant development. Considering that the exclusive emphasis of this analysis was RONSS metabolites and their applications as biostimulants agents or signaling molecules [33].

4 Reactive Nitrogen Species—RNS in Cellular Signaling

Reactive nitrogen species (RNS) have become essential components in cellular signaling and their function in controlling several physiological processes. Nevertheless, the signaling capabilities of RNS have been shown to contribute to many of the same physiological processes that ROS were previously implicated in, including gene expression, differentiation, proliferation, and death [41].

4.1 Nitric Oxide to Reactive Nitrogen Species (RNS)

NO significantly influences cellular signaling, and due to its high reactivity, it also produces a variety of secondary reactive nitrogen species. Such molecules include nitrite (NO_2^-), nitrate (NO_3^-), as well as highly reactive species like peroxynitrite ($ONOO^-$), which is a by-product of the reaction between superoxide (O_2^-) as well as NO. Peroxynitrite is especially significant when oxidative and nitrative stress is present, because, if left unchecked, it can act with lipids, proteins, and DNA, among other macromolecules, producing impairment. However, in addition to being oxidative damage consequences, these reactive compounds are also crucial signaling molecules [42]. Protein nitrosation and nitrosylation, for instance, are important posttranslational changes that control protein activity. A nitric oxide (NO) group is attached to the sulfur atom of a cysteine thiol (–SH) group during the reversible alteration known as nitrosylation, creating S-nitrosothiols (SNO's). The stability, interactions with other molecules, and function of the protein are all changed by this transformation [43].

4.2 Nitrosation and Nitrosylation in Cellular Signaling

The NMDA receptor in the neurological system is regulated by nitrosylation in cellular signaling. The neuronal NOS (nNOS) produces nitric oxide, which controls the NMDA receptor's activation by S-nitrosylation, which, in some circumstances, inhibits its function. NO plays a key role in synaptic plasticity, the process of learning, and memory by dynamically regulating NMDA receptor activation. Redox-dependent NO actions on the NMDA receptors are significant; NO may stimulate the receptor and affect synaptic transmission when it combines with superoxide to create peroxynitrite. Protein nitrosation and nitrosylation, for instance, are important post-translational changes that control protein activity, especially in the thiol group of cysteine residues [44]. Nitrosylation is a reversible process that creates S-nitrosothiols (SNO's) by adding a nitric oxide (NO) group to the sulfur atoms of a cysteine thiol (–SH) group. The protein's stability, function, and relationship with other molecules are all altered by this mutation. Numerous cellular functions, including immunological response, neurotransmission, apoptosis, and cellular metabolism, have been linked to the nitrosylation of proteins. The tiny GTPase p21ras, a crucial modulator of cell division and proliferation, is involved in nitrosylation in signaling. Ras's activity has been demonstrated to be modulated by NO nitrosylation, which impacts how cells react to growth factors and external stimuli. Furthermore, the activation of downstream pathways, including the diacylglycerol (DAG) pathway essential for cell growth and survival, could originate from the nitrosylation of other signal transduction-related proteins, such as Src kin [45, 46].

4.3 Peroxynitrite and Its Function of Protein Regulation

Peroxynitrite is a highly reactive molecule that can alter proteins, lipids, and nucleic acids. It is a by-product of NO and superoxide. Several biological reactions have been linked to the alteration of proteins by oxidizing their tyrosine residues, which results in the creation of nitrotyrosine. Nitrotyrosine proteins may have functional repercussions that impact cellular localization, enzyme activity, and protein–protein interactions. Peroxynitrite-mediated protein modification is anticipated to be characterized by the nitration of tyrosine residues in specific [47]. Cellular processes may become dysregulated due to this alteration, which may impact the activity of important signaling proteins and enzymes. Interestingly, nitrated proteins can be eliminated via degradation routes or other processes that are still being thoroughly studied; thus, they are not irreversible changes. Because nitration is temporary, it is possible to adjust cellular signaling without causing permanent damage [48].

4.4 Redox Balance and Nitric Oxide in Cellular Signaling

In cellular signaling, the cellular redox environment regulates the function of RNS and nitric oxide. The signaling roles of ROS and RNS depend on their concentrations and localization, along with redox buffering mechanisms inside the cell. However, both can cause cellular harm. Recognizing the complex connections between ROS and the RNS in cell signaling has been decoded by studying global oxidative alterations of proteins or redox proteomics [49]. Alkylating agents, mass spectrometry, and biotin-based tagging are some methods used to detect and monitor protein modification by RNS, which can reveal details regarding the basic processes of redox signaling. ROS and RNS are crucial components of redox signaling. In certain situations, the RNS and ROS can control cellular processes like immune response and cell cycle progression, and trigger apoptosis. In other cases, oxidative stress imposed by these species' overproduction might accelerate disease progression [50].

5 Bioimaging Probes of RNS

Reactive nitrogen species (RNS) bioimaging probes are specialized detectors developed to identify and demonstrate these extremely reactive substances' existence, distribution, and motion in living cells and tissues. RNS components such as nitrotyrosine, peroxynitrite ($ONOO^-$), and nitric oxide (NO) are important for cellular signaling and play a part in oxidative stress and illness. Bioimaging probes are designed to react with RNS to study these species specifically. These probes produce measurable signals, including fluorescence or luminescence, which may be recorded using sophisticated microscopy techniques [51]. These probes usually consist of chemiluminescent reporters or fluorophores that, when interacting with specific RNS, alter biochemical and exhibit a noticeable change in optical characteristics. Nitric oxide, for instance, can be detected by probes such as DAF-2 (4,5-diaminofluorescein), which produces a fluorescent signal when NO attaches to it. On the other hand, peroxynitrite can be detected by other probes that exhibit a change in fluorescence emission [52]. These probes are essential for in situ, real-time monitoring of the RNS in biological processes because they enable research to observe variations in RNS levels over time in response to environmental factors, disease progression, or cellular stimulation. Bioimaging probes for the RNS are also used to evaluate the results of therapeutic interventions, offering important new information on the role of the RNS in conditions like cancer, cardiac issues, and neurodegenerative diseases. A significant development in the fields of redox biology and molecular imaging is their capacity to observe RNS dynamics with great specificity and sensitivity [49].

5.1 Reactive Sulfur Species (RSS)

Energy metabolism, antioxidant defense, and cell communication comprise only some of the biochemical functions of sulfur in biological systems. Glutathione (GSH), cysteine, and other sulfur-containing peptides and proteins are examples of sulfur-containing compounds essential for preserving the redox equilibrium within cells. Reactive sulfur species (RSS) have arisen due to the reactivity of sulfur compounds, particularly in the presence of oxidative stress [53]. These species, which include sulfenic acids, disulfides, disulfide-S-oxides, and sulfur-centered radicals, can significantly affect protein activity, cellular health, and cellular functions. RSS is also becoming more widely recognized for its function in pathophysiological processes, signaling, and cellular damage. Depending on their level of concentration and reactivity, these species can have both pro-oxidant and antioxidant qualities [54].

5.2 *RSS's Cellular Signaling and Redox Mechanisms*

Reactive sulfur species (RSS), frequently produced by oxidizing sulfur-containing compounds such as cysteine, glutathione (GSH), and proteins, are essential in cellular signaling and redox processes. Sulfur-centered radicals (thiyl radicals), disulfides and disulfide-S-oxides, and sulfenic acids are among the species that can modify the redox state of enzymes and proteins, influencing cellular processes and protein activity. Many oxidative stimuli, including hydrogen peroxide, peroxynitrite, superoxide, and singlet oxygen, cause the production of RSS [55]. Under physiological applications, thiyl radicals—which arise when thiols undergo one-electron oxidation—are extremely reactive and unstable. However, they are crucial for cellular processes, including redox signaling and protein modification. Sulfenic acids or sulfonyl radicals are generated when these radicals interact with different components of the cell. These radicals then take part in signaling pathways or cause oxidative damage. Additionally, RSS interact with ROS, demonstrating the complex connection between oxygen and sulfur species. Redox-disproportionation, for example, can occur in sulfenic acids and disulfide-S-oxides produced under oxidative stress, possibly resulting in the synthesis of thiols, disulfides, and sulfinic acids that reduce oxidative damage [56]. This biochemical network emphasizes RSS's role in preserving the redox balance inside cells and their participation in vital functions like cellular signaling, oxidative stress response, and protein function modulation. The ability of RSS to modify redox-sensitive enzymes and proteins, influencing significant biological processes and disease pathways, is an example of their regulatory potential [57].

5.3 *Formation and Mechanism of RSS*

Sulfur-containing compounds, primarily thiols (–SH groups), are oxidized through reactions against ROS or other oxidative agents that generate RSS. Protein cysteine residues, the tripeptide glutathione (GSH), and other sulfur-based metabolites are the main sulfur-containing compounds in this context. By oxidizing these substances, reactive intermediates, including thiyl radicals ($RS^{\bullet}$), can be generated [58].

5.4 Sulfur-Centered Radicals ($RS^{\bullet}$)

A thiol group undergoing one-electron oxidation produces the sulfur-centered radical ($RS^{\bullet}$), among the most reactive types of RSS. This highly unstable radical disrupts proteins, lipids, and DNA, among various biological constituents, by creating oxidative stress. Sulfinyl and sulfonyl radicals, two more complicated radical species that can spread oxidative damage in cells, can be produced from thiyl radicals [59].

5.5 *Sulfenic Acids (–SOH)*

Sulfenic acids are created when ROS oxidize the sulfur in thiol groups. These substances are extremely reactive, even if they are more stable than thiyl radicals. By creating covalent connections with cysteine residues, sulfenic acids can alter proteins, changing their function and possibly helping to control their activity. Because they enable the cells to regulate cellular activities and govern redox-sensitive proteins, the inability to reverse sulfenic acid production is essential [60].

5.6 *Disulfide Bonds and Disulfide-S-Oxides*

Disulfides (RSSR) are another significant type of RSS that arises from the oxidation of two thiol groups and their covalent formation by a sulfur-sulfur bond. Protein stability and bending depend on disulfide bonds, but when these bonds undergo further oxidation to disulfide-S-oxides (RSSO), they become increasingly reactive and may cause malfunctioning proteins. These oxidized molecules can interfere with biological functions, particularly if they build up in an oxidative stress environment [61].

5.7 Sulfinic and Sulfonic Acids (–SO_2H, –SO_3H)

Increased oxidation states of sulfur are produced when sulfenic acids undergo further oxidation. Sulfinic and sulfonic acids contribute to cell dysfunction if they collect or impact important cellular proteins, although they are generally less reactive than thiyl radicals or sulfenic acids [62]. ROS triggers numerous metabolic processes and actions that lead to RSS production. RSS is formed by the interaction of ROS with thiol-containing compounds, such as ($O_2^{\bullet-}$), (H_2O_2), (•OH), and ($ONOO^-$). The mechanism of RSS production is the interaction of ROS with glutathione and cysteine. As a significant intracellular antioxidant that preserves redox equilibrium, glutathione (GSH) is a tripeptide with a thiol group. GSH can undergo additional oxidation to produce other RSS, including sulfenic acids and disulfide-S-oxides, or it can be reduced to its disulfide form (GSSG) when it interacts with ROS or another oxidative agent. The thiol peroxyl radical ($RSOO^\bullet$), another reactive species, can be created when thiyl radicals combine with oxygen. Redox cell environments can get even more complex when thiyl radicals react with additional thiols to generate disulfide radical anions ($RSSR^{\bullet-}$) [63, 64].

6 Role of RSS in Oxidative Stress

Macromolecule cells and proteins, lipids, and DNA damage, result from oxidative stress, which happens when the generation of ROS and RSS surpasses the antioxidant capability of the cell. Cellular signaling pathways, especially those about the reaction to oxidative stress, are regulated by RSS. Sulfenic acids and other RSS can influence proteins and enzymes, changing how they react to alterations in the cellular redox state. Sulfenic acids, for instance, accumulate cysteine residues in protein and can affect the action of enzymes involved in apoptosis, redox control, and cell survival [65]. Regulation of oxidative stress responses by transcription factors is one instance of RSS-induced protein modification. Additionally, RSS can alter the activity of signal transduction-related kinases along with phosphatases, which may impact processes like differentiation, apoptosis, and cell proliferation [66].

6.1 Defense Mechanisms of Antioxidants Against RSS

Cells have determined numerous antioxidant defense mechanisms to fight the harmful effects of RSS. By neutralizing excessive ROS and RSS, these mechanisms primarily aim to preserve cellular redox balance and shield cellular components from oxidative damage. Glutathione (GSH) is a significant antioxidant. Glutamate, cysteine, and glycine comprise the tripeptide GSH, which is essential for detoxifying ROS and RSS. Through direct interaction, GSH neutralizes ROS and produces oxidized glutathione, or GSSG [67]. GSH's ability to decrease sulfenic acids and other oxidized sulfur compounds greatly aids the antioxidant defense against RSS. Protein disulfide linkages, including disulfide-S-oxides, can be broken down by thioredoxin, which also helps redox-sensitive proteins function again. Additionally, sulfenic acids and thioredoxin interact to convert them to thiol groups, preventing oxidative damage to cellular proteins. Enzymes known as peroxiredoxins catalyze the decrease of organic and hydrogen peroxides. These enzymes also help cells defend against oxidative stress by lowering sulfenic acids and RSS. Glutathione peroxidase (GP_x), an antioxidant enzyme, incorporates selenium, another chalcogen like sulfur. These enzymes are essential for the decrease of RSS and ROS. GP_x enzymes that contain selenium can directly eliminate hydroperoxides and protect cells from oxidative damage, highlighting the role that selenium and sulfur play in antioxidant defense [68, 69].

6.2 Mitochondria—ROS

The mitochondria, the cell's powerhouse, are required for energy production via oxidative phosphorylation (OXPHOS). Superoxide ($O_2{\bullet}^-$), a precursor to other ROS,

Table 2 Key studies on mitochondrial ROS production mechanisms

S. no	Findings	References
1	Identified complex I and III are the main locations in the ETC where superoxide is produced	[70]
2	Discussed the principles governing superoxide production within the mitochondrial matrix	[71]
3	Investigated how ROS have a dual function in oxidative damage and cellular signaling	[72]
4	Measured the amount of hydrogen peroxide and superoxide produced by mitochondrial regions in different situations	[73]
5	Reviewed the activity of mitochondrial reactive oxygen species in physiological and disease, highlighting their signal activities	[74]
6	Discussed the origins and processes of ROS creation inside mitochondria and gave a thorough overview of mitochondrial production of ROS	[75]
7	Improved our knowledge of ROS homeostasis in disease pathology by developing mathematical models to explain how the electron transport chain produces ROS	[76]

is formed during this process when some electrons escape from the ETC and intermingle with molecular O_2. Although ROS are required for signaling pathways, excessive generation leads to stress related to age and variability of ailments such as cardiovascular disease and neurological problems (Table 2). Mitochondrial ROS are primarily produced by the ETC at Complex I and III [6].

(i) *Mitochondrial mechanisms include*

- Electron Leakage: Superoxide anions ($O_2{\bullet}^-$) are created when oxygen is partially reduced.
- Enzymatic ROS Production: ROS levels are influenced by enzymes such as NADPH oxidase and monoamine oxidase (MAO).
- Fenton Reaction: Most reactive ROS species, hydroxyl radicals (•OH), are created when hydrogen peroxide ($H2O_2$) reacts with iron or copper.
- Reverse Electron Transport (RET): When electrons go backward to Complex I under specific metabolic circumstances, more superoxide is produced [71].

7 Pathophysiological Consequences of Mitochondrial ROS

ROS are vital secondary messengers at low concentrations in several biological activities, such as cellular signaling control of copy factors like HIF-1α and NF-κB, as well as kinases. Redox Homeostasis: ROS regulate the production of antioxidant enzymes. Immune Response: By activating macrophages, ROS aid in pathogen protection. Overproduction of ROS results in oxidative damage, impacts Lipid Peroxidation, modifies membrane integrity, and impairs mitochondrial function. Proteotoxic stress and enzyme inactivation are caused by protein oxidation. DNA Damage: Aging and degenerative disorders are exacerbated by mutations in mitochondrial DNA (mtDNA) [72, 74].

7.1 Mitochondrial ROS in Disease Pathogenesis

Numerous illnesses are linked to dysregulated ROS levels: ROS contribute to atherosclerosis and hypertension by mediating endothelial dysfunction. While regulated ROS levels can cause cancer cells to undergo apoptosis, ROS-induced DNA damage can also promote carcinogenesis. Cells have built-in antioxidant defenses to prevent harm from ROS: Superoxide is changed into hydrogen peroxide by the enzyme superoxide dismutase (SOD). Glutathione System: Uses glutathione reductase and glutathione peroxidase to reduce ROS. Catalase: Produces oxygen and water from hydrogen peroxide. Numerous strategies are being investigated to alter the mitochondria's ROS amount. Antioxidants: MitoQ and SkQ1 are antioxidants that target mitochondria [74]. Lifestyle Interventions: Exercise and calorie restriction improves mitochondrial resilience. Gene therapy: boosting genes that produce antioxidants to prevent damage caused by ROS. Pharmaceutical Agents: Coenzyme Q10 and N-acetylcysteine are small compounds that alter mitochondrial activity.

7.2 Mitochondrial ROS Have a Dual Function in Cellular Activity

As vital signaling molecules and oxidative damaging agents. Developing targeted therapeutics requires an understanding of the symmetry between the production of ROS in the body and in disease. To reduce ROS-related diseases, future studies should concentrate on customized antioxidant techniques and innovative treatments [75].

7.3 Enzymatic Antioxidants Versus Reactive Oxygen Species (ROS)

Derivatives of cellular metabolism, reactive oxygen species (ROS), are mostly formed in the mitochondria throughout oxidative phosphorylation. Low quantities of ROS are necessary for homeostasis and cell signaling, but excessive ROS production causes oxidative stress, which damages cells and causes several illnesses, including cancer, dementia, and cardiovascular conditions. Enzymatic antioxidants have developed in biological systems to combat ROS, neutralizing these reactive molecules and preserving cellular redox equilibrium. The processes, importance, and regulatory features of enzymatic antioxidants in controlling ROS levels are examined below [76, 77]. Enzymatic antioxidants co-ordinate with one another to clean up ROS and prevent oxidative damage (Table 3).

(i) *Interplay Between ROS and Enzymatic Antioxidants*

- ROS Signaling: ROS control apoptosis, immunological responses, and gene expression at the physiological level.
- Oxidative Stress: When ROS production exceeds antioxidant defenses, macromolecular damage results.
- Antioxidant Response: Under oxidative stress, nuclear factor erythroid 2-related factor 2 (Nrf2) upregulates enzymatic antioxidants [78, 79].

(ii) *Implications in Health and Disease*

- Cancer: Tumor growth and treatment resistance are facilitated by the dysregulation of antioxidant enzymes.
- Neurodegeneration: Alzheimer's and Parkinson's disorders are linked to ROS-induced neuronal damage.
- Cardiovascular Diseases: Myocardial infarction and atherosclerosis are accelerated by oxidative stress.
- Diabetes: β-cell dysfunction is caused by ROS generated by hyperglycemia [81].

(iii) *Therapeutic Approaches Targeting Antioxidant Enzymes*

- Pharmacological Activators: Substances that increase SOD, CAT, and GPx activity include curcumin and resveratrol (Table 4).
- Gene therapy: Preventing illness by overexpressing antioxidant enzymes.
- Dietary Antioxidants: Enzymatic antioxidant action is supported by polyphenols, vitamins C, and E [80, 82].

Enzymatic antioxidants preserve cell integrity and functionality and are the defense against ROS. Their exact management is essential for balancing oxidative stress and physiological signaling. Gaining knowledge of these enzymes' molecular processes aids in the development of treatment plans for illnesses linked to oxidative stress [84, 85].

8 Non-Enzymatic Antioxidants Versus Reactive Oxygen Species (ROS)

Reactive oxygen species (ROS), or increased reactive compounds, are the natural products of cellular metabolism, primarily in the mitochondria. Excessive ROS buildup can result in oxidative stress, which can cause cellular malfunction, ageing, and several illnesses, including cancer, cardiovascular problems, and neurodegenerative ailments. However, regulated amounts of ROS are crucial for immune responses

Table 3 Enzymatic antioxidant defense systems

Enzymatic antioxidants	Function	Reaction	Types	Significance	Reference
SOD	Facilitates the conversion of superoxide anion ($O_2\bullet^-$) into molecular oxygen and hydrogen peroxide (H_2O_2)	$2O_2\bullet^- + 2H^+ \rightarrow H_2O_2 + O_2$	(Cu/Zn–SOD)–Cytosolic—SOD1 (Mn–SOD)–Mitochondrial—SOD2 (EC–SOD)–Extracellular—SOD3	Reduces oxidative stress and prevents mitochondrial dysfunction	[78–82]
CAT	Decomposes H_2O_2 into H_2O and O_2, preventing hydroxyl radical (•OH) formation	$2H_2O_2 \rightarrow 2H_2O + O_2$	Peroxisomes, cytoplasm, and mitochondria	Essential for detoxifying high concentrations of H_2O_2	
GPx	Decreases H2O2 and lipid hydroperoxides using glutathione (GSH) as a substrate	$H_2O_2 + 2GSH \rightarrow 2H_2O + GSSG$	GPx1–GPx8, located in cytosol and mitochondria	Prevents lipid peroxidation and maintains redox homeostasis	
Glutathione Reductase (GR)	Decreases glutathione (GSH) from oxidized glutathione (GSSG), sustaining antioxidant cycle	$GSSG + NADPH + H^+ \rightarrow 2\ GSH + NADP^+$	–	Maintains cellular GSH levels essential for redox balance	
Ascorbate Peroxidase (APX)	Reduces hydrogen peroxide using ascorbate as an electron donor	H_2O_2 + ascorbate → 2 H_2O + dehydroascorbate	–	Plays a crucial role in plant defense against oxidative stress	
GST	Helps in detoxification by catalyzing the conversion of decreased glutathione to electrophilic compounds	RX + GSH → R–SG + HX	–	Protects cells against toxic metabolites and oxidative damage	

Table 4 Clinical implications of enzymatic antioxidants and its key findings

S. n0	Key findings	Enzymatic antioxidants investigated	References
1	The continuous production of ROS in plant cellular compartments and emphasizes the role of enzymatic antioxidants like SOD, CAT, and APX in mitigating oxidative stress	(SOD), (CAT), (APX)	[83]
2	The dual role of ROS at low concentrations causing toxicity at high levels. It underscores the importance of enzymatic antioxidants in maintaining ROS balance and preventing oxidative-stress-related damage	General enzymatic antioxidants	[84]
3	The dual role of ROS in cell physiology, serving as sources of cellular damage and important signal molecules. It highlights the significance of enzymatic antioxidants in modulating ROS levels to maintain cellular homeostasis	General enzymatic antioxidants	[85]

and cell communication. Biological systems use enzymatic and nonenzymatic antioxidant defenses to combat ROS [86]. The processes and functions of nonenzymatic antioxidants in defending cells against oxidative damage are explained here. Types of Nonenzymatic Antioxidants, micromolecules known as nonenzymatic antioxidants destroy ROS by giving away hydrogen atoms or electrons. They contain metal-binding proteins, vitamins, carotenoids, polyphenols, and thiols (Table 5 and Fig. 2).

8.1 Sources of Free Radicals and ROS

Nonenzymatic antioxidants neutralize ROS, prevent oxidative-stress-related diseases, and maintain cellular homeostasis as shown in Fig. 3. A balanced diet rich in antioxidants and potential pharmacological interventions offers promising strategies to combat oxidative damage and promote health. Atoms or molecules containing extremely reactive unpaired electrons are known as free radicals. Nitric oxide (NO•), (•OH), and superoxide ($O_2•^-$) are a few examples. A particular free radical and nonradical O_2– containing molecules that take part in oxidative processes is known as ROS. O_2, H_2O_2, and peroxynitrite ($ONOO^-$) are a few examples [8, 88]. ROS and free radicals are double-edged swords that are necessary for cellular activity but may be dangerous when present in excess. Gaining knowledge of their causes, consequences, and regulation helps with illness prevention and treatment

Table 5 Types of nonenzymatic antioxidants and its significant function

S. no	Details	Function	Mechanism	Significance	Reference
1a	Vitamin-based antioxidants Ascorbic acid	Scavenges superoxide, hydroxyl radicals, and singlet oxygen	Donates electrons to neutralize ROS, forming dehydroascorbate, which can be regenerated	Defends lipids, proteins, and DNA from oxidative damage and regenerates vitamin E	[78, 79, 86, 87]
b	Vitamin E (tocopherols and tocotrienols)	Lipid-soluble antioxidant that protects membranes from peroxidation	Neutralizes lipid peroxyl radicals, preventing chain reactions of lipid oxidation	Preserves membrane integrity and reduces oxidative damage in lipoproteins	
c	Vitamin A (retinoids and carotenoids)	Quenches singlet oxygen and scavenges peroxyl radicals	Absorbs ROS energy and stabilizes them	Supports immune function, vision, and cellular differentiation	
2	Polyphenols and flavonoids (quercetin, catechins, resveratrol, curcumin)	Act as hydrogen donors and metal chelators	Neutralize ROS through hydrogen atom transfer and electron donation	Exhibit anti-inflammatory, neuroprotective, and cardioprotective effects	[80]
3	Glutathione (GSH)	Tripeptide antioxidant that detoxifies peroxides and regenerates other antioxidants	Reduces hydrogen peroxide and lipid peroxides via glutathione peroxidase	Maintains redox homeostasis and supports detoxification	[88]
4	Uric acid	Scavenges hydroxyl radicals and peroxynitrite	Directly neutralizes ROS and acts as a metal chelator	Major antioxidant in plasma, protecting against neurodegenerative diseases	[89]
5	Metal chelators (ferritin, ceruloplasmin, transferrin)	Bind transition metals (Fe, Cu) to prevent ROS generation via Fenton reactions	Limit free metal availability, reducing hydroxyl radical production	Protects against neurodegeneration and cardiovascular diseases	[90]

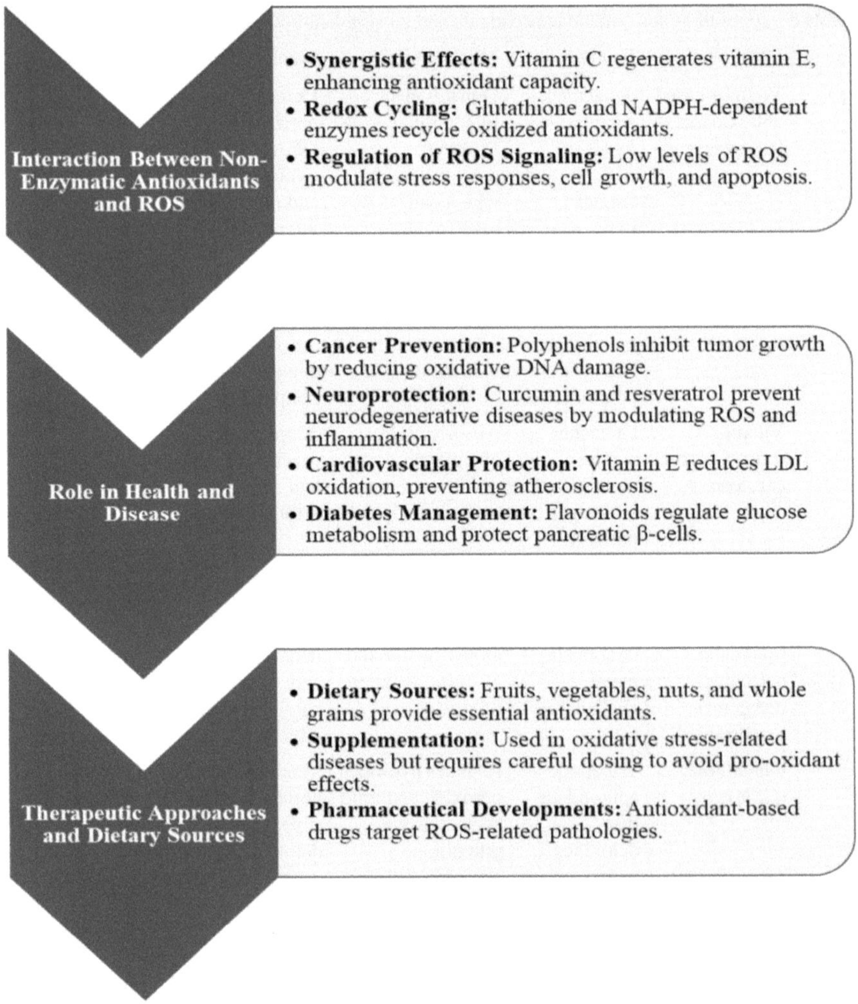

Fig. 2 Different roles played by ROS in man

strategies. Maintaining cellular health and lifespan requires balancing ROS creation and antioxidant defenses [6].

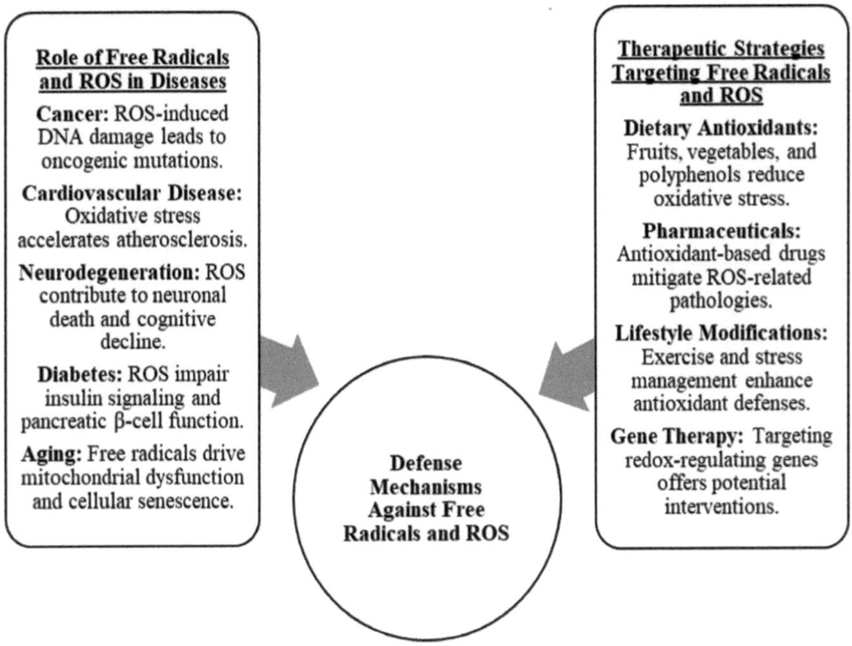

Fig. 3 Defense mechanism role of ROS

9 Conclusion

Over the last two decades, reactive species of ROS have been intensively investigated, and molecules have undergone alterations, causing damage known as oxidative stress to molecular signaling pathways and disrupting normal pathophysiological effects on redox biology. The chemistry and ion levels are elevated, although the assortment of hydrogen peroxide commands redox biology, and the defect of superoxide of –OH radicals increase oxidative stress. Furthermore, ROS-involved redox signaling is required for the multiple cellular pathways seen by the impacts of ROS on appropriate cellular differentiation, degeneration, and tissue proliferation in response to ageing. Meanwhile, redox signaling is not limited to oxidative stress. It also plays an important role in treating and controlling several chronic conditions, such as autoimmune disorders, cancer, and tissue regeneration loss with age. It may be difficult to understand that shifting ions and their interference with redox biology, which is supported by antioxidants, would alter both the normal and pathological states of the pathophysiology of the molecular mechanism. With the current situation, physical sternness is the primary method that triggers increased ROS generation, resulting in positive processes linked with reducing diabetes, ageing, cancer, and various chronic diseases. Finding the redox biology effectors that will maintain the right pathophysiological and biological molecular response, resulting in equilibrated homeostasis against human illness, is critical. As a result, targeted treatment can alleviate pain and promote tissue health.

References

1. Davis, C. G. (1998). *Signal transduction through the RAS pathway of Saccharomyces cerevisiae*. Princeton University.
2. Cortese-Krott, M. M., Koning, A., Kuhnle, G. G., Nagy, P., Bianco, C. L., Pasch, A., Wink, D. A., Fukuto, J. M., Jackson, A. A., van Goor, H., & Olson, K. R. (2017). The reactive species interactome: Evolutionary emergence, biological significance, and opportunities for redox metabolomics and personalized medicine. *Antioxidants and Redox Signaling, 27*(10), 684–712.
3. Li, J., Mara, P., Schubotz, F., Sylvan, J. B., Burgaud, G., Klein, F., Beaudoin, D., Wee, S. Y., Dick, H. J., Lott, S., & Cox, R. (2020). Recycling and metabolic flexibility dictate life in the lower oceanic crust. *Nature, 579*(7798), 250–255.
4. Biswas, M. S., & Mano, J. I. (2021). Lipid peroxide-derived reactive carbonyl species as mediators of oxidative stress and signaling. *Frontiers in Plant Science, 12*, Article 720867.
5. Zhang, Z., Dong, Q., Li, P., Fereja, S. L., Guo, J., Fang, Z., Zhang, X., Liu, K., Chen, Z., & Chen, W. (2021). Highly catalytic selectivity for hydrogen peroxide generation from oxygen reduction on Nd-doped $Bi_4Ti_3O_{12}$ nanosheets. *The Journal of Physical Chemistry C, 125*(45), 24814–24822.
6. Medrano-Macías, J., Flores-Gallegos, A. C., Nava-Reyna, E., Morales, I., Tortella, G., Solís-Gaona, S., & Benavides-Mendoza, A. (2022). Reactive oxygen, nitrogen, and sulfur species (RONSS) as a metabolic cluster for signaling and biostimulation of plants: An overview. *Plants, 11*(23), 3203.
7. Shin, S. Y., Kim, M. H., Kim, Y. H., Park, H. M., & Yoon, H. S. (2013). Co-expression of monodehydroascorbate reductase and dehydroascorbate reductase from Brassica rapa effectively confers tolerance to freezing-induced oxidative stress. *Molecules and Cells, 36*, 304–315.
8. Olson, K. R., Derry, P. J., Kent, T. A., & Straub, K. D. (2023). The effects of antioxidant nutraceuticals on cellular sulfur metabolism and signaling. *Antioxidants and Redox Signaling, 38*(1–3), 68–94.
9. Curieses Andrés, C. M., Pérez de Lastra, J. M., Andrés Juan, C., Plou Gasca, F. J., & Pérez-Lebeña, E. (2024) Antioxidant Metabolism Pathways in Vitamins, Polyphenols, and Selenium: Parallels and Divergences.
10. Dong, C., Fang, W., Yi, Q., & Zhang, J. (2022). A comprehensive review on reactive oxygen species (ROS) in advanced oxidation processes (AOPs). *Chemosphere, 308*, Article 136205.
11. Romero, E., Gómez Castellanos, J. R., Gadda, G., Fraaije, M. W., & Mattevi, A. (2018). Same substrate, many reactions: Oxygen activation in flavoenzymes. *Chemical reviews, 118*(4), 1742–1769.
12. Ross, J. (1995). MRNA stability in mammalian cells. *Microbiological reviews, 59*(3), 423–450.
13. Zhang, K. L., Leng, Y. N., Hao, R. R., Zhang, W. Y., Li, H. F., Chen, M. X., & Zhu, F. Y. (2024). Adaptation of high-altitude plants to harsh environments: Application of phenotypic-variation-related methods and multi-omics techniques. *International Journal of Molecular Sciences, 25*(23), 12666.
14. Lehti-Shiu, M. D., & Shiu, S. H. (2012). Diversity, classification and function of the plant protein kinase superfamily. *Philosophical Transactions of the Royal Society B: Biological Sciences, 367*(1602), 2619–2639.
15. Chen, J., Liu, L., Wang, W., & Gao, H. (2022). Nitric oxide, nitric oxide formers and their physiological impacts in bacteria. *International journal of molecular sciences, 23*(18), 10778.

16. Kumari, A., Pathak, P. K., Bulle, M., Igamberdiev, A. U., & Gupta, K. J. (2019). Alternative oxidase is an important player in the regulation of nitric oxide levels under normoxic and hypoxic conditions in plants. *Journal of Experimental Botany, 70*(17), 4345–4354.
17. Corpas, F. J., & Palma, J. M. (2018). Assessing nitric oxide (NO) in higher plants: An outline. *Nitrogen, 1*(1), 12–20.
18. Han, S., Li, Y., & Gao, H. (2022). Generation and physiology of hydrogen sulfide and reactive sulfur species in bacteria. *Antioxidants, 11*(12), 2487.
19. Selbach, B. P., Pradhan, P. K., & Dos Santos, P. C. (2013). Protected sulfur transfer reactions by the Escherichia coli Suf system. *Biochemistry, 52*(23), 4089–4096.
20. Hanaoka, K., Sasakura, K., Suwanai, Y., Toma-Fukai, S., Shimamoto, K., Takano, Y., Shibuya, N., Terai, T., Komatsu, T., Ueno, T., & Ogasawara, Y. (2017). Discovery and mechanistic characterization of selective inhibitors of H2S-producing enzyme: 3-mercaptopyruvate sulfurtransferase (3MST) targeting active-site cysteine persulfide. *Scientific reports, 7*(1), 40227.
21. Capaldi, F. R., Gratão, P. L., Reis, A. R., Lima, L. W., & Azevedo, R. A. (2015). Sulfur metabolism and stress defense responses in plants. *Tropical plant biology, 8*, 60–73.
22. Bano, K., Kumar, B., Alyemeni, M. N., & Ahmad, P. (2022). Protective mechanisms of sulfur against arsenic phytotoxicity in Brassica napus by regulating thiol biosynthesis, sulfur-assimilation, photosynthesis, and antioxidant response. *Plant Physiology and Biochemistry, 188*, 1–11.
23. Gibson, K. E., Kobayashi, H., & Walker, G. C. (2008). Molecular determinants of a symbiotic chronic infection. *Annual review of genetics, 42*(1), 413–441.
24. Li, Q., Gao, Y., & Yang, A. (2020). Sulfur homeostasis in plants. *International Journal of Molecular Sciences, 21*(23), 8926.
25. Zenda, T., Liu, S., Dong, A., & Duan, H. (2021). Revisiting sulfur—The once neglected nutrient: It's roles in plant growth, metabolism, stress tolerance and crop production. *Agriculture, 11*(7), 626.
26. Mondal, S., Panda, D., & Bose, B. (2022) Seed biopriming with biopesticide: A key to sustainability of agriculture. *New and future developments in microbial biotechnology and bioengineering*, pp. 265–288.
27. Adak, M. K., Das, A., Kundu, A., Chatterjee, M., & Hasanuzzaman, M. (2023). Molecular mechanisms in understanding anoxia tolerance in rice seeds under submergence and their implication in rice biotechnology. *Seeds, 2*(3), 246–258.
28. Boccaccini, A., Zacharaki, V., Sun, J., & Jeong, B. R. (2023). Jingli Yang1, Jinnan Song1 and Byoung Ryong Jeong 1, 2, 3. *Women in Plant Physiology, 2022*, 7.
29. Arif, Y., Hayat, S., Yusuf, M., & Bajguz, A. (2021). Hydrogen sulfide: A versatile gaseous molecule in plants. *Plant Physiology and Biochemistry, 158*, 372–384.
30. Tuteja, N., and Singh, G. S. (Eds.) (2012). *Plant acclimation to environmental stress.* Springer Science and Business Media.
31. Ghosal, K., Sinha, D., & Singh, S. P. (2022). A mechanistic overview of heavy metal detoxification in plants. *Heavy Metals in Plants*, 163–197.
32. Sachdev, S., Ansari, S. A., & Ansari, M. I. (2023). *Reactive oxygen species in plants.* Springer.
33. Dikmen, M. İR. İŞ, Canturk, Z., Tilki, E. K., & Engur, S. (2017). Evaluation of antiangiogenic and antimetastatic effects of Penicillium chrysogenum secondary metabolites. *Indian Journal of Pharmaceutical Sciences, 79*(1), 49–57.
34. Li, L., Wang, Y., & Shen, W. (2012). Roles of hydrogen sulfide and nitric oxide in the alleviation of cadmium-induced oxidative damage in alfalfa seedling roots. *BioMetals, 25*(3), 617–631.
35. Amooaghaie, R., & Enteshari, S. (2017). Role of two-sided crosstalk between NO and H2S on improvement of mineral homeostasis and antioxidative defense in Sesamum indicum under lead stress. *Ecotoxicology and Environmental Safety, 139*, 210–218.
36. Kaya, C., Ashraf, M., Alyemeni, M. N., & Ahmad, P. (2020). Responses of nitric oxide and hydrogen sulfide in regulating oxidative defence system in wheat plants grown under cadmium stress. *Physiologia plantarum, 168*(2), 345–360.

37. Uchida, A., Jagendorf, A. T., Hibino, T., Takabe, T., & Takabe, T. (2002). Effects of hydrogen peroxide and nitric oxide on both salt and heat stress tolerance in rice. *Plant Science, 163*(3), 515–523.
38. Christou, A., Manganaris, G. A., & Fotopoulos, V. (2014). Systemic mitigation of salt stress by hydrogen peroxide and sodium nitroprusside in strawberry plants via transcriptional regulation of enzymatic and non-enzymatic antioxidants. *Environmental and Experimental Botany, 107*, 46–54.
39. Wang, Y., Li, L., Cui, W., Xu, S., Shen, W., & Wang, R. (2012). Hydrogen sulfide enhances alfalfa (Medicago sativa) tolerance against salinity during seed germination by nitric oxide pathway. *Plant and Soil, 351*, 107–119.
40. Antoniou, C., Xenofontos, R., Chatzimichail, G., Christou, A., Kashfi, K., & Fotopoulos, V. (2020). Exploring the potential of nitric oxide and hydrogen sulfide (NOSH)-releasing synthetic compounds as novel priming agents against drought stress in Medicago sativa plants. *Biomolecules, 10*(1), 120.
41. Fubini, B., & Hubbard, A. (2003). Reactive oxygen species (ROS) and reactive nitrogen species (RNS) generation by silica in inflammation and fibrosis. *Free radical Biology and medicine, 34*(12), 1507–1516.
42. KC, S. K., Ghimire, B., Hong, S. H., Oh, J. S., & Szili, E. J. (2025). How to control the plasma jet production of reactive species for medical therapy? A topical review. *Journal of Physics D: Applied Physics.*
43. Negi, M., Kaushik, N., Lamichhane, P., Patel, P., Jaiswal, A., Choi, E. H., & Kaushik, N. K. (2024). Nitric oxide water-driven immunogenic cell death: Unfolding mitochondrial dysfunction's role in sensitizing lung adenocarcinoma to ferroptosis and autophagic cell death. *Free Radical Biology and Medicine, 222*, 1–15.
44. Kumar, A., Saini, V. K., Huaccallo Aguilar, Y., & Reinecke, S. F. (2023). Characterization of reactive species in water induced by cold atmospheric air plasma: Experimental applications for industrial micropollutant removal from wastewater and seed germination. Available at SSRN 5059536.
45. Stomberski, C. T., Hess, D. T., & Stamler, J. S. (2019). Protein S-nitrosylation: Determinants of specificity and enzymatic regulation of S-nitrosothiol-based signaling. *Antioxidants and redox signaling, 30*(10), 1331–1351.
46. Bignon, E., Allega, M. F., Lucchetta, M., Tiberti, M., & Papaleo, E. (2018). Computational structural biology of S-nitrosylation of cancer targets. *Frontiers in oncology, 8*, 272.
47. Ahmad, R., Rasheed, Z., & Ahsan, H. (2009). Biochemical and cellular toxicology of peroxynitrite: Implications in cell death and autoimmune phenomenon. *Immunopharmacology and immunotoxicology, 31*(3), 388–396.
48. Ferrer-Sueta, G., Campolo, N., Trujillo, M., Bartesaghi, S., Carballal, S., Romero, N., Alvarez, B., & Radi, R. (2018). Biochemistry of peroxynitrite and protein tyrosine nitration. *Chemical reviews, 118*(3), 1338–1408.
49. Di Meo, S., Reed, T. T., Venditti, P., & Victor, V. M. (2016). Role of ROS and RNS sources in physiological and pathological conditions. *Oxidative Medicine and Cellular Longevity, 2016*(1), 1245049.
50. Circu, M. L., & Aw, T. Y. (2010). Reactive oxygen species, cellular redox systems, and apoptosis. *Free Radical Biology AND Medicine, 48*(6), 749–762.
51. Geng, Y., Wang, Z., Zhou, J., Zhu, M., Liu, J., & James, T. D. (2023). Recent progress in the development of fluorescent probes for imaging pathological oxidative stress. *Chemical Society Reviews, 52*(11), 3873–3926.
52. Kolanowski, J. L., Kaur, A., & New, E. J. (2016). Selective and reversible approaches toward imaging redox signaling using small-molecule probes. *Antioxidants & Redox Signaling, 24*(13), 713–730.
53. Mukwevho, E., Ferreira, Z., & Ayeleso, A. (2014). Potential role of sulfur-containing antioxidant systems in highly oxidative environments. *Molecules, 19*(12), 19376–19389.
54. Francioso, A., Baseggio Conrado, A., Mosca, L., & Fontana, M. (2020). Chemistry and biochemistry of sulfur natural compounds: Key intermediates of metabolism and redox biology. *Oxidative Medicine and Cellular Longevity, 2020*(1), 8294158.

55. Karimi, M., Ignasiak, M. T., Chan, B., Croft, A. K., Radom, L., Schiesser, C. H., Pattison, D. I., & Davies, M. J. (2016). Reactivity of disulfide bonds is markedly affected by structure and environment: Implications for protein modification and stability. *Scientific reports, 6*(1), 38572.
56. Jacob, C., Battaglia, E., Burkholz, T., Peng, D., Bagrel, D., & Montenarh, M. (2012). Control of oxidative posttranslational cysteine modifications: From intricate chemistry to widespread biological and medical applications. *Chemical Research in Toxicology, 25*(3), 588–604.
57. Moldogazieva, N. T., Mokhosoev, I. M., Feldman, N. B., & Lutsenko, S. V. (2018). ROS and RNS signalling: Adaptive redox switches through oxidative/nitrosative protein modifications. *Free Radical Research, 52*(5), 507–543.
58. Ijaz, M., Ali, Q., Fahad, S., Ashraf, S., Shahid, M., Ahmad, S., & Hasanuzzaman, M. (2019). Role of reactive sulfur species in the oxidative metabolism in plants. *Reactive oxygen, nitrogen and sulfur species in plants: Production, metabolism, signaling and defense mechanisms*, pp. 729–742.
59. Kelly, S. S. (2022). *New strategies for the design of reactive sulfur/nitrogen species releasing systems*. Washington State University.
60. Claiborne, A. L., Miller, H., Parsonage, D., & Ross, R. P. (1993). Protein-sulfenic acid stabilization and function in enzyme catalysis and gene regulation. *The FASEB Journal, 7*(15), 1483–1490.
61. Nagy, P. (2013). Kinetics and mechanisms of thiol–disulfide exchange covering direct substitution and thiol oxidation-mediated pathways. *Antioxidants & Redox Signaling, 18*(13), 1623–1641.
62. Roos, G., & Messens, J. (2011). Protein sulfenic acid formation: From cellular damage to redox regulation. *Free Radical Biology and Medicine, 51*(2), 314–326.
63. Kettenhofen, N. J., & Wood, M. J. (2010). Formation, reactivity, and detection of protein sulfenic acids. *Chemical Research in Toxicology, 23*(11), 1633–1646.
64. Biswas, S., Chida, A. S., & Rahman, I. (2006). Redox modifications of protein–thiols: Emerging roles in cell signaling. *Biochemical Pharmacology, 71*(5), 551–564.
65. Szechyńska-Hebda, M., Ghalami, R. Z., Kamran, M., Van Breusegem, F., & Karpiński, S. (2022). To be or not to be? Are reactive oxygen species, antioxidants, and stress signalling universal determinants of life or death? *Cells, 11*(24), 4105.
66. Sonmez, M. C., Yirmibesoglu, S. S. S., Ozgur, R., Uzilday, B., & Turkan, I. (2024). Roles of reactive carbonyl species (RCS) in plant response to abiotic stress. In *ROS signaling in plants: Methods and protocols* (pp. 101–130). New York, NY: Springer US.
67. Rai, G. K., Kumar, P., Choudhary, S. M., Singh, H., Adab, K., Kosser, R., Magotra, I., Kumar, R. R., Singh, M., Sharma, R., & Corrado, G. (2023). Antioxidant potential of glutathione and crosstalk with phytohormones in enhancing abiotic stress tolerance in crop plants. *Plants, 12*(5), 1133.
68. Chen, T. H., Wang, H. C., Chang, C. J., & Lee, S. Y. (2024). Mitochondrial glutathione in cellular redox homeostasis and disease manifestation. *International Journal of Molecular Sciences, 25*(2), 1314.
69. de Barboza, G. D., Guizzardi, S., Moine, L., & de Talamoni, N. T. (2017). Oxidative stress, antioxidants and intestinal calcium absorption. *World Journal of Gastroenterology, 23*(16), 2841.
70. Brand, M. D. (2010). The sites and topology of mitochondrial superoxide production. *Experimental Gerontology, 45*(7–8), 466–472.
71. Murphy, M. P. (2009). How mitochondria produce reactive oxygen species. *Biochemical Journal, 417*(1), 1–13.
72. Zhang, J., Wang, X., Vikash, V., Ye, Q., Wu, D., Liu, Y., & Dong, W. (2016). ROS and ROS-mediated cellular signaling. *Oxidative Medicine and Cellular Longevity, 2016*(1), 4350965.
73. Boveris, A. (1977). Mitochondrial production of superoxide radical and hydrogen peroxide. *Tissue Hypoxia and Ischemia*, 67–82.
74. Zorov, D. B., Juhaszova, M., & Sollott, S. J. (2014). Mitochondrial reactive oxygen species (ROS) and ROS-induced ROS release. *Physiological Reviews, 94*(3), 909–950.

75. Turrens, J. F. (2003). Mitochondrial formation of reactive oxygen species. *The Journal of Physiology, 552*(2), 335–344.
76. Chenna, S., Koopman, W. J., Prehn, J. H., & Connolly, N. M. (2022). Mechanisms and mathematical modeling of ROS production by the mitochondrial electron transport chain. *American Journal of Physiology-Cell Physiology, 323*(1), C69–C83.
77. Venditti, P., Di Stefano, L., & Di Meo, S. (2013). Mitochondrial metabolism of reactive oxygen species. *Mitochondrion, 13*(2), 71–82.
78. Liu, Z., Ren, Z., Zhang, J., Chuang, C. C., Kandaswamy, E., Zhou, T., & Zuo, L. (2018). Role of ROS and nutritional antioxidants in human diseases. *Frontiers in Physiology, 9*, Article 360203.
79. Galli, F., Piroddi, M., Annetti, C., Aisa, C., Floridi, E., & Floridi, A. (2005). Oxidative stress and reactive oxygen species. *Contributions to Nephrology, 149*(149), 240–260.
80. Liguori, I., Russo, G., Curcio, F., Bulli, G., Aran, L., Della-Morte, D., Gargiulo, G., Testa, G., Cacciatore, F., Bonaduce, D., Abete, P. (2018). Oxidative stress, aging, and diseases. *Clinical Interventions in Aging*, 757–772.
81. Pisoschi, A. M., Pop, A., Iordache, F., Stanca, L., Predoi, G., & Serban, A. I. (2021). Oxidative stress mitigation by antioxidants-an overview on their chemistry and influences on health status. *European Journal of Medicinal Chemistry, 209*, Article 112891.
82. Ali, S. S., Ahsan, H., Zia, M. K., Siddiqui, T., & Khan, F. H. (2020). Understanding oxidants and antioxidants: Classical team with new players. *Journal of food biochemistry, 44*(3), Article e13145.
83. Das, K., & Roychoudhury, A. (2014). Reactive oxygen species (ROS) and response of antioxidants as ROS-scavengers during environmental stress in plants. *Frontiers in Environmental Science, 2*, 53.
84. Lázaro, J. J., Jiménez, A., Camejo, D., Iglesias-Baena, I., Martí, M. D. C., Lázaro-Payo, A., Barranco-Medina, S., Sevilla, F. (2013). Dissecting the integrative antioxidant and redox systems in plant mitochondria. Effect of stress and S-nitrosylation. *Frontiers in plant science, 4*, 460.
85. Sies, H., & Jones, D. P. (2020). Reactive oxygen species (ROS) as pleiotropic physiological signalling agents. *Nature Reviews Molecular Cell Biology, 21*(7), 363–383.
86. Juan, C. A., Pérez de la Lastra, J. M., Plou, F. J., & Pérez-Lebeña, E. (2021). The chemistry of reactive oxygen species (ROS) revisited: Outlining their role in biological macromolecules (DNA, lipids and proteins) and induced pathologies. *International journal of molecular sciences, 22*(9), 4642.
87. Krumova, K., & Cosa, G. (2016). Overview of reactive oxygen species.
88. Teleanu, D. M., Niculescu, A. G., Lungu, I. I., Radu, C. I., Vladâcenco, O., Roza, E., Costăchescu, B., Grumezescu, A. M., & Teleanu, R. I. (2022). An overview of oxidative stress, neuroinflammation, and neurodegenerative diseases. *International Journal of Molecular Sciences, 23*(11), 5938.
89. Mirończuk-Chodakowska, I., Witkowska, A. M., & Zujko, M. E. (2018). Endogenous non-enzymatic antioxidants in the human body. *Advances in Medical Sciences, 63*(1), 68–78.
90. Zandi, P., & Schnug, E. (2022). Reactive oxygen species, antioxidant responses and implications from a microbial modulation perspective. *Biology, 11*, 155.

A Review of Redox-Active Natural Products, Their Synthesis, Isolation, and Biological Activities

R. Parameswari, P. Madhan Kumar, D. Abilash, K. Kasthuri, and R. Babujanarthanam

Abstract Recent study on designing medications from natural ingredients demonstrates a certain resilience to treatment time. The need for pharmaceuticals is urgent, and the use of effective antibiotics as medicine is determined by the desire for recycling as well as environmental friendliness, which is safer for agriculture or other species. Apart from medicine, natural goods developed for nutrition and antiaging, as well as cosmetics to meet societal demands, require study in nature's treasure chest known as green gold. Secondary metabolite products/agents of redox materials generated from plants, fungal microorganisms, and bacteria are the primary sources of fascinating discoveries. While these agents provide commanding instructions to humans, other organisms, and animals via molecular cellular signals, they also protect them with antioxidants, repellents, fungicides, and anticancer agents, as well as promote the use of effective sensors and antibiotics by destroying/killing foreign particles. Interestingly, research on natural products over the decades has demonstrated that some of the more than ten decades of substances identified, such as allicin and other flavonoids, have been investigated in depth in live creatures, both in vivo and in vitro, as well as in numerous clinical trials. Furthermore, classic approaches such as chemo-genetic screening of mRNA profiling, bioanalytical techniques of proteomics, Western blot, PCR, and RT-PCR are employed as potent diagnostic tools for diagnosing intracellular physiological conditions. As a result, in the case of active redox chemicals and notably reactive sulfur species (RSS), the methodologies have weakened the hypothesis of cellular thiolstat, and further study is needed to understand why and how ROS substances function on a regular or frequent basis.

R. Parameswari · P. Madhan Kumar · R. Babujanarthanam (✉)
Department of Biotechnology, Thiruvalluvar University, Serkkadu, Vellore 632115, Tamil Nadu, India
e-mail: babukmg@gmail.com

D. Abilash
Gene Cloning and Technology Lab, Department of Biotechnology, SBST, VIT Vellore, Vellore 632014, Tamil Nadu, India

K. Kasthuri
Department of Biochemistry, Meenakshi Medical College Hospital and Research Institute, Meenakshi Academy of Higher Education and Research, Kanchipuram 631552, Tamil Nadu, India

K. Kasinathan et al. (eds.), *Recent Advances in Oxidative Stress Associated Chronic Diseases Volume 1*, https://doi.org/10.1007/978-981-96-9670-3_2

1 Introduction

Natural products derived from plants and other microbes have satisfied the demands of nutrition, agriculture, pharmacology, medicine, and antiaging cosmetics through functional foods. Nature's riches, on the other hand, are given with a wide range of materials, beginning with mixed crudes and oils and progressing to pure goods. Among these, various materials as substances and redox-active materials are of particular interest. Furthermore, these redox chemicals have been extant for a long time, but only a few revivals have occurred due to environmental variables [1]. As a result, we may all recognize these changes as advancements: (i) a modernized society requires nutritious meals for individuals who are beginning to age, particularly those aged 50 and more., (ii) the need surrounding the environment and the usage of eco-friendly agents used to protect plants, also known as green pesticides, which are compatible with the food chain, conveniently available, and affordable by all sorts of society, (iii) day-by-day emerging pathogenic microbes such as bacteria, virus, and fungi needs to design modified antibiotics with respect to molecular structure as antimicrobial agents, (iv) multiresistant drugs need to overcome the newly emerging pathogenic against protection, (v) selective foods that will be used to isolate were contains selective competitive compounds that act against cancer cells and their proliferation along with existing normal cells, results in provided, (vi) the use of modern technology transforms formerly inaccessible natural goods into newer, more bioavailable, and solubilized products in response to demand, (vii) the use of updated nanotechnology in natural goods to provide targeted medicine delivery in a short period of time in order to treat or meet the respect demand/need (viii) in order to detect the flow and monitor the intracellular redox signals, focused novel staining and diagnosis approaches were identified as intracellular diagnostics for RSI [2–4]. Surprisingly, developing natural products is simple and yields better results than artificial products, whereas using active redox secondary metabolites, synthesized based on their origin, physiochemical properties, and optimized biological activities, can also be monitored by recent developmental research [5]. As a result, this chapter focuses on natural goods derived from redox-active compounds, and it is time to conduct multiple research investigations in academic and industrial settings to accelerate the growth of such products.

2 Why Natural Products?

In the recent twenty-first century, the new history of study topics is *"gold rush"* and *"green gold"* for distinct chunks of chemical compounds that were buried in this evolution, and describing it is the key effort now. The perception of redox-active compounds was derived from commonly available garlic plants, particularly allicin, or various fruits rich in phenols, polyphenols, flavonoids, and other exotic components of lichens, cacti, and carnivorous plants, all of which release redox agents and are known as secondary active redox substances. Few substances were as easily available as garlic, selenium, and allium of onions, which have been recognized after being grown in massive quantities across the world. On the other hand, some parts of the planet are difficult to develop owing to weather and nutritional requirements, resulting in a decrease in the growth of plants and other microbes that produce biologically active products [6]. Furthermore, in certain unusual places of the environment, organisms may hide, which may be poisonous, and on Earth, they must be frozen to prevent microbial outbreaks during natural catastrophes such as geography. In any event, the quest for new products from natural molecules as new sources results in scientifically and commercially viable resources. Oxidative stress or oxidative-stress-related cause damage rises with age, and it may be prevented or controlled with a balanced nutritional diet high in chemodigestive/compromised agents such as antioxidants and vitamins, as well as competitive agents, particularly when separated [7]. Certainly, cudarflavone B, a redox-active molecule found as a neuroprotective agent, was investigated in most clinical studies. Furthermore, how antioxidants function in favor of molecular pathways and, controversially, against redox species must be investigated and given specific attention. Furthermore, it is unexpected that ROS species are required, as redox ions maintain membrane charges and guard against harmful chemicals. Because of its electronic supply to the cell membrane, it protects the membrane from electrical heat shock caused by pathogenic impediment entering the host tissue. Reducing equivalents of redox species simply transfer or take electrons, and they primarily operate as antioxidants, particularly flavonoids, without interfering with or contributing to molecular signaling [8–10].

3 Reactive Oxygen Species (ROS)

Treating diseases effects on one organ/one symptom toward an approach based on the patient's pathophysiology system condition, which includes individual history, lifestyle, microbiome, and genetic inheritance blueprint, are the major deciding factors in deciding and treating diseases in the practice of medicine in the paradigmatic shift. Undeniably, illnesses diagnosed/symptoms with the same characteristics in clinical and multifactorial genesis include a wide range of cells, molecules, and organs. Thus, multiple entrance of complexity locations presents the difficulty of biological reasons and their related appropriate treatment/therapy [11]. Each therapy requires an understanding of the multidisciplinary approach of drug biology into pharmacokinetic parameters. Recent research based on existing illness modules discovered that further mathematical techniques are required as core hubs for disease treatments. Even cardiovascular diseases, diabetes, cancer, asthma, neurological illnesses, obesity, and cancer have few known causes. It is extremely perplexing since a single symptom represents the cause of several diseases, resulting in a significant worldwide burden to classify diseases and their severity. As a result, death and morbidity rates increase. Despite being discovered as a common component, the reactive species interactome (RSI) caused oxidative stress, prompting some to create the term "redox diseases." The phrase oxidative stress refers to an imbalance between antioxidants and oxidants that favors oxidants, resulting in the disruption of redox signals and dominance/damage to molecular membranes. At the first stage of illnesses, it is thought that they formed or were triggered [12]. Later, even if the use of specific antioxidants against redox imbalances did not provide the expected results, it was proposed that further research into the underlying pathophysiology of redox illnesses be conducted. The redox code in current heredity enables the identification of reproduction and the genetic code-based physiological therapy described by Jones and Sies; precise regulatory enzymes and their associated cofactor elements determine molecular function and define, allow, and safeguard it. Based on this idea, reactive oxygen species (ROS), an endogenous oxidative signaling that regulates enzymatic molecules, operate as a switch to change cysteine redox residues [13]. The modification of cysteine residues on its functional residual site results in a disruption of its enzymatic activity-associated protein structure, which leads to inefficient transcription. Adjusting the enzymatic environment causes undesirable changes to the intended molecules, resulting in unfitness as shown in Fig. 1. Furthermore, redox species form a larger network than ROS. It is a kind of free radical, like nitric oxide (NO), that is produced endogenously by the enzyme nitric oxide synthases (NOS), an efficient second messenger molecule that plays a variety of physiological roles [14]. The chemical interaction between NO and ROS reduces the production of reactive nitrogen species (RNS). According to molecular signaling, both ROS and RNS targeted cysteine thiols and their oxidative physiology alterations, which included the modification of sulfonic acid and sulfane, as well as sulfur-containing polysulfide molecules. As a result of their equivalency, ROS and RNS chemicals are both classified as reactive sulfur species. RSS, like ROS and RNS, is a pathologic oxidative damage-causing agent implicated in signaling systems, but it is not as well understood as ROS and RNS [15, 16].

4 Redox Materials

Materials, which are capable of oxidation–reduction (redox) processes, which indicate the transferring of electrons between various chemical species, are referred to as redox-active materials. These materials are essential for many scientific, technological, and industrial activities, because they can engage in electron transfer, which makes them appropriate for applications such as biosensing, energy storage, catalysis, and environmental remediation. Redox-active materials are distinguished by their large surface area, redox potential, conductivity, or stability, as well as their capacity to reversibly receive or lose electrons, allowing them to go through several redox cycles without suffering significant damage. Redox-active materials frequently break down into several groups, each with its own special characteristics, such as metal oxides, carbon-based materials, conductive polymers, and hybrid materials [17]. High redox activity and stability are characteristics of metal oxides like TiO_2, FeO_4, and MnO_2, which have uses in photocatalysis for eliminating pollutants and

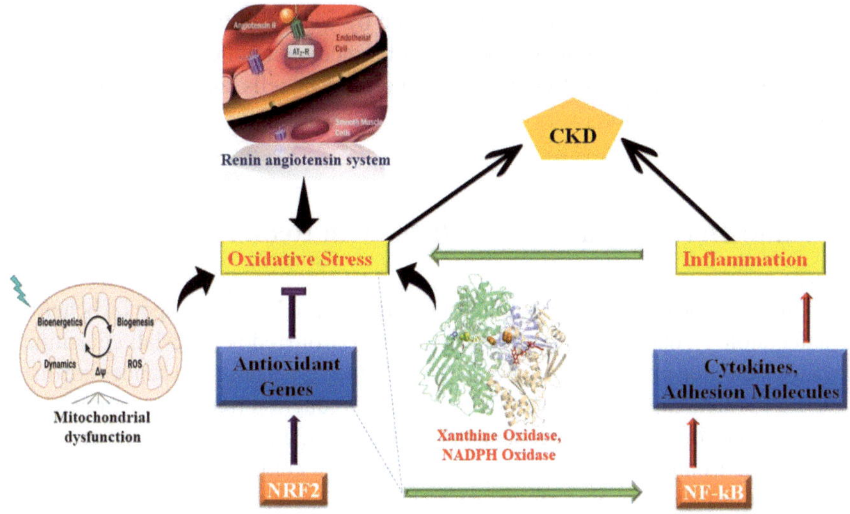

Fig. 1 Redox-active material and their interaction with cellular signaling

storage of energy in lithium-ion batteries. Carbon-based materials, such as graphene, graphene oxide, and carbon nanotubes, are perfect for electrochemical sensors high supercapacitors because of their large surface area and electrical conductivity. Due to their conductivity as well as biocompatibility, polymers with conductivity like polyaniline (PANI) and polypyrrole (PPy) are frequently utilized in biosensors and energy storage devices. On the other hand, hybrid materials like metal–organic frameworks (MOFs) combine the qualities of both inorganic and organic components, making them adaptable for energy storage, environmental remediation, along with catalysis [18]. These materials' redox activity mechanisms differ based on their composition and structure. Iron cycles between +2 and +3 oxidation states in FeO_4, an example of how metal oxides engage in redox reactions by changing their oxidation state of metal ions. The redox activity of carbon-based materials is enhanced by functional groups such as carboxyl (–COOH) and hydroxyl (–OH), which promote electron transport through conjugated π-electron systems. Conductive polymers, like PANI, go through redox reactions when they are doped and undoped, changing from conducting to insulating states. By combining the redox properties of their metal centers and organic linkers, hybrid materials allow for a variety of electron transport pathways. Pt/C catalysts are essential for hydrogen fuel cells, while redox-active materials such as TiO_2 are employed in photocatalytic water dissolution and pollutant degradation [19]. Redox-active materials are utilized in electrochemical sensors in the sensing field to detect gasses, biomolecules, and contaminants; graphene-based sensors are utilized to detect glucose. Additionally, conductive polymers are important for biosensing, which is the detection of cells, proteins, and DNA. The redox characteristics of these materials are advantageous for environmental remediation applications, such as the removal of arsenic using FeO_4 nanoparticles and the possibility of producing hydrogen and oxygen by water splitting using redox-active materials. Conductive polymers find use in biomedical applications such as tissue engineering to encourage differentiation and growth of cells and deliver medications for controlled absorption [20].

5 Nano-Derived Natural Products with Redox Materials

Combining the advantageous qualities of natural materials with the special benefits of nanotechnology, nano-derived organic compounds from redox materials constitute an exciting new area in material science and biotechnology. These substances are distinguished by their capacity to engage in oxidation–reduction (redox) reactions, which promote electron transport mechanisms vital to a wide range of industrial, biological, and environmental applications. Redox-active natural compounds provide new opportunities for developing multifunctional, sustainable, and cost-effective systems in areas including biosensing, energy storage, environmental remediation, and drug delivery [21]. The ability to use nanostructured materials to improve the performance of biologically derived substances, such as vitamins, enzymes, and plant-based compounds, while utilizing their natural properties—often attaining higher strength, activity, as well as versatility—makes this integration particularly captivating. Bioactive chemicals, many of which include redox characteristics, are abundant in natural products [22]. Natural compounds with anti-inflammatory, anticancer, antimicrobial, and antioxidant qualities include vitamins, polyphenols, flavonoids, alkaloids, and terpenoids. In addition to retaining their bioactivity, these natural compounds acquire new properties like controlled release, more effectively targeting, as well as increased stability when they are integrated into nanomaterials like nanoparticles, nanocomposites, or nanoemulsions. These properties are crucial for successful use in biomedical and environmental contexts [23]. Redox-active natural materials by offering a barrier that prevents these substances from deteriorating, so increasing their useful life and durability. For example, liposomes, micelles, or polymeric nanoparticles are examples of nanocarriers that can encapsulate polyphenolic chemicals, which are well-known for their strong antioxidant capabilities. Targeted cancer therapies and antioxidant-based treatments benefit greatly from this encapsulation, because it increases the chemicals' solubility and bioavailability, enabling them to more effectively reach tissues or cells. These natural polyphenols maintain their redox characteristics, and in certain situations, the combined effects of the nanoparticles utilized for administration increase their effectiveness [24]. Nanomaterial-derived redox-active natural compounds are also important for restoration of the environment, because they can be applied to the immediate issue of chemical contamination. Integrating certain plant-derived chemicals, such as tannins or anthocyanins, with nanomaterials, like iron oxide nanoparticles, can improve their redox properties. Heavy metals and organic pollutants can be eliminated, or wastewater can be treated with these materials. Iron oxide nanoparticles and natural polyphenols, for instance, can function as catalysts to reduce toxic metals like chromium (VI) into less dangerous types [25]. Due to their capacity to scavenge free radicals and alter signaling pathways, redox-active natural compounds like resveratrol, curcumin and quercetin have demonstrated promise in the treatment of several illnesses, including cancer, inflammation, as well as cardiovascular conditions. These substances have greater therapeutic efficacy when they are encapsulated in nanoparticles because they may be transported to the target site more efficiently. These chemicals' redox-active properties also make them suitable for use in biosensors that identify disease-related biomarkers. For example, redox-active natural products functionalized into gold nanoparticles can be employed in electrochemical sensors to identify proteins, glucose, or DNA, providing extremely sensitive and specific tools for diagnosis [26, 27].

6 Nano-Derived Natural Products Isolation

Improving that the bioactive components are effectively extracted, purified, and combined with nanomaterials for increased efficacy requires several meticulous and crucial stages in the isolation of nano-derived natural products. The first step is accumulating natural sources, which may originate from a variety of living things, such as bacteria, fungus, plants, animals, and marine life. The selection of raw materials is crucial since the bioactive components they contain influence the end product's therapeutic potential and utilization [28]. Several extraction techniques are used in this stage to obtain bioactive chemicals. Solvent extraction, which dissolves and extracts the required chemicals using solvents including ethanol, methanol, or hexane, and steam distillation or cold pressing, which are mostly used to extract essential oils from plants, are the most often used extraction methods. The process of steam distillation employs steaming to extract aromatic oils, which helps protect delicate components, while cold pressing, specifically, includes manually pressing material from plants to get oils without heat. Purification and separation are necessary since the crude mixture that results from extraction frequently contains a variety of undesirable chemicals in addition to the desired bioactive molecules. To separate the extract's bioactive ingredients from other contaminants, this procedure is essential [29]. The extract is purified using methods like filtration, centrifugation, and chromatography—more especially, high—performance liquid chromatography, or HPLC. HPLC ensures that target bioactive molecules are isolated with great precision by effectively separating compounds according to their chemical properties [30]. Encapsulation and functionalization of nanomaterials are done to improve the bioactive chemicals' stability, controlled release, and bioavailability. The organic substance may be affixed to the surfaces of nanomaterials or enclosed in nanoparticles. Dendrimers, liposomes, micelles, and biodegradable polymer nanoparticles such as chitosan and PLGA [poly(lactic-co-glycolic acid)] are examples of frequently utilized nanomaterials. These nanomaterials aid in the transportation of bioactive substances across biological membranes, prevent degradation of the compounds, and ensure the targeted or sustained release of their therapeutic agents [31]. Furthermore, hydrophobic natural compounds—which are more difficult to work with in conventional formulations—are made more soluble and stable by the incorporation of nano-sized carriers. By combining cutting-edge extraction, purification, along with nanotechnology techniques, the isolation process of nano-derived natural products is a multi-step process that optimizes the bioactivity, stability, as well as therapeutic potential of natural compounds. This results in the creation of more potent drug delivery systems, supplements, along with therapeutic agents [32].

7 Encapsulation of Nano-Derived Redox Natural Products

Natural product encapsulation in nanomaterials is an essential technique for improving the pharmacokinetic characteristics of bioactive substances, guaranteeing their stability, solubility, and regulated release. Several kinds of nanomaterials are used in this procedure; each has special benefits based on the type of natural product and its intended use. Both hydrophilic and hydrophobic natural substances can be encapsulated in liposomes, which are vesicular structures made of lipid bilayers. The lipid-based nanoparticles are perfect for drug administration, because they enhance the solubility, prevent degradation, and permit regulated release of the active ingredients [33]. Conversely, polymeric nanoparticles are made of biodegradable, biocompatible, and multipurpose polymers like chitosan and poly(lactic-co-glycolic acid) (PLGA). The hydrophilic outer shell of micelles, which are self-assembled structures formed by surfactants, assures water solubility, while the hydrophobic center can enclose naturally occurring chemicals that are poorly soluble. Essential oils and polyphenols are examples of hydrophobic substances whose solubility and bioavailability can be improved by micelles. Nanocapsules resemble nanoparticles in their core–shell structure, but the bioactive ingredient is enclosed in the core and encased in a biodegradable polymeric shell [34]. The encapsulated natural compounds are further protected by the improved stability and regulated release that Nanocapsules provide. Lastly, functional groups give dendrimers—branched, tree-like macromolecules—the ability to encapsulate natural materials. These macromolecules are ideal for applications requiring controlled distribution as well as specific targeting, because they have a large surface area, allow for exact control over their shape and size, and may be created for the targeted delivery of bioactive substances. The techniques used to encapsulate natural products derived from nanotechnology differ based on the solubility and desired qualities of the compounds [35]. Solvent evaporation is a widely used process in which a solvent is used to dissolve the natural product and polymer material, and the solvent is then evaporated to create nanoparticles that encapsulate the bioactive chemical. Producing polyphenol and other natural chemical nanoparticles is a common use for this method. Compounds that need aqueous conditions to be stable benefit greatly from this technique. The process of creating an emulsion with a polymer and natural materials is known as emulsion polymerization. The natural ingredient is then trapped within the polymer matrix by controlling the polymerization process, which makes it the perfect technique for hydrophobic natural compounds that require stabilization and slow-release encapsulation [36]. Spray drying is another popular method that involves spraying a natural product and polymer solution into a heated chamber. This causes the solvent to evaporate quickly and creates nanoparticles or microparticles that encapsulate the bioactive ingredient. This process works especially well for making fine powders of natural ingredients that have been encapsulated. Natural compounds with limited solubility can be encapsulated using the solvent diffusion method. A solvent is used to dissolve the natural product, which is subsequently dispersed into a polymer matrix. The solvent then evaporates, leaving behind nanoparticles that encapsulate the component [37]. For substances that are hard to dissolve in water, this method works well for producing nanocarriers. This technique is highly beneficial for the formation of micelles and liposomes because the lipids and surfactants are amphiphilic, which allows them to self-assemble into aggregates that can contain both hydrophilic and hydrophobic substances. Self-assembly has the virtue of being straightforward and producing extremely stable nanostructures with minimum outside assistance. Every encapsulation technique has advantages, and the best one is chosen depending on the natural product's physicochemical characteristics, the intended release profile, and the parameters of the specific application [38]. Overall, employing nanoparticles to encapsulate natural goods is a potent way to increase their bioavailability, shield them from deterioration, and allow for tight oversight over their release—all of which are very useful for applications in pharmaceuticals, nutraceuticals, and cosmetics. Through the enhancement of bioactive chemicals' solubility, stability, and targeted treatment, these methods have great potential for creating innovative treatments and products that capitalize on the potential of natural materials [39].

8 Redox Materials and Its Biological Functions

Materials identified as redox materials are materials that are capable of initiating oxidation–reduction (redox) reactions, in which electrons migrate between atoms, molecules, or ions. These substances are essential for many biological processes, because they can affect a broad range of biochemical activities, such as signaling, energy production, and cellular regulation. Redox processes are essential to cellular respiration in biological systems, which is the process by which cells oxidize resources like glucose and fatty acids to make adenosine triphosphate (ATP), which is the source of energy [40]. Redox processes are crucial to the production of ATP by the mitochondria, which are referred to as the cell's powerhouse. Electron transport is facilitated by redox materials, including proteins, enzymes, and cofactors (such as cytochromes, NADH, and NADPH), which allow for the effective conversion of energy inside cells. Redox materials also have a role in controlling cellular signaling pathways. Signaling molecules such as transcription factors, which regulate gene expression, can be influenced by the equilibrium between both reduction and oxidation states in a cell. In redox signaling, for instance, reactive oxygen species (ROS), which are extremely reactive molecules that include oxygen, function as secondary messengers and influence physiological reactions like inflammation, stress reactions, and apoptosis [41, 42]. Overproduction of ROS can result in oxidative stress, which damages proteins, lipids, and DNA and causes diseases including cancer, neurodegeneration, as well as cardiovascular disorders, even though ROS are essential for regular cellular processes. By neutralizing excess ROS and eliminating cellular damage, antioxidants, a different type of redox-active chemicals, combat oxidative stress. These antioxidants are crucial for preserving redox equilibrium in cells and include substances like glutathione, vitamin C, and vitamin E [43]. Diseases can arise from disturbances in the delicate balance between the two processes, which is strictly controlled to guarantee healthy cellular activity. Redox materials are also used in detoxification procedures, where they aid in the body's removal and neutralization of toxic compounds. Enzymes like cytochrome P450, for example, aid in the metabolism of medicines and xenobiotics, enabling their removal through oxidation–reduction processes. Immune responses also exhibit the function of redox materials. ROS are produced by immune cells like neutrophils and macrophages as a component of their natural defense against infections. For example, to prevent tissue damage, the immune system must manage ROS levels, even though their controlled synthesis promotes in the destruction of pathogens and viruses [44]. Redox materials perform a dual function in this context, supporting immunological defense and preserving the equilibrium of the immune system. Redox mechanisms are also essential for DNA synthesis and repair. By repairing oxidative DNA damage, redox-sensitive enzymes maintain genomic stability and stop mutations that can cause illnesses like cancer. The therapeutic potential of redox materials has also been studied in several biomedical applications. Redox-active nanoparticles, for example, are being studied for drug delivery systems [45]. By undergoing redox reactions, these nanoparticles can release medications in a controlled way by reacting to the redox conditions found in diseased tissues. Modification of the redox pathway is being studied in cancer therapy to target cancer cells specifically, since they frequently have different redox states than healthy cells. Diabetes, cardiovascular disease, and neurodegenerative illnesses are among the conditions linked to oxidative stress that are being investigated for treatment with redox-based therapies such antioxidant supplements and redox-modulating medications [46].

9 Applications of Redox Polymers in Medicine

9.1 Redox Polymers in Biosensors

One of redox polymers' most common application areas has historically been sensors and biosensors. These materials have been essential to the development of chemical and specific optical sensors, as well as electrochemical biosensors. Electrochemical glucose biosensors are arguably the most widely used devices. The redox polymer plays a crucial role in the efficient immobilization of enzymes onto electrodes, acting as a mediator to facilitate the electronic flux between the electrode surface and the protein and raise the sensor's sensitivity [47]. Several redox polymers have been analyzed, including conducting polymers including polypyrrole, osmium-containing redox polymers, ferricyanide-containing redox polymers, along with ferrocene-modified redox polymers. Studies of the implanted redox polymers using cyclic voltammetry typically reveal a rise in current as the amount of glucose within the media increases, which makes them appropriate for use as glucose biosensors. Polymer preparation that results in the creation of biosensors with enhanced selectivity and sensitivity to various analytes. A ferrocene media tor is used by the electrochemical transistor to transfer electrons between the PEDOT:PSS gate electrode and the enzyme glucose [48]. Polymers with redox activity are used similarly, and sometimes without the requirement for a particular enzyme, in electrochemical biosensors to detect dopamine, and ascorbic acid, hydrogen peroxide, ethanol, glutamic acid, atmospheric carbon monoxide, and DNA hybridization. The electrochemical DNA (E-DNA) sensor is one of the most exciting recent developments [49]]. A surface-confined DNA stem-loop tagged with an electrically active redox polymer makes up an E-DNA sensor. Prussian Blue, hexaammineruthenium (III) chloride, methylene blue, ferrocene, boron, and enzymes are the most often utilized redox moieties in sensors. The redox pair close to the electrode generates a significant faradaic current when there is no target. The redox molecule along with electrode becomes significantly more distant because of the stem-loop breaking down during hybridization using the complementary nucleic acid. The current that is ascribed to the target decreases measurably as the distance increases [50].

9.2 Redox Polymers in Drug Delivery

Redox active surroundings or certain reagents are used to initiate the regulated delivery of a medication or gene in a redox-responsive polymeric delivery system. For instance, the redox polymers can act as a trigger to deliver the treatment in a targeted manner by using the clear distinctions in redox capabilities between the cytosol or the intracellular fluid with extracellular fluid. Redox polymers may be subject to quick breakage within the cell, over a period of minutes to hours, even though they must be adequately stable in the bloodstream. As a result, a medication or gene molecule may be trapped or enclosed in a nanocarrier that has one or more cleavage redox linkages [51]. The intended therapeutic substance will be delivered by reducing or oxidizing these linkage(s) inside the cell. To achieve effective intracellular drug and gene release, the notable variation in GSH levels has been identified as the perfect and pervasive internal stimulus for the quick disruption of reduction-sensitive nano-carriers within cells, specifically in the cytosol, endo/lysosome, and cell nucleus. The primary redox active groups include the chemical processes that enable their reversible or non-reversible action [48]. The drug release is caused by the redox process, which breaks a chemical bond and frequently causes the polymer's chain to split or the cross-linking bonds in a micro-nanoparticle to break down. In certain situations, ultimately, the redox reaction results in a chemical shift that accelerates the medication by altering its ionic character or solubility [52].

9.3 Disulfide Bonds in Redox Polymer

The gold standard of redox polymers is considered to have disulfide bonds. Depending on the need, thiol groups and the other redox-responsive groups may be added at various precisely defined positions. Therefore, this disulfide bond may be at the side chain or in the main backbone as a cross-linker point. To promote the breakdown of the polymer backbone, for instance, a disulfide link should be inserted to the main backbone when rapid biodegradability is essential. Another option is to perform a redox sensible crosslinking using the chemicals that include disulfides [53]. In this case, the primary backbone will lose its three-dimensional structure when reductive redox is present, but it will not be dependent upon redox conditions. In a reduction environment, polyurethane micelles containing the broad-spectrum anticancer medication doxorubicin (DOX) showed increased drug release. Modification processes are another method for adding disulfide bonds to linear polymers. In this way, the polymer is end-capped with a group, like pyridyl disulfide, which may interact with sulfhydryl groups to generate disulfide linkages [48].

10 Biomedical Applications of Redox-Active Nanoparticle

10.1 Antibacterial Activity

Nanoceria's emerged multiple applications in a variety of sectors, most notably biomedicine, have made it one of the important metal oxide nanoparticles. Significant antibacterial efficacy against Gram-negative as well as Gram-positive bacteria has been demonstrated for nanoceria. Bacterial cells are killed by nanoceria through two different mechanisms: direct and indirect contact. Direct bacterial cell adsorption of nanoceria damages the cell wall, which further causes intracellular ROS to be produced. Nanoceria enters the bacterial cell and deactivates the bacterial enzymes that produce hydrogen peroxide (ROS) because of this contact, rendering the cellular proteins inactive [54]. Bacterial cells die as a result of the intracellular ROS that are so produced, further damaging RNA, protein, and DNA. Oxidative stress may result from the decrease of nanoceria from Ce^{4+} to Ce^{3+} on the cell's surface during adsorption through bacterial wall membrane oxidation. The process of indirect interaction typically occurs in bacteria that are enclosed in polysaccharides. These bacteria have a polysaccharide layer associated with their cell envelope, which prevents nanoparticles from making contact directly with the cell wall. Outside the bacterial cell, ROS are produced. They then enter the cell through the membrane and harm the bacterial cells by breaking down proteins and nucleic acids, which eventually results in cell death [55]. One of the mechanisms that cause cyanobacteria to produce ROS involves oxidative photosynthesis. Superoxide anions and hydroxyl radicals are produced when the ROS produced by this mechanism reacts with the Ce^{3+} site on the surface of CNP, which comes in close contact with cyanobacteria. Lipid peroxidation caused by these ROS damages membrane integrity and further disrupts the photosynthetic process, eventually resulting in cell death [56].

10.2 Antioxidant and Anticancer Properties

The capability of nanoceria to alternate between two separate states of oxidation, Ce^{3+} and Ce^{4+}, depending on the environment, indicates they have strong antioxidant activity. This special redox characteristic of nanoceria can shield numerous tissues and organs against cellular damage brought on by reactive oxygen species (ROS) or free radicals. The ability of nanoceria to protect a pulmonary-like cellular system from oxidative damage caused by $KBrO_3$. They utilized BEAS-2B, a human epithelium lung cell line, and then treated the cells with nanoceria prior to exposing them to $KBrO_3$ [57]. They demonstrated that by lowering intracellular ROS generation, nanoceria avoided cell death. Nanoceria improved the antioxidant defense system by raising total antioxidant capacity and total thiol content, which reduced caspase 3 activities, oxidative DNA damage, and lipid peroxidation. Nanoceria have been shown to have strong anticancer potential in addition to their antibacterial and antioxidant effects. By causing ROS to develop, it kills cancer cells while offering cryoprotection against ROS to healthy cells. Because cancer cells produce more lactic acid and glycolysis than healthy cells do, their environments become more acidic [58]. Nanoceria lose their antioxidant cytoprotective effect in low pH environments and function as prooxidants, which generate ROS and oxidative stress, ultimately leading to cell death and apoptosis. The biomedical field has utilized the significant use of nanoceria as a targeted and effective medication and gene delivery system. Nanoceria's intrinsic cytotoxicity to cancer cells demonstrates a synergistic anticancer impact in this drug-delivery application in cancer cells. Due to their active intracellular absorption, the disintegration of the nanoceria lid in a strongly reducing cellular environment, and the resulting expulsion of the encapsulated drug, MSN loaded with camptothecin and capped with nanoceria exhibit dosage and time-dependent cytotoxicity in BxPC-3 cells [59].

10.3 Iron and Titanium Oxide Nanoparticles

For biomedical applications ranging from targeted therapies and tissue engineering to MRI imaging as well as cell tracking, iron oxide nanoparticles (IONs) are among the most studied and potentially adaptable nanomedicines. The intriguing characteristic of IONs—that is, their ability to exhibit magnetism only when exposed to an external magnetic field and to stop when the field is removed—is used in most of these applications. Superparamagnetic ions (SPIONs) can be directed to tissues or organs via an external magnetic field, or they might generate the heat whenever an alternating magnetic field is utilized [60]. ROS production from Fe states can cause oxidative stress via a few previously mentioned ways. Numerous studies have demonstrated that SPIONs induce oxidative stress, which includes aggravating oxidative DNA damage in human lymphocytes, interfering with lysosomal along with mitochondrial function in rat cardiomyocytes, and causing human macrophage apoptosis through excessive stimulation in the c Jun N-terminal kinase (JNK) signaling cascade [61]. Titanium dioxide (TiO_2) and its nanoconstituents (TiO_2 NMs) are gaining attention for use in a wide range of biomedical applications. Due to its high strength, endurance, low density, strong resistance to corrosion and wear, and, most importantly, biocompatibility, this substance has been demonstrated to be a great candidate for use in dentistry as implants for dentistry and orthopedics. This material's biocompatibility and bioinertness have been linked to the development of a native oxide layer on the surface of TiO_2 upon exposure to air. The primary reason for TiO_2 NPs' successful application as cytotoxic reagents in human cervical adenocarcinoma, hepatocarcinoma, non-small cell lung cancer, breast cancer, and leukemia cell lines is their efficient generation of ROS. Because of its photocatalytic qualities, TiO_2 and its nanocomponents are good options for both single- and multi-drug delivery therapies. Smaller particles' cytotoxic effects have been shown to trigger emphysema, epithelial cell death, and lung inflammation [62].

10.4 Gold Nanomaterials

Gold-based nanomaterials (Au NMs) are extremely promising for a variety of applications in nanomedicine because of their optical qualities, biocompatibility, high stability, and multifunctionality. Photo-induced cancer treatments, targeted therapeutic administration, cell and tissue imaging, and electrochemical sensing are a few examples of these uses. Additionally, due to their versatility, AuNMs have lately been employed as theragnostic sensors to target, diagnose, administer a therapeutic function, and track therapy response in extremely specialized single clinical operations. Although gold nanoparticles (AuNPs) are generally not regarded as redox active, their presence in biological systems can result in additional interactions that could trigger certain biological redox reactions [63]. As redox catalysts, AuNPs can facilitate improved electron transfer for a variety of electroactive biological species, making them a perfect platform for electrochemical biosensors. AuNPs are used in electrochemical sensing, which involves the direct electrochemistry of redox proteins, primarily enzymes, electrochemical genosensors (DNA), aptamer antibodies sensors, and immunosensors to identify clinically meaningful biomarkers that are pertinent to the diagnosis and/or tracking of human diseases. AuNPs' involvement in the production of ROS and oxidative-stress-induced DNA damage [64]. Lipid peroxidation, glutathione peroxidase reduction, and a substantial increase in 8-hydroxydeoxyguanosine levels are all caused by AuNPs and are indicators of oxidative-stress-induced DNA damage. A new approach to the construction of gene carriers based on the production of trace levels of ROS following mild light irradiation has recently been made possible by photo-induced endosome/lysosome disruption. ROS produced by radiation allows for the selective regulation of endosome/lysosome escape without damaging the loaded gene or causing cell death. Because they can produce ROS, gold nanoclusters have better properties than earlier photo-controlled delivery systems [65].

10.5 Redox Cancer Chemotherapeutics

ROS production is responsible for at least some of the effectiveness of cancer chemotherapy medications. ROS stress can occur via inhibiting antioxidants or by acting as a pro-oxidant. Because cancerous and healthy cells have different metabolisms, it is now believed that cancer cells are more sensitive to ROS stress. Cancer cells employ oxidative glycolysis and a drastically altered metabolic route to produce adenosine triphosphate (ATP) through oxidative phosphorylation. ROS, particularly O^{2-} and H_2O_2, must be produced and maintained at relatively high levels in tumors. Probably, it is well-established that the majority of cancer cells have higher-than-normal ROS levels [66]. Higher ROS levels will cause cell death, while relatively low ROS levels promote cell survival and proliferation. Cancer cells' metabolic alterations and protein translation result in elevated ROS levels, which they use to live by producing additional antioxidants such as reduced glutathione (GSH) and thioredoxin (TRX) in the presence of NADPH, which lowers ROS concentrations to levels that are sustainable. ROS can cause mutagenesis in cancer cells despite this survival response. Redox chemotherapy has become a promising new avenue for anti-cancer chemotherapy, particularly for tumors like pancreatic cancer and melanoma that have few other treatment options [66]. ROS are known to have the ability to cause tumors. Antioxidants and pro-oxidants regulate the steady-state levels of ROS within cells. However, cancer cells have higher levels of ROS production and loss, which raises the cells' initial ROS concentration. Both cell types will experience an increase in ROS concentrations after receiving redox-inducing chemotherapy, but normal cells' reserve antioxidants will prevent the ROS concentration from rising to an amount where it triggers apoptosis. Cancer cells have a greater basal ROS content, which causes them to have no extra antioxidant capacity and to undergo apoptosis [67].

11 Conclusion

Surprisingly, secondary metabolites of redox-active chemicals have been extant for several millennia, with most of them being recognized and ingested by humans today. Data and updated research have radically altered the trend over the previous several years, depending on the concept and technique. Crude plant extracts and related pure chemicals were extracted using physiochemically modified techniques, resulting in compounds and nanoparticles. Furthermore, antioxidants and antibiotics in refined form from various microbial sources have been observed in current molecular signaling processes of the redox modulation concept known as cellular thiolstat, which has developed as a novel trend. As intracellular diagnostics, sophisticated analytical techniques such as HPLC, different electrophoresis, GC–MS, physical, chemical, and green chemistry-derived nanoparticles, and emerging nanotechnology are all used to diagnose reducing equivalents, imbalanced pathophysiology, and homeostatic status in living organisms.

References

1. Rinalducci, S., Murgiano, L., & Zolla, L. (2008). Redox proteomics: Basic principles and future perspectives for the detection of protein oxidation in plants. *Journal of Experimental Botany, 59*(14), 3781–3801.
2. Mantravadi, P. K., Kalesh, K. A., Dobson, R. C., Hudson, A. O., & Parthasarathy, A. (2019). The quest for novel antimicrobial compounds: Emerging trends in research, development, and technologies. *Antibiotics, 8*(1), 8.
3. Thiele-Bruhn, S. (2021). The role of soils in provision of genetic, medicinal and biochemical resources. *Philosophical Transactions of the Royal Society B, 376*(1834), 20200183.
4. Martins, A., Vieira, H., Gaspar, H., & Santos, S. (2014). Marketed marine natural products in the pharmaceutical and cosmeceutical industries: Tips for success. *Marine Drugs, 12*(2), 1066–1101.
5. L Larghi, E., BJ Bracca, A., A Arroyo Aguilar, A., A Heredia, D., L Pergomet, J., O Simonetti, S., & S Kaufman, T. (2015). Neocryptolepine: A promising indoloisoquinoline alkaloid with interesting biological activity. Evaluation of the drug and its most relevant analogs. *Current topics in medicinal chemistry, 15*(17), 1683–1707.
6. Sajjad, W., Rafiq, M., Din, G., Hasan, F., Iqbal, A., Zada, S., Ali, B., Hayat, M., Irfan, M., & Kang, S. (2020). Resurrection of inactive microbes and resistome present in the natural frozen world: Reality or myth? *Science of The Total Environment, 735*, Article 139275.
7. Alum, E. A., Urom, S. M. O. C., & Ben, C. M. A. (2016). Microbiological contamination of food: The mechanisms, impacts and prevention. *International Journal of Scientific and Technology Research, 5*(3), 65–78.
8. Pisoschi, A. M., Pop, A., Iordache, F., Stanca, L., Predoi, G., & Serban, A. I. (2021). Oxidative stress mitigation by antioxidants-an overview on their chemistry and influences on health status. *European Journal of Medicinal Chemistry, 209*, Article 112891.
9. Ahlawat, J. (2022). *Nanotechnology-based approaches to mitigate environmental pollutants'-induced neurological disorders* (Doctoral dissertation, The University of Texas at El Paso).
10. Ahlawat, J., & Narayan, M. (2022). Multifunctional carbon quantum dots prevent soluble-to-toxic transformation of amyloid and oxidative stress. *ACS Sustainable Chemistry and Engineering, 10*(14), 4610–4622.

11. Pollard, C., Smith, S. C., & Theodorescu, D. (2010). Molecular genesis of non-muscle-invasive urothelial carcinoma (NMIUC). *Expert Reviews in Molecular Medicine, 12*, Article e10.
12. Langley, R. R., & Fidler, I. J. (2007). Tumor cell-organ microenvironment interactions in the pathogenesis of cancer metastasis. *Endocrine Reviews, 28*(3), 297–321.
13. Bityutsky, V. S., Tsekhmistrenko, S. I., Tsekhmistrenko, O. S., Tymoshok, N. O., Spivak, M. Y. (2020). Regulation of redox processes in biological systems with the participation of the Keap1/Nrf2/ARE signaling pathway, biogenic selenium nanoparticles as Nrf2 activators. *Regulatory Mechanisms in Biosystems, 11*(4).
14. Carroll, D. W. (2018). Targeting the cellular redox environment: A novel approach for the treatment of hematopoietic neoplasms.
15. Cho, C. Y. (1999). *Role of oxidative stress in two models of insulin resistance within primary rat adipocytes* (Doctoral dissertation).
16. Medrano-Macías, J., Flores-Gallegos, A. C., Nava-Reyna, E., Morales, I., Tortella, G., Solís-Gaona, S., & Benavides-Mendoza, A. (2022). Reactive oxygen, nitrogen, and sulfur species (RONSS) as a metabolic cluster for signaling and biostimulation of plants: An overview. *Plants, 11*(23), 3203.
17. Calbo, J., Golomb, M. J., & Walsh, A. (2019). Redox-active metal–organic frameworks for energy conversion and storage. *Journal of Materials Chemistry A, 7*(28), 16571–16597.
18. Bachi, A., Dalle-Donne, I., & Scaloni, A. (2013). Redox proteomics: Chemical principles, methodological approaches and biological/biomedical promises. *Chemical Reviews, 113*(1), 596–698.
19. Ma, Y., Wang, X., Jia, Y., Chen, X., Han, H., & Li, C. (2014). Titanium dioxide-based nanomaterials for photocatalytic fuel generations. *Chemical Reviews, 114*(19), 9987–10043.
20. National Academies of Sciences, Engineering, and Medicine, (2022). *New directions for chemical engineering*. National Academy of Sciences, Washington, DC (United States).
21. Ansari, S. P., & Dar, R. A. (2025). Quinones as redox-active materials for energy applications. In *Quinone-based compounds in drug discovery* (pp. 209–228). Academic Press.
22. Sims, C. M., Hanna, S. K., Heller, D. A., Horoszko, C. P., Johnson, M. E., Bustos, A. R. M., Reipa, V., Riley, K. R., & Nelson, B. C. (2017). Redox-active nanomaterials for nanomedicine applications. *Nanoscale, 9*(40), 15226–15251.
23. Elnour, A. A. M., Mirghani, M. E. S., Musa, K. H., Kabbashi, N. A., & Alam, M. Z. (2018). Challenges of extraction techniques of natural antioxidants and their potential application opportunities as anti-cancer agents. *Health Science Journal, 12*(5), 596.
24. Ashok, A., Andrabi, S. S., Mansoor, S., Kuang, Y., Kwon, B. K., & Labhasetwar, V. (2022). Antioxidant therapy in oxidative stress-induced neurodegenerative diseases: Role of nanoparticle-based drug delivery systems in clinical translation. *Antioxidants, 11*(2), 408.
25. Jain, A. K., Mehra, N. K., & Swarnakar, N. K. (2015). Role of antioxidants for the treatment of cardiovascular diseases: challenges and opportunities. *Current Pharmaceutical Design, 21*(30), 4441–4455.
26. Daraban, B. S., Popa, A. S., & Stan, M. S. (2024). Latest perspectives on Alzheimer's disease treatment: The role of blood-brain barrier and antioxidant-based drug delivery systems. *Molecules, 29*(17), 4056.
27. Pingarrón, J. M., Yanez-Sedeno, P., & González-Cortés, A. (2008). Gold nanoparticle-based electrochemical biosensors. *Electrochimica Acta, 53*(19), 5848–5866.
28. Shahidi, F., Varatharajan, V., Peng, H., Senadheera, R. (2019). Utilization of marine by-products for the recovery of value-added products. *Journal of Food Bioactives, 6*.
29. Vinatoru, M., Toma, M., & Mason, T. J. (1999). Ultrasonically assisted extraction of bioactive principles from plants and their constituents. *Advances in Sonochemistry, 5*, 209–248.
30. Mahato, N., Sinha, M., Sharma, K., Koteswararao, R., & Cho, M. H. (2019). Modern extraction and purification techniques for obtaining high purity food-grade bioactive compounds and value-added co-products from citrus wastes. *Foods, 8*(11), 523.
31. Shukla, S. K., Shukla, S. K., Govender, P. P., & Giri, N. G. (2016). Biodegradable polymeric nanostructures in therapeutic applications: Opportunities and challenges. *RSC Advances, 6*(97), 94325–94351.

32. Bhatia, S., & Bhatia, S. (2016). Nanoparticles types, classification, characterization, fabrication methods and drug delivery applications. *Natural polymer drug delivery systems: Nanoparticles, plants, and algae*, pp. 33–93.
33. McClements, D. J. (2015). Encapsulation, protection, and release of hydrophilic active components: Potential and limitations of colloidal delivery systems. *Advances in Colloid and Interface Science, 219*, 27–53.
34. Cimino, C., Maurel, O. M., Musumeci, T., Bonaccorso, A., Drago, F., Souto, E. M. B., Pignatello, R., & Carbone, C. (2021). Essential oils: Pharmaceutical applications and encapsulation strategies into lipid-based delivery systems. *Pharmaceutics, 13*(3), 327.
35. Enrico, C. (2019). Nanotechnology-based drug delivery of natural compounds and phytochemicals for the treatment of cancer and other diseases. *Studies in Natural Products Chemistry, 62*, 91–123.
36. Kumar, S., Dilbaghi, N., Saharan, R., & Bhanjana, G. (2012). Nanotechnology as emerging tool for enhancing solubility of poorly water-soluble drugs. *Bionanoscience, 2*, 227–250.
37. Meena, J., Gupta, A., Ahuja, R., Panda, A. K., Bhaskar, S. (2020) Inorganic particles for delivering natural products. *Sustainable agriculture reviews 44: Pharmaceutical technology for natural products delivery Vol. 2 impact of nanotechnology*, pp. 205–241.
38. Vriezema, D. M., Comellas Aragonès, M., Elemans, J. A., Cornelissen, J. J., Rowan, A. E., & Nolte, R. J. (2005). Self-assembled nanoreactors. *Chemical Reviews, 105*(4), 1445–1490.
39. Xing, P., & Zhao, Y. (2016). Multifunctional nanoparticles self-assembled from small organic building blocks for biomedicine. *Advanced Materials, 28*(34), 7304–7339.
40. Zhu, J., & Thompson, C. B. (2019). Metabolic regulation of cell growth and proliferation. *Nature reviews Molecular Cell Biology, 20*(7), 436–450.
41. Ježek, P., Holendová, B., & Plecitá-Hlavatá, L. (2020). Redox signaling from mitochondria: Signal propagation and its targets. *Biomolecules, 10*(1), 93.
42. Möller, M. N., Orrico, F., Villar, S. F., López, A. C., Silva, N., Donzé, M., Thomson, L., & Denicola, A. (2022). Oxidants and antioxidants in the redox biochemistry of human red blood cells. *ACS Omega, 8*(1), 147–168.
43. Fedoreyeva, L. I. (2024). ROS as signaling molecules to initiate the process of plant acclimatization to abiotic stress. *International Journal of Molecular Sciences, 25*(21), 11820.
44. Knight, J. A. (2000). Free radicals, antioxidants, and the immune system. *Annals of Clinical and Laboratory Science, 30*(2), 145–158.
45. Paiva, C. N., & Bozza, M. T. (2014). Are reactive oxygen species always detrimental to pathogens? *Antioxidants and Redox Signaling, 20*(6), 1000–1037.
46. Blagov, A. V., Summerhill, V. I., Sukhorukov, V. N., Zhigmitova, E. B., Postnov, A. Y., & Orekhov, A. N. (2024). Potential use of antioxidants for the treatment of chronic inflammatory diseases. *Frontiers in Pharmacology, 15*, 1378335.
47. Cash, K. J., Ricci, F., & Plaxco, K. W. (2009). An electrochemical sensor for the detection of protein—small molecule interactions directly in serum and other complex matrices. *Journal of the American Chemical Society, 131*(20), 6955–6957.
48. Casado, N., Hernandez, G., Sardon, H., & Mecerreyes, D. (2016). Current trends in redox polymers for energy and medicine. *Progress in Polymer Science, 52*, 107–135.
49. Luong, J. H. T., Male, K. B., Hrapovic, S. (2007). Carbon nanotube-based electrochemical biosensing platforms: Fundamentals, applications, and future possibilities. *Recent Patents on Biotechnology, 1*(2), 181–191.
50. Tsekeli, T. R. (2021) *The development of electrochemical (APTA) sensors based on carbon nanofiber composite and silver nanocomposite platforms and their applications on the detection of endocrine disrupting bisphenol A*. University of Johannesburg (South Africa).
51. Garcia-Garcia, A., Muñana-González, S., Lanceros-Mendez, S., Ruiz-Rubio, L., Alvarez, L. P., & Vilas-Vilela, J. L. (2024). Biodegradable natural hydrogels for tissue engineering, controlled release, and soil remediation. *Polymers, 16*(18), 2599.
52. Testa, B., & Krämer, S. D. (2007). The biochemistry of drug metabolism–an introduction: Part 2. Redox reactions and their enzymes. *Chemistry and Biodiversity, 4*(3), 257–405.

53. Chivukula, P. (2012) *Design, synthesis, and evaluation of biodegradable hydrogels and novel polymeric nanocarriers for the treatment of colorectal cancer* (Doctoral dissertation, Department of Pharmaceutics and Pharmaceutical Chemistry, University of Utah)
54. Barker, E., Shepherd, J., & Asencio, I. O. (2022). The use of cerium compounds as antimicrobials for biomedical applications. *Molecules, 27*(9), 2678.
55. More, P. R., Pandit, S., Filippis, A. D., Franci, G., Mijakovic, I., & Galdiero, M. (2023). Silver nanoparticles: Bactericidal and mechanistic approach against drug resistant pathogens. *Microorganisms, 11*(2), 369.
56. Bidle, K. D., & Falkowski, P. G. (2004). Cell death in planktonic, photosynthetic microorganisms. *Nature Reviews Microbiology, 2*(8), 643–655.
57. Kargozar, S., Baino, F., Hoseini, S. J., Hamzehlou, S., Darroudi, M., Verdi, J., Hasanzadeh, L., Kim, H. W., & Mozafari, M. (2018). Biomedical applications of nanoceria: New roles for an old player. *Nanomedicine, 13*(23), 3051–3069.
58. Ray, S. D., Bagchi, D. (2004). Roles of polyphenols, flavonoids, and oligomeric proanthocyanidins in cancer chemoprevention. *Phytopharmaceuticals in cancer chemoprevention* (pp. 311–348). CRC Press USA.
59. Thakur, N., Das, J., & Sil, P. C. (2021). Emerging role of redox-active nanoceria in cancer therapeutics via oxidative stress. In *Handbook of oxidative stress in cancer: Therapeutic aspects* (pp. 1–23). Singapore: Springer Singapore.
60. Nelson, N. R., Port, J. D., & Pandey, M. K. (2020). Use of superparamagnetic iron oxide nanoparticles (SPIONs) via multiple imaging modalities and modifications to reduce cytotoxicity: An educational review. *Journal of Nanotheranostics, 1*(1), 105–135.
61. Nabavinia, M., & Beltran-Huarac, J. (2020). Recent progress in iron oxide nanoparticles as therapeutic magnetic agents for cancer treatment and tissue engineering. *ACS Applied Bio Materials, 3*(12), 8172–8187.
62. Ahamed, M., Khan, M. M., Akhtar, M. J., Alhadlaq, H. A., & Alshamsan, A. (2017). Ag-doping regulates the cytotoxicity of TiO_2 nanoparticles via oxidative stress in human cancer cells. *Scientific reports, 7*(1), 17662.
63. Shafiee, A., Ghadiri, E., Kassis, J., Pourhabibi Zarandi, N., & Atala, A. (2018). Biosensing technologies for medical applications, manufacturing, and regenerative medicine. *Current Stem Cell Reports, 4*, 105–115.
64. Zhu, C., Yang, G., Li, H., Du, D., & Lin, Y. (2015). Electrochemical sensors and biosensors based on nanomaterials and nanostructures. *Analytical Chemistry, 87*(1), 230–249.
65. Liu, H. M., Cheng, M. Y., Xun, M. H., Zhao, Z. W., Zhang, Y., Tang, W., Cheng, J., Ni, J., & Wang, W. (2023). Possible mechanisms of oxidative stress-induced skin cellular senescence, inflammation, and cancer and the therapeutic potential of plant polyphenols. *International Journal of Molecular Sciences, 24*(4), 3755.
66. Li, Y., Jia, A., Wang, Y., Dong, L., Wang, Y., He, Y., Wang, S., Cao, Y., Yang, H., Bi, Y., & Liu, G. (2019). Immune effects of glycolysis or oxidative phosphorylation metabolic pathway in protecting against bacterial infection. *Journal of Cellular Physiology, 234*(11), 20298–20309.
67. Graves, D. B. (2012). The emerging role of reactive oxygen and nitrogen species in redox biology and some implications for plasma applications to medicine and biology. *Journal of Physics D: Applied Physics, 45*(26), Article 263001.

Oxidative Stress and Pathophysiology of Chronic Disease

Sathyabama Balaji, Muthu Senthil Pandian, R. Ganesamoorthy, R. Parameswari, and R. Babujanarthanam

Abstract Oxidative stress, sometimes known as "*good stress*," is differentiated by a low to medium concentration of oxidants. Biochemical modifications, such as hydroxylation, carboxylation, peroxidation, or modulation of signaling pathways, such as the mitogen-activated protein kinase (MAPK) cascade, nuclear factor-κB (NF-κB), nuclear factor erythroid 2-related factor 2 (Nrf2), and phosphoinositide 3-kinase, can all contribute to regulation. ROS/RNS, which are produced by the two mitochondria and NADPH oxidases, are elevated in oxidative stress (OS), commonly known as "*bad stress*," a detrimental condition. OS is a common denominator with chronic illnesses, even though many of them are thought to have many origins. This chapter discusses the relevance of operating systems. OS can help to explain some of the organism's pathogenic circumstances. Any discrepancy between RS and antioxidants results in "OS" leading to various pathological diseases, including diabetes. Most data suggest that OS influences diabetes pathogenesis by affecting enzyme system levels. Diabetes complications caused by OS may lead to stroke. The OS indicators are also evaluated quantitatively and qualitatively. OS impacts the pathophysiology of cancer, heart disease, diabetes, COVID-19, and the kidney. They collaborate to diminish the detrimental effects of OS on small-molecule antioxidants in a minimum state (vitamins C and E, flavonoids, carotenoids, melatonin, ergothioneine, and others) and antioxidant enzymes. Vitamin E is the first line of defense against lipid peroxidation and one of the most powerful molecular-weight antioxidants.

S. Balaji · M. S. Pandian (✉)
Department of Physics, SSN Research Centre, Sri Sivasubramaniya Nadar College of Engineering, Chennai, Tamil Nadu, India
e-mail: senthilpandianm@ssn.edu.in

R. Ganesamoorthy
Department of Chemistry, Vinayaka Mission's Kirupananda Variyar Arts and Science College, Vinayaka Mission's Research Foundation (DU), Salem, Tamilnadu, India

R. Parameswari
Department of Biochemistry, Thiruvalluvar University, Serkkadu, Vellore 632115, India

R. Babujanarthanam
Department of Biotechnology, Thiruvalluvar University, Serkkadu, Vellore 632115, India

K. Kasinathan et al. (eds.), *Recent Advances in Oxidative Stress Associated Chronic Diseases Volume 1*, https://doi.org/10.1007/978-981-96-9670-3_3

1 Introduction

Sies originally coined the term "*oxidative stress*" (OS). Reactive oxygen species (ROS) or reactive nitrogen species (RNS) generation and decay are out of equilibrium, which is their distinguishing feature [1]. ROS are extremely reactive molecules due to their chemical makeup and can result from the breakdown of nitrogen or oxygen. ROS and RNS can be free radicals (FR) like hydroxyl radical (OH·), nitric oxide (NO·), and superoxide radical ($O_2^{\cdot -}$). However, peroxynitrite (ONOO–) and hydrogen peroxide (H_2O_2), another nonfree radical, are also present [2]. ROS cause enzymatic processes inside the mitochondria, causing the electron transport chain to reduce oxygen [3]. ROS can also originate in peroxisomes and the endoplasmic reticulum [4, 5]. The quantity of ROS in the cell influences various biological activities, including immunity, apoptosis, protein phosphorylation, and transcription factor activation [6, 7]. Superoxide dismutase (SOD) is a kind of metallic enzyme that transforms $O_2^{\cdot -}$ into H_2O_2 and oxygen [8]. SOD has three types, also known as mammals: mitochondrial SOD, cytoplasmic SOD, and extracellular SOD [9]. The nucleic acids, membrane lipids, and cellular proteins hinder proper cellular activity, and excessive ROS emission causes bodily injury [10]. Diabetes mellitus is characterized by hyperglycemia (*high blood glucose*) and inadequate insulin synthesis or action by the pancreas throughout the body. Chronically high blood glucose levels have been associated with both macrovascular and microvascular issues, including blindness, kidney disease, heart disease, and stroke [11]. This chapter will discuss the importance of ROS production in the blood vessels. Following that, we intended to study the OS, with a specific emphasis on ROS. Many chronic illnesses, including neurological disorders, cancer, renal diseases, cardiovascular disease (CVD), diabetes, and COVID-19, are linked with OS.

2 Oxidative Stress (OS)—An Overview

In living cells, both enzymatic and nonenzymatic redox processes regularly create reactive oxygen species and reactive nitrogen species, forming a family of reactive compounds. Free radicals, also known as primary radicals, are solitary chemical entities that possess one or more unpaired electrons. Examples include the less reactive superoxide anion ($O_2^{\bullet -}$), hydroxyl ($OH^{\bullet}$), nitrogen dioxide radicals ($NO_2^{\bullet}$), and nitric oxide radicals ($NO^{\bullet}$). They are very reactive, eagerly seeking more unpaired electrons. A new nonradical molecule is formed when two unpaired electrons engage and form a covalent bond. When FR impacts nonradical molecules, it typically initiates a chain reaction that forms a new (*secondary*) radical molecule. Because of oxidative damage was eventually sustained in various tissues and organs as shown in Fig. 1 [12]. Secondary radicals include ozone (O_3), hydrogen peroxide (H_2O_2), singlet oxygen ($^1O_2$), dinitrogen trioxide (N_2O_3), hypochlorous acid (HClO), nitrous acid (HNO_2), and peroxynitrite (ONO^-). Measuring the amount of FR in circulation is difficult due

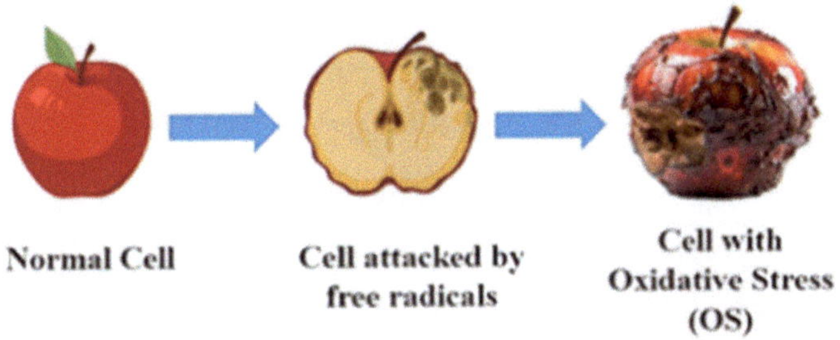

Fig. 1 Normal cell to oxidative stress

Table 1 Oxidative stress (OS)-induced chronic diseases

Organ	Chronic disease
Mind	AD, memory loss, and so on
Eyes	Cataract and retinal diseases
Heart	Arteriosclerosis, ischemia, cardiomyopathy, hypertension, and heart failure
Multi-organs	Cancer, diabetes, inflammation, infection, and aging

to its volatility, short half-life, and impact on chronic illness. Biomarkers OS are divided into 4 types: (1) ROS creation, (2) protecting the antioxidant, (3) initiating ROS alterations, (4) ROS damage caused by functional markers downstream [13]. OS-induced chronic diseases are given in Table 1.

2.1 *Free Radicals (FRs)*

Free radicals, also called radicals (FRs), are transitory chemical entities that can react with any number of unpaired electrons. They are also unavoidable because of the information exchanged throughout the regular migration and differentiation processes. According to Bansal and Bilaspuri, fFR oxidizes cell components and molecules by transferring unpaired electrons, resulting in cell damage. They are generally extremely reactive and unstable as shown in Fig. 2 [14]. A brief list of Free radicals (FRs) and nonradical oxidants (NROs) is given in Table 2.

2.2 *The Biomarkers in OS*

As previously stated, OS is notified by an imbalance in the ratio of oxidants to antioxidants, which favors oxidants and potentially leads to irreversible chemical

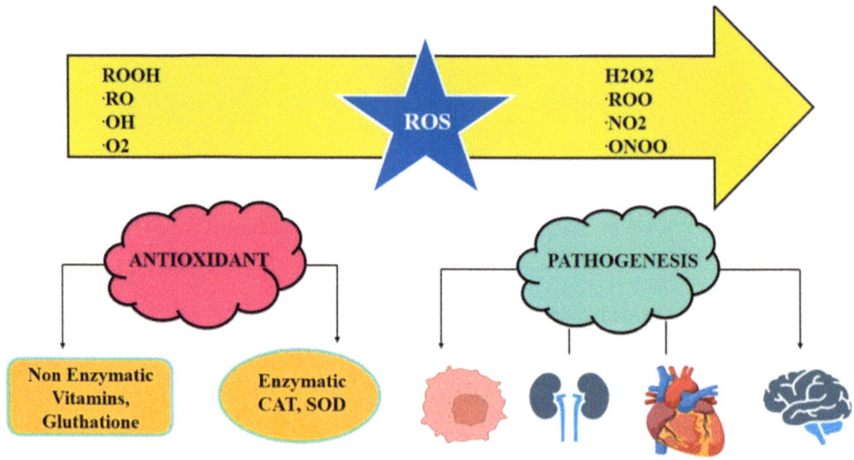

Fig. 2 Free radicals on human being

Table 2 List of free radicals (FRs) and nonradical oxidants (NROs)

FRs and NROs	Examples
ROS	Hydroxyl radical (·OH), peroxyl radical (ROO·), superoxide radical anion (O_2^-), per hydroxyl radical (HOO·), alkoxyl radical (RO·), hydrogen peroxide (H_2O_2), singlet oxygen ($^1O_2$)
RNS	Nitric oxide (NO·), nitrosyl anion (NO^-), nitrous acid (HNO_2), peroxynitrite anion ($ONOO^-$), nitrogen dioxide (NO_2)
RCS	Nitryl chloride ($ClNO_2$) & hypochlorous acid (HClO)
RSS	Thiyl radical (RS·), persulfides (i.e., GSSH, CSSH), hydrogen sulfide (H_2S)

alterations. A biomarker of OS needs to satisfy the following: relevance (a), precision for specific illness (b), a reflection of disease stage (c), and excellent consistency in biomarker measures. OS-related indicators are divided into various categories: (a) ROS production indicators, (b) shielding antioxidant markers, (c) ROS-induced modification biomarkers, and (d) ROS-induced damage caused by downstream functional markers [15] as shown in Fig. 3. The biomarkers of OS and those related to human disease are given in Table 3.

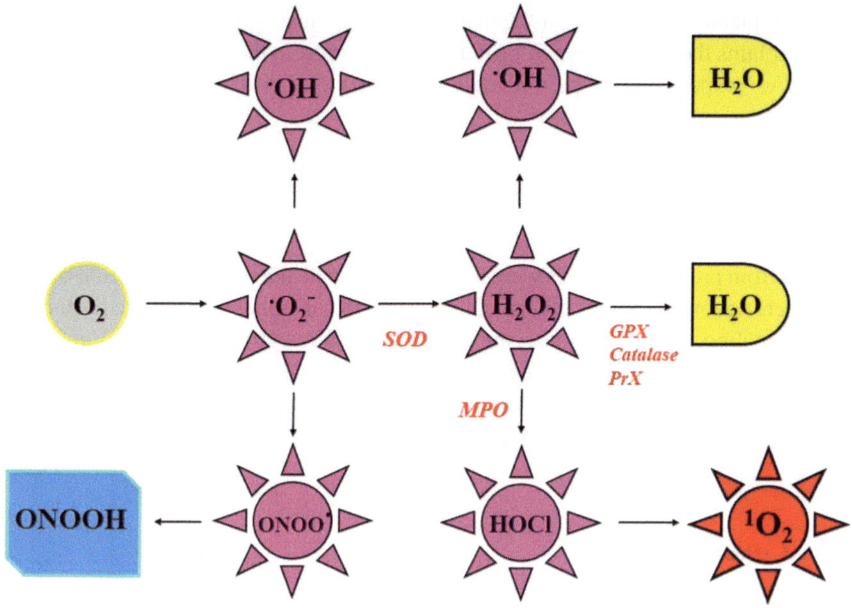

Fig. 3 Formation and detoxification of ROS

Table 3 Biomarkers of OS and those related to human diseases

OS to biomarkers	Disease	Reference
MDA	AD, atherosclerosis, and cancer	[16]
4-Hydroxynonenal	CVD, atherosclerosis, and AD	[17]
F2-isoPs (F2-isoprostanes)	CVD, atherosclerosis, and heart failure	[18]
AGEs 8-oxo-G	Atherosclerosis, and AD	[19]
Oxo-LDL	CVD, atherosclerosis, and AD	[20]
Acrolein	CVD, atherosclerosis, and AD	[21]
Cysteine/cystine	Atherosclerosis and kidney failure	[22]
GSH/GSSG	CVD, AD, and cancer	[23]
NO_2-Tyr	CVD, atherosclerosis, AD, and cancer	[24]
3-Cl-Tyr	CVD	[25]

3 Pathophysiology: An Overview

Pathophysiology is recognized as a flexible sense that helps to prevent injuries and effectively notifies us of threats to our bodies' integrity. It results from a complex interaction between psychological and neurological activities in the central and peripheral nerve systems. Acute pain has a strong adaptive function, but chronic disease typically does not. It has been proposed that the process of "*central sensitization*" explains the phenomena of prolonged discomfort in the absence of appropriate "*objective*" experiences of pain stimuli, which is linked with chronic disease [26].

3.1 Cancer

Results from previous decades were used to investigate the two-fold function of ROS in cancer. According to Reczek and Chandel, ROS can either (i) promote protumorigenic signaling, which promotes cancer cell survival and growth, or (ii) promote anti-tumorigenic signaling [27]. Cancer cells generate more ROS than healthy ones. They restore and balance ROS while increasing antioxidant capability. Malignant cells are susceptible to ROS-targeting therapy because of their altered redox equilibrium [28].

3.2 Cancer and OS

ROS can damage DNA and promote cancer development [29]. Hydroxy free radical is the major ROS that targets nuclear and DNA strands [30]. Various enzyme activities enable cells to repair damaged DNA. In case the DNA damage is not repaired on time, carcinogenesis occurs due to alterations caused by base deletion. The more oxidative damage in DNA, the greater the likelihood of poor DNA repair. Aging has two effects: reduced DNA repair and the accumulation of oxidative damage. Oxidative alterations in DNA cause chromosome abnormalities and inhibit DNA replication and cytotoxicity [31]. Although acute free radicals attack DNA, free radical interactions with other cell molecules can also induce mutagenesis. LPO has the potential to induce cancer [32]. Antioxidants have shown promise in the treatment of type 1 and 2 diabetes. Few biological processes like, oxidation of glucose is linked to higher levels of oxygen and nitrogen free radicals (ROS/RNS), which worsen diabetes and its complications. Numerous research outputs presented a link between OS and diabetes, as well as potential complications in the heart, liver, kidneys, and eyes [33]. Thus, OS appears more problematic in metabolic diseases, notably type 2 diabetes. Adducts can be formed when MDA interacts with guanine bases. We still do not understand the processes contributing to OS-induced carcinogenesis. Thus, OS alterations in gene expression, cell proliferation, and apoptosis play an important role in tumour genesis and development. Innovative procedures [34] enable OS to change protein and gene transcript phrases linked with cell proliferation and development. Cancer and its antioxidants were given in Table 4 (Fig. 4).

Table 4 Cancer and its antioxidants benefits

Antioxidants	Benefits
Polyphenols—tea catechins	It contains a lot of epigallocatechin-3-gallate (EGCG), which can stop cancer cells from growing and developing. It prevents the NF-κB transcription. By controlling pathways crucial to tumor cell migration through Metallo-proteinase 2, 7, and 9, it can stop cancer metastases
Quercetin	Pro-oxidant properties
Luteolin	4 OH group acts as an anticancer property mediated by OS suppression
Apigenin	It displays anticancer activity
Melatonin, carotenoids, and vitamins C and E	Potential in cancer prevention (carotenoid lycopene)
Vitamin E	Against chemotherapy, the sensitivity of cancer cells will increase

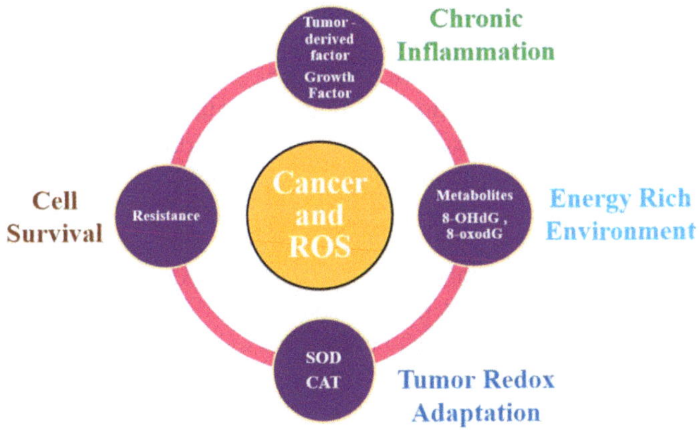

Fig. 4 Role of cancer and OS in the human body

4 Chronic Kidney Disease (CKD)

OS is a prevalent component in chronic kidney disease (CKD), which can lead to inflammation and cardiovascular issues. Uremia occurs when the kidneys malfunction and excrete debris rather than urine into the circulation. The OS mechanism causes the uncoupling of endothelial nitric oxide synthase (eNOS) [35]. Causes of OS in CKD and its many routes. There are four types of stress: oxidative (OS) induced by ROS (•OH), nitrosative (•NO_2), chlorinated (HOCl), and carbonyl stress (AGEs) [36].

4.1 CKD and OS

Even tiny variations in the redox balance of medulla and renal cortex might negatively impact since an acceptable level of ROS is required for normal kidney function. These include fibrotic disorders, inflammation, higher urine protein levels (*proteinuria*), affects blood passage over Na^+ storage [37]. OS has been shown to cause kidney failure and other issues associated with type 1 and type 2 diabetes (diabetic nephropathy). Mitochondrial dysfunction, endothelial dysfunction, abnormal NADPH oxidase (NOX) activity, endothelial nitric oxide synthase (eNOS) uncoupling, reduced antioxidant enzyme activity, and lower molecular weight antioxidants have all been associated with higher OS [38]. Several ways activate NADPH oxidases (NOX). The aggressive uremic toxin indoxyl sulfate (IS) is well known to activate NOX4, resulting in increased ROS generation. Long-term angiotensin II receptor blocking (ATII-R, such as Azilsartan or Losartan) in people with diabetes or chronic renal diseases may prevent, cure, or relieve kidney failure symptoms by relaxing veins and arteries and reducing blood pressure. Numerous research has investigated OS's impact on the development of kidney disease and the subsequent loss of renal function. Both the medulla and the renal cortex can create or regulate ROS levels. This can have a range of consequences, including changes in renal blood flow owing to sodium/fluid retention, inflammation, fibrotic changes, and the onset of proteinuria. Reduced OS is at least partially caused by increased eNOS and SOD transcription, as well as decreased NOX activity [39] as shown in Fig. 5. The benefits of antioxidants in CKD were presented in Table 5. Patients with chronic renal diseases have higher Xanthine oxidase (XO) activity. It has recently been proposed that chronic renal diseases and abnormal xanthine oxidase levels are associated with hyperuricemia, a disease defined by abnormally high blood urate levels. Patients with hypertension who underwent XO inhibitor medication showed a minor improvement in renal function [40]. The kidneys contain high quantities of SOD, the principal detoxification enzyme. Patients with chronic renal diseases have been observed to have diminished SOD function [41].

5 Cardiovascular Diseases (CVDs)

Atherosclerosis and blood vessel alterations that limit blood flow are two of the leading causes of CVD. Heart failure, coronary artery disease, hypertension, and vascular diseases are the most common pathologies linked with CVD.

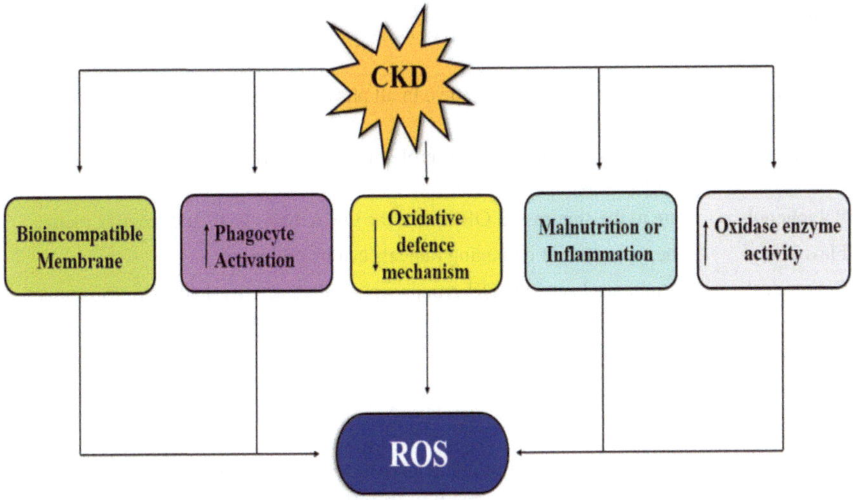

Fig. 5 Pathogenic factors leading to ROS

Table 5 Benefits of antioxidants in CKD

Antioxidant	Benefits
Flavonoid quercetin berberine	OS used in diabetic nephropathy
Bardoxolone methyl (BARD)	Anti-inflammatory
Coenzyme Q (ubiquinone)	Patients with renal lithiasis can be protected from renal injury
Curcumin	Nondiabetic CKD proteinuria patients have reduced lipid peroxidation
Gentamicin (aminoglycoside antibiotic)	Prevents bacterial infections

5.1 CVD and OS

Certain kinases are required for the redox signaling that regulates heart function. The H_2O_2-treated Ca/calmodulin-dependent kinase II has one of the most significant changes for excitation–contraction coupling. Furthermore, p38 MAPK used to increase glucose absorption, whereas JNKs (c-Jun N-terminal kinase) are linked to insulin resistance [42]. Several neurotransmitters, hormones, and stress signals activate PKA, an important regulator of cardiac shape and function. Elevated PKA activity promotes cardiac contraction and the Ca^{2+} cycle. Nuclear factor E2-related factor 2 (Nrf2) trans binds to the antioxidant response element (ARE) after entering the nucleus. ARE regulates the activity of a variety of antioxidant enzymes, including SOD, glutathione peroxidase, thioredoxin reductase, and others [43]. Furthermore, OS is associated with risk factors for CVD, such as hyperhomocysteinaemia, dyslipidemia, and diabetes mellitus [DM]. Antioxidants in CVD was given in Table 6. As

Table 6 Antioxidants in CVD

Antioxidants	Benefits
Vitamin E	Anti-inflammatory properties. In all phases of the disease, it leads to chronic inflammation
Vitamin C	In patients with hypercholesterolemia, cholesterol level is reduced due to non-antioxidant properties
Polyphenols	Due to the suppressed OH levels, the OH reduces ROS and reactive metals
Flavonoids	The formation of damaging radicals can be prevented
Resveratrol	It contains 3 separate OH groups, antioxidant activity, and ROS-scavenging ability. Activates eNOS, which may prevent hypertension
Vitamins B and D	It is an intermediary in metabolism and is indicative of hyperhomocysteinemia

a result, it has become more obvious that novel antioxidant therapies, such as gene transfer of antioxidant enzymes, may be beneficial in treating CVD. Early on, cellular redox reactions were commonly referred to as "*redox signalling*" and "*redox control*" indicating their critical role in intracellular regulation and communication. Later, the concept of an OS was expanded to encompass the critical function of redox signaling pathways. A variety of endogenous and exogenous causes can cause OS. It is widely assumed that OS, which is created by an imbalance between RNS and excessive ROS, is a key contributor to the genesis of many physiological illnesses, including CVDs. The generation of ROS, predominantly superoxide radical anion $O_2\bullet^-$, is regulated by membrane-bound electron-transporting proteins named NADPH (NOX). However, hydrogen peroxide is also formed [44]. In Fig. 6 the most often expressed cardiovascular systems are NOX 1, 2, 4, and 5. Cardiomyocytes, fibroblasts, and endothelial cells have NOX2 and 4. Angiotensin II (AngII), a peptide hormone that regulates vascular NOXs, this can subsequently stimulate Angiotensin II receptor type 1 (AT1), aiding it to self-strengthen [45]. AngII also regulates electrolyte balance and blood pressure. NOS has three isoforms: neural (NOS1), induced (NOS2), and endothelial (eNOS). Each of the three kinds can be found in distinct organs and cells. Endothelial cells and striated muscle produce NOS1 and NOS3, which have a Ca-dependent effect. The NOS2 polypeptide is particularly active in the heart during pathological conditions such as inflammation [46].

6 Neurodegenerative (NDG)

Each group of neurons in the brain is equally sensitive to OS. For example, neurons with multiple connections and longer axons require more power for overtime plasticity or axonal movement [47]. Specific neuron groups are more vulnerable to degeneration due to increased ATP demand and dysfunctional mitochondria.

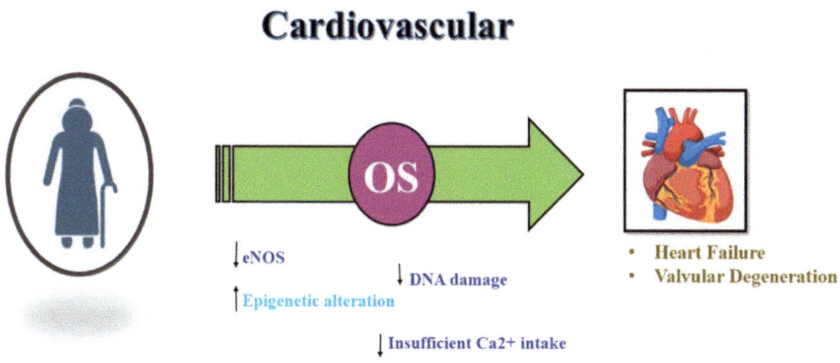

Fig. 6 OS in cardiovascular disease

6.1 NDG and OS

The NDG disorders associated with OS include multiple sclerosis [48], depression [49], Parkinson's disease [50], and Alzheimer's disease [51]. Ageing is the main link between OS and NDG diseases. OS causes oxidative damage and increases mitochondrial dysfunction as it accumulates with ageing. The hippocampus in animal models of Alzheimer's disease shows impaired mitochondrial complex IV activity. High OS creates neurotoxic chemicals and causes mitochondrial oxidative damage. ROS promotes the creation of toxic peptide β-amyloid, leading to the NDG phase of Alzheimer's disease [52]. Furthermore, β-amyloid promotes the production of H_2O_2 in neocortical neurons, increasing OS. In Parkinson's disease models derived from microglial cells, NDG development of dopaminergic neurons is associated with dysregulated NADPH stimulation. OS is also linked to the NDG and inflammatory processes associated with depression and multiple sclerosis. Multiple sclerosis relates to increased detection of oxidative damage to DNA (8-OHdG), carbonated proteins, and inhibition of the GPx enzyme. Individuals with unipolar depression, on the other hand, have higher MDA levels, lower ascorbic acid levels, and lower SOD enzyme activity. Because lipids are necessary for brain neuronal function, they are also more abundant in the plasma membrane, which operates as a barrier between the extracellular and intracellular compartments. They undergo lipid peroxidation and are more susceptible to free radical assaults. Lipid peroxidation decreases membrane fluidity, increasing membrane leakiness. This allows chemicals that cannot pass through the barrier to enter the intracellular system via specific channels (e.g., K+, Ca^{2+}). As a result, these chemicals harm receptors, enzymes, and membrane proteins. Because polyunsaturated fatty acids (PUFAs) are prevalent in neuronal membrane lipids, their side chains are more vulnerable to ROS/RNS, which can result in OS [53] (Fig. 7). The destruction of brain cells is a defining feature of Alzheimer's disease, a complicated condition. This is the primary symptom of dementia, characterized by a decline in cognitive function. Amyloid and cholinergic hypotheses are the primary ideas proposed to explain AD's complex etiology. Furthermore, RNA damage can

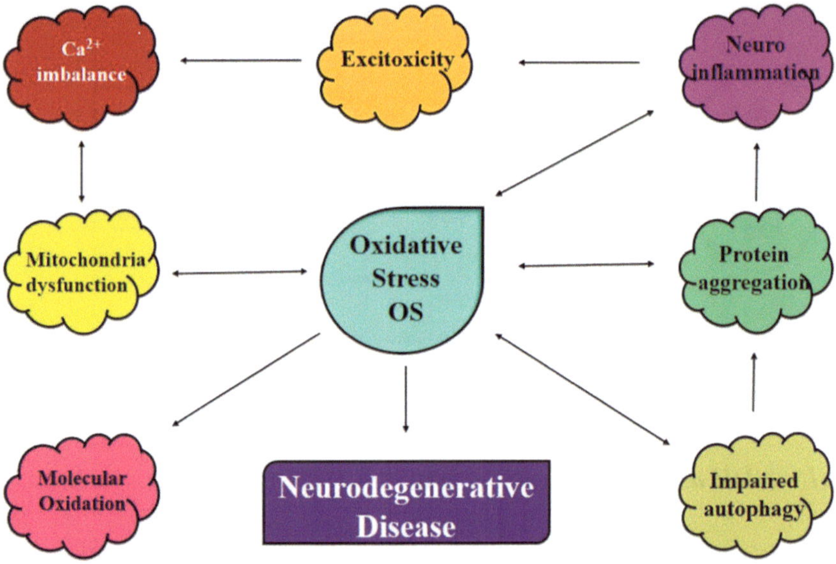

Fig. 7 Role of OS in neurodegenerative diseases

interfere with protein synthesis and translation. The length of amyloid-β peptides appears to impact its toxicity [54].

7 Diabetes

Hyperglycemia, or high blood glucose, and inadequate levels of insulin or function by the pancreas within the body are hallmarks of diabetes mellitus, a set of metabolic illnesses. Diabetes is often categorized into two types: Type I (IDDM) and Type II (NIDDM) [55].

7.1 Pathway of Diabetes

When people eat, their blood glucose levels rise, triggering insulin release. This causes an increase in the mobility, biological transformation, and storage of fat and muscular tissue. When fasting, the liver supplies glucose to the blood, which the brain consumes without insulin. In addition to storing glucose, insulin inhibits glucagon release and lowers blood fatty acid levels. This lowers the quantity of glucose generated by the liver. Intracellular hypoglycemia and extracellular hyperglycemia are caused by insufficient insulin or insulin resistance, which results in inadequate glucose absorption in the tissue. Extracellular hyperglycemia causes hyperglycemic coma and osmotic diuresis. Intracellular hypoglycemia promotes glucogenesis and gluconeogenesis. This causes fat breakdown (*resulting in diabetic ketoacidosis*) and lowers protein synthesis and gamma globulins (*producing cachexia, polyphagia, and delayed wound healing*) [56].

7.2 OS in Diabetes

Data demonstrating the importance of OS in the pathogenesis of type I and type II diabetes are currently available. An increased lipid peroxidation produces free radicals, damaging enzymes and cellular machinery and enhancing insulin resistance due to OS. A recent study indicates that the lipid and apolipoprotein components of LDL contribute to oxidative damage in the symptoms of diabetes I and II. The cross-linking of apo-B monomers, which is started by hydroxyl radicals, produces intractable clumps [57]. The major cause of OS in diabetes mellitus is mitochondria. Data demonstrating the importance of OS in the pathogenesis of type I and type II diabetes are currently available. The free radicals produced may damage the enzymes and cellular machinery and enhance insulin resistance due to OS. A recent study indicates that the lipid and apolipoprotein components of LDL contribute to oxidative damage in the symptoms of diabetes I and II. The cross-linking of apo-B monomers, which is started by hydroxyl radicals, produces intractable clumps [57]. The major cause of OS in diabetes mellitus is mitochondria. The oxygen is converted into O free radical (O·). A major ROS later turned into other RS such as $ONOO^-$, OH, and H_2O_2. ROS/RNS regulates insulin responsiveness in two ways. (i) RNS and ROS are produced in response to insulin for physiological function; (ii) They negatively impact insulin signaling, increasing the risk of type II diabetes [58] as shown in Fig. 8.

8 OS in Diabetes Complications

Numerous studies have found a relationship between diabetes and OS by evaluating numerous biomarkers, including lipid peroxidation products and DNA damage biomarkers. Because of their propensity to damage proteins, lipids, and DNA, FRs are thought to play a crucial role in late-stage diabetes complications. OS causes a variety of clinical conditions, including cancer, diabetes, and rheumatoid arthritis. Numerous experimental discoveries have shown a link between diabetes and OS by evaluating various biomarkers, including lipid peroxidation products and DNA damage indicators. OS causes various medical conditions, including cancer and diabetes mellitus.

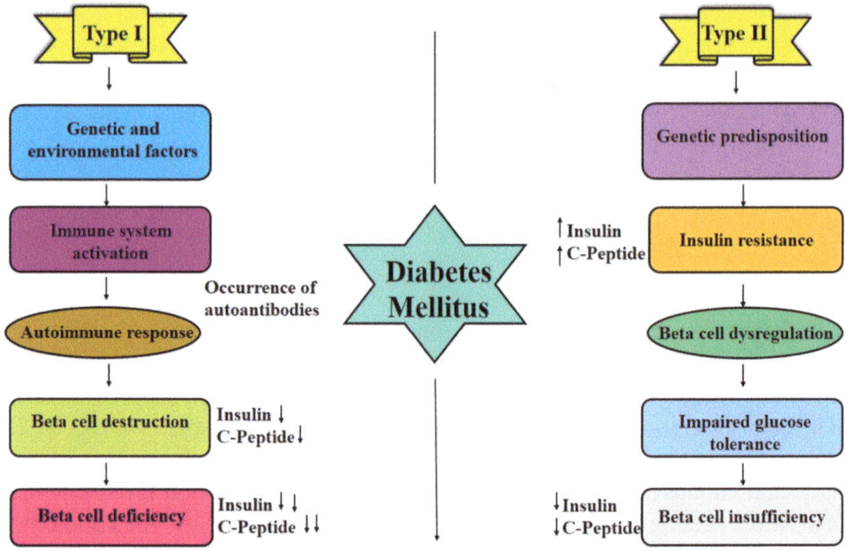

Fig. 8 Flow chart of diabetes mellitus

8.1 COVID-19

OS plays a superiority in the genesis and progression of illnesses, as well as a variety of viral diseases such as rhinoviruses, and COVID-19. Chronic OS typically affects all people who have viral infections. As the illness progresses, the body's antioxidant capacity dwindles, the oxidative state deteriorates, and antioxidants like glutathione and ascorbate are exhausted. Furthermore, there is undeniable evidence of substantial stress in the form of enhanced nitrosative synthase-2 (NOS2) expression and a high concentration.

8.2 COVID-19 and OS

When analyzing data from diverse research studies, it is vital to remember that OS levels change throughout the disease's course and are affected by the severity of COVID-19 [59]. The amounts of hydrogen peroxide were similar in all hospitalized patients, independent of illness severity. Superoxide anion levels increased considerably with disease severity. Nitric oxide levels were much higher in COVID-19 patients during hospitalization than in regulators. In contrast, nitric oxide levels were lower in severe COVID-19 than in milder or moderate COVID-19. Nitric oxide levels were shown to be lower in instances of fatal COVID-19, but there was no significant difference between moderate and severe COVID-19 patients and healthy controls [60]. Compared to healthy controls, there was an increase in OS among COVID-19 patients ranging from mild to severe. Severe COVID-19 was linked to reduced vitamin D levels, elevated plasma peroxide levels, and a higher OS index. Severe COVID-19 patients with fatal outcomes had a longer OS, whereas survivors had a shorter OS. Alwazeer et al. found that as the severity of COVID-19 increased, so did total oxidant status. In contrast, it was revealed that the severe, serious, and mild categories of hospitalized pediatric and adult patients had similar levels of total antioxidant and total oxidant status [61].

8.3 Challenges

This page compiles data from various meta-analyses that investigate the pathways of chronic disorders. Proteins, lipids, nucleic acids, and other biological elements are degraded due to excessive FRs [62]. OS has been related to several conditions that are now being discussed, including autoimmune diseases, cancer, CVD, and Alzheimer's disease. Air pollution, ionizing radiation, prolonged physical activity, diseases, and ingesting excessive polyunsaturated fatty acids are all risk factors for the generation of FR [63]. However, antioxidants—substances that inhibit FR—are supposed to solve these problems. "*Antioxidants*" refers to a substance's ability to provide electrons (Fig. 9).

Certain substances, depending on the chemical makeup of their environment, can act as prooxidants in certain cases and antioxidants in others. Antioxidants such

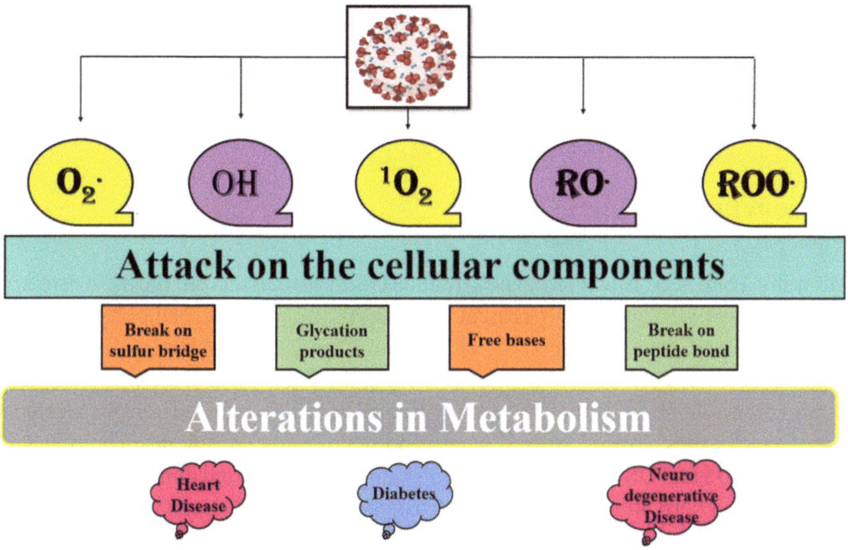

Fig. 9 Role of free radicals and OS in COVID-19

as polyphenols can occasionally have low blood concentrations without obvious action [65]. Most studies on extra benefits used only one kind of vitamin E, although our foods have eight distinct types [66]. Several complementary fronts must be addressed to continue increasing the benefit-risk window of anti-inflammatory treatment in chronic illnesses. Current research on this subject is critical to determine if certain illnesses, including cancer and atherosclerosis, can eliminate the inflammatory trigger. For example, innovative treatment techniques may be available to treat atherosclerosis by lowering plasma levels and avoiding the retention of atherogenic lipoproteins. Although these goals are theoretically possible, they have historically proven difficult to reach, and they are not now feasible in other chronic disorders, such as neurological disease [67]. Given the discovery that chronic inflammatory disorders may diminish the number or function of lipid mediators of inflammation resolution, therapeutic administration of these molecules might be an additional method. Finally, it is vital to consider noninvasive approaches for monitoring the pro-resolving or anti-inflammatory effects of drugs targeting chronic disorders' inflammatory part. This is especially important in situations like atherosclerosis, where clinical effects are often delayed, erratic, and deadly. Finally, during the last 20 years, a wealth of data has emerged demonstrating that inappropriate, nonresolving inflammation is the underlying cause of numerous prevalent chronic diseases when the primary pathophysiologic route is not infection, fundamental abnormalities in inflammatory control, or autoimmune. Even though this insight might lead to a slew of new therapeutic breakthroughs, the promise is hindered by the fact that the inflammatory response is a complex network required for appropriate physiology. The issues of need, compensation, and redundancy usually result in a narrow risk–benefit window. However, a better knowledge of inflammatory signaling provides promise for innovative therapy options that can successfully address these difficulties, particularly in endogenous regulatory networks and swelling resolution [68]. Furthermore, studies usually employ healthy people for whom the body's OS is not a significant factor in predicting sickness risk.

9 Conclusion

Adopting a balanced lifestyle can help reduce OS. Antioxidants are the most effective barrier against OS, and diet is critical. OS heavily impacts the pathophysiology of potentially dangerous disorders. Elevated prooxidant factor levels may cause persistent alterations in cellular architecture or enzymatic function and structural mistakes in mitochondrial DNA. It may be useful as an adjuvant therapy for various ailments, including diabetes, cancer, heart disease, and neurological problems. The observed abnormal ROS and OS levels indicate a substantial degree of cell component damage, nonfunctional derivative formation, and structural abnormalities in macromolecules in severe and extreme instances of COVID-19 [64]. It is possible to conclude that SARS-CoV-2's high fatality rate is due to both severe oxidative damage during infection and illness-related inflammation. Antioxidants have previously shown promise in treating diabetes. Substantial levels of oxygen and nitrogen free radicals (ROS/RNS) deteriorate diabetes and its complications. The bulk of research has shown a link between OS and diabetes, as well as potential complications in the heart, liver, kidneys, and eyes. Thus, OS appears more problematic in metabolic diseases, notably type 2 diabetes [69]. Substantial research needs to be done to decode the benefits and drawbacks of antioxidant consumption in cancer therapy. Oxidation and reduction processes do not take place in a vacuum. Antioxidants as meals (or supplements) should be used cautiously, because they are also prone to oxidation. Because there is accumulating evidence of some detrimental consequences, ingesting significant levels of antioxidants has become increasingly important. Their chemical ingredients are less toxic and more effective than those already on the market; thus, studying them as possible preventive and therapeutic agents is fascinating. To summarize, OS relates to humans, and more study in this field might be critical to understanding and treating several illnesses in future generations.

References

1. Fujii, H., Nakai, K., & Fukagawa, M. (2011). Role of oxidative stress and indoxyl sulfate in progression of cardiovascular disease in chronic kidney disease. *Therapeutic Apheresis and Dialysis, 15*(2), 125–128. https://doi.org/10.1111/j.1744-9987.2010.00883.x
2. Phaniendra, A., Jestadi, D. B., Periyasamy, L. (2015). Free radicals: properties, sources, targets, and their implication in various diseases. *Indian journal of clinical biochemistry, 30* (2015), 11–26. https://doi.org/10.1007/s12291-014-0446-0.
3. Brand, M. D. (2010). The sites and topology of mitochondrial superoxide production. *Experimental gerontology, 45*(7–8), 466–472. https://doi.org/10.1016/j.exger.2010.01.003
4. Santos, C. X. C., Tanaka, L. Y., Wosniak Jr, J., Laurindo, F. R. M. (2009). Mechanisms and implications of reactive oxygen species generation during the unfolded protein response: roles of endoplasmic reticulum oxidoreductases, mitochondrial electron transport, and NADPH oxidase. *Antioxidants & Redox Signaling, 11*(10), 2409–2427. https://doi.org/10.1089/ars.2009.2625.
5. Fransen, M., Nordgren, M., Wang, B., Apanasets, O. (2012). Role of peroxisomes in ROS/RNS-metabolism: implications for human disease. *Biochimica et Biophysica Acta (BBA)-Molecular Basis of Disease, 1822*(9), 1363–1373. https://doi.org/10.1016/j.bbadis.2011.12.001.
6. Rajendran, P., Nandakumar, N., Rengarajan, T., Palaniswami, R., Gnanadhas, E. N., Lakshminarasaiah, U., Gopas, J., & Nishigaki, I. (2014). Antioxidants and human diseases. *Clinica chimica acta, 436*, 332–347. https://doi.org/10.1016/j.cca.2014.06.004.
7. Small, D. M., Coombes, J. S., Bennett, N., Johnson, D. W., & Gobe, G. C. (2012). Oxidative stress, anti-oxidant therapies and chronic kidney disease. *Nephrology, 17*(4), 311–321. https://doi.org/10.1111/j.1440-1797.2012.01572.x
8. Yasui, K., & Baba, A. (2006). Therapeutic potential of superoxide dismutase (SOD) for resolution of inflammation. *Inflammation Research, 55*, 359–363. https://doi.org/10.1007/s00011-006-5195-y
9. Kim, S.-H., Kim, S.-H., Lee, J.-H., Lee, B.-H., Yoon, H. J., Shin, D. H., Park, S. S., Jang, S. B., Park, J.-S., & Jee, Y.-K. (2015). Superoxide dismutase gene (SOD1, SOD2, and SOD3) polymorphisms and antituberculosis drug-induced hepatitis. *Allergy, Asthma & Immunology Research, 7*(1), 88–91. https://doi.org/10.4168/aair.2015.7.1.88.
10. Mirończuk-Chodakowska, I., Witkowska, A. M., & Zujko, M. E. (2018). Endogenous non-enzymatic antioxidants in the human body. *Advances in Medical Sciences, 63*(1), 68–78. https://doi.org/10.1016/j.advms.2017.05.005.

11. Czerska, M., Mikołajewska, K, Zieliński, M., Gromadzińska, J., & Wąsowicz, W. (2015). Today's oxidative stress markers. *Medycyna pracy, 66*(3).
12. Birben, E., Sahiner, U. M., Sackesen,C., Erzurum, S., & Kalayci, O. (2012). Oxidative stress and antioxidant defense. *World Allergy Organization Journal, 5*, 9–19. https://doi.org/10.1097/WOX.0b013e3182439613.
13. Bansal, A. K., & Bilaspuri, G. S. (2011). Impacts of oxidative stress and antioxidants on semen functions. *Veterinary Medicine International* (1), 686137. https://doi.org/10.4061/2011/686137.
14. Frijhoff, J., Winyard, P. G., Zarkovic, N., Davies, S. S., Stocker, R, Cheng, D., Knight, A.R., et al. (2015). Clinical relevance of biomarkers of oxidative stress. *Antioxidants & Redox Signaling, 23*(14), 1144–1170. https://doi.org/10.1089/ars.2015.6317.
15. Mohideen, K., Sudhakar, U., Balakrishnan, T., Almasri, M. A., Al-Ahmari, M. M., Al Dira, H. S., Suhluli, M., et al. (2021). Malondialdehyde, an oxidative stress marker in oral squamous cell carcinoma—A systematic review and meta-analysis. *Current Issues in Molecular Biology, 43*(2), 1019–1035. https://doi.org/10.3390/cimb43020072.
16. Perkovic, M. N., Jaganjac, M., Milkovic, L., Horvat, T., Rojo, D., Zarkovic, K., Ćorić, M., et al. (2023). Relationship between 4-hydroxynonenal (4-HNE) as systemic biomarker of lipid peroxidation and metabolomic profiling of patients with prostate cancer. *Biomolecules, 13*(1), 145. https://doi.org/10.3390/biom13010145.
17. Davies, S. S., & Jackson Roberts II, L. (2011). F2-isoprostanes as an indicator and risk factor for coronary heart disease. *Free Radical Biology and Medicine, 50*(5), 559–566. https://doi.org/10.1016/j.freeradbiomed.2010.11.023.
18. Al-Taie, A., Sancar, M., Izzettin, F. V. (2021). 8-Hydroxydeoxyguanosine: A valuable predictor of oxidative DNA damage in cancer and diabetes mellitus. In *Cancer*, pp. 179–187. Academic Press. https://doi.org/10.1016/B978-0-12-819547-5.00017-1.
19. Martin-Perez, M., Urdiroz-Urricelqui, U., Bigas, C., & Benitah, S. A. (2022). The role of lipids in cancer progression and metastasis. *Cell metabolism, 34*(11), 1675–1699. https://doi.org/10.1016/j.cmet.2022.09.023.
20. Chen, H. J., Lin, W. P., Chiu, S. D., & Fan, C. H. (2014). Multistage mass spectro metric analysis of human hemoglobin glutathionylation: Correlation with cigarette smoking. *Chemical Research in Toxicology, 27*(5), 864–872. https://doi.org/10.1021/tx5000359
21. Zhu, J. W., Yuan, J. F., Yang, H. M., Wang, S. T., Zhang, C. G., Sun, L. L., Yang, H., Zhang, H. (2012). Extracellular cysteine (Cys)/cystine (CySS) redox regulates metabotropic glutamate receptor 5 activity. *Biochimie, 94*(3), 617–627. https://doi.org/10.1016/j.biochi.2011.09.013.
22. Alisik, M., Alisik, T., Nacir, B., Neselioglu, S., Genc-Isik, I., Koyuncu, P., & Erel, O. (2021). Erythrocyte reduced/oxidized glutathione and serum thiol/disulfide homeostasis in patients with rheumatoid arthritis. *Clinical Biochemistry, 94*, 56–61. https://doi.org/10.1016/j.clinbiochem.2021.04.023
23. Nascimento, J. A. M., Nascimento, M. L., Mendes, C. H. S., & Oliveira, S. C. B. (2021). Voltammetric studies of 3-nitro-tyrosine electro-oxidation at solid electrodes and its interaction with DNA. *Electroanalysis, 33*(1), 216–225. https://doi.org/10.1002/elan.202060256.
24. Kettle, A. J., Albrett, A. M., Chapman, A. L., Dickerhof, N., Forbes, L. V., Khalilova, I., & Turner, R. (2014). Measuring chlorine bleach in biology and medicine. *Biochimica et Biophysica Acta (BBA)-General Subjects, 1840*(2), 781–793. https://doi.org/10.1016/j.bbagen.2013.07.004.
25. Sharifi-Rad, M., Anil Kumar, N. V., Zucca, P., Varoni, E. M., Dini, L., Panzarini, E., Rajkovic, J., et al. (2020). Lifestyle, oxidative stress, and antioxidants: back and forth in the pathophysiology of chronic diseases. *Frontiers in physiology, 11*, 552535. https://doi.org/10.3389/fphys.2020.00694.
26. Reczek, C. R., & Chandel, N. S. (2017). The two faces of reactive oxygen species in cancer. *Annual review of cancer biology, 1*(1), 79–98. https://doi.org/10.1146/annurev-cancerbio-041916-065808
27. Valko, M., Rhodes, C. J. B., Moncol, J., Izakovic, M. M., & Mazur, M. (2006). Free radicals, metals and antioxidants in oxidative stress-induced cancer. *Chemico-Biological Interactions, 160*(1), 1–40. https://doi.org/10.1016/j.cbi.2005.12.009.

28. Valko, M., Izakovic, M., Mazur, M., Rhodes, C. J., & Telser, J. (2004). Role of oxygen radicals in DNA damage and cancer incidence. *Molecular and Cellular Biochemistry, 266*, 37–56. https://doi.org/10.1023/B:MCBI.0000049134.69131.89
29. Valavanidis, A., Vlachogianni, T., & Fiotakis, C. (2009). 8-hydroxy-2′-deoxyguanosine (8-OHdG): A critical biomarker of oxidative stress and carcinogenesis. *Journal of Environmental Science and Health Part C, 27*(2), 120–139. https://doi.org/10.1080/10590500902885684
30. Chatterjee, N., & Walker, G. C. (2017). Mechanisms of DNA damage, repair, and mutagenesis. *Environmental and Molecular Mutagenesis, 58*(5), 235–263. https://doi.org/10.1002/em.22087
31. Klaunig, J. E. (2018). Oxidative stress and cancer. *Current Pharmaceutical Design, 24*(40), 4771–4778. https://doi.org/10.2174/1381612825666190215121712
32. Tan, S.-N., Sim, S.-P., & Khoo, A.-B. (2018). Oxidative stress-induced chromosome breaks within the ABL gene: A model for chromosome rearrangement in nasopharyngeal carcinoma. *Human genomics, 12*, 1–21. https://doi.org/10.1186/s40246-018-0160-8
33. Döppler, H., & Storz, P. (2017). Mitochondrial and oxidative stress-mediated activation of protein kinase D1 and its importance in pancreatic cancer. *Frontiers in Oncology, 7*, 41. https://doi.org/10.3389/fonc.2017.00041
34. Himmelfarb, J., Stenvinkel, P, Alp Ikizler, T., & Hakim, R. M. (2002). The elephant in uremia: oxidant stress as a unifying concept of cardiovascular disease in uremia. *Kidney International, 62*(5), 1524–1538. https://doi.org/10.1046/j.1523-1755.2002.00600.x.
35. Ling, X. C., & Kuo, K.-L. (2018). Oxidative stress in chronic kidney disease. *Renal Replacement Therapy, 4*(1), 1–9. https://doi.org/10.1186/s41100-018-0195-2.
36. Nistala, R., Whaley-Connell, A., & Sowers, J. R. (2008). Redox control of renal function and hypertension. *Antioxidants & Redox Signaling, 10*(12), 2047–2089. https://doi.org/10.1089/ars.2008.2034
37. Galvan, D. L., Green, N. H., & Danesh, F. R. (2017). The hallmarks of mitochondrial dysfunction in chronic kidney disease. *Kidney International, 92*(5), 1051–1057. https://doi.org/10.1016/j.kint.2017.05.034
38. Finch, J. L., Suarez, EduB., Husain, K., Ferder, L., Cardema, M. C., Glenn, D. J., Gardner, D. G., Liapis, H., & Slatopolsky, E. (2012). Effect of combining an ACE inhibitor and a VDR activator on glomerulosclerosis, proteinuria, and renal oxidative stress in uremic rats. *American Journal of Physiology-Renal Physiology, 302*(1), F141–F149. https://doi.org/10.1152/ajprenal.00293.2011
39. Kohagura, K., Tana, T., Higa, A., Yamazato, M., Ishida, A., Nagahama, K., Sakima, A., Iseki, K., & Ohya, Y. (2016). Effects of xanthine oxidase inhibitors on renal function and blood pressure in hypertensive patients with hyperuricemia. *Hypertension Research, 39*(8), 593–597. https://doi.org/10.1038/hr.2016.37
40. Tbahriti, H. F., Kaddous, A., Bouchenak, M., & Mekki, K. (2013). Effect of different stages of chronic kidney disease and renal replacement therapies on oxidant-antioxidant balance in uremic patients. *Biochemistry Research International* (1), 358985. https://doi.org/10.1155/2013/358985.
41. Bennett, A. M., Lawan, A. (2020). Improving obesity and insulin resistance by targeting skeletal muscle MKP-1. *Journal of Cell Signal, 1*(4), 160–168. https://doi.org/10.33696/Signaling.1.025.
42. Lismont, C., Revenco, I., & Fransen, M. (2019). Peroxisomal hydrogen per oxide metabolism and signaling in health and disease. *International Journal of Molecular Sciences, 20*(15), 3673. https://doi.org/10.3390/ijms20153673
43. Yoboue, E. D., Sitia, R., & Simmen, T. (2018). Redox crosstalk at endoplasmic reticulum (ER) membrane contact sites (MCS) uses toxic waste to deliver messages. *CELL Death & Disease, 9*(3), 331. https://doi.org/10.1038/s41419-017-0033-4.
44. Santillo, M., Colantuoni, A., Mondola, P., Guida, B., & Damiano, S. (2015). NOX signaling in molecular cardiovascular mechanisms involved in the blood pressure homeostasis. *Frontiers in Physiology, 6*, 194. https://doi.org/10.3389/fphys.2015.00194
45. Förstermann, U., & Sessa, W. C. (2012). Nitric oxide synthases: Regulation and function. *European Heart Journal, 33*(7), 829–837. https://doi.org/10.1093/eurheartj/ehr304

46. Salehi, B., Sestito, S., Rapposelli, S., Peron, G., Calina, D., Sharifi-Rad, M., Sharopov, F., Martins, N., & Sharifi-Rad, J. (2018). Epibatidine: A promising natural alkaloid in health. *Biomolecules, 9*(1), 6. https://doi.org/10.3390/biom9010006
47. Gilgun-Sherki, Y., Melamed, E., & Offen, D. (2004). The role of oxidative stress in the pathogenesis of multiple sclerosis: The need for effective antioxidant therapy. *Journal of Neurology, 251*, 261–268. https://doi.org/10.1007/s00415-004-0348-9
48. Michel, M., Tanja, D. P., & Thome, J. (2012). The role of oxidative stress in depressive disorders. *Current Pharmaceutical Design, 18*(36), 5890–5899. https://doi.org/10.2174/138161212803523554.
49. Hwang, O. (2013). Role of oxidative stress in Parkinson's disease. *Experimental Neurobiology, 22*(1), 11. https://doi.org/10.5607/en.2013.22.1.11
50. Tönnies, E., & Trushina, E. (2017). Oxidative stress, synaptic dysfunction, and Alzheimer's disease. *Journal of Alzheimer's Disease, 57*(4), 1105–1121. https://doi.org/10.3233/JAD-161088
51. Du, H., Guo, L., Yan, S., Sosunov, A. A., McKhann, G. M., & Yan, S. S. (2010). Early deficits in synaptic mitochondria in an Alzheimer's disease mouse model. *Proceedings of the National Academy of Sciences, 107*(43), 18670–18675. https://doi.org/10.1073/pnas.1006586107
52. Allan Butterfield, D. (2002). Amyloid β-peptide (1–42)-induced oxidative stress and neurotoxicity: Implications for neurodegeneration in Alzheimer's disease brain. A review. *Free Radical Research, 36*(12), 1307–1313. https://doi.org/10.1080/1071576021000049890
53. Tasset, I., Agüera, E., Sánchez-López, F., Feijóo, M., Giraldo, A. I., Cruz, A. H., Gascón, F., & Túnez, I. (2012). Peripheral oxidative stress in relapsing–remitting multiple sclerosis. *Clinical Biochemistry, 45*(6), 440–444. https://doi.org/10.1016/j.clinbiochem.2012.01.023
54. Bajpai, A., Verma, A. K., Srivastava, M., & Srivastava, R. (2014). Oxidative stress and major depression. *Journal of Clinical and Diagnostic Research: JCDR,* 8(12), CC04. https://doi.org/10.7860/JCDR/2014/10258.5292.
55. Xu, Y., Tang, G., Zhang, C., Wang, N., & Feng, Y. (2021). Gallic acid and diabetes mellitus: Its association with oxidative stress. *Molecules, 26*(23), 7115. https://doi.org/10.3390/molecules26237115
56. Tang, Z., Wang, P., Dong, C., Zhang, J., Wang, X., & Pei, H. (2022). Oxidative stress signaling mediated pathogenesis of diabetic cardiomyopathy. *Oxidative Medicine and Cellular Longevity, 2022*(1), 5913374. https://doi.org/10.1155/2022/5913374
57. Zhao, M., Xie, Y., Gao, W., Li, C., Ye, Q., & Li, Y. (2023). Diabetes mellitus promotes susceptibility to periodontitis—novel insight into the molecular mechanisms. *Frontiers in Endocrinology, 14*, 1192625.
58. Yousef, H., Khandoker, A. H., Feng, S. F., Helf, C., & Jelinek, H. F. (2023). Inflammation, oxidative stress and mitochondrial dysfunction in the progression of type II diabetes mellitus with coexisting hypertension. *Frontiers in Endocrinology, 14*, 1173402. https://doi.org/10.3389/fendo.2023.1173402
59. Chernyak, B. V., Popova, E. N., Prikhodko, A. S., Grebenchikov, O. A., Zinovkina, L. A., & Zinovkin, R. A. (2020). COVID-19 and oxidative stress. *Biochemistry (Moscow), 85*, 1543–1553. https://doi.org/10.1134/S0006297920120068
60. Fodor, A., Tiperciuc, B., Login, C., Orasan, O.H., Lazar, A. L., Buchman, C., Hanghicel, P., et al. (2021). Endothelial dysfunction, inflammation, and oxidative stress in COVID-19—mechanisms and therapeutic targets. *Oxidative Medicine and Cellular Longevity*, (1), 8671713. https://doi.org/10.1155/2021/8671713.
61. Alwazeer, D., Fuh-Ching Liu, F., Wu, X. Y., & LeBaron, T. W. Combating oxidative stress and inflammation in COVID-19 by molecular hydrogen therapy: mechanisms and perspectives. *Oxidative Medicine and Cellular Longevity,* (1), 5513868. https://doi.org/10.1155/2021/5513868.
62. Knopman, D. S., Lundt, E. S., Therneau, T. M., Albertson, S M., Gunter, J. L., Senjem, M. L., Schwarz, C. G., et al. (2021). Association of initial β-amyloid levels with subsequent flortaucipir positron emission tomography changes in persons without cognitive impairment. *JAMA Neurology, 78*(2), 217–228. https://doi.org/10.1001/jamaneurol.2020.3921.

63. Poprac, P., Jomova, K., Simunkova, M., Kollar, V., Rhodes, C. J., & Valko, M. (2017). Targeting free radicals in oxidative stress-related human diseases. *Trends in Pharmacological Sciences, 38*(7), 592–607. https://doi.org/10.1016/j.tips.2017.04.005
64. Georgieva, E., Ananiev, J., Yovchev, Y., Arabadzhiev, G, Abrashev, H., Abrasheva, D., Atanasov, V., et al. (2023). COVID-19 complications: oxidative stress, inflammation, and mitochondrial and endothelial dysfunction. *International Journal of Molecular Sciences, 24*(19), 14876. https://doi.org/10.3390/ijms241914876.
65. Fernández-García, E., Carvajal-Lérida, I., Jarén-Galán, M., Garrido-Fernández, J., Pérez-Gálvez, A., & Hornero-Méndez, D. (2012). Carotenoids bioavailability from foods: From plant pigments to efficient biological activities. *Food Research International, 46*(2), 438–450. https://doi.org/10.1016/j.foodres.2011.06.007
66. Firuzi, O., Miri, R., Tavakkoli, M., & Saso, L. (2011). Antioxidant therapy: current status and future prospects. *Current Medicinal Chemistry, 18*(25), 3871–3888. https://doi.org/10.2174/092986711803414368.
67. Tabas, I., & Glass, C. K. (2013). Anti-inflammatory therapy in chronic disease: Challenges and opportunities. *Science, 339*(6116), 166–172. https://doi.org/10.1126/science.1230720
68. Evans, M., Lewis, R. D., Morgan, A. R., Whyte, M. B., Hanif, W., Bain, S. C., Sarah Davies, S., et al. (2022). A narrative review of chronic kidney disease in clinical practice: current challenges and future perspectives. *Advances in Therapy, 39*(1), 33–43. https://doi.org/10.1007/s12325-021-01927-z.
69. Asmat, U., Abad, K., & Ismail, K. (2016). Diabetes mellitus and oxidative stress—A concise review. *Saudi Pharmaceutical Journal, 24*(5), 547–553. https://doi.org/10.1016/j.jsps.2015.03.013

The Impact of Natural and Synthetic Antioxidants in Disease Development in Humans

S. Diviya Priya, S. Subhashini, R. Ramesh, and L. Anandaraj

Abstract Antioxidant in general called as the defense system and a complex mechanism of chemical reactions that protects all the living organisms in various ways including developing a protective shield against diseases causing pathogens, fighting against the malicious threats in the cells of the body, creating a self-operated mechanism to monitor with the ROS induced stress related diseases and ensuring the diagnosis of oxidative stress induced communicable, non-communicable diseases and chronic illness. Overall, the antioxidant plays a multifunctional role, and these are available as natural and synthetic antioxidants, which varies in their properties and functions. A variety of antioxidants were extracted from the plant species, fruits, seed coatings, vegetables, etc., and naturally occurring antioxidants including vitamins, C, A and E, carotenoids, beta carotene, lutein, zeaxanthin, phytocompounds derived from plants used in the pharmacological sector in developing antioxidant-aided drug systems to cure various diseases, also in the food preservation, cosmetics etc., It acts as free radical terminators and retards the chain initiation mechanism. Similarly, the synthetic antioxidants are manmade and found to have numerous applications in the development of ROS and self-mechanism strategy in the development of diseases and protection system. Synthetic antioxidants include several chemical additives and some of them have also the negative impact on the human immune system because of the adaptability variations. Here in this chapter the role of the natural and the synthetic antioxidants in the disease development, diagnostics, and protection against chronic illness were discussed in detail.

Keywords The role of ROS in stress-related diseases · Various functions are performed by antioxidants · Plant phytochemicals · Systematic use of antioxidants in drug development

S. Diviya Priya · R. Ramesh (✉) · L. Anandaraj
Department of Physics, Sacred Heart College (Autonomous), (Affiliated to Thiruvalluvar University), Tirupattur 635601, Tamil Nadu, India
e-mail: rramesh@shctpt.edu

S. Subhashini
School of Physiotherapy, Dhanalakshmi Srinivasan University, Tiruchirappalli 621112, Tamil Nadu, India

K. Kasinathan et al. (eds.), *Recent Advances in Oxidative Stress Associated Chronic Diseases Volume 1*, https://doi.org/10.1007/978-981-96-9670-3_4

1 Introduction

Antioxidant is defined in a variety of ways. Vitamin E and other chain-breaking antioxidant inhibitors of lipid peroxidation are frequently implicitly limited to using this phrase. Reactive oxygen species (ROS) play a vital role when generated in living systems. Ageing has been associated with chronic disease, and research indicates that there is no proof that antioxidant supplements cause this condition [1]. Both nutrients and nonnutrients are necessary for the growth and development of living things. The primary oxidative radicals harmful to human bodies are macronutrients, which can be made up of carbon, hydrogen, oxygen, and other elements. These have the potential to harm the tissue system and its functions. Globally, about 10% of people have chronic kidney disease (CKD). Vitamins are nutritional components in metabolic processes and should be taken in tiny amounts daily. Animals have them, a crucial part of the animal's oxidative stress equilibrium is examined in poultry. Despite being used as an antioxidant in food for the past century [2], BHA (butylated hydroxyl anisole) has negative side effects, including the development of cancer. There are more antioxidants in plant extracts than in BHA, according to some research. Natural plant extract serves as a substitute for artificial preservers and enhances the end product's significance [3]. Phytocompounds have gained more attention because of the negative effects that medications and other synthetic chemicals have on animal health as well as the quality and safety of products. As a supplement to animal feed, plant polyphenols are becoming more popular due to their usage in animal nutrition [4]. Clean, safe, and green agriculture as well as a sustainable, eco-friendly method of raising livestock would also benefit farmers and society as a whole. It is becoming more and more fashionable to alter animal nutrition in order to create healthier animal-based products with a higher proportion of polyunsaturated fatty acids (PUFAs) than saturated fatty acids. Numerous additional compounds have been propounded to have higher antioxidant properties [5].

In general, the necessary condition for the typical physiological functions of every organism is as follows: the equilibrium between the free radicals and the antioxidants is must. If the level of the radials overwhelms the ability of the vital functions including digestion, healing, sexual activity etc., For the regulation of these systems of cellular stress is produced [6]. In this case, the application of external antioxidants be involved in chief role in coping and balancing the cellular stress. Natural antioxidants including carnosine, metallothionein, phytic acid, taurine and its compounds, uric acid, bilirubin, creatinine dihydrolipoic acid, ubiquinol flavonoids, polyamines and other compounds of phenolics are of plant origin [7]. Some of the compounds like acetyl cysteine and nonsteroidal anti-inflammatory medicines have antioxidant properties when consumed by humans [8]. On the other hand, the animal kingdom had evolved with a self-built mechanism to reduce or to avoid the generation of RS oxidative metabolism and the antioxidant defense systems that can limit the antioxidant activity of the RS production [9]. Chronic diseases seem to be increasing in the industrialized nations and according to the data, about 30 million adults in the globe suffer from one or more chronic conditions that limit their activity of routine life;

this number will grow steadily into the future and the physicians are trying their best to take care of the patients with chronic diseases [10]. Naturally occurring synthetic antioxidants are very important for human health and longevity. The superoxide anion is considered as one of the primary ROS systems, which is responsible for the physiological, which in turn generates the secondary ROS via direct interaction with very many molecules through enzymes [11]. Overproduction of the ROS, RNS, and RSS in the biological systems leads to an imbalance in the biological systems creating oxidative stress due to damage in the DNA, proteins, and lipids. Reports reveal that ROS-initiated free radical injury is linked to the ageing process, as well as several diseases such as cancer, hyperglycemia leading to diabetes, cardiothoracic diseases, cataracts, neurodegenerative disorders, etc., In this chapter, naturally occurring antioxidants, synthetic antioxidants, and their properties in the development of diseases are discussed in detail.

2 Natural Antioxidants of Plant Species

Angiosperms are thought to include between 250,000 and 300,000 species in the field of botany. Less than 400 of these species representing 33 of the 51 orders and 89 of the 279 families have been recognized or raised for human consumption [12]. Many edible portions of the plant species including, leaves, flowers, seeds roots, and trunks have natural antioxidants, which include the same or the combination of the compounds that may or may not be present in these plants and some of the antioxidants can be isolated through various processes [13]. Dare edible, although, in most species, just one component of the plant is edible; however, they are not always found in the edible sections. All higher plants and their diverse components contain natural antioxidants, including usually, these antioxidants fall into the phenolic or polyphenolic compound category. Some of the compounds with antioxidant properties found in the edible parts of plants include flavonoids, tocopherols, derivatives of cinnamic acid, polyfunctional organic acids, phosphatides, etc. The flavonoid group includes chalcones, catechins, isoflavones, flavones, flavanols, and flavones [14]. Chlorogenic acid, ferulic acid and caffeic acids are the derivatives of cinnamic acid. It has long been known that using natural chemicals can help prevent lipids and other organic materials from oxidizing. It was about a century ago that one of the first and most thorough studies in this area took place, which greatly stimulated interest in antioxidation. Researchers looked at the antioxidant qualities of more than 500 synthetic and natural substances in this study [15]. Most of the later research was on substances unrelated to food, like plastics, rubber, and gasoline intended to stop oxidation. However, the foundation for investigating compounds that might regulate oxidation in food systems was established by these groundbreaking investigations. It was shown in the early 1930s that adding some quantity of the flours of soy, sesame, and oats to lard might successfully lessen its rancidity. The process of adding the defatted soy flour to the melted lard had the most noticeable antioxidant effect. Natural antioxidants in food can come from (a) naturally occurring chemicals in one or more food components; (b) molecules produced by reactions during processing; and (c) naturally occurring food additives [16]. Several of the natural antioxidants are algae, oil seeds, amla, citrus fruits, cocoa powder, extracts from shell plants, products of heated proteins, soy products, Oat flour, microbial products, amino acids, peptides, tannins, flavonoids, Melanoidin Tocopherols, etc.,

2.1 Polyphenolic Resveratrol

It is one of natural substances present in the several fruits including grapes, mulberries, soy, Raspberries. Additionally, it can be found in the rootstock of well-known plant species Japanese knotweed (*Polygonum cuspidatum*). This antioxidant is found to be abundant in red wine. It possesses a good antioxidant, antiproliferative effects against several fungi and UV-radiation [17]. Research indicates that resveratrol may support cardiovascular health and has been explored for its possible advantages in managing diabetes.

2.2 Procyanidins and Cinnamon

Foods that are high in procyanidins include cinnamon, apples, cranberries, avocados, aduki beans, prunus dulcis, and peanuts. Among these, the spice species called cinnamon had undergone the majority thorough research, likely due to its diverse types of procyanidins [18]. While cinnamon is commonly utilized as a spice to enhance the flavor of various dishes, it has also been advocated in both Ayurvedic and Chinese medicinal practices, potentially owing to its high content of tannins. Cinnamon cassia (*Cinnamomum cassia*), which occurs abundantly in the compound called cinnamaldehyde, was the subject of extensive studies aimed at improving the health of individuals with diabetes [19].

2.3 Carnitine

L-carnitine participates in a critical role in the transformation on free fatty acids (FFAs) into energy [20]. It facilitates the smooth transport of FFA inside the matrix of mitochondrial system, where it undergoes metabolism for energy production through the citric acid cycle. This compound is present in various food sources, including nuts, fishes, dairy products, red meat, and seeds. Additionally, it helps in reducing cellular stress and has been the subject of extensive research regarding its potential benefits for individuals with type 2 diabetes [21].

2.4 Alpha Lipoic Acid

Lipoic acid is synthesized as a minimal quantity within the human body. It serves as a cocatalyst for several enzymes, particularly those associated with two oxoacid dehydrogenase complexes, including pyruvate dehydrogenase, 2-oxoglutartate dehydrogenase complex, oxoacid, and acetoin complex [22]. Notably, these complexes take part in energy metabolism of carbohydrates. Additionally, Alpha-lipoic acid functions as an antioxidant agent and can be found in various foods such as carrots, spinaches, broccoli, yams, potatoes, and red meat. Although the concentration of alpha-lipoic acid in these foods is found to be relatively low, and the bioavailability is restricted because it coalesces with the chemical compound lysine. Clinical studies have investigated its application in the sense of dietary supplement, particularly for its effective and ability advantages in managing obesity and diabetes [23].

3 Antioxidants Defense System

Antioxidant defense mechanisms play a vital role in preserving redox equilibrium inside the body system, particularly during the pregnancy, when cellular oxidative stress can profoundly affect both maternal and fetal well-being [24]. These mechanisms include a combination of enzymatic and nonenzymatic elements that collaborate to counteract reactive oxygen species (ROS) and reduce oxidative harm. A thorough understanding of these protective systems is critical for tackling complications associated with oxidative stress, including pregnancy-induced hypertension (PIH) [25].

3.1 Enzymatic Antioxidants

Enzymatic antioxidants are proteins that facilitate reactions to stabilize the active ROS species and mitigate oxidative disorder. Some of the enzymatic antioxidants were discussed as follows.

3.2 Superoxide Dismutase (SOD)

This enzyme agitates the transformation of radicals of superoxide into a compound called hydrogen peroxide (H_2O_2) and dilute molecular oxygen [26]. Various isoforms of SOD exist in various forms such as cytosolic SOD, mitochondrial SOD, and extracellular SOD, each serving specific functions within different cellular environments.

Increased levels of SOD are essential for safeguarding molecular cells in opposition to cellular stress, particularly inside the endothelial cells that line blood vessels [27].

3.3 Catalase

Catalase further detoxifies hydrogen peroxide generated by the SOD and in turn it converts hydrogen peroxide into its constituent water molecule and oxygen [28]. This catalyst compound is primarily located in the liver and RBCs of the blood stream, where it acts a vital part in preventing several harmful accumulations of H_2O_2 at elevated concentrations.

3.4 Glutathione Peroxidase (GPx)

GPx comprises a group of enzymes that reduce hydrogen peroxide and lipid peroxides by utilizing glutathione (GSH) as a cofactor. Through this mechanism, GPx protects cellular membranes and other essential components from oxidative damage. Sufficient levels of GPx are critical for sustaining cellular health, especially during times of heightened oxidative stress, such as during pregnancy [29].

3.5 Nonenzymatic Antioxidants

Similar to enzymatic ones, nonenzymatic antioxidants are essential in the neutralization of ROS and for the preservation of redox equilibrium. Prominent nonenzymatic antioxidants consist of the following.

3.6 Glutathione

This tripeptide, made up of glutamine, cysteine, and glycine, is recognized as a vital and effective intracellular antioxidants compound. The compound Glutathione is available in two forms: oxidized (GSSG) and reduced (GSH), with the reduced form serving as the active antioxidant. Antioxidants play a vital role in cellular protection by directly stabilizing the radicals generated by ROS and other antioxidant compounds like vitamin C and vitamin E, and it enhances the overall efficiency of the cells [30].

3.7 *Vitamins C and E*

These macronutrients are one of the essential dietary antioxidant compounds. Vitamin C, called as ascorbic acid, is soluble in water and it donates the electrons to stabilize the free radicals, thus preventing oxidative damage in aqueous environments. Conversely, Vitamin E, or also called as tocopherol, is fat soluble and it protects the membrane cells in the generation of peroxides from lipids by hunting free lipid radicals. The combined effects of these vitamins significantly bolster the antioxidant defense system, especially during pregnancy when oxidative stress is elevated [31]. Other phytonutrients, including flavonoids, polyphenolic compounds, and carotenoids, exhibit antioxidant properties. These plant-derived compounds contribute to the reduction of cellular stress by hunting the free radicals, several chelating metal ions, and modulating cellular signaling pathways. Consuming foods rich in these antioxidants, particularly fruits and vegetables, is crucial for maintaining a healthy diet during pregnancy [32].

3.7.1 Types of Antioxidants Investigated

Numerous antioxidants have been examined for their potential advantages in managing PIH.

(i) *Omega-3 Fatty Acids:* Renowned primarily for their anti-inflammatory effects, omega-3 fatty acids also demonstrate antioxidant properties. They may assist in regulating oxidative stress levels and improving endothelial function, positioning them as a potential complementary treatment for PIH.

(ii) *Coenzyme Q10 (CoQ10):* CoQ10 is a crucial antioxidant involved in mitochondrial energy production. It has been shown to increase antioxidant defenses and enhance endothelial function. Some studies indicate that CoQ10 supplementation may lead to reduced blood pressure and lower oxidative stress in pregnant women with hypertensive disorders [33].

(iii) *L-Arginine:* While primarily recognized as a precursor to nitric oxide, L-arginine also exhibits antioxidant characteristics. Supplementation with L-arginine can boost NO production, resulting in enhanced vascular function and decreased pressure in the blood in women with PIH.

3.8 *Polyphenols*

Approximately 8,000 polyphenols have been identified in various plants and fruits, with around 500 of this exhibiting biological activity and being sourced from natural products, processed foods, or juices [34]. These compounds are chemically classified as polyphenolic compounds substances and are distinguished by the presence of minimum of two phenyl rings and several hydroxyl groups. Certain polyphenols share structural similarities with 17β-E and may partially exhibit estrogenic activity by interacting with estrogen receptors. Polyphenols are recognized for their effective antioxidant and anti-inflammatory properties, which helps in bone protection [1]. They play a significant role in the prevention and management of osteoporosis and bone inflammatory conditions by mitigating ROS and OS. By inhibiting oxidative stress, polyphenols prevent the differentiation of bone precursor cells into non-bone tissue cells. Additionally, they can diminish osteoclast separation and its activity, thereby improving osteoblast segregation and overall growth in bones. Regular consumption of these compounds is associated with an increase in peak bone mass, and the intake of polyphenolic antioxidants helps prevent bone loss linked to a higher incidence of fractures. A study in Europe has reported that Mediterranean diet and an enhanced intake of calcium correlated with a reduced level of fracturesin adolescent males [35]. Numerous researches conducted on both humans and animals have demonstrated positive impacts of polyphenol intake on bone-related health: all these studies suggest that polyphenol substances may serve as therapeutic one rather than a protective one. Furthermore, there is conflicting evidence regarding the impact of polyphenol-rich foods on disorders in bone, especially at the stage of postmenopausal individuals. Specifically, it created a positive effect on bone-related health in osteogenic problems in women at postmenopausal stage, the same cannot be conclusively stated for other conditions [36].

3.9 Flavonoids

The most notable flavonoids include anthocyanins, which are commonly found in various fruits, particularly blueberries and bilberries, as well as in vegetables, grains, tea products, legumes, wine, chocolate associated food products, and wine. Latest studies indicate that the compound anthocyanins in blueberries subsist arabinosides, galactosides and glycosides [37]. The significant radical-scavenging properties of flavonoids can be attributed to the presence of a 3′,4′-dihydroxy group, also known as catechol structure, in the B ring, which exhibits the property of electron-donating characteristics that acts as a target for radicals, along with three OH group in the C ring. Anthocyanins and their sugar-less forms, anthocyanidins, was found to be improve bone health in human beings and animals in spite of its potent antioxidant properties, which aid in the development of bone mass [38]. Specifically, flavonoids demonstrate various beneficial effects, including the reduction process of bone resorption in regulating down the osteoclastic activity and enhancing osteoblastic activity via the inhibition of RANKL signaling. Recent findings have shown that the fruit extract of blueberry has a high content of anthocyanins and reveals both antioxidant and anti-osteoclastogenic properties in MLO-Y4 osteocytes and can restore osteogenic disparity in SaOS-2 osteoblasts, both of which are affected by oxidative stress-stimulated glutathione exhaustion and starvation [39]. BJ has been found to hinder the RANKL and signaling of sclerostin by blocking the activation of ERK1/2, JNK and NF-κB, while also promoting the activity of SIRT1, thereby supporting the bone cells formation step and the mineralization processes through the activation of

ALP and RUNX-2. Furthermore, it has been proved that blueberry polyphenols can stimulate the SIRT1 exaggeration in both cellular and mammalian models, regardless of oxidative stress conditions. Additional research indicates that diets rich in blueberries contribute to increased bone mass density and growth [40].

3.10 Nonflavonoids

Several compounds like xanthones, stilbenes, phenolic acids, and tannins are categorized as nonflavonoid phenolics and found to possess excellent antioxidant properties in developing bone-related health. In the case of phenolic group, a carboxyl group is attached to the benzene ring, whereas in the other compound, including stilbene, xanthones and tannins have two aromatic rings that are fused through C–C bonds [41]. Hydroxybenzoic acid, anacardic acid, curcumin, and hydroxycinnamic acids are found to improve bone health drastically as. A yellow-colored polyphenol curcumin obtained from turmeric plant possesses antioxidant and anti-inflammatory properties. It has a long history of use in treating various ailments, including age-related conditions like osteoporosis. Research indicates that curcumin may offer protection against glucocorticoid-stimulated osteoporosis by protecting osteoblasts from apoptosis [42]. Furthermore, it has been found to suppress osteoclast genesis by obstructing the movement of bone-marrow-derived macrophages and diminishing the synthesis of CCL3, an inflammatory chemokine, thereby hindering their maturation into osteoclasts. Recent studies have suggested that curcumin can be used alongside alendronate to enhance mineral density of the bone in postmenopausal women who are suffering from osteoporosis [43].

4 Chronic Diseases Due to Natural and Synthetic Antioxidants

Chronic illnesses are prevalent in health-care systems across the globe, including in the United States. The role of patients in managing their own diseases is crucial for mitigating the impact of these conditions. Various factors within an individual's social and physical environments can either facilitate or hinder their management efforts [44]. Implementing interventions aimed at enhancing patient management can lead to favorable outcomes, such as improved condition monitoring, reduced symptoms, better physical and psychosocial well-being, and decreased utilization of health-care services. Effective programs are often grounded in theoretical frameworks, with self-regulation emerging as a promising approach for developing such interventions. However, significant gaps persist in our understanding and enhancement of patient-driven disease management [45]. These gaps arise from a focus on clinical environments for program implementation, a lack of attention to external factors influencing effective management, limitations in the design of existing studies, an emphasis on short-term rather than long-term evaluations, and an inadequate assessment of the distinct contributions of various program elements. The management of chronic diseases remains a primary focus for health-care systems. This emphasis arises from the lack of primary prevention and curative options for numerous conditions, coupled with the global trend of increasing longevity, which often brings about chronic health issues [46]. Implementing effective primary prevention strategies, such as modifying dietary habits, enhancing physical activity, and reducing smoking, poses significant challenges. Similarly, the application of secondary prevention measures—aimed at preventing and managing the consequences of diseases—also proves to be difficult [47]. This chapter examines the various factors that empower individuals with chronic illnesses to maintain control over their health conditions. For the purposes of this discussion, optimal disease management by the patient is defined as achieving the highest possible level of functioning while minimizing symptoms, relative to the severity of the condition as shown in Fig. 1. Globally, the primary contributors to mortality include heart disease, cancer, and stroke, even in regions heavily affected by infectious diseases. Additionally, several other conditions, while not as directly linked to premature death, impose significant burdens in terms of human suffering and economic impact [48]. Conditions such as arthritis, diabetes, and asthma exemplify this, with HIV/AIDS also being notable. Despite being an infectious disease, advancements in treatment have allowed HIV/AIDS to be managed as a chronic condition.

4.1 Obesity and Antioxidants

Obesity is a multifaceted and complex disease. Since 1980, overweight and obesity have doubled in prevalence worldwide, with about one-third of the world's population falling into one of these categories [49]. Although prevalence rates are typically higher among women and older people, this increase in obesity rates has been seen across all age groups and genders, irrespective of geographic location, ethnicity, or socioeconomic position. Although this tendency has been similar across different countries and areas, there are considerable differences in the absolute rates of obesity and overweight. Obesity rates in a few developed countries seem to have levelled off recently [50]. BMI is frequently used in epidemiological research to classify people as overweight or obese, although it has low sensitivity and there is significant variation in body fat percentage among people with the same BMI. Include Name age, and sex factors for example, individuals of Asian descent typically exhibit a higher percentage of body fat than their Caucasian counterparts at the same BMI [51]. Additionally, Increased fat accumulation in the tissues of visceral and ectopic locations (*tissues of liver and muscular tissues*) visceral adipose tissue and ectopic locations (liver and muscle) is mainly associated with highly risk of cardiometabolic disease [52]. as well as in cases where there is a higher in the mass ratio such as lean-to-fat and the research findings reports that obesity can be more prevalent and

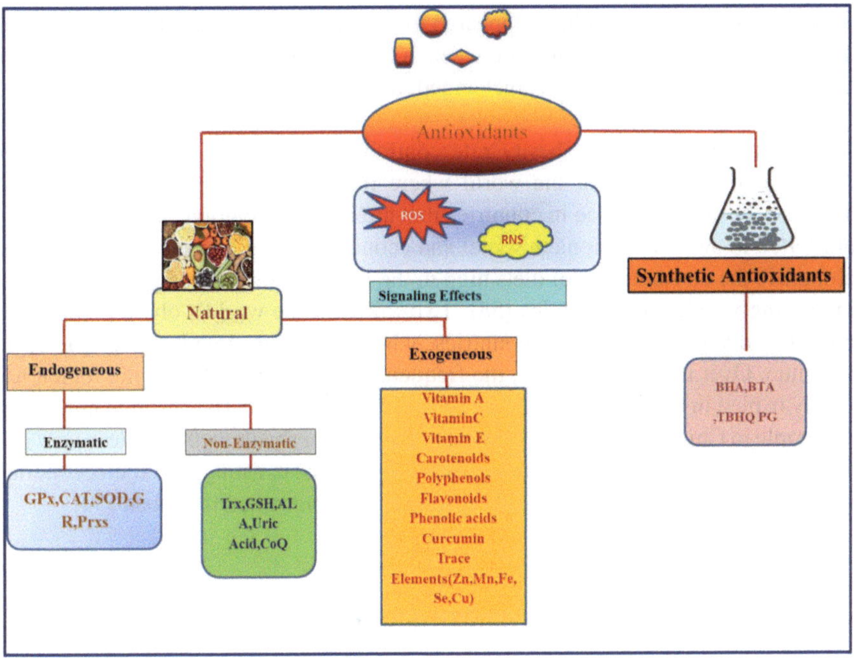

Fig. 1 Broad view of natural and synthetic antioxidants

needs more attention than the suggested in the epidemiological studies [53]. Dietary antioxidants, inflammation diet combination and body composition are interrelated with reciprocal relations in a bidirectional nature which affects health conditions a lot. Obesity is due to the result of caloric value extension at the biological level and undergoes partitioning that favors accumulation rather than oxidation [54]. Also, the availability ratio and the level of cellular stress and the ROS in the foods that lack antioxidants leads to alterations (*increased ghrelin due to the reductions in leptin changes their signal*) in the hormonal secretion which causes the increase in weight gain and results in a low-level caloric diet. Reports on experimental case studies revealed that flavonoids inhibit digestion and movement of postprandial glucose in the small intestine by enhancing the disposal level of glucose in the tissues and hence regenerating affected beta cells by improving insulin secretion in the pancreas [55]. Another naturally occurring compound curcumin found in Curcuma longa belongs to the ginger family was found to have good anti-obesity properties and prevent preadipocyte and rectify negative effects caused by obesity. Obesity can manifest at any stage of life. Studies on the prevalence of obesity have shown that it is becoming more common in people and children of all ages, irrespective of socioeconomic status, geography, or race. Obesity has become customary among people of middle age in areas of wealthy urban places and specifically for women in countries of low income [56]. Conversely, in high-income countries, it impacts individuals of all ages and genders, although its occurrence is notably higher among disadvantaged populations. Obesity rates have appeared to stabilize over the past decade in various developed nations. For instance, between 2005 and 2015, the rates of obesity in the US and the UK stayed largely unchanged, averaging between 30 and 34% and 23 and 24%, respectively [57].

In contrast, the global obesity rate continues to rise, with accelerating trends observed in other parts of the world where most of the population resides. The surge in obesity prevalence in countries that are in the developing stage is primarily attributed to swift alterations in social and economic conditions and demographics, alongside the consumption of diets high in the form of fats and energy coupled with increasingly sedentary lifestyles [58]. Two-thirds of the world's obese populations currently dwell in developing countries, where the rising trends are predicted to continue. Over the past 35 years, the frequency of obesity in the Pacific region has nearly tripled in China, from 0.6% in 1980 to 5.3% in 2015. The absolute number of fat people in China is among the greatest in the world, even though the rate is very low than in the US. Differences in gender, lifestyle patterns in eating, and socioeconomic position play the main determinants among rural women and the urban [59].

4.2 Male and Female Infertility and Antioxidants

The importance of the mind–body connection to fertility has been emphasized in several studies; yet the psychological aspects of infertility have not received enough attention. Women and their partners experience physical and psychological stress because of fertility treatments, which include hormone therapy, medical monitoring, and in vitro fertilization (IVF). A lower chance of a healthy pregnancy is linked to psychological factors such as despair, anxiety, and stress-related changes in heart rate and cortisol levels [60]. When pregnancy does not happen right away, couples who are trying to conceive frequently experience feelings of disappointment and frustration. Furthermore, self-esteem, body image, and views of masculinity or femininity may all be significantly impacted if difficulties continue and one couple is diagnosed with infertility [61]. Three possible connections between psychological factors and infertility have been put forth: (i) Psychological factors may act as risk factors for infertility in the future, (ii) psychological distress may result from the diagnosis and treatment of infertility, and (iii) there may be a reciprocal relationship between psychological factors and infertility. This chapter looks at the data about these three associations and suggests a different method of treating infertility that involves stress evaluation either before or in addition to reproductive treatments. Regarding the interaction between psychological variables and infertility, three types of correlations have been put up: (i) Psychological variables that increase the likelihood of infertility in the future, (ii) psychological suffering that results from the diagnosis and treatment of infertility, and (iii) there is a reciprocal interaction between psychological issues and infertility. In the present scenario, infertility is one of the major problems among men and women in general. Fertility is a process that is very complex, and it includes the decidualization of the uterine endometrium, which is essential for pregnancy and regulates the elevated levels of the hormones of the ovaries and the changes in the estrogen-E2 and progesterone-P4 needed for the implantation of the embryo in the phase change of menstrual cycle [62]. This will take place only at certain conditions of physiological levels of signals generated by the ROS for the chemical interaction in the uterine endometrium during the remarkable changes in the oxygen level tension during fertilization and pregnancy. It is essential that for the proper functioning of the physiological process of uterine balance between oxidants and antioxidants is mandatory. Retinoids are another compound that acts as an embryonic signalling pathway during the 3rd trimester and the deficiency of vitamin A is associated with premature birth risk and anemia [63]. Folate is one of the crucial water-soluble vitamins which is considered one of the necessary compounds for the biosynthesis of nucleotides, WBC's, metabolism of amino acids and DNA repair. In comparison with vitamins E and C, folate acts as an effective antioxidant to produce quality oocyte and maturation, embryogenesis, fetal growth and the development of organs during pregnancy. In the female reproductive system, melatonin controls the endocrinological system of the reproductive at childhood and it can control the ovulation process the level of melatonin decreases with the increase of age and can cause reproductive disorders like endometriosis and PCOS [64]. PCOS is mainly due to serum prolinase activity and may result in menstrual irregularities and cardiovascular risk at a certain age.

Suggesting a two-way relationship in which psychosocial distress and infertility may have an impact on one another concurrently or successively. Many couples consider fertility therapy to be a major psychological burden, and many feel that the medical system does not sufficiently relieve this mental burden [65]. Furthermore, studies show that approximately half of women receiving fertility therapy say that their infertility has been the most stressful event of their lives, compared to only 15% of men who say the same. The second hypothesis is supported by a comparative international survey of women waiting for IVF treatment, which found that their rates of depressive symptoms were four times greater than those of a control group of women without reproductive problems [66]. Furthermore, their self-reported scores were significantly lower than those of the control group in domains like memory/concentration, anxiety, and attractiveness. Investigations indicated that women facing infertility exhibited depression levels that were twice as elevated as those in a control group. Additionally, it was observed that women who had been dealing with infertility for a duration of 2 to 3 years reported the highest levels of depression, in contrast to those who had experienced infertility for less than one year or for more than six years [67]. Male infertility has been significantly associated with elevated scores across all three subscales of burnout, with particular emphasis on cognitive fatigue. Potential explanations for this correlation may include gender differences in coping mechanisms related to infertility. Infertility treatment can vary based on the gender of the affected partner, and there is a reciprocal relationship between psychological factors and infertility.

4.3 Miscarriage and Antioxidants

Although many pregnancies are ended in the early stages of pregnancy, medical professionals still do not have adequate tools to determine the likelihood for miscarriage. Therefore, the purpose of this prospective cohort analysis was to determine the risk variables for first-trimester miscarriage in women who did not show any obvious signs of possible loss of pregnancy [68]. The investigation comprised of 1,098 women who were between weeks 4 and 12 of pregnancy and had no overt indications of a pregnancy hazard. Pregnancy outcomes were tracked after a demographic, medical history, and psychological, and biological information was gathered at the time of recruitment. There were 55 occurrences of miscarriage and 809 pregnancies that progressed successfully among the patients with sufficient data [69]. This clinical issue is exacerbated by the trend of women postponing pregnancy until their 30s. Consequently, even a single miscarriage can significantly threaten the chances of successful reproduction for individual couples [70]. Traditionally, clinical evaluation and treatment are not provided until three or more miscarriages occur consecutively, which may be too late for older women. The cost of accessing preventive services appears to be disproportionately high for women, particularly given that the risk of miscarriage is believed to increase with the number of prior losses, potentially raising the risk of subsequent pregnancies by 40%. Therefore, both clinicians and expectant mothers would greatly benefit from the establishment of standardized risk assessment tools for broader clinical application [71]. In summary, the subtle yet significant risk of miscarriage during the first trimester poses a serious concern for both expectant mothers and their health-care providers. This study offers important insights for clinicians by reclassifying groups of seemingly low-risk women considering increased, often overlooked, miscarriage risks. By combining the results of this study with data already available from epidemiology and basic science, it is proposed that human miscarriage should not be seen as a single occurrence but rather as the result of complex interactions between risk factors related to demographics, medical history, physiology, and psychology [72]. Notably, advancing age, low level of BMI, and little serum progesterone were identified as independent risk factors, all of which can be easily assessed during a routine office visit. These findings will assist clinicians in recognizing pregnant women who may need additional monitoring or therapeutic measures, such as progestogen supplementation, in the early stages of pregnancy [73, 74].

4.4 Antioxidants—Stress and Diabetes Mellitus

Stress is a significant factor that may contribute to chronic hyperglycemia in individuals with diabetes. It has been well-documented that stress can profoundly influence metabolic processes. The fight or flight response primarily triggers energy mobilization, leading to the release of various hormones that can elevate blood glucose levels

[75]. A relative or absolute lack of insulin may make it difficult for diabetics to digest the stress-induced spikes in glucose, even though this reaction is adaptive in a healthy person. Furthermore, diabetic patients may have abnormalities in the regulation of these stress hormones. The evidence regarding the impact of stress on type I diabetes remains inconsistent. Some retrospective studies in humans have indicated that stress may trigger the onset of type I diabetes, while animal research has demonstrated that different types of stressors can either induce or inhibit the development of experimental models of the disease [76]. Autonomic neuropathy, which is common in type I diabetes, may be connected to this heterogeneity. On the other hand, stronger evidence suggested that stress plays a crucial role in type II diabetes. While little study has been done on the effects of stress on the development and course of type II diabetes in humans, a huge data on animal experiments revealed that the stress is a constant cause of hyperglycemia in this type of diabetes [77]. Additionally, there is mounting evidence that the path physiology of type II diabetes involves the autonomic nerve system in both human and animal models. Human type I diabetes has been linked to stress, according to certain theories. According to research, people with diabetes have a higher chance of experiencing stress. In humans, stress has been shown to cause hyperglycemia and hypoglycemia or have no significant effect on glycemic control in those already diagnosed with diabetes and significant family loss before the manifestation of symptoms [78]. Nevertheless, these studies often suffer from inadequate controls and frequently depend on participants' recollections of specific life events. In one study, diabetic patients were compared with their non-diabetic siblings of similar age and a control group from the same neighborhood. The findings revealed that the diabetic individuals experienced a greater number of severe life events in the three years leading up to their diagnosis compared to either control group [79]. Severe life events were categorized based on the level of immediate or prolonged threat posed to the individual. However, this study is constrained by its limited sample size and does not provide detailed examples of the severe life events encountered. While these findings are not definitive, notably, concordance for type I diabetes is present in just 50% of identical twins. Despite both twins showing signs of autoimmune irregularities, the disease manifests in only half of the pairs. This suggests that an environmental trigger may be essential for the full expression of the disease. Consequently, stress might influence the onset of type I diabetes by either directly or indirectly activating this autoimmune irregularity. Nonetheless, definitive evidence supporting this mechanism remains elusive. In the last decade, there has been ongoing speculation regarding the effectiveness of the autonomic central nervous system in the pathophysiology of type II diabetes [80]. This concept was initially proposed by Claude Bernard, who demonstrated that hyperglycemia could be induced in healthy, nondiabetic rabbits through lesions in the hypothalamic region. More recently, various studies have indicated that hyperglycemia can also be triggered by chemical stimulation of the brain using morphine and various endogenous neuropeptides, with the condition being reversible through bilateral adrenalectomy [81]. Additionally, hyperglycemia has been observed following a slow intravenous infusion of epinephrine and during stress conditions that lead to prolonged sympathetic activation. Autonomic activity that may contribute to metabolic decomposition

could be influenced by stress or by the impact of dietary fats and carbohydrates on sympathetic outflow [82].

4.5 Malnutrition

Dietary compounds are the real nutrients that have physiological effects which include vitamins, minerals, amino acids, fatty acids, fibers and some herbal extracts that act as potential antioxidants in maintaining nutritional level metabolism. People with metastatic cancer are affected with cachexia and sarcopenia. A study reports that patients with metastatic cancer are severely malnutrition with symptoms like weight loss syndrome which involves complex metabolic processes [83]. The reduced form of NAD(H), NADP(H) and NADPH is very essential for the reduction of the oxidized glutathione. Also, these forms can be used as decreasing equivalent for the lipooxygenases, and it acts as a prompt source for the free radicals hence there is a deficiency in the nicotinic acid that causes a contaminant decrease in GSH in many internal organs. The impact of malnutrition on physiological functions significantly influences clinical outcomes. Observations made by surgeons in the 1930's indicated that patients who were either starved or underweight experienced a higher rate of postoperative complications and mortality. Numerous studies have since corroborated this initial finding. Surgical patients suffering from malnutrition exhibit complication and mortality rates that are three to four times greater than those of adequately nourished patients, often resulting in extended hospital stays and costs that can be up to 50% higher. Similar trends have been noted among medical patients, particularly the elderly. Distinguishing the harmful effects of malnutrition from the effects of the underlying disease can be challenging, as each can act as both a cause and a consequence of the other. Nevertheless, substantial evidence demonstrates that nutritional support markedly enhances outcomes for these patients; thus, it is essential to identify malnutrition through appropriate screening measures [84]. Malnutrition is a prevalent issue that is often overlooked and inadequately addressed by both patients and health-care providers. It serves as both a contributing factor and a result of various health conditions, affecting individuals in both institutional settings and the wider community as shown in Fig. 2. In the United Kingdom, about 5% of people are very low weight and have a BMI of 20 kg/m^2. In addition, people who are obese who lose weight accidentally, bringing their level of BMI to normal range were also found to be malnutrition. Following acute medical occurrences, such as small bowel infarction, some patients may be at risk, which can impair their capacity to meet metabolic needs over the long and short term. Malnutrition rates are at least twice as high among the elderly and those with chronic illnesses, and they can even triple among those residing in institutes with utmost care facilities [85]. In wealthy countries, malnutrition is still a major problem, especially for those who are poor, socially isolated, and abusing drugs. However, the bulk of adult malnutrition is associated with several illnesses and can be brought on by (i) reduced meal intake; (ii) reduced ability to

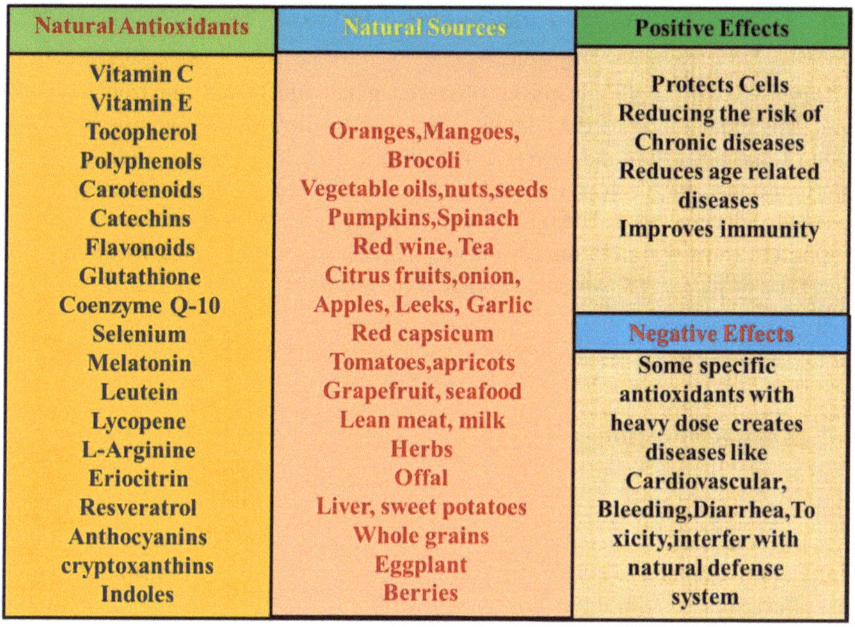

Fig. 2 Antioxidants and its sources: beneficial and harmful effects

absorb micronutrients and/or macronutrients; (iii) increased losses or altered dietary requirements; (iv) increased energy use in specific medical disorders.

4.6 Muscle Function

Weight loss brought on by a decrease in muscle and fat mass of the various organ masses is the main symptom of malnutrition. Fluctuations in the diet have a greater impact on the functions of the muscle in the absence of fluctuations in the composition of the muscle [86]. Once dietary assistance is provided there will be an appreciable improvement in muscular function due to the downregulation of the membranes of the cells in a pumping mechanism called as reductive adaption. After only a short time of famine, this behavior may occur. However, if food is insufficient to meet the body's needs for a long period, the body starts to use up functional stores in tissues like muscle, fat, and bone, which causes changes in body composition. This eventually has direct effects on tissue function, resulting in a reduction in functional ability and a metabolic state that is both fragile and steady [87]. Stressors like diseases and trauma can trigger rapid decomposition. It is important to remember that abrupt or imbalanced increases in energy consumption can also result in malnutrition. Patients at risk for decomposition and refeeding syndrome.

4.7 Cardiorespiratory Function and Disease

A decrease in cardiac muscle mass is observed in individuals suffering from malnutrition. This reduction in cardiac output subsequently affects renal function by diminishing renal perfusion and lowering the glomerular filtration rate. Deficiencies in micronutrients and electrolytes, such as thiamine, can also impair cardiac function, especially during the refeeding process. Additionally, inadequate function of the diaphragm and respiratory muscles leads to decreased cough pressure and impaired clearance of secretions, which can hinder recovery from respiratory tract infections [88].

4.8 Gastrointestinal Diseases

Proper nutrition is essential for maintaining gastrointestinal (GI) function. Chronic malnutrition leads to alterations in pancreatic exocrine function, intestinal blood circulation, villous structure, and intestinal permeability [89]. The colon becomes less effective at reabsorbing water and electrolytes, resulting in the secretion of ions and fluids in both the small and large intestines. This condition can lead to diarrhea, which is linked to a significantly elevated mortality rate among patients suffering from severe malnutrition.

4.9 Immunity and Wound Healing

Immune function is significantly impacted, leading to an increased susceptibility to infections because of compromised cell-mediated immunity, as well as diminished cytokine, complement, and phagocyte activity. Additionally, delayed wound healing is a well-documented issue among malnourished surgical patients. Malnutrition is frequently underestimated by health-care providers, yet it is prevalent and has extensive implications for physiological functioning. It correlates with heightened morbidity and mortality rates among hospitalized patients, thereby substantially increasing health-care expenses [90]. The adoption of a straightforward screening tool can effectively identify at-risk patients, enabling the initiation of appropriate treatment, which can lead to substantial improvements in clinical outcomes and a reduction in health-care costs. It is imperative for every physician to acknowledge that adequate nutritional care is essential to exemplary clinical practice. By addressing educational gaps among all health-care professionals and leveraging clinical leadership, significant advancements in nutritional care can be achieved [91].

5 The Challenges of Treatment and Prevention

Traditionally, public health initiatives have concentrated on the individual, advocating for healthy dietary choices and consistent physical activity. Consequently, the conventional understanding of the rising obesity rates, the challenges in weight loss, or the recurrence of weight gain post initial loss is often attributed to the shortcomings of the individual, the intervention, or a combination of both. However, the escalating prevalence of obesity does not seem to stem from a lack of motivation among individuals within the population [92]. In fact, in the United States, where data on increasing obesity rates is robust, there is evidence indicating that a significant portion of the population is actively engaged in efforts to manage their weight. Recent analyses suggest that a shift in perspective is necessary—one that considers the environmental factors influencing individuals' food choices and their levels of physical activity. There is a growing agreement that effectively addressing the obesity crisis necessitates a comprehensive approach that encompasses all societal levels, benefiting both the population at large and individuals [93]. This approach involves achieving a balance in intervention strategies that range from personalized healthcare (downstream investments) to the implementation of policies and legislation that impact entire populations on a macroscale (upstream investments). At present, a substantial number of resources is allocated to downstream initiatives in comparison to upstream interventions. There is an increasing recognition that enhanced upstream investment is essential to confront the obesogenic environment [94]. Furthermore, the novel concept of integration in chronic disease prevention has been introduced into public health practices. This concept advocates for the coordination of activities to simultaneously address multiple chronic diseases that share common risk factors. For instance, a program that integrates the three primary healthy living strategies—nutrition, physical activity, and mental well-being—has the potential to effectively combat cardiovascular disease. Risk factors have been recognized in children as young as five years old who are overweight. A critical element of the childhood obesity epidemic is the rising incidence of Type 2 diabetes among pediatric populations, along with the potential for related macrovascular and microvascular complications. Additionally, the mental health consequences of enduring obesity are likely to contribute to further morbidity. In conclusion, the overall expected effect on the prevalence of chronic diseases poses significant challenges for the sustainability of health-care systems. Historically, public health initiatives have concentrated on the individual, advocating for nutritious dietary choices and consistent physical activity [95]. Consequently, the conventional understanding of the rising rates of obesity, difficulties in weight loss, or relapse following initial weight reduction is often attributed to shortcomings in the individual, intervention methods, or the combination of both. Nevertheless, the escalating prevalence of obesity seems to stem from a lack of motivation among individuals within the population. A significant portion of the population is actively engaged in efforts to manage their weight. Recent evaluations indicate that a fundamental change in perspective is essential to comprehend and address the issue. This shift should consider the context of the individuals in decision-making regarding the consumption of food and involvement in physical activity regularly.

6 The Determinants of Antioxidants Associated with Chronic Diseases.

Health promotion focuses on addressing both the fundamental determinants and the direct causes of poor health. The primary factors contributing to significant dental conditions, such as dental caries and periodontal disease, include diet, plaque accumulation, smoking, lesions in oral, and oral cancer, joint dysfunction, alcohol consumption, as well as stress and physical trauma to the teeth. Given that these factors are prevalent in various other chronic illnesses [96–98].

7 Comorbidity of Chronic Diseases

In the research presented in this document, we acquired independent evaluations of chronic illnesses, specifically the diagnosis recorded by the doctors, alongside assessments of status reported by the patients themselves. Utilizing this information, our objective was to evaluate both the individual and combined impacts of coexisting chronic conditions on physical and mental functioning. For the sake of prevention, we assert that there is a significant necessity to enhance understanding of the connections between specific pairs of somatic diseases and mental functioning, given that patients with chronic illnesses are recognized to have an elevated risk of psychiatric disorders, although the psychological implications of comorbidity remain inadequately understood [99]. Prior to analyzing the effects of specific combinations of diseases and the influence of various chronic diseases with and without comorbidity. The results indicate that individuals diagnosed with chronic diseases experience poorer physical functioning compared to the general population in the United States, irrespective of the chronic condition examined. Additionally, those with comorbidities tend to rate their physical functioning as subpar [100]. Among the participants, those with arthritis, irrespective of whether they had additional comorbidities, reported the most significant decline in physical functioning. In terms of mental functioning, the average scores on the Mental Component Summary (MCS) were approximately 50, indicating that emotional and mental health is unaltered. However, low-level MCS scores were observed in patients with thyroid dysfunction. Notably, unspecified comorbidity did not correlate with greater impairment in mental functioning. The notable result obtained among the patients that were conducted on PCS suggests the presence of synergistic effects among pairs of chronic diseases [101]. Specifically, the following combinations demonstrated synergism in physical functioning: diabetes coupled with CVD, CRD with diabetes, Arthritis with CVD, CVD and CRD.

Also, some reports revealed multimorbidity with major chronic diseases including diabetes, CVD, and CRD, signifying a deterioration in physical health status. Given the prevalence of these disease pairs within the population, general practitioners and medical specialists must recognize this heightened risk of physical disability [102].

8 Qualitative Research in Antioxidant-Induced Diseases in Humans

Qualitative research concerning chronic illness has not yet been distinctly established as a specific area within medical sociology. Chronic illness was first introduced into sociological discourse in the 1960s without clear guidelines regarding its research foundations. This lack of specification allowed for its use as a critique of the medical field, coinciding with significant transformations in modern society, including the early stages of widespread medicalization. Concurrently, qualitative research began to emerge, drawing upon interactionist and ethnomethodological perspectives to understand social reality [103]. Chronic illness was first brought to the forefront of sociological discourse through Mechanic's critical yet widely accepted challenge to Parsons' concept of the sick role, which was deemed inadequate for analyzing conditions beyond acute illness. More than a decade later, Parsons managed to articulate how his framework of the doctor-patient role system could accommodate patients' potential for complete recovery while pursuing the highest possible level of normalcy. However, the theoretical implications of this discussion remain unresolved to this day [104]. Similarly, Scheff, in his work Being Mentally Ill (1966), emphasizes the effectiveness of the social impact on the sick can have lifelong consequences, leading to irreversible deprivation and self-devaluation, often initiated by an accidental public crisis that results in the labelling of otherwise inconspicuous yet confused behaviors [105]. A significant driving force underlying this issue is the perception that modern medicine has inadequately addressed the needs of the disabled and marginalized populations. The ultimate objective is to "*humanize health care*" transforming the pervasive feelings of alienation and helplessness experienced by patients due to the prevailing professional authority in clinical settings into a compelling justification for a fundamental reorganization of health-care systems [106].

9 Influence of Economic Factors and Environmental Pollutants in Oxidative-Stress-Associated Diseases

At the macroeconomic level, it is frequently presumed economic growth across the globe leads to increased income, which in turn enhances various health outcomes in developing nations. However, while increased economic investment and elevated incomes for certain segments of the population have alleviated some health issues in these countries, the prevalence of chronic diseases has intensified, See Table 1 [107]. Key factors influencing this situation include worldwide trade, foreign investments, and urbanization. Access to healthy food by all the people belongs to the low and middle classes. Unavailability of clean water, high exposure to environmental pollution, stressful working conditions, and limited access to health-care centers. All these lead to the generation of free radicals and decreased antioxidant defenses inside the body [108–111]. Investment policies can yield economic advantages; however, they may also promote unhealthy behavioral risks. Notably, many nonfinancial corporations in the world ranked by foreign assets in 2000 are linked to heavy risk factors of getting oxidative-stress-associated diseases, especially those who are into the tobacco, food, and automobile sectors. These corporations invest significantly in marketing their products, which, in the absence of regulation, can lead to an increase in risk factors. Marginalized populations have the highest risk leading to increased oxidative stress because of low-level access to healthy lifestyle choices [112]. Certain companies and their trade organizations have made concerted efforts to obstruct the implementation of regulations and undermine effective advocacy aimed at promoting tobacco control. Furthermore, sugar lobbyists have successfully propagated the narrative that "*sugar is not detrimental to health, but reduced sugar intake negatively impacts economies*," a viewpoint that has been embraced by some governments, despite contrary evidence [113–115].

10 Regulating and Governing Risk Factors Associated with Oxidative Stress-Related Diseases

The adverse effects of the development of the economic system are mitigated through coordinated efforts across various sectors to address chronic diseases at their source. Policymakers should promote the engagement of transnational corporations in enhancing health outcomes while maintaining profitability by implementing innovative business models, like those being pursued by Unilever [116]. Additionally, they ought to advocate for the establishment of guidelines and incentives that foster healthy foreign investments in developing nations, as well as the integration of chronic disease considerations into accountability frameworks. Every nation's government and associated sectors play an essential role in the formulation and execution of effective control of risk factors initiatives, World Health Organization's members must develop a framework comprising of tobacco control, food policy including intake of fruits and vegetables, whole grains, nuts, flavonoid-rich foods [117–119]. Other ways of creating awareness about limiting smoking, healthy blood pressure, maintaining of healthy weight, and regular physical exercises may prevent the risk factors leading to cardiovascular diseases shortly. Governments have the capacity to influence economic incentives for both businesses and individuals. By broadening the markets for fruits and vegetables, developing nations can benefit from enhanced export revenues and higher consumption levels. Additionally, increasing

Table 1 Disease developments due to antioxidants

S. no	Antioxidants	Adverse effects and disease development	References
1	Anthocyanin	Damages kidney, causes tumors, thyroid imbalance, affects medications, reduces the absorption of nutrients, gastrointestinal issues, skin diseases	[134]
2	Vitamin C	Headache, kidney stones, insomnia, heartburn, swelling of esophagus	[135]
3	Vitamin E	Bleeding, fatigue, blurred vision, intestinal cramps, interferes with some drugs	[135]
4	Tocopherol	Impaired reflexes, weak muscles Lou Gehrig's disease, angina, atherosclerosis, eczema, colon cancer	[136]
5	Polyphenols	Mutagenetic effects, carcinogenic effects, genotoxic effects, prooxidative activity, tumors, thyroid hormone imbalance	[137]
6	Carotenoids	Carotenoderma, canthaxanthin retinopathy, lung cancer, interferes with some blood tests	[138]
7	Flavonoids	Drug interactions, contact dermatitis, hemolytic anemia, estrogenic-related disease, nervousness, liver toxicity	[139, 140]
8	Glutathione	Allergy, digestion problems, abdominal cramps, kidney dysfunction, Steven Johnson syndrome, toxic epidermal necrolysis	[141]
9	Coenzyme Q10	Hypertension, dizziness, insomnia, irritability, nausea, rash	[142, 143]
10	Selenium	Gastrointestinal issues, nail issues, nervous system issues, respiratory issues, garlic breath	[144]
11	Leutein	Vision loss, eye pain, irregular heartbeats, shortness of breath, fluttering in the chest	[145, 146]
13	Melatonin	Constipation, tremors, mood swings, decreased appetite, diarrheal, abdominal pain, drowsiness	[147]
14	Lycopene	Nausea, vomiting, stomach upset, low blood pressure, bleeding	[148]
15	L-Arginine	Bloating, gout, headache, allergy, asthma, abdominal pain	[149]
16	Resveratrol	Gastrointestinal issues, liver dysfunction, slowing the process of blood clotting	[150]

the excise tax on tobacco to a degree that significantly curtails consumption will not only boost government revenue but also alleviate the health-related costs associated with tobacco use. Unlike the methodologies applied in oral-related health issues, the prevailing perspective on general health and disease posits that socio-political factors predominantly influence health outcomes [120]. Focusing solely on individual health behaviors, while neglecting the necessity for robust social health policies, represents not only a simplification of the issue but also a shirking of accountability. An emphasis on lifestyle choices frequently masks the wider determining factors of health and the critique in prioritizing the lifestyle of the individual is a cause and remedy for the issues associated with health and is especially pertinent in the field of dentistry.

The predominant emphasis of numerous oral health policies is on altering individual behaviors. This strategy tends to shift focus away from the fundamental factors that contribute to diseases with connection to oral [121].

11 Restructuring of Health Systems Globally

Restructuring health systems is necessary to properly handle chronic disease, which includes the diagnosis and also prevention at various stages. Government backing for this change is crucial if significant health gains are to be made [122]. Based on several possibilities, the UK government's Treasury has put forth a set of precise health goals and targets for the country in a model that has worldwide impliocations. In the world's poorest countries, the advent of new funding sources for the treatment of infectious chronic diseases including AIDS/HIV and tuberculosis offers a significant way to create a well-organized health system [123]. Many academics have highlighted the possibility of secondary prevention of diabetes and cardiovascular diseases in underdeveloped nations; yet these potentials must be upgraded for practical measures. We are in a situation to create a protocol for the global health and wealth of the nation [124].

12 Limitations of the Lifestyle Approach

Lifestyle is often perceived as a deliberate choice made by individuals regarding their behavior. Conversely, some view lifestyle as a reflection of the social and cultural contexts that shape and limit behavior, alongside the personal choices made by individuals. However, seemingly straightforward actions are intertwined with more intricate lifelong habits and social conditions linked to lifestyle [125]. In the MRFIT study, conditions were optimal for lifestyle modifications. Nevertheless, even among a highly motivated and carefully selected group of participants, only minimal adjustments in dietary and smoking habits were observed over a six-year timeframe. The strategies employed in MRFIT were grounded in the most advanced research on behavior modification. The inability to achieve significant changes in critical health-related behaviors through these traditional approaches should serve as a cautionary note for dental health educators [126]. The main start of lifestyle style is often perceived as a new set of personal behaviors that individuals consciously select. However, some view lifestyle as a reflection of the social and cultural contexts that shape and limit behavior, alongside the personal choices made by individuals. Seemingly simple actions are intertwined with more intricate lifelong habits and social conditions linked to lifestyle. The conditions in which individuals live play a main circular place and role in the maintenance of their lifestyles [127–129]. Blane (1985) compellingly argues that the emphasis placed on individual behaviors as causal factors has been overstated. Instead, the true agents of causation are social structures, and these behaviors should be seen as markers of underlying causes that are more closely related to them. The fact that mortality and illness are consistently observed to be gradients in all societies emphasizes the importance of social structure. Disparities in health are not just apparent between the rich and the poor or between the disadvantaged and the general population. Higher social class people are typically healthier than those in lower social classes. Despite changes in life expectancy and average income throughout time, this gradient has been constant [130]. Interestingly, the gradient extends across the social hierarchy and is continuous without a clear threshold. Even if they are at the same level of social economic status and education, the people who are at the higher status are healthier than that right below them. This trend persists as one moves down the employment ladder, with worse health outcomes at each level. Additionally, this gradient can be seen in a variety of health conditions and death causes. Consequently, a compelling rationale for forming partnerships among various sectors engaged in the promotion of health is to prevent redundancy and enhance the quality of treatment of chronic diseases [131]. Additionally, it is important to notice that most populations that are most burdened by various diseases are often those who are deprived and socially marginalized. The resolution of chronic disease issues relies on collaborative approaches. The strategies designed to address these challenges are outlined in the Ottawa Charter for Health Promotion. Key components include community engagement and support, environmental modifications, legislative action, enhancement of personal skills, and empowering individuals to take an active role in society, thereby collectively addressing the systems that influence their health [132]. Effective management of dental diseases is predominantly achievable through social policy initiatives. Oral health professionals are tasked with persuading policymakers and the public to adopt the necessary social measures to tackle both general and oral health concerns, as well as to engage in the execution of these policies [133].

13 Conclusion

The documented potential for patient management to play a significant role in disease control is substantial, and specific areas requiring further research have been identified. Interventions aimed at enhancing patient management have demonstrated associations with critical health outcomes, including improved condition monitoring, reduced symptoms, better physical functioning, and psychological well-being, as well as decreased health-care utilization. Additionally, patients have been engaged and supported across various clinical and some nonclinical environments. Given the success of numerous existing models, their broader implementation appears justified. Utilizing established programs as designed, rather than making extensive adaptations without subsequent evaluation, it is likely to enhance the likelihood of achieving the intended results. Some interventions have proven to be more theoretically sound and effective than others, and the outcomes of different programs have varied. Most evaluation periods have been relatively brief compared to the length of most chronic illnesses. In instances where long-term evaluations have been conducted, the necessity for periodic "booster" interventions over time has become apparent. Research conducted in the last ten years has enhanced the theoretical frameworks for comprehending how patients manage chronic diseases. Interventions designed to support patients have evolved, increasingly aligning with established models or fundamental principles of behavior change. Recent investigations have evaluated chronic disease interventions that incorporate aspects of self-regulation. The core principles of self-regulation provide insight into the formulation of patients' management strategies and may represent a favored focus for future interventions. The management of diseases by patients suffering from various chronic conditions has been improved through the direct implementation of interventions that employ diverse strategies and educational methods. Likewise, initiatives aimed at enhancing clinician training, patient education, and modifications in practice organization seem to have yielded some positive results in disease management. Nevertheless, there is a notable lack of studies that have definitively pinpointed the specific components of these multi-faceted programs that are most influential in achieving desired outcomes. Furthermore, there has been minimal research conducted on settings outside of clinical environments, such as workplaces and schools, as potential avenues for engaging individuals with chronic diseases and aiding them in better management of their conditions. These alternative entry points may represent valuable opportunities for improving disease control. The expenses associated with the implementation of effective programs aimed at improving disease management for patients, as well as enhancing the capabilities of clinicians and health-care organizations, have not been sufficiently evaluated in relation to their benefits and cost savings. Considering the global trend of individuals living longer with chronic illnesses, there is a need for research to increase innovative and effective functions and policies that can efficiently empower a significant number of patients to manage their conditions successfully. Both natural and synthetic antioxidants have the potential compounds in turn develop diseases in humans. Preventive and precautionary methodology in daily routine life including eating food, regular physical exercise, a healthy and sound mind, positive thinking, and scientific analysis and tools will guide mankind to lead a healthy life in the future.

References

1. Carocho, M., & Ferreira, I. C. (2013). A review on antioxidants, prooxidants and related controversy: Natural and synthetic compounds, screening and analysis methodologies and future perspectives. *Food and Chemical Toxicology, 51*, 15–25.
2. Dai, J., & Mumper, R. J. (2010). Plant phenolics: Extraction, analysis and their antioxidant and anticancer properties. *Molecules, 15*, 7313–7352.
3. Scalbert, A., & Williamson, G. (2000). Dietary intake and bioavailability of polyphenols. *Journal of Natural, 130*, 2073S-2085S.
4. Chen, L. H. (1989). Interaction of vitamin E and ascorbic acid (review). *In Vivo, 3*, 199–209.
5. Du, J., Cullen, J. J., & Buettner, G. R. (2012). Ascorbic acid: Chemistry, biology and the treatment of cancer. *Biochimica et Biophysica Acta, 1826*, 443–457.

6. Rietjens, I. M., Boersma, M. G., Haanm, L. D., Spenkelink, B., Awad, H. M., & Cnubben, N. H. et al. (2002) The pro-oxidant chemistry of the natural antioxidants vitamin C, vitamin E, carotenoids and flavonoids. *Environmental Toxicology Pharmacology, 11*, 321–33.
7. Kang, M. J., Lee, S. S., & Koh, H. C. (2012). Prooxidant properties of ascorbic acid in the nigrostriatal dopaminergic system of C57BL/6 mice. *Toxicology, 294*, 1–8.
8. Heo, J. H., et al. (2013). The possible role of antioxidant vitamin C in Alzheimer's disease treatment and prevention. *American Journal of Alzheimer's Disease and Other Dementias, 28*, 120–125.
9. Putchala, M. C., Ramani, P., Sherlin, H. J., Premkumar, P., & Natesan, A. (2013). Ascorbic acid and its pro-oxidant activity as a therapy for tumours of oral cavity—a systematic review. *Archives of Oral Biology, 58*, 563–574.
10. Bast, A., & Haenen, G. R. (2013). *Ten misconceptions about antioxidants. Trends PharmacolSci., 34*, 430–436.
11. Halliwell, B. (1996). Commentary: Vitamin C: Antioxidant or pro-oxidant in vivo? *Free Rad Research, 25*, 439–54 (1996)
12. Halliwell, B. (2006). Reactive species and antioxidants. Redox biology is a fundamental theme of aerobic life. *Plant Physiology, 141*, 312–22.
13. Halliwell, B. (1991). Drug antioxidant effects. *Drugs, 42*, 569–605.
14. Fransen, M., Nordgren, M., Wang, B., & Apanasets, O. (2012). Role of peroxisomes in ROS/RNS-metabolism: Implications for human disease. *Biochim BiophysActa, 1822*, 1363–1373.
15. Schrader, M., & Fahimi, H. D. (2006). *Peroxisomes and oxidative stress. BiochimBiophysActa, 63*, 1755–1766.
16. Rodrigo, R., Guichard, C., & Charles, R. (2007). Clinical pharmacology and therapeutic use of antioxidant vitamins. *Fundamental & Clinical Pharmacology, 21*, 111–127.
17. Kohen, R., & Nyska, A. (2002). Oxidation of biological systems: Oxidative stress phenomena, antioxidants, redox reactions, and methods for their quantification. *Toxicology Pathology, 30*, 620–50.
18. Prior, R. L., Wu, X., & Schaich, K. (2005). Standardized methods for the determination of antioxidant capacity and phenolics in foods and dietary supplements. *Journal Agriculture Food Chemical, 53*, 4290–02.
19. Kasote, D. M. (2013). Flaxseed phenolics as natural antioxidants. *International Food Research Journal, 20*, 1797–1804.
20. Ou, B., Woodill-Hampsch, M., & Prior, R. L. (2001). Development and validation of an improved oxygen radical absorbance capacity assay using fluorescein as the fluorescent probe. *Journal of Agriculture and Food Chemistry, 49*, 4619–4626.
21. Barbehenn, R. V., & Peter, C. C. (2011). Tannins in plant-herbivore interactions. *Phytochemistry, 72*, 1551–1565.
22. Alam, M. N., Bristi, N. J., & Rafiquzzaman, M. (2013). Review on in vivo and in vitro methods. Evaluation of antioxidant activity. *Saudi Pharm Journal, 13*(21), 143–52.
23. Winkel-Shirley, B. (2002). Biosynthesis of flavonoids and effects of stress. *Current Opinion in Plant Biology, 5*, 218–223.
24. Bennett, R. N., & Wallsgrove, R. M. (1994). Secondary metabolites in plant defence mechanisms. *New Phytologist, 127*, 617–633.
25. Espinoza, A., San, M. A., López-Climent, M., Ruiz-Lara, S., Gómez-Cadenas, A., & Casaretto, J. A. (2013). Engineered drought-induced biosynthesis of α-tocopherol alleviates stress-induced leaf damage in tobacco. *Journal of Plant Physiology, 170*, 1285–1294.
26. Gao, Q., & Zhang, L. (2008). Ultraviolet-B-induced oxidative stress and antioxidant defense system responses in ascorbate-deficient vtc1 mutants of Arabidopsis thaliana. *Journal of Plant Physiology, 165*, 138–148.
27. Apel, K., & Hirt, H. (2004). Reactive oxygen species: Metabolism, oxidative stress, and signal transduction. *Annual Review of Plant Biology, 55*, 373–399.
28. Szalai, G., Kellos, T., Galiba, G., & Kocsy, G. (2009). Glutathione as an antioxidant and regulatory molecule in plants under abiotic stress conditions. *Journal of Plant Growth Regulation, 28*, 66–80.

29. Suzuki, N., Koussevitzky, S., Mittler, R., & Miller, G. (2012). ROS and redox signalling in the response of plants to abiotic stress. *Plant, Cell and Environment, 35*, 259–270.
30. Shao, H. B., Chu, L. Y., Shao, M. A., Jaleel, C. A., & Mi, H. M. (2008). Higher plant antioxidants and redox signaling under environmental stresses. *Comptes Rendus Biologies, 331*, 433–441.
31. Korkina, L. G. (2007). Phenylpropanoids as naturally occurring antioxidants: From plant defense to human health. *Cellular and Molecular Biology, 53*, 15–25.
32. Galant, A., Preussm, M. L., Cameron, J. C., & Jez, J. M. (2011). Plant glutathione biosynthesis: Diversity in biochemical regulation and reaction products. *Frontiers in Plant Science, 2*, 45.
33. Meyer, A. J., & Hell, R. (2005). Glutathione homeostasis and redox-regulation by sulfhydryl groups. *Photosynthesis Research, 86*, 435–457.
34. Lü, J. M., Lin, P. H., Yao, Q., & Chen, C. (2010). Chemical and molecular mechanisms of antioxidants: Experimental approaches and model systems. *Journal of Cellular and Molecular Medicine, 14*, 840–860.
35. del Río, L. A., Sandalio, L. M., Corpas, F. J., Palma, J. M., & Barroso, J. B. (2006). Reactive oxygen species, reactive nitrogen species in peroxisomes. Production, scavenging, and, role in cell signaling. *Plant Physiology, 141*, 330–335.
36. Badarinath, A. V., Rao, K. M., Chetty, C. M. S., Ramkanth, V., Rajan, T. V. S., & Gnanaprakash, K. (2010). A review on in-vitro antioxidant methods: Comparisons, correlations and considerations. *International of Journal PharmTech Research, 2*, 1276–1285.
37. Chand, S., & Dave, R. (2009). In vitro models for antioxidant activity evaluation and some medicinal plants possessing antioxidant properties: An overview. *African Journal of Microbiology Research, 3*, 981–996.
38. Tuteja, N., Ahmad, P., Panda, B. B., & Tuteja, R. (2009). Genotoxic stress in plants: Shedding light on DNA damage, repair and DNA repair helicases. *Mutation Research, 681*, 134–149.
39. Hemre, G. I., Lock, E. J., Olsvik, P. A., Hamre, K., Espe, M., Torstensen, B. E., Silva, J., Hansen, A. C., Waagbo, R., Johansen, J. S., Sanden, M., & Sissener, N. H. (2016). Atlantic salmon (Salmo salar) require increased dietary levels of B-vitamins when fed diets with high inclusion of plant based ingredients. *Peer Journal, 4*, Article e2493.
40. Bergin, A., Carvell, J., & Butler, M. (2022). Applications of bio-capacitance to cell culture manufacturing. *Biotechnology Advances, 61*, Article 108048.
41. Chen, S., Li, Z., Zhang, S., Zhou, Y., Xiao, X., Cui, P., Xu, B., Zhao, Q., Kong, S., & Dai, Y. (2022). Emerging biotechnology applications in natural product and synthetic pharmaceutical analyses. *Acta Pharmaceutica Sinica B, 12*, 4075–4097.
42. De Bont, R., & van Larebeke, N. (2004). Endogenous DNA damage in humans: A review of quantitative data. *Mutagenesis, 19*, 169–185.
43. Da Pozzo, E., De Leo, M., Faraone, I., Milella, L., Cavallini, C., Piragine, E., et al. (2018). Antioxidant and antisenescence effects of bergamot juice. Oxidation Medical Cell Longevity, 9395804.
44. Chen, X., Guo, C., & Kong, J. (2012). Oxidative stress in neurodegenerative diseases. *Neural Regeneration Research, 7*, 376–385.
45. Chen, X.-F., Wang, L., Wu, Y.-Z., Song, S.-Y., Min, H.-Y., & Yang, Y. (2018). Effect of puerarin in promoting fatty acid oxidation by increasing mitochondrial oxidative capacity and biogenesis in skeletal muscle in diabeticrats. *Nutrition & Diabetes, 8*, 1–13.
46. Cardoso, B. R., Hare, D. J., Bush, A. I., & Roberts, B. R. (2017). Glutathioneperoxidase 4: A new player in neurodegeneration? *Molecular Psychiatry, 22*, 328–335.
47. Cadet, J., & Wagner, J. R. (2013). DNA base damage by reactive oxygen species, oxidizing agents, and UV radiation. *Cold Spring Harbor Perspectives in Biology, 5*, Article a012559.
48. Buj, R., & Aird, K. M. (2018). Deoxyribonucleotide triphosphate metabolism in cancer and metabolic disease. *Frontiers in Endocrinology, 9*, 177.
49. Bedard, K., & Krause, K. H. (2007). The NOX family of ROS-generating NADPHoxidases: Physiology and pathophysiology. *Physiological Reviews, 87*, 245–313.
50. Battin, E. E., & Brumaghim, J. L. (2009). Antioxidant activity of sulfurand selenium: A review of reactive oxygen species scavenging, glutathioneperoxidase, and metal-binding antioxidant mechanisms. *Cell Biochemistry and Biophysics, 55*, 1–23.

51. Battelli, M. G., Polito, L., & Bolognesi, A. (2014). Xanthine oxidoreductasein atherosclerosis pathogenesis: Not only oxidative stress. *Atherosclerosis, 237*, 562–567.
52. Battelli, M. G., Bolognesi, A., & Polito, L. (2014). Pathophysiology of circulating xanthine oxidoreductase: New emerging roles for a multi-tasking enzyme. *Biochimica et Biophysica Acta, 1842*, 1502–1517.
53. Bartosz, G. (2009). Reactive oxygen species: Destroyers or messengers? *Biochemical Pharmacology, 77*, 1303–1315.
54. Barrera, G. (2012). Oxidative stress and lipid peroxidation products in cancer progression and therapy. *ISRN Oncology*, 137289–137289.
55. Banafsheh, A. A., & Sirous, G. (2016). Studies on oxidants and antioxidants with a brief glance at their relevance to the immune system. *Life Sciences, 146*, 163–173.
56. Aroor, A. R., Mandavia, C., Ren, J., Sowers, J. R., & Pulakat, L. (2012). Mitochondria and oxidative stress in the cardiorenal metabolic syndrome. *Cardiorenal Medical, 2*, 87–109.
57. Antonioni, A., Fantini, C., Dimauro, I., & Caporossi, D. (2019). Redoxhomeostasis in sport: Do athletes really need antioxidant support? *Research in Sports Medical, 27*, 147–165.
58. Abramov, A. Y., Jacobson, J., Wientjes, F., Hothersall, J., Canevari, L., & Duchen, M. R. (2005). Expression and modulation of an NADPH oxidase inmammalian astrocytes. *Journal of Neuroscience, 25*, 9176–9184.
59. Ezzati, M., & Riboli, E. (2013). Behavioral and dietary risk factors for noncommunicable diseases. *New England Journal of Medicine, 369*(10), 954–964.
60. Barlow, C. E., LaMonte, M. J., FitzGerald, S. J., Kampert, J. B., Perrin, J. L., & Blair, S. N. (2006). Cardiorespiratory fitness is an independent predictor of hypertension incidence among initially normotensive healthy women. *American Journal of Epidemiology, 163*(2), 142–150.
61. Münzel, T., Camici, G. G., Maack, C., Bonetti, N. R., Fuster, V., & Kovacic, J. C. (2017). Impact of oxidative stress on the heart and vasculature. *Journal of the American College of Cardiology., 70*(2), 212–229.
62. Mozaffarian, D., Hao, T., Rimm, E. B., Willett, W. C., & Hu, F. B. (2011). Changes in diet and lifestyle and long term weight gain in women and men. *New England Journal of Medicine, 364*, 2392–2404.
63. Islam, F. M., Wu, J., Jansson, J., & Wilson, D. P. (2012). Relative risk of cardiovascular disease among peopleliving with HIV: A systematic review and meta-analysis. *HIV Medicine, 13*(8), 453–468.
64. Doughan, A. K., & Dikalov, S. I. (2007). Mitochondrial redox cycling of mitoquinone leads to superoxide production and cellular apoptosis. *Antioxidants & Redox Signaling, 9*, 1825–1836.
65. Timmers, S., Konings, E., Bilet, L., Houtkooper, R. H., Van De Weijer, T., Goossens, G. H., Hoeks, J., Van Der Krieken, S., Ryu, D., & Kersten, S. (2011). Calorie restriction-like effects of 30 days of resveratrol supplementation on energy metabolism and metabolic profile in obese humans. *Cell Metabolism, 14*, 612–622.
66. Tomé-Carneiro, J., Gonzálvez, M., Larrosa, M., Yáñez-Gascón, M. J., García-Almagro, F. J., Ruiz-Ros, J. A., García-Conesa, M. T., Tomás-Barberán, F. A., & Espín, J. C. (2012). One-year consumption of a grape nutraceuticalcontaining resveratrol improves the inflammatory and fibrinolytic status of patients in primary prevention of cardiovascular disease. *American Journal of Cardiology, 110*, 356–363.
67. Mankowski, R. T., You, L., Buford, T. W., Leeuwenburgh, C., Manini, T. M., Schneider, S., Qiu, P., & Anton, S. D. (2020). Higher dose of resveratrol elevated cardiovascular disease risk biomarker levels in overweight olderadults–A pilot study. *Experimental Gerontology, 131*, Article 110821.
68. Yasui, K., & Bab, A. (2006). Therapeutic potential of superoxidedismutase (SOD) for resolution of inflammation. *Inflammation Research, 55*(9), 359–363.
69. Rajendran, P., Nandakumar, N., & Rengarajan, T. (2014). Antioxidants and human diseases. *Clinica Chimica Acta, 436*, 332–347.
70. Chen, S., Li, Z., Zhang, S., Zhou, Y., Xiao, X., Cui, P., Xu, B., Zhao, Q., Kong, S., & Dai, Y. (2022.) Emerging biotechnology applications in natural product and synthetic pharmaceutical analyses. *Acta Pharmaceutica Sinica B, 12*, 4075–4097.

71. Tenney, L., McKenzie, L. M., Matus, B., Mueller, K., & Newman, L. S. (2019). Effect of an opioid management program for Colorado workers' compensation providers on adherence to treatment guidelines for chronic pain. *American Journal of Industrial Medicine, 62*(1), 21–29.
72. Turk, D. C., Wilson, H. D., & Cahana, A. (2011). Treatment of chronic non-cancer pain. *Lancet, 377*(9784), 2226–2235.
73. Els, C., Jackson, T. D., Kunyk, D., Lappi, V. G., Sonnenberg, B., Hagtvedt, R., Sharma, S., Kolahdooz, F., & Straube, S. (2017). Adverse events associated with medium- and long-term use of opioids for chronic non-cancer pain: an overview of Cochrane Reviews. *Cochrane Database System Review, 10*(10), CD012509.
74. Padayatty, S. J., Sun, H., Wang, Y., Riordan, H. D., Hewitt, S. M., Katz, A., Wesley, R. A., & Levine, M. (2004). Vitamin C pharmacokinetics: Implications for oral and intravenous use. *Annals of Internal Medicine, 140*(7), 533–537.
75. Walsh, Z., Gonzalez, R., Crosby, K., Thiessen, M. S., Carroll, C., & Bonn-Miller, M. O. (2017). Medical cannabis and mental health: A guidedsystematic review. *Clinical Psychology Review, 51*, 15–29.
76. Samii, A., Ryan-Dykes, P., Tsukuda, R. A., Zink, C., Franks, R., & Nichol, W. P. (2006). Telemedicine for delivery of health care in Parkinson's disease. *Journal of Telemedicine and Telecare, 12*, 16–18.
77. Miller, G. E., Chen, E., Sze, J., Marin, T., Arevalo, J. M., Doll, R., Ma, R., & Cole, S. W. (2008). A functional genomic fingerprint of chronic stress in humans: Blunted glucocorticoidand increased NF-kappaB signaling. *Biological Psychiatry, 64*, 66–72.
78. Slavich, G. M., & Irwin, M. R. (2014). From stress to inflammation and major depressive disorder: A social signal transduction theory of depression. *Psychological Bulletin, 140*, 774–815.
79. Barrett, L. F., & Simmons, W. K. (2015). Interoceptive predictions in the brain. *Natural Review Neuroscience, 16*, 419–429.
80. Masoli, M., Fabian, D., & Holt, S. (2004). Global Initiative forAsthma (GINA) program: The global burden of asthma: Executive summary of the GINA Dissemination Committeereport. *Allergy, 59*, 469–478.
81. Ward, B. W., Schiller, J. S., & Goodman, R. A. (2014). Peer reviewed: Multiple chronic conditions among us adults: A 2012 update. *Preview Chron Disease,* 11.
82. Fortin, M., Stewart, M., Poitras, M. E., Almirall, J., & Maddocks, H. (2012). A systematic review of prevalence studies on multimorbidity: Toward a more uniform methodology. *The Annals of Family Medicine., 10*(2), 142–151.
83. Subramanian, S. V., Corsi, D. J., Subramanyam, M. A., & Davey, S. G. (2013). Jumping the gun: Theproblematic discourse on socioeconomic status and cardiovascular health in India. *International Journal of Epidemiology, 42*(5), 1410–1426.
84. Baker, M. A., Lin, H. H., Chang, H. Y., & Murray, M. B. (2012). The risk of tuberculosis disease among persons with diabetes mellitus: A prospective cohort study. *Clinical Infectious Diseases, 54*(6), 818–825.
85. Jakovljevic, M., Bjedov, S., Jaksic, N., & Jakovljevic, I. (2020). COVID-19 pandemia and public and global mental health from the perspective of global health security. *PsychiatrDanub., 32*, 6–14.
86. Veitch, N. C. (2004). Horseradish peroxidase: A modern view of a classic enzyme. *Phytochemistry, 65*, 249–259.
87. Ning, G., & Bloomgarden, Z. (2013). Diabetes in China: prevalence, diagnosis, andcontrol. *Journal of Diabetes, 5*(4),372.
88. Turner, J. E., Lira, V. A., & Brum, P. C. (2017). New insights into the benefits of physicalactivity and exercise for aging and chronic disease. *Oxidative Medical CellLongev.*, 2503767.
89. Cohen, I. G., & Mello, M. M. (2018). HIPAA and protecting health information in the21st century. *JAMA, 320*(3), 231–232.
90. Wang, D. D., & Hu, F. B. (2018). Precision nutrition for prevention and management of type 2 diabetes. *The Lancet Diabetes and Endocrinology, 6*(5), 416–426.

91. Topol, E. J. (2019). High-performance medicine: The convergence of human and artificial intelligence. *Nature Medicine, 25*(1), 44–56.
92. Stylianou, E. (2019). Epigenetics of chronic inflammatory diseases. *Journal of InflammRe-search, 12*, 1–14.
93. van der Harst, P., & Verweij, N. (2018). Identification of 64 novel genetic loci provides an expanded view on the genetic architecture of coronary artery disease. *Circulation Research, 122*(3), 433–443.
94. Park, S. L., Cheng, I., & Haiman, C. A. (2018). Genome-wide association studies of cancer in diverse populations. *Cancer Epidemiology, Biomarkers & Prevention, 27*(4), 405–417.
95. Wang, H. J., Zakhari, S., & Jung, M. K. (2010). Alcohol, inflammation, and gut-liverbraininteractions in tissue damage and disease development. *World Journal of Gastroen-terology, 16*(11), 1304–1313.
96. West, R. (2017). Tobacco smoking: Health impact, prevalence, correlates and interventions. *Psychology & Health, 32*(8), 1018–1036.
97. Marsit, C. J. (2015). Influence of environmental exposure on human epigeneticregulation. *Journal of Experimental Biology, 218*, 71–79.
98. Booth, F. W., Roberts, C. K., & Laye, M. J. (2012). Lack of exercise is a major cause of chronic diseases. *Comprehensive Physiology, 2*(2), 1143–1211.
99. Sanli, E., & Kabaran, S. (2019). Maternal obesity, maternal over nutrition and fetal programming: Effects of epigenetic mechanisms on the development of metabolic disorders. *Current Genomics, 20*(6), 419–427.
100. Cavalli, G., & Heard, E. (2019). Advances in epigenetics link genetics to the environment and disease. *Nature, 571*(7766), 489–499.
101. Ghishan, F. K., & Kiela, P. R. (2017). Vitamins and minerals in inflammatory boweldisease. *Gastroenterology Clinics of North America, 46*(4), 797–808.
102. Vickers, M. H. (2014). Early life nutrition, epigenetics and programming of laterlife disease. *Nutrients, 6*(6), 2165–2178.
103. Nielsen, F. H. (2014). Effects of magnesium depletion on inflammation in chronic disease. *Current Opinion in Clinical Nutrition and Metabolic Care, 17*(6), 525–530.
104. Kundakovic, M., & Champagne, F. A. (2015). Early-life experience, epigenetics, and the developing brain. *Neuropsychopharmacology, 40*(1), 141–153.
105. Zmora, N., Suez, J., & Elinav, E. (2019). You are what you eat: Diet, health and the gutmicrobiota. *Nature Reviews. Gastroenterology & Hepatology, 16*(1), 35–56.
106. Shi, Z. (2019). Gut microbiota: An important link between western diet and chronic diseases. *Nutrients, 11*(10), 2287.
107. Katan, M. B., & Ludwig, D. S. (2010). Extra calories cause weight gain–but how much? *JAMA, 303*(1), 65–6.
108. Man, A. W., & Wang, Y. (2017). Age-associated arterial remodelling. *EC Cardiology, 4*, 137–164.
109. Moran, L. J., Hutchison, S. K., Norman, R. J., & Teede, H. J. (2011). Lifestyle changes in women with polycystic ovary syndrome. *Cochrane Database System Review, 7*, CD007506.
110. Foster, G. D., Makris, A. P., & Bailer, B. A. (2005). Behavioral treatment of obesity. *American Journal of Clinical Nutrition, 82*(1), 230S-235S.
111. Bernstein, C. N. (2017). The brain-gut axis and stress in inflammatory bowel disease. *Gastroenterology Clinics of North America, 46*, 839–846.
112. Scheiermann, C., Kunisaki, Y., & Frenette, P. S. (2013). Circadian control of the immune system. *Nature Reviews Immunology, 13*, 190–198.
113. Upala, S., & Sanguankeo, A. (2015). Effect of lifestyle weight loss intervention on disease severity in patients with psoriasis: A systematic review and meta-analysis. *International Journal of Obesity, 39*, 1197–1202.
114. Kurian, A. K., & Cardarelli, K. M. (2007). Racial and ethnic differences incardiovascular disease risk factors: A systematic review. *Ethnicity and Disease, 17*(1), 143–152.
115. Elia, M., & Stratton, R. J. (2000). How much under nutrition is there in hospitals? *British Journal of Nutrition, 84*, 257–259.

116. Kim, D. D., & Basu, A. (2016). Estimating the medical care costs of obesity in the United States: Systematic review, meta-analysis, and empirical analysis. *Value Health, 19*(5), 602 613.
117. W. H. O. Expert Consultation. (2004). Appropriate body-mass index for Asian populations and its implications for policy and intervention strategies. *Lancet. 363*(9403), 157–163.
118. Zhou, B. F. (2002). Predictive values of body mass index and waist circumference for risk factors of certain related diseases in Chinese adults–study on optimal cut-off points of body mass index and waist circumference in Chinese adults. *Biomedical and Environmental Sciences, 15*(1), 83–96.
119. Ding, C., Chan, Z., & Magkos, F. (2016). Lean, but not healthy: The 'metabolically obese, normal weight' phenotype. *Current Opinion in Clinical Nutrition and Metabolic Care, 19*(6), 408–417.
120. Wild, S., Roglic, G., Green, A., Sicree, R., & King, H. (2004). Global prevalence of diabetes: Estimates for 2000 and projections for 2030. *Diabetes Care, 27*, 1047–1053.
121. Basnayar, B., & Rajapasha, L. C. (2004). Cardiovascular and infectious diseases in South Asia: The double whammy. *BMJ, 328*,781.
122. Gajalakshmi, V., Peto, R., Kanaka, T. S., & Jha, P. (2003). Smoking and mortality from tuberculosis and other diseases in India. *Lancet, 362*, 507–515.
123. Sklar, P., & Henry, M. (2003). HIV Infection and cardiovascular disease—is there really a link? *Nengl Journal Medical, 349*, 2065–2067.
124. Bettcher, D. W., Yach, D., & Guindon, G. E. (2000). Global trade and health: Key linkages and future challenges. *BullWorld Health Organ., 78*, 521–534.
125. Sung, R. Y., Yu, C. W., Chang, S. K., Mo, S. W., Woo, K. S., & Lam, C. W. (2002). Effects of dietary intervention and strength training on blood lipidlevel in obese children. *Archives of Disease in Childhood, 86*, 407–410.
126. Gijsen, R., Hoeymans, N., Schellevis, F. G., Ruwaard, D., Satariano, W. A., & Bos, G. A. M. van den. (2001). Causes and consequences of comorbidity: A review. *Journal of Clinical Epidemiology, 54*, 661–674,
127. Heijmans, M., Foets, M., Rijken, M., Schreurs, K., de Ridder, D., & Bensing, J. (2001). Stress in chronic disease: Do the perceptions of patients and their general practitioners match? *British Journal of Health Psychology, 6*, 229–242.
128. Sambamoorthi, U., Tan, X., & Deb, A. (2015). Multiple chronic conditions and health care costs among adults. *Expert Review Pharmacoeconomics Outcomes Research, 15*(5), 823–832.
129. Bhattacharya, R., Ajmera, M., Bhattacharjee, S., & Sambamoorthi, U. (2014). Use of antidepressants and statins and short-term risk of new-onset diabetes among high risk adults. *Diabetes Research and Clinical Practice, 105*(2), 251–260.
130. Thorpe, K., Jain, S., & Joski, P. (2017). Prevalence and spending associated with patients who have abehavioral health disorder and other conditions. *Health Affairs, 36*, 124–132.
131. Powers, B. W., Chaguturu, S. K., & Ferris, T. G. (2015). Optimizing high-risk care management. *JAMA, 313*, 795–796.
132. Muhrer, J. C. (2010). Detecting and dealing with substance abuse disorders in primary care. *Journal of Nurse Practitioners, 6*, 597–605.
133. Ducat, L., Philipson, L. H., & Anderson, B. J. (2014). The mental health comorbidities of diabetes. *JAMA, 312*, 691–692.
134. Goncalves, A. C., Nunes, A. R., Falcao, A., Alves, G., & Silva, L. R. (2021). Dietry Effects of Anthocynanins in Human health: A comprehensicve review. *Pharmaceuticals(Basel), 18*, 14(7).
135. Lykkesfeldt, J. (2020). On the effect of Vitamin C intake on human health: How to(mis)interpretethe clinical evidence. *Redox Biology, 34*, 101532
136. Kopańska, M., Batoryna, M., Banaś-Ząbczyk, A., Błajda, J., Lis, M. W. (2022). The effect of α-tocopherol on the reduction of inflammatory processes and the negative effect of acrylamide. *Molecules*, *27*(3), 965.
137. Duda-Chodak, A., Tarko, T., Satora, P., & Sroka, P. (2015). Interaction of dietary compounds, especially polyphenols, with the intestinal microbiota: A review. *European Journal of Nutrition, 54*, 325–341.

138. Crupi, P., Faienza, M. F., Naeem, M. Y., Corbo, F., Clodoveo, M. L., & Muraglia, M. (2023). Overview of the potential beneficial effects of carotenoids on consumer health and well-being. *Antioxidants, 12*(5), 1069.
139. Barreca, M. M., Alessandro, R., & Corrado, C. (2023). Effects of flavonoids on cancer, cardiovascular and neurodegenerative diseases: Role of NF-κB Signaling pathway. *International Journal of Molecular Sciences, 24*(11), 9236.
140. Adinew, G. M., Taka, E., Mendonca, P., Messeha, S. S., & Soliman, K. F. A. (2021). The Anticancer effects of flavonoids through miRNAs modulations in triple-negative breast cancer. *Nutrients, 13*(4), 1212.
141. Alam, P., Balawi, T. H., Altalayan, F. H. Hatamleh, A., Muhammad, A, & Ahamad, P. (2021). Silicon attenuates the negative effects of chromium stress in tomato plants by modifying antioxidant enzyme activities, ascorbate–glutathione cycle and glyoxalase system. *Acta Physiology Plant, 43*, 110.
142. Gasmi, A., Bjorklund, G., Mujawdiya, P. K., Semenova, Y., Piscopo, S., & Peana, M. (2022). Coenzyme Q10in aging and disease. *Critical Reviews in Food science Nutrition, 64912*, 3907–3919.
143. Rauchova, H. (2021). Coenzyme Q10 effects in neurological diseases. *Physiological Research, 70*(14), S683–S714.
144. Genchi, G., Lauria, G., Catalano, A., Sinicropi, M. S., & Carocci, A. (2023). Biological activity of selenium and its impact on human health. *International Journal of Molecular Sciences, 24*(3), 2633.
145. Ahn, Y. J., & Kim, H. (2021). Lutein as a modulator of oxidative stress-mediated inflammatory diseases. *Antioxidants, 10*(9), 1448.
146. Zhang, Y., Ding, H., Xu, L., Zhao, S., Hu, S., Ma, A., & Ma, Y. (2022). Lutein can alleviate oxidative stress, inflammation, and apoptosis induced by excessive alcohol to ameliorate reproductive damage in male rats. *Nutrients, 14*(12), 2385.
147. Tennoune, N., Andriamihaja, M., & Blachier, F. (2022). Production of indole and indole-related compounds by the intestinal microbiota and consequences for the host: the good, the bad, and the ugly. *Microorganisms., 10*(5), 930.
148. Ding, S., Jiang, H., Fang, J., & Liu, G. (2021). Regulatory effect of resveratrol on inflammation induced by Lipopolysaccharides via reprogramming intestinal microbes and ameliorating serum metabolism profiles. *Immunology, 12*, 777159.
149. Kurhaluk, N. (2023). The effectiveness of L-arginine in clinical conditions associated with hypoxia. *International Journal of Molecular Sciences, 24*(9), 8205.
150. Ye, X., Li, H., Anjum, K., Zhong, X., Miao, S., Zheng, G., Liu, W., & Li, L. (2022). Dual role of indoles derived from intestinal microbiota on human health. *Frontiers Immunology, 13*, 902526.

A Comprehensive Overview of the Role of Oxidative Stress and Antioxidants in Cardiovascular Diseases

Harsath Ram Govindan Radhakrishnan, Pandian Chithamparam, and Azhaguchamy Muthukumaran

Abstract Oxidative stress significantly influences the emergence of several diseases in humans. The two main factors, including the production of reactive oxygen species (ROS) and intracellular defense responses, constitute the maintenance of intracellular oxidative equilibrium. Elevated levels of oxidative stress can potentially modify lipid, protein, and DNA, leading to inflammatory responses and apoptosis. Studies demonstrate that oxidative stress is a significant factor in the occurrence of numerous types of cardiovascular conditions, including atherosclerosis, myocardial infarction, hypoxia, hypertension, and atrial fibrillation. Multiple treatment approaches are available to address cardiovascular disorders associated with oxidative stress. The study examined widely recognized antioxidants like dietary supplements and more peculiar antioxidants. Furthermore, innovative treatment approaches employing cardiac patches and miRNA have become available to address various cardiovascular conditions. This chapter provides a complete overview of oxidative stress and its consequences. Then, the discussion will delve into the correlation between oxidative stress and various heart diseases. Finally, we turned our attention to the clinical implications of oxidative stress in cardiovascular disorders.

Keywords Oxidative stress · Reactive oxygen species · Antioxidants · Cardiovascular diseases · Cardiac patch

1 Introduction

Endocrinologist Hans Selye first used the term "*stress*" in the biological realm in 1936 to describe an organism's inadequate physiological response to any psychological, emotional, or physical obstacles [1]. The general adaptation syndrome is a three-stage response pattern seen in humans and animals: (a) the alarm stage: This occurs when we first perceive something as stressful, leading the body to trigger the

H. R. G. Radhakrishnan · P. Chithamparam · A. Muthukumaran (✉)
Department of Biotechnology, Kalasalingam Academy of Research and Education, Krishnankoil, Tamilnadu, India
e-mail: a.muthukumaran@klu.ac.in

K. Kasinathan et al. (eds.), *Recent Advances in Oxidative Stress Associated Chronic Diseases Volume 1*, https://doi.org/10.1007/978-981-96-9670-3_6

fight-or-flight reaction; (b) the resistance stage: If the stress continues, the body stays active and increases its metabolic rate to cope with it [2]. The body cannot maintain this level forever and will eventually deplete its resources; and (c) the exhaustion stage: Long-term exposure to stress wears down the body resources and causes immunological suppression, resource depletion, and loss of bodily functioning. This might result in a wide range of health problems and disorders, including depression, diabetes, cardiovascular disease, and gastrointestinal issues [3]. Oxidative stress is a condition in which the production of free radicals exceeds the ability to counteract them through antioxidants [4] (Fig. 1). Excessive production of reactive oxygen species (ROS) in biological systems can be harmful due to their potential reaction with DNA, proteins, and lipids [5]. Studies reveal that reactive oxygen species (ROS) trigger signaling cascades, leading to various pathologies, cell injury, and death. ROS and intermediate free radicals serve as the main contributors to oxidative damage [6]. ROS are commonly regarded as key signaling molecules that are critical for the maintenance of physiological functions, such as cell maturation, division, and survival [5]. ROS, produced by cells including immune, endothelial, and inflammatory cells, play a dual role in redox signaling and oxidative stress or injury. While oxidative stress generates high levels of ROS that damage biomolecules, redox signaling responds to low levels of ROS by activating signaling pathways [7]. Mitochondria are the primary source of ROS production during both physiological and pathological conditions. At the same time, O_2 can be produced by inflammatory and endothelial cells, lipoxygenases (LOX), and cyclooxygenases (COX) during the arachidonic acid metabolism and by cellular respiration [8]. Free radicals and nonradicals are the two categories of ROS. Unpaired electrons in molecules give rise to free radicals, which are reactive and share their unpaired electrons to produce nonradicals. ROS might come from two distinct sources: endogenous and exogenous [9]. Enzymatic endogenous sources of ROS include xanthine oxidoreductase, cytochrome p450 oxidase, NADPH oxidase, arachidonic acid, and monoamine oxidase, and nonenzymatic sources include Fenton's and Haber's reactions [10]. ROS can also arise from other endogenous sources, including the endoplasmic reticulum, lysosomes, mitochondria, receptor-mediated generation, and peroxisomes. Cigarette smoke, environmental agents, exhaust gas, heavy metals, industrial pollutants, radiation, narcotics, and xenobiotics are some of the exogenous sources for ROS formation.

Typically, ROS can cause various adverse effects, among the most significant being structural alterations to lipids, proteins, and nucleic acids within cellular compartments. For instance, polyunsaturated fatty acids (PUFA) are susceptible to free radical attack: this initiates a chain reaction in lipid oxidation, namely, ROS-induced damage [11]. Cells exposed to ROS are highly and vulnerable to frequent DNA damage. They may activate nucleases, and hydroxyl radicals may react directly with the DNA, giving rise to a specific pattern of chemical modifications [12]. ROS significantly contributes to the immune function by maintaining redox balance and activating various cell signaling pathways [13]. Overproduction of reactive oxygen species damages membranes, lipids, proteins, nucleic acids, and organelles and may initiate cell apoptotic processes. However, polysaccharides like hyaluronic acid can be subjected to oxidative degradation. Heart-related diseases are the primary root

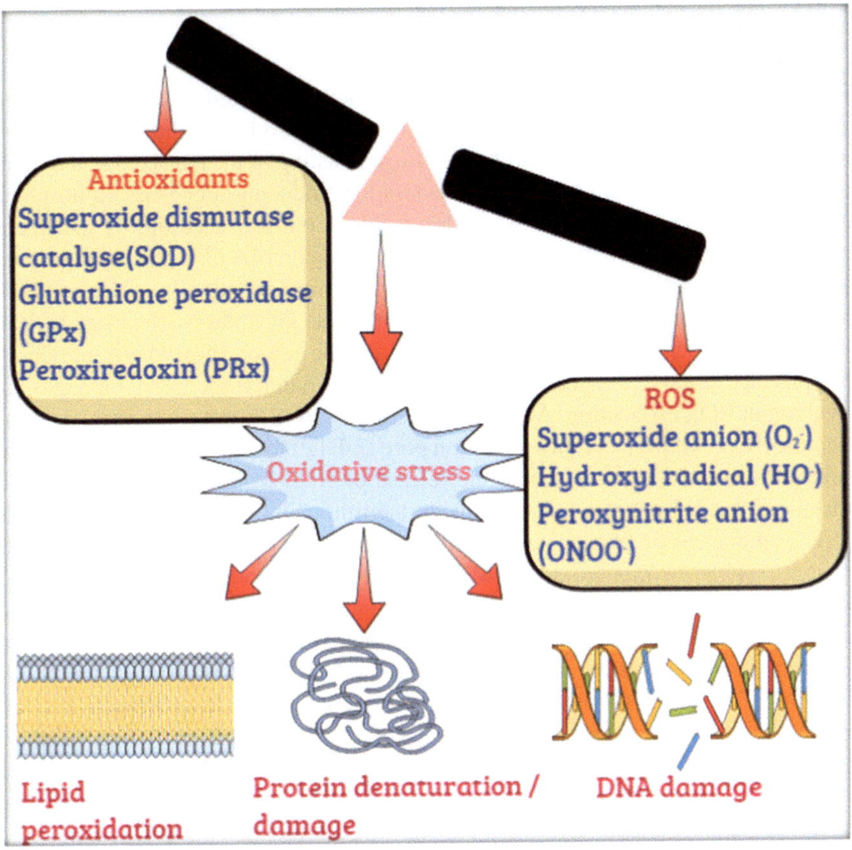

Fig. 1 Schematic diagram representing the overview of oxidative stress

cause of both illness and death around the world. Research indicates that oxidative stress significantly contributes to the emergence of cardiovascular diseases, including arteriosclerosis, cardiac arrest, arrhythmia, and myocardial infarction (6). Extensive research has been conducted on antioxidant therapy for the early detection and management of cardiovascular diseases [14]. The dysfunction of ROS-clearance enzymes, like superoxide dismutase (SOD), leads to increased levels of oxidative stress. Innovative strategies and novel antioxidants have recently been developed to counteract ROS overproduction [15]. This chapter highlights the underlying causes of oxidative stress and their therapeutic significance in relation to cardiovascular diseases (CVDs).

2 Production of ROS

Mitochondria are mainly responsible for producing endogenous reactive oxygen species (ROS). These are primarily oxidative phosphorylation and electron transport chain (ETC) by-products. Mitochondria, consisting of the matrix, and the inner and outer membranes, are the primary centers for producing adenosine triphosphate (ATP) [4]After entering the mitochondria, acetyl-CoA produced during the Krebs cycle is transferred from complexes I, III, and IV to complex V, where ATP production is carried out. Superoxide is made when the mitochondrial respiratory chain gets disrupted under certain clinical circumstances. This causes electrons to leak into the oxygen [16]. Complexes I and III are the major sites where mitochondria produce reactive oxygen species (ROS). More than ten additional enzymes are involved in ROS production. An overproduction of ROS in mitochondria can cause the mitochondrial permeability transition pore (mPTP) to open, which in turn destabilizes the membrane, and enable the release of ROS from mitochondrial matrix into the cytosol. Enzymes located outside the mitochondria, such as nicotinamide adenine dinucleotide phosphate (NADPH) oxidase (NOX), uncoupled nitric oxide synthase (NOS), myeloperoxidase (MPO), lipoxygenase, and xanthine oxidase (Xo), contribute to the production of ROS [17]. Furthermore, NOX is generally seen as a significant factor in ROS. The NOX family consists of more than seven members. The cardiovascular system exhibits the presence NOX_1, NOX_2, NOX_4, and NOX_5 enzymes. The ability to regulate NOX_2, commonly referred to as gp91phox, is the most extensively researched. Other members of the NOX family, including p22phox, p40phox, p47phox, and p67phox, facilitate NOX_2 activation. Ser303, 304, and 328 phosphorylation enable p47phox to become active [18]. When p22phox and p47phox bind together, NOX_2 becomes active and may access p40phox and p67phox. Activation of NOX_2 permits NADPH to connect with the intracellular C-terminus, thereby transferring electrons to oxygen-producing ROS (4). Furthermore, NOX-derived ROS activate secondary oxidase systems, including Xo activation, mitochondrial dysfunction, and NOS uncoupling [19]. Environmental factors may influence the production of both exogenous and endogenous ROS. Exogenous sources of ROS encompass alcohol, smoking, xenobiotics, UV radiation, and environmental pollutants [20]. By interacting with endogenous substances or enzymes, these exogenous ROS sources increase the production of ROS. The smoke emitted from cigarette can trigger the activation of NOX and increase ROS production. Ultraviolet radiation facilitates the generation of ROS through its interaction with water. On the other hand, alcohol can reduce the expression of cytoprotective enzymes and antioxidants.

3 Antioxidants

The antioxidant enzymes in the cell regulate redox equilibrium. SOD, thioredoxin (Tr_x), peroxiredoxin (PR_x), glutathione peroxidase (GP_x), and catalase are examples of antioxidant enzymes [18]. The degradation of ROS plays a key role in defending against oxidative damage. The superoxide dismutase (SOD) is the sole enzyme that can transform the superoxide anion into H_2O_2 and O_2 [17]. There are three known isoforms of SOD: copper-zinc SOD in the cytoplasm and nucleus (Cu/ZnSOD), extracellular SOD in extracellular fluids (ECSOD), and manganese SOD in the mitochondrial matrix (MnSOD) [21]. Catalase, an enzyme in the peroxisomes of all mammalian cells, can transform hydrogen peroxide into water, except in the case of RBCs and vascular endothelial cells. GP_x catalyzes the conversion of peroxides and organic hydroperoxides into oxygen, water, or alcohol via glutathione. There are eight known GPX isoforms. The most common isoform, GP_x1, is present in the mitochondria, nucleus, and cytoplasm [22]. The cytoplasm and nucleus both contain GP_x2. The cytosol is where GP_x3 is usually located. GP_x4 is situated at the membrane with the mitochondria, cytosol, and nucleus [23]. PR_x utilizes NADPH to facilitate the formation of organic superoxide peroxides, hydroperoxides, and peroxynitrite. The PR_x family has six isoforms with cysteine residues. PR_x1–4 posses a standard 2-Cys structure, PR_x5 has an unusual 2-Cys structure, and PRX6 is recognized for its 1-Cys structure [24]. The Tr_x antioxidant family, which includes Tr_x and Tr_xR, is crucial for safeguarding cells from oxidative stress. The inhibition of ribonucleotide reductase and the regulation of a redox-sensitive transcription factor may facilitate the repair of DNA and proteins. Additionally, Tr_xR and NADPH are capable of sustaining Tr_x levels in a reduced form.

4 Effects of Oxidative Stress

ROS within cells is highly active and unstable and could lead to changes in proteins, lipids, and nuclei, altering protein function and signal pathway regulation (Fig. 2). Lipids, especially PUFAs and cholesterol, are crucial oxidative stress substrates due to their role in cell membranes. Moreover, lipids are key cellular metabolites [25]. The method of lipid oxidation includes three primary steps: initiation, elongation, and termination. The process starts when an unsaturated lipid molecule loses one of its hydrogen atoms, producing free radicals. During the propagation process, the combination of both lipid radicals and oxygen gives rise to peroxyl radicals. The generated peroxyl radicals, subsequently, interact with other lipids to produce an increased number of lipid peroxyl radicals [26]. The exhaustion of the hydrogen source leads to the production of numerous nonradical products. Lipid peroxidation yields essential compounds known as active aldehydes, which significantly contribute to the physiopathology of various disorders. Examples of these products include malondialdehyde (MDA) and 4-hydroxynonenal (4HNE) [27]. The detection of active aldehydes in the blood stream can serve as an indicator of disease progression. Proteins are susceptible to oxidant damage when their sulfhydryl group (–SH) is oxidized with sulfur, particularly in amino acids like Met and Cys. Furthermore, synthesizing protein carbonyls, such as ketones and aldehydes, is a standard process. Direct oxidation of side-chain amino acids produces protein carbonyls less frequently than lipid oxidation-derived aldehydes like MDA, 4-oxo-2-nonenal, 4-HNE, 2,4-decadienal, and acrolein [28]. Protein oxidation products vary due to the diverse amino acid side chains in proteins, offering numerous targets for various radical reactions. Arginine, proline, threonine, and lysine are the amino acid residues most susceptible to oxidation [29]. In proteins, radical reactions can take place through three different mechanisms: (1) detachment of hydrogen atoms from positions adjacent to electron-delocalizing groups like guanidine group in arginine, –SH, hydroxy groups (in serine and threonine), and the carboxyl and amide (in aspartic acid, glutamic acid, asparagine, and glutamine); (2) detachment of electrons from areas with high-electron density [30]. Posttranslational modifications associated with protein oxidation change the charge, hydrophilicity/hydrophobicity, structure, folding, and composition of amino acids and proteins [31]. These alterations, such as enzymatic activity, can impair their functions and result in aggregation, cross-linking, or conformational changes. Additionally, cardiovascular, pulmonary, or neurological diseases are more likely due to the oxidized proteins. For instance, cysteine may come into contact with oxidants $O_2^{-\cdot}$, $HO^{\cdot}$, $^{\cdot}NO$, and H_2O_2 during oxidative stress. Sulfenic acids, formed by thiol oxidation in cysteine's side chain, are more reactive and possess electrophilic and nucleophilic properties [27, 31]. Thus, irreversible alteration results from the lack of ability of thioredoxin or glutathione to reduce the sulfenic and sulfonic acids released. Compounds including 5-hydroxymethyluracil, 5-hydroxyuracil, and 6-hydroxy-5,6-dihydrocytosine are produced when ROS reacts with nucleoside base double bonds. It can cause G-T transversions with the most extensively studied compound, 8-oxo-deoxyguanosine [19, 30]. Moreover, oxidative stress might affect histone methylation and DNA, altering chromatin structure and genome function. Lipid peroxidation has the potential to alter DNA directly. Translation synthesis, homologous recombination, and nucleotide excision repair are methods for repairing DNA damage [15, 31]. Inaccuracies in transcription and translation may result from an imbalance between damaged and repaired DNA. Evidence suggests that ROS also attacks the DNA in mitochondria. Mitochondrial DNA is a double-stranded, closed-circular DNA that regulates mitochondrial activity [24, 28]. When ROS damages mitochondrial DNA, there may be a reduction in the number of mitochondrial copies and transcripts and a release of proapoptotic factors and mitochondrial DNA.

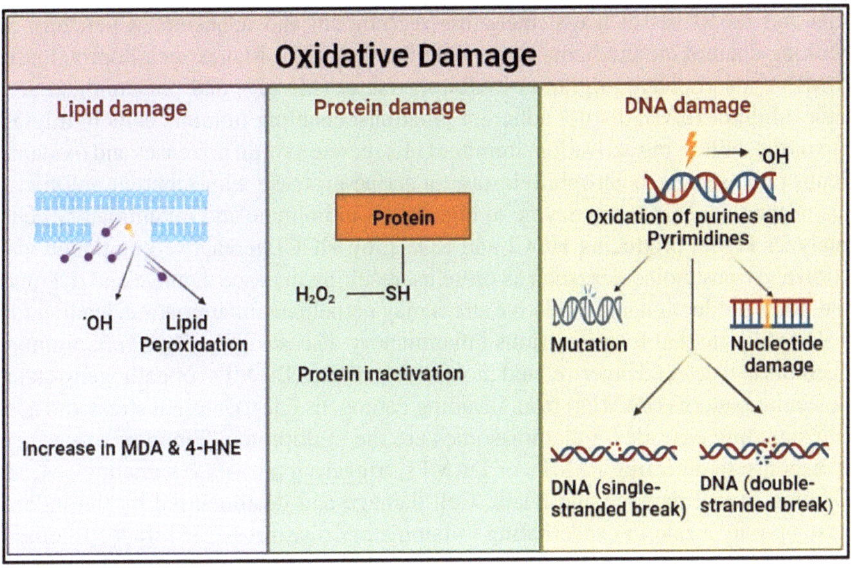

Fig. 2 Oxidative-stress-induced oxidative damage

5 Inflammation: Main Cause of Oxidative Stress

Inflammation is an essential component in the pathogenesis of numerous widely occurring diseases. Inflammation is when immune cells enter blood vessel walls, penetrate tissues, and release ROS that damage the tissue [32]. The catalytic component gp91phox (NOX_2) regulates NADPH oxidase 2, the primary pathway by which PMN produces ROS. The natural or catalytic production of O_2^- transforms into H_2O_2 (6). When MPO catalyzes H_2O_2, it results in the formation of highly reactive radicals like hydroxyl radical (·OH) or hypochlorous acid (HOCl). Both of these radicals possess the potential to inflict tissue damage and promote the oxidation of proteins [33]. Inflammatory cells generate ROS, which induce oxidative stress to eradicate pathogens and activate pathways that promote inflammation. The presence of ROS activates various kinases, including PKC, JNK, and p38 MAPK, induces the activation of transcription factors, producing chemokines and proinflammatory cytokines [34] (Fig. 3). Upon activation, these cytokines and chemokines attach to their specific receptors, including the Vascular endothelial growth factor (VEGF) receptor, Epidermal growth factor receptor (EGFR), and platelet-derived growth factor receptors, which are known to produce ROS. In addition to promoting the release of cytokines like MIP-1/2, MCP-1/2, IL-1, IL-6, TNF-α, and Regulated upon Activation, Normal T-cell Expressed and presumably Secreted (RANTES), ROS triggers the expression of PAI-1, MMP-2, MMP-9, Inducible Nitric Oxide Synthase (iNOS), and MMIF, through NF-κB, plays a significant role in immune cell recruitment and adhesion [26, 34]. CAMs control the interaction between immune cells and blood vessel walls, including recruitment and adhesion, which rely on cytokine-chemokine gradients to regulate the flow [35]. Matrix metalloproteinases (MMPs), macrophage migration inhibitory factor (MMIF), and plasminogen activator inhibitor (PAI), disrupt adherent junctions, enabling immune cells to migrate into tissue. Following activation, immune cells secrete several proteases and oxidants. Neutrophils secrete neutrophil elastase, a serine protease with inherent antimicrobial properties. MPO, an enzyme at high levels in immune and inflammatory cells, catalyzes H_2O_2, producing HOCl and ROS [36]. HOCl, a reactive compound, can oxidize various molecules, such as proteins, resulting in tissue damage and dysfunction. ROS production and oxidative stress may perpetuate inflammation, leading to a prolonged state that fosters various inflammatory diseases [37]. Therefore, immune receptors can detect danger-related molecular patterns (DAMPs) or pathogen-related molecular patterns (PAMPs) from invading pathogens due to internal stress and activate signaling cascades. Inflammasomes are the multiprotein complexes that form in response to detecting PAMPs or DAMPs, triggering a caspase signaling cascade resulting in cell damage and death. Cell damage and death caused by the inflammasome play a role in exacerbating inflammatory diseases [3, 35]. Indeed, inflammasomes have been associated with various inflammatory conditions ranging from neurodegenerative diseases like Parkinsonism, dementia, and encephalomyelitis disseminata to metabolic issues such as obesity, type 2 diabetes, and atherosclerosis [38]. While the main pathway for inflammasome activation is typically through sensing PAMPs and DAMPs, numerous studies have suggested that ROS and various oxidants can also trigger the inflammasome complex.

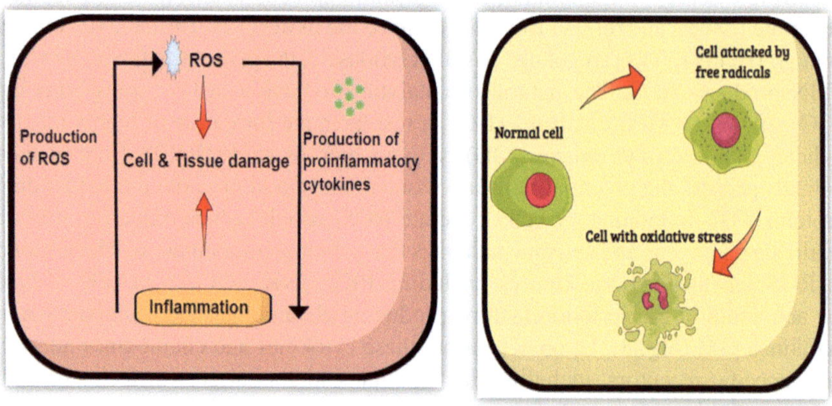

Fig. 3 Schematic representation of oxidative stress and inflammation

6 Oxidative Stress and Programmed Cell Death

Free radical damage or inflammation can cause cell death through necrosis, apoptosis, aging, or autophagy, which breaks down oxidized or defective proteins [39]. ROS can induce cell death in two main ways: either by causing harm to cell membranes or cellular structures or by triggering the caspases (enzymes responsible for cell death) through lipid peroxidation products such as 4-hydroxy-2-nonenal (HNE) [11, 38]. Accumulated HNE in cells can trigger apoptosis by activating either "*mitochondrial*" pathway or the "*death receptor*." Lipid-peroxide-induced cytokines activate caspases 8 and 10, activating the death receptor pathway [12, 34]. These enzymes then break down intracellular proteins, triggering apoptosis. When ROS and lipid peroxidation products lead to membrane permeabilization, they activate the mitochondrial pathway, which in turn causes caspase-dependent and caspase-independent programmed cell death. Furthermore, HNE and ROS can worsen the situation by activating transcription factors AP-1 and NF-κB, leading to the propagation of proinflammatory signaling cascades [26, 37]. Redox signals and ROS play a role in cellular defense mechanisms like autophagy. The autophagy mechanism breaks down complexes and adducts of peroxidized proteins, such as HNE-protein Michael adducts [24, 37]. Autophagy is the process of removing mutated proteins and organelles by forming a vacuole or an autophagosome through the fusion of membrane surfaces. The vacuole is produced and then uses the typical lysosomal breakdown process to dispose of its contents. Studies indicate that ROS (H_2O_2 and superoxide anion) play a critical role in autophagy, because antioxidant treatment can partially or fully reverse the process [38]. ROS activates autophagosomes, which break down damaged proteins, or autophagy execution, which activates lysosomal cysteine proteases for protein degradation. Low levels of ROS are thought to serve as secondary messengers in signal transduction pathways involved in cell division, transformation, and programmed cell death. The process consists in activating signaling pathways that trigger growth-stimulatory genes like c-fos, c-jun, and c-myc, alongside enzymes that facilitate apoptosis, such as caspases [6, 38].

7 Relationship Among Oxidative Damage, Inflammation, and Disease Progression

Oxidative stress and inflammation have the potential to induce cellular integrity. Perhaps the vascular endothelium, where inflammation occurs, is the one cell type particularly affected by this damage [40]. Beyond its function of supplying nutrients to various organs and forming a barrier between tissues and blood circulation, the endothelium also acts as a site for the adhesion of immune cells and other types of cells. These immune cells attach themselves to the vessel wall upon activation in response to inflammatory signals (11). After sticking to the surface, these cells can exit the blood vessels into the nearby tissue and produce ROS, oxidant radicals, and

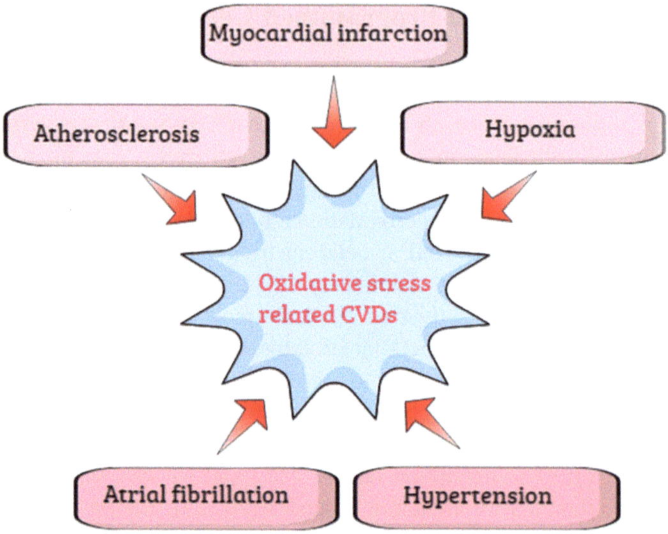

Fig. 4 Oxidative stress-induced CVDs in humans

metabolites that harm endothelial cells and tissues. Therefore, inflammation results in impaired endothelial function [41]. Endothelial dysfunction initiates a proinflammatory environment characterized by upregulation of chemokines, cytokines, adhesion molecules, and chemoattractants due to a positive feedback loop (10). Endothelial dysfunction significantly influences the early stages of diabetes and atherosclerosis. Apart from the endothelium, various cell types, like epithelial and neuronal cells, can experience oxidative stress and contribute to a range of diseases [42]. In summary, the presence of siginificant levels of ROS, abnormal expression of inflammatory chemokines and cytokines, activation of NF-κB, and elevated COX-2 are among the molecular factors contributing to inflammation-induced cardiovascular diseases, as outlined below (Fig. 4).

7.1 Atherosclerosis

A high-fat diet, cholesterol deposition, and hypercholesterolemia in the arterial wall are deemed to increase the risk of arteriosclerosis. A strong relationship has been established among low-density lipoproteins (LDL), total cholesterol, and the ratio of total cholesterol to HDL concerning this health risk [43]. The early phase of arteriosclerosis lesions in large, elastic, and muscular arteries is characterized by the development of a fatty streak, which is marked by the accumulation of lipids within smooth muscle cells, T-lymphocytes, macrophages, and ingested lipoprotein proteoglycan complexes in complex foam cells. Smooth muscle cells and macrophages produce collagen, elastin, and proteoglycans, which move to the intima and form fibrous plaques [17, 41]. Later, qualitative changes in fibrous plaques can lead to ulceration, thrombosis, and/or hemorrhage, resulting in arterial occlusions. Peroxidation of polyunsaturated fatty acids (PUFAs) leads to ischemic necrosis, causing significant consequences due to the production of ROS and endogenous peroxides [44]. Increases in malondialdehyde (MDA) concentration and polymorphonuclear leukocytes' ability to produce oxygen-free radical (OFR) in the blood and aortic tissue are associated with hypercholesterolemic atherosclerosis. Evidence suggests that both lipid-rich, very-low-density lipoprotein (VLDL) and low-density lipoprotein (LDL) might improve monocyte adherence to endothelial cells over a while. The scavenger receptor mechanism takes up modified LDL forms like acetyl or oxidized LDL, which leads to an increase in cholesterol levels and the subsequent formation of foam cells [45]. Excess LDL cholesterol can cause atherosclerosis, according to clinical and epidemiological studies. All primary artery wall cells can oxidatively modify LDL, contributing significantly to in vivo atherosclerosis [46]. Macrophages serve as scavengers because of their remarkable ability to retain modified LDL and β-VLDL that arises from the dietary sources: macrophages function as scavengers. Studies show macrophages have a constrained number of receptors for the uptake of native LDL, while they demonstrate a strong ability to internalize chemically modified LDL via the acetyl-LDL receptor. (saturable receptor). Minimally oxidized LDL (MM-LDL), which originates in the subendothelial compartment, can be internalized by the conventional LDL receptor through apoB and does not engage with macrophages in the same manner as typical LDL. The unsaturated acyl chains of phospholipids and cholesteryl esters in mid-oxidized LDL undergo oxidation to short-chain aldehydes, hydroperoxides, and isoprostanes, all of which possess potent biological effects [47]. Monocyte chemotactic protein-1 (MCP-1), produced by low-density lipoprotein (LDL), enables monocytes to adhere to the endothelium and migrate to the subendothelial region, producing M-CSF. Monocytes can grow and eventually differentiate into macrophages with the help of M-CSF [48]. Macrophages have the ability to transform MM-LDL into a more oxidized variant that evades detection by the LDL receptor. This process facilitates its uptake through the scavenger receptor pathway, resulting in the accumulation of cholesterol. A significant amount of cholesterol (up to 50%) is oxidized, yielding 7-ketocholesterol and other oxysterols that attach to scavenger receptors such as CD36 and/or SR-A [49]. The oxidation of unsaturated fatty groups produces a complex mixture of fragmented, derivatized, and cross-linked apoB, with significant aldehyde modification resulting in hydroxynonenal and malondialdehyde [50]. The distinctive features of oxidatively modified LDL (ox-LDL) include its chemoattractant effect on monocytes, inhibits arterial macrophage movement, and shows cytotoxicity, these properties potentially lead to endothelial dysfunction and atherogenesis by impacting arterial gene expression. Oxidatively modified LDL (ox-LDL) has unique properties, including being a chemoattractant for monocytes, an inhibitor of macrophage mobility in the arterial wall, and being cytotoxic, potentially promoting endothelial dysfunction and atherogenesis by affecting gene expression in the arterial wall. High-density lipoprotein (HDL) correlates with reduced likelihood of developing atherosclerosis. ApoA1, an HDL apolipoprotein, aids in cholesterol efflux from peripheral vasculature deposits, transporting it to the liver for excretion. The antiatherogenic effect of HDL is primarily due to its ability to inhibit the oxidation of LDL, which in turn reduces cellular absorption by the monocyte-macrophage system [45]. Ceruloplasmin and other proteins are present on the lipoprotein's surface, giving HDL its chelation characteristics and serving as part of its antioxidant mechanism and ability to accept hydroperoxides from oxidized membranes, potentially facilitating excretion or detoxification. In vitro studies have shown that HDL may absorb hydroperoxides from oxidized membranes, opening up a channel for excretion or detoxification. Furthermore, it has been demonstrated that Apo A-I stops the chain reactions of lipid peroxidation by transforming lipid hydroperoxides into redox-inactive molecules. Enzyme proteins linked with HDL that have antioxidant activity also contribute to HDL's function [51]. Paraoxonase 1 (PON 1), an HDL-associated esterase/lactonase, displays anti-inflammatory and antioxidant characteristics. LDL particle degrades oxidized fatty acids, preventing proatherosclerotic functions and decreasing monocyte binding to the endothelium.

ROS plays a critical factor in developing atherogenesis and endothelial dysfunction. Several oxidants, such as mitochondrial respiratory chain/oxidative phosphorylation, nitric oxide synthase, myeloperoxidase, NADPH oxidase, lipoxygenase/cyclooxygenase, and xanthine oxidase, are produced via enzymatic processes within the vessel wall [45]. TNF-α, IL-1, Angiotensin II, VEGF, and thrombin are stimulating factors that trigger endothelial NADPH oxidase to release ROS into blood vessels. C-reactive protein (CRP), a cardiovascular risk factor, stimulates superoxide production in smooth muscle cells. This may affect NADPH oxidase regulation in human aortic endothelium through nitric oxide synthases (NOS) and protein kinases [50]. Active eNOS has an oxygenase domain (containing heme, tetrahydrobiopterin, Zn, L-arginine, and BH4) and a reductase domain (binding sites for FMN, FAD, and NADPH), joined by a hinge region that calmodulin binds to. Normally, enzymes in the oxygenase domain transfer electrons from a heme group to L-arginine, generating L-citrulline and NO, with BH4 acting as a cofactor [49]. A decline in the availability of BH4 or L-arginine causes eNOS to move from a coupled to an uncoupled state, causing vascular dysfunction due to the rapid reaction of NO^* with O_2^{*-}. Activated macrophages and smooth muscle cells in blood vessels express iNOS in atherosclerotic lesions. Myeloperoxidase (MPO), an enzyme that contains heme, facilitates the conversion of chloride ions (Cl^-) into hydrochlorous acid (HOCl). Chlorinated biomolecules act as distinct indicators of enzymatic-facilitated oxidation reactions. The $MPO/H_2O_2/Cl^-$ system may produce free amino acid or protein-bound tyrosyl radicals, 3-chlorotyrosine, chlorohydrins from fatty acids and cholesterol, and α-chloro fatty acid aldehydes. LDL oxidation is one example of a secondary oxidation process in which tyrosyl radicals might participate.

The $MPO/H_2O_2/Cl^-$ system, along with HOCl, oxidizes nitrite to produce nitryl chloride (NO_2Cl), a nonradical oxidant, as well as the radical $*NO_2$. These entities promote nitration and the transformation of tyrosine into 3-nitrotyrosine [50]. MPO significantly contributes to the formation of nitrating species within the biological systems, and the formation of 3-nitrotyrosine being reliant on the availability of NO_2. The study found that xanthine-oxidase-induced oxidative stress in hypercholesterolemic rabbits leads to atherosclerosis. In the vascular wall, lipoxygenase

(LPO) converts polyunsaturated fatty acids (PUFAs) into hydroperoxy fatty acid derivatives, the primary ROS production source. It has also been demonstrated that the increased susceptibility to atherosclerosis is linked to mitochondrial ROS.

7.2 Myocardial Infarction

A myocardial infarction (MI), commonly known as cardiac arrest, arises when a specific area of the heart muscle does not receive enough blood, ultimately leading to the death of heart cells [52]. The rupture of atherosclerotic plaques in the arteries, which clog the coronary arteries, is the leading cause of this problem. If not treated promptly, the lack of oxygen due to ischemia can lead to harm or death of the heart muscle tissue. Furthermore, myocardial infarction is associated with reperfusion and tissue ischemia, which influence both hypoxia and hypoxia-reoxygenation (H/R). When blood flow resumes after an interval of ischemia or oxygen deprivation, tissue may endure damage known as a reperfusion injury. The inflammatory response exacerbates reperfusion injury. Newly formed white blood cells release inflammatory substances such as free radicals and cytokines when responding to tissue damage [53]. Redox signaling can indirectly initiate apoptosis in reactive species. Moreover, white blood cells can attach to the walls of small blood vessels, thereby blocking them and causing increasing ischemia. When ischemia lasts more than 60 min, the breakdown of ATP metabolism results in hypoxanthine formation [54]. The xanthine oxidase transforms molecular oxygen into highly reactive superoxide and hydroxyl radicals [55]. Xanthine oxidase is responsible for the production of uric acid, which can function as both a scavenger of reactive species like peroxy nitrite and a prooxidant. During reperfusion, the overproduction of nitric oxide interacts with superoxide, resulting in the formation of highly reactive compound known as peroxy nitrite [56]. These unbound molecules cause additional harm to the glycosaminoglycans, proteins, and lipids of the cell membrane. They can also use redox signaling to initiate several biological processes. Shortly after reperfusion begins, chemical reactions occur, activating the mitochondrial permeability pore (MPT pore) in the cardiac cell mitochondrial membrane. This causes water to enter the mitochondria, leading to their dysfunction and collapse. The release of calcium then disrupts the energy production in the mitochondria, ultimately leading to cell death. Therefore, ensuring the protection of mitochondria is an effective way to keep the heart safe. Cyclophilin D, produced in response to high calcium levels, facilitates the opening of the MPT pore by interacting with other pore components. Several studies have discussed using bone marrow mesenchymal stem cell transplantation as a therapeutic approach for treating the heart after a heart attack. Nevertheless, the method's effectiveness is significantly impeded by the low survival rate of donor MSCs in damaged hearts, particularly under oxidative stress conditions. Numerous reports show how high-density lipoprotein (HDL) reduces the risk of ischemic diseases due to its ability to reverse transport cholesterol. According to another study, pretreating with HDL increased rates of MSC survival, enhanced cardiac remodeling, and improved myocardial function compared to the MSC control group. Research indicates that HSF1, a transcription factor responsible for activating heat shock proteins (HSPs), protects against heart diseases like ischemia/reperfusion injury and heart attacks [57]. HSF1 can inhibit the apoptosis of cardiomyocytes caused by different stimuli, while cytotoxic oxidative stress, which is cytotoxic, can ultimately result in apoptosis. HSPs and HSF1 offer protection against oxidative damage and reduce ischemia/reperfusion injury by inhibiting JNK activity. Reactive oxygen species formation within cells is inhibited by HSF1, which reduces JNK activity and prevents cardiomyocytes from apoptosis [58]. A recent research experiment involved transferring ASK1, HSF1, or both into cultured cardiomyocytes of neonatal rats, followed by exposure to ROS and H_2O_2 to study JNK activity and apoptosis.

7.3 Hypoxia

Hypoxia refers to a reduction in oxygen delivery to a tissue or organ, and if it impacts the brain specifically, it is called cerebral hypoxia. This situation is triggered by incidents that greatly obstruct the brain's capacity to obtain or utilize oxygen. Excessive oxygen deficiency causes nerve cells to die through apoptosis, leading to stagnant anoxia [59]. A stroke arises when insufficient blood flow and less oxygen reaches the brain. During a few days, or maybe a week, following an initial stroke, there is extensive self-destruction. Oxidative stress greatly impacted the progression of acute ischemic stroke. A contributing factor to the severity of a stroke might be the generation of free radicals and oxidative damage. Findings from a study showed that within 48 h after an acute stroke, patients had considerably higher serum concentrations of GSH, NO, and MDA, in contrast to the control group. During this period, there was a negative correlation between the "*neurological deficit score*" and NO and MDA levels, with GSH levels acting as a compensatory mechanism. Oxidative stress, a critical factor in ischemic stroke, affects cerebral blood vessels and can lead to significant brain damage following a heart attack. Research involving animal models of ischemic stroke demonstrates that superoxide and its derivatives produce vasodilation through altered vascular response, activation of K^+ channels, disruption of the blood–brain barrier, and localized destructive lesions [60]. In addition to other factors, hypoxia is a common stress factor that is known to promote BBB collapse. Signs of oxidative stress in hypoxic or ischemic conditions suggest that it may contribute to the breakdown of barriers. The pathophysiological process of oxygen deprivation injury is critical to the BBB's collapse [61]. The research discovered that applying antioxidant therapy to RBE4 monolayers under conditions of oxygen deprivation enhances barrier efficiency and survival of cells. Due to its capacity to produce ROS, free radical damage plays a significant role in the EC barrier's collapse [62]. When subjected to oxygen deprivation stress, RBE4 monolayers infused with antioxidants resulted in barrier efficiency and cell viability. Additionally, HIF-1 plays a pivotal role in the activation of various genes that are responsive to hypoxic conditions, establishing it as a key regulator of genes associated with low oxygen levels. Its regulatory influence extends to genes associated with angiogenesis, cellular proliferation, energy metabolism, erythropoiesis, and vasomotor regulation [63]. Controlling HIF-1 activation and buildup could be a promising treatment strategy for cerebral ischemia, as these genes may aid in brain cell healing after ischemia and reperfusion. Researchers have discovered that the neuroprotective effect of HIF-1 expression of its gene product, erythropoietin, protects cells from hypoxic and ischemic injuries via various mechanism. HIF-1-regulated VEGF, which protects against blood-flow-related damage and cell death, promotes cell survival by blocking p53 activation, PARP cleavage, caspase activation, and cytochrome c release.

HIF-1 likely has a role in the deterioration of tissues and cells. Reports indicate that HIF-1 can facilitate apoptosis in hypoxic or ischemic situations. Hypoxic conditions have shown HIF-1-induced apoptosis in embryonic stem (ES) cells. According to the research, an inducible subunit, HIF-1α, accumulates and ensures the stability of the active wild-type p53 under hypoxic conditions. The observed apoptosis in hypoxia-induced ES cells is due to increased p53 protein levels [64]. According to experimental research, HIF-1α can cause cell death during severe ischemia but also improve survival after mild events. Moreover, NOX_4 is the primary vascular isoform induced during acute stroke, making it a crucial target for effective therapy. Ischemia increases NOX4 expression in the brains of mice and human brains. After experiencing sudden and long-term cerebral ischemia, mice without NOX4 were protected from damage caused by blood–brain barrier breakdown, neuronal degeneration, and oxidative stress [62]. In $NOX_4^{-/-}$ mice, oxidative stress restoration significantly impacted their stroke-prevention behavior. The enzyme NOX_4 significantly contributes to oxidative stress associated with brain ischemia and particular forms of human stroke, indicating its potential as a promising avenue for the development of innovative stroke therapies.

7.4 Hypertension

Hypertension is commonly recognized as a significant risk factor for cardiovascular disease. An acute increase in blood pressure is often caused by heightened levels of vasorelaxation factors. Findings from laboratory investigations reveal that oxidative stress levels in hypertensive models are distinct from those found in the control group [65]. NOX families, along with Xo, NOS, and mitochondria, are considered significant contributors to the increase in ROS, much like atherosclerosis. Endothelial cells are the target of oxidative stress, which regulates blood pressure [66]. The balance between endothelium-derived hyperpolarizing (EDHFs) and endothelium-derived relaxing (EDRFs) factors controls vascular tonicity. Nitric oxide (NO) belongs to group of EDRFs, whereas ROS is considered a member of endothelium-derived hyperpolarizing factors (EDHF). ROS can raise the quantity of endogenous endothelium NOS antagonists, such as asymmetric dimethyl arginine (ADMA), and lower NO bioavailability [67]. Factors involved in gene expression, including activated protein-1 (AP-1) and p53 and tyrosine kinases, MAPKs, and Phosphoinositol-3-kinase/Akt kinase (PI3K/Akt), are some of the cell phosphorylation pathways that influence endothelial function. Oxidative stress is capable of initiating and governing numerous pathways along with the factors that influence gene expression. Chayama et al. studied oxidative stress in renovascular hypertension patients by examining their urinary excretion of MDA-modified LDS and 8-hydroxy-2-deoxyguanosine [68]. The study revealed that hypertensive patients exhibited elevated levels of oxidative stress indicators relative to the control group of non-hypertensive individuals. Beevers' analysis showed that hypertension patients exhibit increased levels of lipid hydroperoxides.

7.5 *Atrial Fibrillation (AF)*

An erratic and fast heartbeat is a sign of atrial fibrillation, a frequent clinical arrhythmia. An elevated incidence of atrial fibrillation (AF) has been associated with decreased plasma antioxidant capacity in rats. According to Quyyumi et al., AF patients had higher glutathione redox potentials, cystine levels, and cystine/glutathione ratios, all signs of increased oxidative stress. The enzymes Xo and $NOX_{2/4}$, both elevated in fibrillating regions, produce ROS [69]. ROS may induce AF through many different pathways. First, in cardiomyocytes, the increase in ROS controls ionic leakage [70]. It causes the sarcoplasmic reticulum (SR) to release more Ca^{2+} and enhance the L-type Ca^{2+} and Na^+ currents. This leads to decreased conduction velocity and an increased action potential duration. Moreover, oxidative stress triggers collagen deposition, leading to cardiac fibrosis [71]. The electrical interaction between myocytes is disrupted in cardiac fibrosis. Ultimately, Oxidative stress can lead to atrial fibrillation (AF) by altering DNA, iron-related proteins, and posttranslational changes. Angiotensin II and hypoxia regulate the mRNA splicing of the Na^+ voltage-gated channel alpha subunit 5 (SCN5A), resulting in disruptions in Na^+ current. ROS can oxidize CaMKII, potentially regulating Na^+ current through the ryanodine receptor, a crucial Ca^{2+}-regulating protein in the SR [72].

8 Therapeutic Strategies for Oxidative-Stress-Related CVDs

8.1 Antioxidants

8.1.1 Dietary Supplements

Research indicates that in different animal models, ROS may induce atherothrombosis and other vascular ailments, while also contributing to the development of cardiovascular diseases (CVDs) [73]. Specific ROS inhibitors are beneficial for treating cardiovascular disease. Ascorbic acid (vitamin C), α-tocopherol (vitamin-E), and β-carotene are the most important antioxidants that benefit the cardiac system while using ROS inhibitors [74]. Vitamin C can detoxify both endogenous and exogenous reactive oxygen ROS and modified lipids and proteins. Vitamin C controls smooth-muscle-related widening of blood vessels, endothelial cell growth, and programmed cell death, which is essential to the onset of cardiovascular diseases [75]. Moreover, large-scale randomized trials examining vitamin C have shown no significant impact on cardiovascular disease (CVD). Antioxidants did not demonstrate cardioprotective effects and failed to reduce clinical events in high-risk or disease-established individuals, according to a comprehensive evaluation of randomized controlled trials. Tocopherols and tocotrienols, which make up vitamin E, have four isoforms each: delta (δ), gamma (γ), beta (β), and alpha (α). The quality of vitamin E is determined by the effectiveness of α-tocopherol among the eight types of molecules [75]. Lipid radicals and the lipid-soluble antioxidant α-tocopherol interact to shield the cell membrane. Vitamin E disrupts the NOX enzyme assembly, thereby preventing the release of superoxide. According to experimental research, vitamin E can lower the risk of cardiovascular problems and coronary heart disease. A reddish-orange pigment is found in a wide variety of fruits and vegetables. According to research, consuming vitamins C and E significantly reduced coronary heart disease incidence, which included 400,000 participants. This phytonutrient effectively reduces singlet oxygen levels. In order to avoid lipid oxidation in cell membranes, beta-carotene is converted into an active form of vitamin A, known as bioactive retinol. According to epidemiological studies, higher beta-carotene levels lead to a lower incidence of heart attacks and CVDs. Novel and more potent ROS inhibitors have been developed due to the inadequacy of existing antioxidants in providing cardioprotective properties. While mitochondria are essential in causing CVDs, there have been no studies on targeting ROS inhibitors to mitochondria in CVD patients [76]. Mitochondrial-based antioxidant therapy might be more effective than conventional antioxidant therapy in treating cardiovascular diseases.

8.1.2 Experimental Therapeutic Approach Using Antioxidants

Three categories of novel antioxidants include (a) chemical compounds for the functional restoration of ROS-induced damage, (b) compounds that block the occurrence of oxidative stress, and (c) stimulants of endogenous protective antioxidant systems [77]. Nuclear factor erythroid 2-related factor 2 (NRF2) activators activate the body's natural defense system against antioxidants. The gene transcription factor NRF2 is in charge of identifying the promoter of the antioxidant sensitivity component. High blood pressure and atherosclerosis are more prevalent among individuals with reduced NRF2 expression and activity [78]. Dimethyl fumarate (DMF) is the most extensively researched drug targeting NRF2. DMF minimized the size of the infarction following I/R damage, according to in vivo studies. Studies conducted in vitro have also suggested that it protects cardiomyocytes from ischemia and perfusion damage. In a mouse model of apolipoprotein E deficiency with streptozotocin-induced hyperglycemia, DMF retarded the atherosclerosis progression [79].

NOX, MPO, and Xo are the targets of allopurinol, a potential therapeutic agent in the second class of drugs that work to prevent oxidative stress. According to clinical studies, allopurinol lowers oxidative stress in endothelial cells, which may positively impact high blood pressure, ischemia, and cardiac arrest [80]. Allopurinol's influence on blood pressure appears slightly less in a meta-analysis of ten clinical investigations. It significantly reduces hospital-acquired deaths and cardiac issues in individuals undergoing first-line percutaneous coronary intervention or coronary artery bypass surgery [81]. GKT137831 is recognized as the first NOX inhibitor currently being studied in clinical settings. In apo-E KO mice, GKT137831 treatment had a strong antiatherosclerotic effect. In an antioxidant model of atherosclerosis, MPO inhibitors seem to influence the composition of atherosclerotic lesions and impact cardiac remodeling in murine models. Finally, compounds primarily work on the nitric oxide-cyclic guanosine monophosphate NO-cGMP signaling pathway to facilitate the functional repair of ROS-induced damage. In order to disrupt NO-cGMP signaling, ROS may harm the NO receptor, disconnect NOS, or chemically scavenge NO. CXL-1427, a second-generation HNO donor molecule, is one of the medications in this class of pharmaceuticals [82]. In a Phase 2a dose-escalation trial, hospitalized patients diagnosed with heart failure with reduced ejection fraction (HFrEF) revealed a favorable safety profile and positive blood flow responses. This category comprises L-arginine and L-citrulline, which aid in NOS recoupling. In an extensive review of 11 randomized, double-blind, placebo-controlled trials, it was demonstrated that oral supplementation with L-arginine led to a significant reduction in blood pressure, specifically 2.66 mmHg in diastolic and 5.39 mmHg in systolic readings. Administering L-citrulline decreased diastolic blood pressure by 3.77 mmHg and systolic blood pressure by 7.54 mmHg, according to a meta-analysis of 15 randomized controlled trials, including 420 people.

8.1.3 Role of Clinical Drugs as Antioxidants

Carvedilol, melatonin, metformin, and proprotein convertase subtilisin/kexin type 9 (PCSK9) inhibitors represent a selection of medications utilized in clinical settings that are known to exhibit antioxidant characteristics [83]. Carvedilol is an antioxidant-rich drug that blocks α-1, β-1, β-2adrenergic receptors. This drug exerts a substantial impact on the therapeutic approach to heart failure. Studies conducted on animals indicate that targeting oxidative stress may provide a protective effect against atherosclerosis, heart failure, and myocardial infarction. Metformin is a medication that enhances the expression of antioxidant genes and inhibits the NOX pathway, thereby reducing the ROS production [84]. Metformin can also assist in controlling the ratio of NO to ADMA, since it is a structural analog of ADMA. There is evidence of protective effects in HF and CAD from clinical trials and animal research. PCSK9 inhibition in animal models of atherosclerosis decreases ROS production in smooth muscle cells of the arterial wall, achieved through the downregulation of lectin-like oxidized LOX-1 expression [74].

8.1.4 Role of MiRNAs

Small noncoding RNAs, commonly known as miRNAs, seem to control the expression of genes by targeting specific messenger RNAs [85]. The regulation of the oxidative stress response depends mainly on miRNAs, and hypoxia causes a significant increase in the hypoxia-induced RNA miRNA-210. miRNA-210, a pivotal microRNA associated with anoxic environments, improves heart functionality by promoting angiogenesis and inhibiting apoptosis. HIF-1α regulates the expression of miRNA-210. HIF-1 α mutant mice continued to exhibit lower levels of miRNA-210 than the normal mice [86]. The study found that individuals who passed away from a sudden cardiac arrest exhibited higher levels of miRNA-210 in their damaged heart tissues than the subjects in the control group. MiRNA-210, administered directly into an animal model of myocardial infarction, results in increased cardiac function. Furthermore, within the cohort of patients with severe coronary artery disease, there was a significant relationship between the levels of circulating miRNA-210 and mortality due to cardiovascular conditions. miRNA-1, the most common miRNA in myocardial muscles, is crucial for muscle cell proliferation and differentiation. In rat cardiomyocytes, the presence of H_2O_2 can significantly enhance the expression of miRNA-1. miRNA-1 overexpression led to elevated levels of ROS and reduced production of SOD. A significant correlation was found between the total levels of miRNA-1 and infarct size in a rat model of coronary artery disease. Furthermore, miRNA-1 treatment improved heart function in mice following a cardiac arrest [87]. Additionally, researchers found an association between the circulating troponin T and blood levels miRNA-1 in individuals with coronary artery disease. Under hypoxic conditions, there is additional evidence of overexpression of other miRNAs, including miRNA-17, miRNA-21, and miRNA-132 [88]. Moreover, several miRNAs play a vital role in controlling atherosclerotic plaque progression. A study involving

100 participants found that patients suffering from cardiovascular disease had higher levels of miRNA-133, particularly in symptomatic plaques, relative to the control group. The study suggests that targeting NOS with miRNA-133 suppression can effectively prevent endothelial dysfunction [89]. Endothelial cells tend to be influenced by miRNA-92a, which controls the expression of endothelial NOS. Through promoting angiogenesis and endothelial cell proliferation, miRNA-92a increases plaque stability and decreases inflammation. However, much more effort remains essential to making miRNAs a viable therapeutic alternative owing to their poor stability and bioavailability.

8.1.5 Cardiac Patches

Cardiac patches are implantable devices designed to treat cardiovascular diseases. Restoring compromised cardiomyocytes (CMs) is their primary objective. Fibrotic scar tissue restores damaged cardiac myocyte cells in the ischemic region due to cardiac disorders such as myocardial infarction and coronary artery disease [90]. However, the loss of contractile ability in this situation indicates heart dysfunction and ultimately leads to cardiac arrest. In this context, cardiac patches made of biologically active scaffold material are one of the most efficient therapeutic options for the regeneration of heart tissue [91]. Cardiac patches consist of a therapeutic ingredient, such as cells or bioactive molecules such as growth factors (HGF or VEGF), miRNA, and extracellular vesicles like exosomes, and a scaffold material replicating the characteristics of a typical native myocardium (Fig. 5). Substrate scaffold materials must sustain biological activity, resemble host tissue mechanically, and inhibit undesirable immune reactions [92]. Synthetic and natural polymeric scaffolds could provide an array of mechanical qualities, good biocompatibility, and readily adjustable degradation rates. When it requires treating damaged heart tissue, cardiac patches constitute a more potent therapeutic option than other injections of growth factors or stem cells, which have short half-lives and poor stability [93]. Bioactive substances packed in acellular cardiac patches include microRNAs, growth factors, and extracellular vesicles. Naturally occurring bioactive substances are known as growth factors that impact cell growth by stimulating proliferation or differentiation. Polymeric scaffolds made from poly (2-ethyl-2- oxazoline) (PEOz) and poly(l-lactide-co-caprolactone) (PLCL) were electrospun with a blend of vascular endothelial growth factor (VEGF) and essential fibroblast growth factor primary fibroblast growth factor (bFGF) [94]. The study found that growth-factor-infused scaffolds enabled cells to respond at a molecular level, thereby reducing scar tissue formation. The team of scientists has engineered a cardiac patch that incorporates several electrodes, which encompass the electrical stimulation of cells and tissues, as well as the long-term release of bioactive materials within cardiac tissue [95]. In this investigation, in vitro cell migration was facilitated by releasing stromal-cell-derived factor-1 (SDF-1) from the cardiac scaffold. In addition, SDF-1 release might improve vascularization. Apoptotic bodies, exosomes, and microvesicles are the three main types of extracellular vesicles, with a 30–150 nm diameter. Cell signaling and intercellular processes may depend on the vesicles secreted by cells in the extracellular environment. As a result, cardiac patches filled with extracellular vesicles remain a promising alternative for treating cardiac disorders [96]. To address the poor stability and short half-life of vesicles released from therapeutic cells following injection, Liu et al. created hydrogels filled with extracellular vesicles for cardiac patches. Cardiomyocytes (CMs) generated from pluripotent stem cells became the subject of this study due to their ability to produce extracellular vesicles. The study found that the application of cardiac patches to rat hearts 24 h post infarction reduces arrhythmic load and cardiomyocyte death rates, facilitating ejection fraction recovery. Hamada et al. prepared extracellular-vesicle-loaded cardiac patches using poly (glycerol-co-sebacate) acrylate ethylene glycol (PGSA-g-EG), a photocurable adhesive polymer, to enhance the therapeutic effect of cell therapy [94]. Studies show that the proposed platform releases active EVs over 14 days. Damaged cardiomyocytes can regenerate using cardiac patches. Recent studies in the field have focused mainly on cell treatment instead of active compound delivery. The main drawback is the implantation procedure. Cardiac patch implants necessitate a major surgical procedure [90]. Experts are currently involved in resolving this issue. It is now possible to implant cardiac patches with less invasive techniques. Not all types of patches allow for this technique. Finally, cardiac patches often contain biomolecules, crucial when designing industrial-scale devices [95].

8.1.6 Drawbacks

Although oxidative stress is a significant factor in several heart-related diseases, the antioxidant therapy employed in medical settings remains limited. Antioxidant treatments in animal models seemed more beneficial, but their effects in human clinical studies remain disputable. The potential cause is that nonspecific ROS suppression is ineffective or undesirable due to disruption of crucial ROS-mediated cellular signaling [97]. For patients suffering from myocardial infarction, targeted reduction of ROS overproduction, such as cardiac patches, may offer a more potent form of antioxidant treatment. Furthermore, selecting a proper time for administering antioxidants becomes essential owing to the disease's rapid progression [98]. Oxidative stress may have specific roles at various phases of a disease. It follows that administering the antioxidant therapy at the right time is essential.

9 Conclusion

This chapter discusses the key findings and research pertaining to the contributions of ROS and other oxidants in inflammation, oxidative stress, and developing certain disease conditions. Oxidative stress is a key factor in the onset and development of heart-related disorders. Numerous treatment approaches aimed at reducing oxidative stress have been introduced. The findings of clinical trials on humans differ from those of animals regarding the therapeutic effect of antioxidant treatment in different

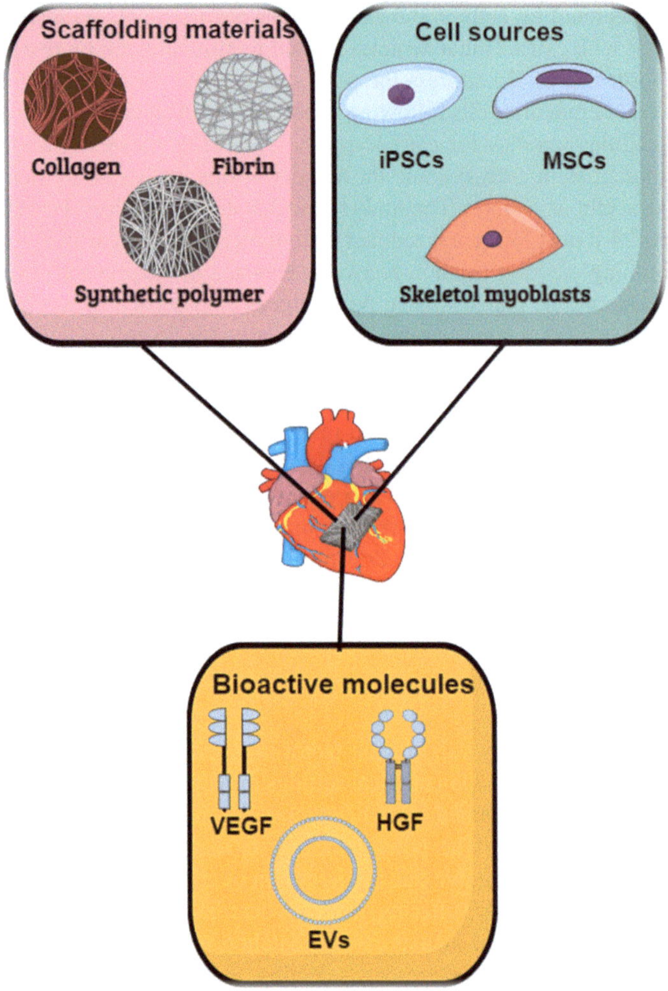

Fig. 5 Scaffolding materials used in the production of cardiac patches

heart-related diseases. Therefore, improved antioxidant therapy and a better understanding of oxidative stress in cardiovascular disease are necessary for more effective treatments.

References

1. Thanan, R., Oikawa, S., Hiraku, Y., Ohnishi, S., Ma, N., Pinlaor, S., et al. (2014). Oxidative stress and its significant roles in neurodegenerative diseases and cancer. *International Journal of Molecular Sciences, 16*(1), 193–217.

2. Preiser, J. C. (2012). Oxidative stress. *Journal of Parenteral and Enteral Nutrition, 36*(2), 147–154.
3. Rahman, T., Hosen, I., Islam, M. M. T., & Shekhar, H. U. (2012). Oxidative stress and human health. *Advances in Bioscience and Biotechnology, 03*(07), 997–1019.
4. Klran, T. R., Otlu, O., & Karabulut, A. B. (2023). Oxidative stress and antioxidants in health and disease. *Journal of Laboratory Medicine, 47*(1):1–11.
5. Cui, H., Kong, Y., & Zhang, H. (2012). Oxidative stress, mitochondrial dysfunction, and aging. *Journal of Receptors and Signal Transduction, 2012*, 1–13.
6. Schiavone, S., Jaquet, V., Trabace, L., & Krause, K. H. (2013). Severe life stress and oxidative stress in the brain: From animal models to human pathology. *Antioxidants Redox Signal, 18*(12), 1475–1490.
7. Bansal, M., & Kaushal, N. (2014). Oxidative stress mechanisms and their modulation.
8. Pizzino, G., Irrera, N., Cucinotta, M., Pallio, G., Mannino, F., Arcoraci, V., et al. (2017). Oxidative stress: Harms and benefits for human health. *Oxidative Medicine and Cell Longevity, 2017*.
9. Vona, R., Gambardella, L., Cittadini, C., Straface, E., & Pietraforte, D. (2019). Biomarkers of oxidative stress in metabolic syndrome and associated diseases. *Oxidative Medicine and Cell Longevity, 2019*.
10. Melov, S. (2002). Animal models of oxidative stress, aging, and therapeutic antioxidant interventions. *International Journal of Biochemistry & Cell Biology, 34*(11), 1395–1400.
11. Chatterjee, S. (2016). *Oxidative stress, inflammation, and disease [Internet]. Oxidative stress and biomaterials* (pp. 35–58). Elsevier Inc. Available from: https://doi.org/10.1016/B978-0-12-803269-5.00002-4
12. Teleanu, D. M., Niculescu, A. G., Lungu, II., Radu, C. I., Vladâcenco, O., Roza, E., et al. (2022). An overview of oxidative stress, neuroinflammation and neurodegenerative diseases. *International Journal of Molecular Sciences, 23*(11).
13. Jomova, K., Raptova, R., Alomar, S. Y., Alwasel, S. H., Nepovimova, E., Kuca, K., et al. (2023). Reactive oxygen species, toxicity, oxidative stress, and antioxidants: chronic diseases and aging. *Archives of Toxicology, 97*, 2499–2574. Available from: https://doi.org/10.1007/s00204-023-03562-9
14. Griendling, K. K., Camargo, L. L., Rios, F. J., Alves-Lopes, R., Montezano, A. C., & Touyz, R. M. (2021). Oxidative stress and hypertension. *Circulation Research, 128*(7), 993–1020.
15. Reddy, V. P., Zhu, X., Perry, G., & Smith, M. A. (2009). Oxidative stress in diabetes and Alzheimer's disease. *Journal of Alzheimer's Disease, 16*(4), 763–774.
16. Li, S., Tan, H. Y., Wang, N., Zhang, Z. J., Lao, L., Wong, C. W., et al. (2015). The role of oxidative stress and antioxidants in liver diseases. *International Journal of Molecular Sciences, 16*(11), 26087–26124.
17. Rottenberg, H., & Hoek, J. B. (2017). The path from mitochondrial ROS to aging runs through the mitochondrial permeability transition pore. *Aging Cell, 16*(5), 943–955.
18. Tejero, J., Shiva, S., & Gladwin, M. T. (2019). Sources of vascular nitric oxide and reactive oxygen species and their regulation. *Physiological Reviews, 99*(1):311–379.
19. Pick, E. (2014). Role of the Rho GTPase Rac in the activation of the phagocyte NADPH oxidase: Outsourcing a key task. *Small GTPases, 5*.
20. Bedard, K., & Krause, K. H. (2007). The NOX family of ROS-generating NADPH oxidases: Physiology and pathophysiology. *Physiological Reviews, 87*(1), 245–313.
21. Cramer-Morales, K., Heer, C. D., Mapuskar, K. A., & Domann. F. E. (2015). SOD2 targeted gene editing by CRISPR/Cas9 yields Human cells devoid of MnSOD. *Free Radical Biology and Medicine, 89*, 379–386. Available from: https://doi.org/10.1016/j.freeradbiomed.2015.07.017
22. Miao, L., & St. Clair, D. K. (2009). Regulation of superoxide dismutase genes: Implications in disease. *Free Radical Biology and Medicine, 47*(4), 344–356. Available from: https://doi.org/10.1016/j.freeradbiomed.2009.05.018
23. Margis, R., Dunand, C., Teixeira, F. K., & Margis-Pinheiro, M. (2008). Glutathione peroxidase family—An evolutionary overview. *FEBS Journal, 275*(15), 3959–3970.

24. Lu, J., & Holmgren, A. (2014). The thioredoxin antioxidant system. *Free Radical Biology and Medicine, 66*, 75–87. Available from: https://doi.org/10.1016/j.freeradbiomed.2013.07.036
25. Gianazza, E., Brioschi, M., Fernandez, A. M., & Banfi, C. (2018). Lipoxidation in cardiovascular diseases. *Redox Biology, 23*, 101119. Available from: https://doi.org/10.1016/j.redox.2019.101119
26. Del Re, D. P., Amgalan, D., Linkermann, A., Liu, Q., & Kitsis, R. N. (2019). Fundamental mechanisms of regulated cell death and implications for heart disease. *Physiological Reviews, 99*(4), 1765–1817.
27. Ayala, A., Muñoz, M. F., & Argüelles, S. (2014). Lipid peroxidation: Production, metabolism, and signaling mechanisms of malondialdehyde and 4-hydroxy-2-nonenal. *Oxidative Medicine and Cellular Longevity, 2014.*
28. Dalle-Donne, I., Rossi, R., Colombo, R., Giustarini, D., & Milzani, A. (2006). Biomarkers of oxidative damage in human disease. *Clinical Chemistry, 52*(4), 601–623.
29. England, K., O'Driscoll, C., & Cotter, T. G. (2004). Carbonylation of glycolytic proteins is a key response to drug-induced oxidative stress and apoptosis. *Cell Death and Differentiation, 11*(3), 252–260.
30. Marnett, L. J. (2000). Oxyradicals and DNA damage. *Carcinogenesis, 21*(3), 361–370.
31. Zhang, X., Wu, X., Hu, Q., Wu, J., Wang, G., Hong, Z., et al. (2019). Mitochondrial DNA in liver inflammation and oxidative stress. *Life Sciences, 236.*
32. Morgan, M. J., & Liu, Z. G. (2011). Crosstalk of reactive oxygen species and NF-κB signaling. *Cell Research, 21*(1), 103–115. Available from: https://doi.org/10.1038/cr.2010.178
33. Steven, S., Frenis, K., Oelze, M., Kalinovic, S., Kuntic, M., Jimenez, M. T. B., et al. (2019). Vascular inflammation and oxidative stress: Major triggers for cardiovascular disease. *Oxidative Medicine and Cell Longevity, 2019.*
34. Lei, X. G., Zhu, J. H., Cheng, W. H., Bao, Y., Ho, Y. S., Reddi, A. R., et al. (2015). Paradoxical roles of antioxidant enzymes: Basic mechanisms and health implications. *Physiological Reviews, 96*(1), 307–364.
35. Sinha, K., Das, J., Pal, P. B., & Sil, P. C. (2013). Oxidative stress: The mitochondria-dependent and mitochondria-independent pathways of apoptosis. *Archives of Toxicology, 87*(7), 1157–1180.
36. Filomeni, G., De Zio, D., & Cecconi, F. (2015). Oxidative stress and autophagy: The clash between damage and metabolic needs. *Cell Death & Differentiation [Internet], 22*(3):377–388. https://doi.org/10.1038/cdd.2014.150
37. Fulda, S. (2016). Regulation of necroptosis signaling and cell death by reactive oxygen species. *Biological Chemistry, 397*(7), 657–660.
38. Dixon, S. J., Lemberg, K. M., Lamprecht, M. R., Skouta, R., Zaitsev, E. M., Gleason, C. E., et al. (2012). Ferroptosis: An iron-dependent form of nonapoptotic cell death. *Cell, 149*(5), 1060–1072. Available from: https://doi.org/10.1016/j.cell.2012.03.042
39. Kim, Y. J., Kim, E. H., & Hahm, K. B. (2012). Oxidative stress in inflammation-based gastrointestinal tract diseases: Challenges and opportunities. *Journal of Gastroenterology and Hepatology, 27*(6), 1004–1010.
40. Caliri, A. W., Tommasi, S., & Besaratinia, A. (2021). Relationships among smoking, oxidative stress, inflammation, macromolecular damage, and cancer. *Mutation Research/Reviews in Mutation Research, 787*, 108365. Available from: https://doi.org/10.1016/j.mrrev.2021.108365
41. Kumar, K., Sonal, Singh, N., Singh, D., Jaggi, A. S., Singh, J., et al. (2021). Role of oxidative stress and inflammation in complications of diabetes. *Diabetes and Diabetic Complications: Current Status and Future Prospectives*, 259–287.
42. Kuang, F., Liu, J., Tang, D., & Kang, R. (2020). Oxidative damage and antioxidant defense in ferroptosis. *Frontiers in Cell and Developmental Biology, 8*(September), 1–10.
43. Scicchitano, P., Cortese, F., Gesualdo, M., De Palo, M., Massari, F., Giordano, P., et al. (2019). The role of endothelial dysfunction and oxidative stress in cerebrovascular diseases. *Free Radical Research, 53*(6), 579–595. Available from: https://doi.org/10.1080/10715762.2019.1620939

44. Lassègue, B., Sorescu, D., Szöcs, K., Yin, Q. Q., Akers, M., Zhang, Y., et al. (2001). Novel gp91phox homologues in vascular smooth muscle cells: Nox1 mediates angiotensin II-induced superoxide formation and redox-sensitive signaling pathways. *Circulation Research, 88*(9), 888–894.
45. Ellmark, S. H. M., Dusting, G. J., Ng Tang Fui, M., Guzzo-Pernell, N., & Drummond, G. R. (2006). The contribution of Nox4 to NADPH oxidase activity in mouse vascular smooth muscle. *Cardiovascular Research, 65*(2), 495–504.
46. Görlach, A., Brandes, R. P., Nguyen, K., Amidi, M., Dehghani, F., & Busse, R. (2000). A gp91phox containing NADPH oxidase selectively expressed in endothelial cells is a major source of oxygen radical generation in the arterial wall. *Circulation Research, 87*(1), 26–32.
47. Ago, T., Kitazono, T., Ooboshi, H., Iyama, T., Han, Y. H., Takada, J., et al. (2004). Nox4 as the major catalytic component of an endothelial NAD(P)H oxidase. *Circulation, 109*(2), 227–233.
48. Griendling, K. K., & FitzGerald, G. A. (2003). Oxidative stress and cardiovascular injury: Part II: Animal and human studies. *Circulation, 108*(17), 2034–2040.
49. Jalife, J., & Kaur, K. (2015). Atrial remodeling, fibrosis, and atrial fibrillation. *Trends in Cardiovascular Medicine, 25*(6), 475–484. Available from: https://doi.org/10.1016/j.tcm.2014.12.015
50. Samman Tahhan, A., Sandesara, P. B., Hayek, S. S., Alkhoder, A., Chivukula, K., Hammadah, M., et al. (2017). Association between oxidative stress and atrial fibrillation. *Heart Rhythm, 14*(12), 1849–1855. Available from: https://doi.org/10.1016/j.hrthm.2017.07.028
51. Ahmad, K. A., Yuan Yuan, D., Nawaz, W., Ze, H., Zhuo, C. X., Talal, B., et al. (2017). Antioxidant therapy for management of oxidative stress induced hypertension. *Free Radical Research, 51*(4), 428–438.
52. Hausenloy, D. J., & Yellon, D. M. (2013). Myocardial ischemia-reperfusion injury: A neglected therapeutic target. *The Journal of Clinical Investigation, 123*(1), 92–100.
53. Eltzschig, H. K., & Eckle, T. (2011). Ischemia and reperfusion-from mechanism to translation. *Nature Medicine, 17*(11), 1391–1401. Available from: https://doi.org/10.1038/nm.2507
54. Binder, A., Ali, A., Chawla, R., Aziz, H. A., Abbate, A., & Jovin, I. S. (2015). Myocardial protection from ischemia-reperfusion injury post coronary revascularization. *Expert Review of Cardiovascular Therapy, 13*(9), 1045–1057.
55. Guzy, R. D., & Schumacker, P. T. (2006). Oxygen sensing by mitochondria at complex III: The paradox of increased reactive oxygen species during hypoxia. *Experimental Physiology, 91*(5), 807–819.
56. Kahles, T., & Brandes, R. P. (2013). Which NADPH oxidase isoform is relevant for ischemic stroke? The case for Nox 2. *Antioxidants & Redox Signaling, 18*(12), 1400–1417.
57. Berry, C. E., & Hare, J. M. (2004). Xanthine oxidoreductase and cardiovascular disease: Molecular mechanisms and pathophysiological implications. *Journal of Physiology, 555*(3), 589–606.
58. Li, H. S., Zhou, Y. N., Li, L., Li, S. F., Long, D., Chen, X. L., et al. (2018). HIF-1α protects against oxidative stress by directly targeting mitochondria. *Redox Biology, 25*.
59. McGarry, T., Biniecka, M., Veale, D. J., & Fearon, U. (2018). Hypoxia, oxidative stress and inflammation. *Free Radical Biology and Medicine, 125*, 15–24. Available from: https://doi.org/10.1016/j.freeradbiomed.2018.03.042
60. Park, A. M., & Suzuki, Y. J. (2007). Effects of intermittent hypoxia on oxidative stress-induced myocardial damage in mice. *Journal of Applied Physiology, 102*(5), 1806–1814.
61. Farías, J. G., Herrera, E. A., Carrasco-Pozo, C., Sotomayor-Zárate, R., Cruz, G., Morales, P., et al. (2016). Pharmacological models and approaches for pathophysiological conditions associated with hypoxia and oxidative stress. *Pharmacology & Therapeutics, 158*, 1–23. Available from: https://doi.org/10.1016/j.pharmthera.2015.11.006
62. Wu, Q., Wu, W., & Kuca, K. (2020). From hypoxia and hypoxia-inducible factors (HIF) to oxidative stress: A new understanding of the toxic mechanism of mycotoxins. *Food and Chemical Toxicology, 135*, 110968. Available from: https://doi.org/10.1016/j.fct.2019.110968
63. Al-Kharashi, A. S. (2018). Role of oxidative stress, inflammation, hypoxia and angiogenesis in the development of diabetic retinopathy. *Saudi Journal of Ophthalmology, 32*(4), 318–323. Available from: https://doi.org/10.1016/j.sjopt.2018.05.002

64. Kim, R. J., An, S. H., Gwark, J. Y., & Park, H. B. (2021). Antioxidant effects on hypoxia-induced oxidative stress and apoptosis in rat rotator cuff fibroblasts. *European Cells and Materials, 41*, 680–693.
65. Rodrigo, R., Prat, H., Passalacqua, W., Araya, J., Guichard, C., & Bächler, J. P. (2007). Relationship between oxidative stress and essential hypertension. *Hypertension Research, 30*(12), 1159–1167.
66. Konukoglu, D., & Hafize, U. (2012). Athlete's heart and left heart disease cesare. *Panminerva Medica, 54*(1), 39–44.
67. Dinh, Q. N., Drummond, G. R., Sobey, C. G., & Chrissobolis, S. (2014). Roles of inflammation, oxidative stress, and vascular dysfunction in hypertension. *Biomed Research International, 2014*.
68. Ren, X., Wang, X., Yuan, M., Tian, C., Li, H., Yang, X., et al. (2018). Mechanisms and treatments of oxidative stress in atrial fibrillation. *Current Pharmaceutical Design, 24*(26), 3062–3071.
69. Kalani, R., Judge, S., Carter, C., Pahor, M., & Leeuwenburgh, C. (2006). Effects of caloric restriction and exercise on age-related, chronic inflammation assessed by C-reactive protein and interleukin-6. *The Journals of Gerontology Series A: Biological Sciences and Medical Sciences, 61*(3), 211–217.
70. Morita, N., Lee, J. H., Xie, Y., Sovari, A., Qu, Z., Weiss, J. N., et al. (2011). Suppression of re-entrant and multifocal ventricular fibrillation by the late sodium current blocker ranolazine. *Journal of the American College of Cardiology, 57*(3), 366–375.
71. Anzai, K., Ogawa, K., Kuniyasu, A., Ozawa, T., Yamamoto, H., & Nakayama, H. (1998). Effects of hydroxyl radical and sulfhydryl reagents on the open probability of the purified cardiac ryanodine receptor channel incorporated into planar lipid bilayers. *Biochemical and Biophysical Research Communications, 249*(3), 938–942.
72. Yoo, S., Aistrup, G., Shiferaw, Y., Ng, J., Mohler, P. J., Hund, T. J., et al. (2018). Oxidative stress creates a unique, CaMKII-mediated substrate for atrial fibrillation in heart failure. *JCI Insight, 3*(21), 1–22.
73. Adhami, M., Martin, N. K., Maguire, C., Courtenay, A. J., Donnelly, R. F., Domínguez-Robles, J., et al. (2023). Drug loaded implantable devices to treat cardiovascular disease. *Expert Opinion on Drug Delivery, 20*(4), 507–522. Available from: https://doi.org/10.1080/17425247.2023.2190580
74. Picardo, M., & Dell'Anna, M. L. (2010). Oxidative stress. *Vitiligo*, 231–237.
75. Generoso, G., & Bittencourt, M. S. (2017). Vitamin A: An enhanced vision of the relationship between apolipoproteins and cardiovascular risk? *Atherosclerosis, 265*, 256–257. Available from: https://doi.org/10.1016/j.atherosclerosis.2017.08.020
76. Farías, J. G., Molina, V. M., Carrasco, R. A., Zepeda, A. B., Figueroa, E., Letelier, P., et al. (2017). Antioxidant therapeutic strategies for cardiovascular conditions associated with oxidative stress. *Nutrients, 9*(9), 1–23.
77. Cuadrado, A., Manda, G., Hassan, A., Alcaraz, M. J., Barbas, C., Daiber, A., et al. (2018). Transcription factor NRF2 as a therapeutic target for chronic diseases: A systems medicine approach. *Pharmacological Reviews, 70*(2), 348–383.
78. Meili-Butz, S., Niermann, T., Fasler-Kan, E., Barbosa, V., Butz, N., John, D., et al. (2008). Dimethyl fumarate, a small molecule drug for psoriasis, inhibits Nuclear Factor-κB and reduces myocardial infarct size in rats. *European Journal of Pharmacology, 586*(1–3), 251–258.
79. Okafor, O. N., Farrington, K., & Gorog, D. A. (2017). Allopurinol as a therapeutic option in cardiovascular disease. *Pharmacology & Therapeutics, 172*, 139–150. Available from: https://doi.org/10.1016/j.pharmthera.2016.12.004
80. Agarwal, V., Hans, N., & Messerli, F. H. (2013). Effect of Allopurinol on blood pressure: A systematic review and meta-analysis. *The Journal of Clinical Hypertension, 15*(6), 435–442.
81. Altenhöfer, S., Radermacher, K. A., Kleikers, P. W. M., Wingler, K., & Schmidt, H. H. H. W. (2015). Evolution of NADPH oxidase inhibitors: Selectivity and mechanisms for target engagement. *Antioxidants Redox Signaling, 23*(5), 406–427.
82. Yu, L., Yang, G., Zhang, X., Wang, P., Weng, X., Yang, Y., et al. (2018). Megakaryocytic leukemia 1 bridges epigenetic activation of NADPH oxidase in macrophages to cardiac ischemia-reperfusion injury. *Circulation, 138*(24), 2820–2836.

83. Sun, H., Gusdon, A. M., & Qu, S. (2016). Effects of melatonin on cardiovascular diseases: Progress in the past year. *Current Opinion in Lipidology, 27*(4), 408–413.
84. Tsai, C. M., Kuo, H. C., Hsu, C. N., Huang, L. T., & Tain, Y. L. (2014). Metformin reduces asymmetric dimethylarginine and prevents hypertension in spontaneously hypertensive rats. *Translational Research, 164*(6), 452–459. Available from: https://doi.org/10.1016/j.trsl.2014.07.005
85. Guan, Y., Song, X., Sun, W., Wang, Y., & Liu, B. (2019). Effect of hypoxia-induced MicroRNA-210 expression on cardiovascular disease and the underlying mechanism. *Oxidative Medicine and Cell Longevity, 2019.*
86. Wang, L., Jia, Q., Xinnong, C., Xie, Y., Yang, Y., Zhang, A., et al. (2019). Role of cardiac progenitor cell-derived exosome-mediated microRNA-210 in cardiovascular disease. *Journal of Cellular and Molecular Medicine, 23*(11), 7124–7131.
87. Karakas, M., Schulte, C., Appelbaum, S., Ojeda, F., Lackner, K. J., Münzel, T., et al. (2017). Circulating microRNAs strongly predict cardiovascular death in patients with coronary artery disease-results from the large AtheroGene study. *European Heart Journal, 38*(7), 516–523.
88. Polyakova, E. A., Zaraiskii, M. I., Mikhaylov, E. N., Baranova, E. I., Galagudza, M. M., & Shlyakhto, E. V. (2021). Association of myocardial and serum miRNA expression patterns with the presence and extent of coronary artery disease: A cross-sectional study. *International Journal of Cardiology, 322*, 9–15. Available from: https://doi.org/10.1016/j.ijcard.2020.08.043
89. Lee, D. Y., & Chiu, J. J. (2019). Atherosclerosis and flow: Roles of epigenetic modulation in vascular endothelium. *Journal of Biomedical Science, 26*(1), 1–17.
90. McMahan, S., Taylor, A., Copeland, K. M., Pan, Z., Liao, J., & Hong, Y. (2020). Current advances in biodegradable synthetic polymer based cardiac patches. *Journal of Biomedical Materials Research Part A, 108*(4), 972–983.
91. Qiao, L., Hu, S., Liu, S., Zhang, H., Ma, H., Huang, K., et al. (2019). MicroRNA-21-5p dysregulation in exosomes derived from heart failure patients impairs regenerative potential. *The Journal of Clinical Investigation, 129*(6), 2237–2250.
92. Sharma, G. N., Gupta, G., & Sharma, P. (2018). A comprehensive review of free radicals, antioxidants, and their relationship with human ailments. *Critical Reviews in Eukaryotic Gene Expression, 28*(2), 139–154.
93. Liu, B., Lee, B. W., Nakanishi, K., Villasante, A., Williamson, R., Metz, J., et al. (2018). Cardiac recovery via extended cell-free delivery of extracellular vesicles secreted by cardiomyocytes derived from induced pluripotent stem cells. *Nature Biomedical Engineering, 2*(5), 293–303. Available from: https://doi.org/10.1038/s41551-018-0229-7
94. Hamada, T., Dubois, J. L. N., Bellamy, V., Pidial, L., Hagège, A., Pereira, M. N., et al. (2020). In vitro controlled release of extracellular vesicles for cardiac repair from poly(glycerol sebacate) acrylate-based polymers. *Acta Biomaterials, 115*, 92–103. Available from: https://doi.org/10.1016/j.actbio.2020.08.015
95. Montgomery, M., Ahadian, S., Davenport Huyer, L., Lo Rito, M., Civitarese, R. A., Vanderlaan, R. D., et al. (2017). Flexible shape-memory scaffold for minimally invasive delivery of functional tissues. *Nature Materials, 16*(10), 1038–1046.
96. Gabisonia, K., Prosdocimo, G., Aquaro, G. D., Carlucci, L., Zentilin, L., Secco, I., et al. (2019). MicroRNA therapy stimulates uncontrolled cardiac repair after myocardial infarction in pigs. *Nature, 569*(7756), 418–422.
97. Lee, D., Bae, S., Hong, D., Lim, H., Yoon, J. H., Hwang, O., et al. (2013). H2O2 - responsive molecularly engineered polymer nanoparticles as ischemia/reperfusion-targeted nanotherapeutic agents. *Science and Reports, 3*, 1–8.
98. Bittner, V. (2008). A randomized factorial trial of vitamins C and E and beta carotene in the secondary prevention of cardiovascular events in women. *Southern Medical Journal, 101*(4), 339–340.

Recent Advances in Mechanisms of In Vitro and In Vivo Evaluation of Antioxidant Activity

Kamalaharshini Mohanarangam, Asma Shereen Mohammed Yaseen, Ayesha Baje Babu, Maria Cynthiya Charles Raj, Smriti Narendra Kumar, and Johnson Retnaraj Samuel Selvan Christyraj

Abstract Antioxidants are pivotal in mitigating oxidative stress by neutralizing reactive oxygen species (ROS), thus reducing cellular damage and potentially preventing chronic diseases. Although in vitro assays are preliminary tools for assessing antioxidant activity, they are limited in replicating living organisms' complex pharmacokinetic and pharmacodynamic profiles. This chapter emphasizes the importance of in vivo testing for accurately evaluating antioxidant compounds' bioavailability, metabolism, and physiological relevance. In vivo models allow for a comprehensive analysis of pharmacokinetic parameters, including absorption, distribution, metabolism, and excretion (ADME), which are crucial in determining therapeutic efficacy. Moreover, in vivo studies reveal the influence of factors such as dose–response relationships, tissue specificity, and interindividual variability, which are often overlooked under in vitro settings. The application of antioxidants extends across fields such as pharmaceuticals, nutraceuticals, and preventive medicine, where their roles in reducing inflammation, slowing aging processes, and potentially lowering the risk of conditions like cardiovascular diseases and neurodegenerative disorders are actively explored. Incorporating in vivo studies yields crucial insights into antioxidants' practical efficacy and safety profiles, facilitating the development of clinically relevant formulations and optimized therapeutic strategies. The comparative analysis supports a translational approach that connects in vitro and in vivo findings, advancing the development of effective antioxidant interventions for disease prevention and health optimization.

Keywords Antioxidants · Oxidative stress · ROS · In vivo methods · In vitro antioxidant assay · Neurodegenerative disease · Free radicals

Kamalaharshini Mohanarangam, Asma Shereen Mohammed Yaseen, Ayesha Baje Babu these authors contributed equally.

K. Mohanarangam · A. S. Mohammed Yaseen · A. Baje Babu · M. C. Charles Raj · S. Narendra Kumar · J. R. S. Selvan Christyraj (✉)
Regeneration and Stem Cell Biology Lab, Centre for Molecular and Nanomedical Sciences, International Research Centre, Sathyabama Institute of Science and Technology, Chennai, Tamil Nadu, India
e-mail: johnnbt@sathyabama.ac.in

K. Kasinathan et al. (eds.), *Recent Advances in Oxidative Stress Associated Chronic Diseases Volume 1*, https://doi.org/10.1007/978-981-96-9670-3_7

1 Introduction

Antioxidants have been widely explored in nutrition, health, food science, and biology. Free radicals are produced in the body as a result of oxidation and are fought by some substances known as antioxidants [1, 2]. As a result, these compounds shield our body from the harmful effects of free radicals, which are produced when oxidation occurs [3]. They boost antioxidant defense and scavenge free radicals and Reactive Oxygen Species (ROS) [3]. Since free radicals and oxidants can be both harm and benefit to the organism, they perform two functions such as toxic and valuable molecules [4, 5]. Free radicals can be created by both natural in situ cell metabolism and external sources, such as pollution, cigarette smoke, radiation, and medications. Food contains antioxidants, mainly fruits, vegetables, and whole plant-based foods. Several vitamins, including C and E, are effective antioxidants. Since these molecules are well known for protecting food and pharmaceutical products from oxidative degradation and the organism from oxidative-stress-mediated disease processes, antioxidants have gained increasing attention. The therapeutic effectiveness of antioxidants, especially those derived from natural sources, is limited by their low bioavailability, which results from their quick metabolism and poor absorption by the body [6]. Evaluating antioxidant activity facilitates evaluating antioxidant capacity in both in vitro and in vivo settings to determine if the chemicals can achieve therapeutic levels in the body and target tissue/cells. While strong antioxidant activity may be observed in vitro, these results may not necessarily indicate that similar advantages will be achieved in therapeutic use (in vivo). Similarly, thorough testing is necessary to confirm that antioxidants have diverse effects on living things, particularly disease models they are meant to treat [7]. This stage ensures that the medicinal substances are adequate and appropriate for clinical settings. Since various compounds may exhibit antioxidant activity—from the well-known vitamin C to the less famous but consistently documented peptides obtained from animal or vegetable sources—tests for antioxidant activity require different methodologies beyond a complete description [8, 9]. Analyzing antioxidant activity, for example, entails assessing the rate of reaction or the potential impact of antioxidants on the rate of autoxidation of the substrate they protect. In-depth assays for biological systems and foods have been created [10]. Antioxidant activity and capacity can be measured in vitro using assays with various mechanisms, as shown in Fig. 1, including (i) DPPH (2,2-diphenyl-1-picrylhydrazyl) Assay, (ii) ABTS (2,2′-azinobis (3-ethylbenzothiazoline-6-sulfonic acid) Assay, (iii) FRAP (Ferric Reducing Antioxidant Power) Assay, (iv) ORAC (Oxygen Radical Absorbance Capacity) Assay, (v) TBARS (Thiobarbituric Acid Reactive Substances) Assay. Several tests, including (i) Superoxide Dismutase (SOD) Assay, (ii) the Glutathione Peroxidase (GPx) Assay, (iii) Glutathione (GSH) Assay, (iv) malondialdehyde (MDA), and (v)8-hydroxy-2′-deoxyguanosine (8-OHdG) Assay, are used to evaluate antioxidant capability in vivo [11, 12].

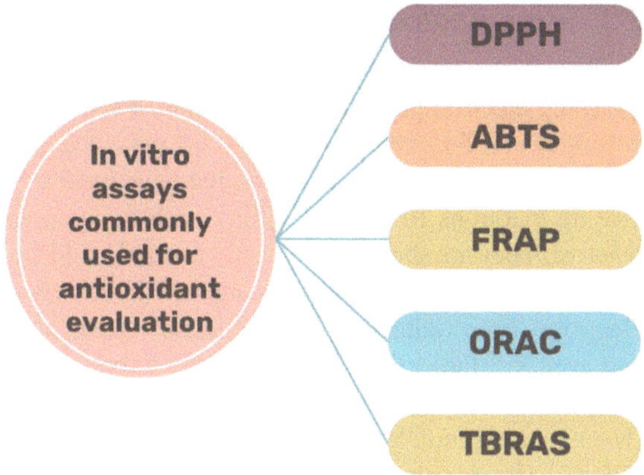

Fig. 1 Types of in vitro assays commonly employed for antioxidant evaluation

2 Principles of Antioxidant Activity

Antioxidants are compounds that scavenge free radicals to protect cells from oxidative damage. Similarly, free radical scavenging, metal ion chelation, and the enhancement of antioxidant enzymes such as catalase (CAT), glutathione peroxidase (GPx), and superoxide dismutase (SOD) are the primary mechanisms of antioxidant action. The principles of antioxidant activity include a variety of defense reactions against oxidative damage occurring in the cells. Besides altering the expression of genes linked to the antioxidant response, certain antioxidants reduce oxidative stress in the cells. Recent research emphasizes how polyphenols from plants like *Syzygium cumini* and Veronica can lower ROS and improve cellular health [13].

2.1 Oxidative Stress

Oxidative stress usually occurs when there is a disproportion between the levels of oxidants and antioxidants, which is often skewed in favor of the oxidants. This imbalance leads to molecular as well as cellular damage and disruption in redox signaling and control. Consumption of fats, carbohydrates, and saturated fatty acids, especially trans-fatty acids, in excessive amounts, triggers specific intracellular reactions that may cause oxidative stress by activating the polyol and hexosamine pathways, glyceraldehyde oxidation, protein kinase C, and superoxide generation through oxidative phosphorylation [14]. Recent research reveals that endothelial destruction, vascular failure, cardiac remodeling, renal failure, stimulation of the sympathetic nervous system, immune cell activation, and systemic inflammation that results in hypertension and heart disorders in the human body have all been linked with chronic inflammation and oxidative stress [15].

2.2 Free Radicals

By using electron paramagnetic resonance spectroscopy, Barry Commoner and his colleagues (1954) were first showed the presence of free radicals in the living organisms. Free radicals are substances that have one or more unpaired electrons and can take up electrons from surrounding molecules. Hence, they are considered highly reactive. They are short-lived, highly unstable, and are considered unselectively reactive to the nearby molecules because of their high electrophilicity due to unpaired electrons. Free radicals can significantly affect the homeostasis of cells, or they can even kill the cell if they are highly expressed in selective tissues or cells. For example, certain studies have shown that a type of ROS, that is, hydrogen peroxide (H_2O_2), can stimulate cell proliferation by activating the Cyclin-dependent kinase 2 (CDK2) protein (plays a significant role in cell division). This discovery is important for studying cancer, since free radicals promote uncontrolled tumor cell proliferation [16].

2.3 Reactive Oxygen Species

Reactive oxygen species play a vital role in cellular proliferation, differentiation, migration, apoptosis, and cell death. Reactive Oxygen Species (ROS) and Reactive Nitrogen Species (RNS) are needed in low to medium quantities to maintain important physiological functions and regulate crucial transcription factors. Similarly, disrupted redox homeostasis is caused by excess formation of ROS, which results in oxidative stress and ROS (reactive oxygen species)-mediated damage to all vital biomolecules, including proteins, DNA, and lipids. ROS are produced as a by-product of either cell signal pathways or cell-mediated immunity mechanisms, and they originate from two primary sources: oxidative phosphorylation and action of cellular response to xenobiotics, bacteria, and cytokines. Exogenous small molecular antioxidants scavenge ROS, such as hydroxyl radicals, peroxynitrite, and other radicals that react quickly with protein, membrane lipids, and DNA [17].

2.4 Mechanism of Antioxidant Capacity

A molecule that lowers in vitro does not always act as an antioxidant in an in vivo system. This is because of free radical's ability to diffuse and spread quickly. It could be difficult for the antioxidants to exist at the movement and place where oxidative damage is produced, because some have a concise life span. On top of that, the reactions occurring between antioxidants and free radicals fall under second-order responses. Still, they also depend on factors related to the chemical makeup

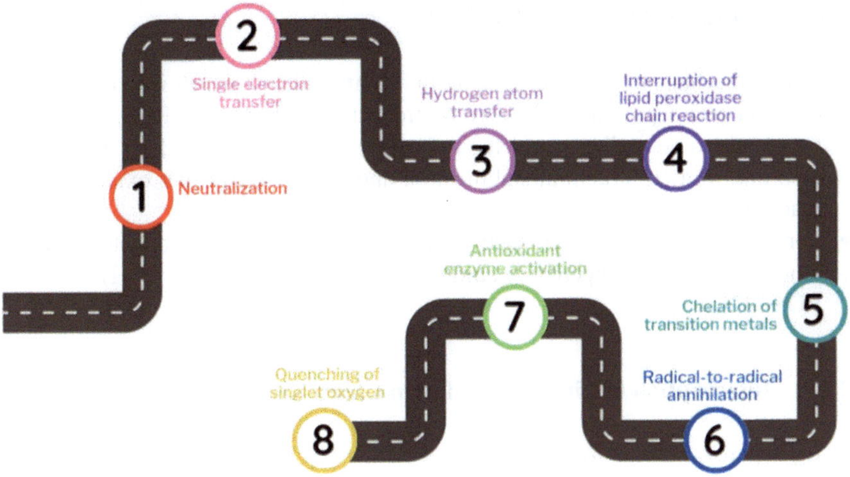

Fig. 2 Road map of antioxidant capacity

of both reagents, the reaction conditions, and the medium, besides depending on the concentration of free radicals and antioxidants as shown in Fig. 2 [18].

2.4.1 Neutralization of Free Radicals

- Free radicals receive electrons from antioxidants as they have an unpaired electron or hydrogen atom. These antioxidants transform free radicals into less reactive moieties [19].
- For example, superoxide radicals (O_2^-) and hydroxyl radicals (·OH), which are two strongly reactive molecules, receive electrons from Vitamin C (ascorbic acid).

2.4.2 Single Electron Transfer (SET)

- In this procedure, a single electron is donated to free radicals by antioxidants, which not only neutralize the radicals into harmless substances but stabilize them. Rather than becoming reactive, antioxidants scavenge ROS and free radicals by reducing them.
- For example, during lipid peroxidation, lipid radicals are formed. Vitamin C (ascorbic acid) donates one of its electrons to lipid radicals, simultaneously preventing the spread of lipid radicals in the cell/tissue [20].

2.4.3 Hydrogen Atom Transfer (HAT)

- In this mechanism, the hydrogen atom of antioxidants, which is made up of an electron and a proton, is donated to free radicals and other ROS. This donation leads to neutralization of the radical by fulfilling the need for an electron, resulting in their structural stabilization and preventing further oxidative chain reactions [21].
- For example, polyphenols such as quercetin and catechins donate hydrogen atoms to neutralize superoxide anions (O_2^-), hydroxyl radicals (·OH), and other ROS [22].

2.4.4 Interruption of Lipid Peroxidation Chain Reaction

- Lipid peroxidation is a chain of reactions started by free radicals, especially in cell membranes where oxidation of polyunsaturated fatty acids occurs. Antioxidants stop the chain reaction by preventing further oxidation of fatty acids by donating an electron or hydrogen atom to lipid radicals [23, 24].
- For example, β-carotene, a type of carotenoid, acts as a potent scavenger of singlet oxygen ($^1O_2$) and can suppress lipid peroxidation by donating one hydrogen atom to lipid radicals, breaking the oxidative damage chain [25].

2.4.5 Chelation of Transition Metals

- Copper (Cu^{2+}) and iron (Fe^{2+}) are well-known transition metals that catalyze the production of ROS through the Fenton reaction, in which Highly reactive hydroxyl radicals (·OH) are formed from hydrogen peroxide (H_2O_2).
- Antioxidants such as flavonoids and phenolic acids get bound to metal ions, which inhibit the metal's capacity to facilitate the formation of free radicals by forming stable complexes [26, 27].
- For example, a type of flavonoid found in green tea, that is, catechins, lowers the production of ROS and the associated oxidative-stress-chelating metal ions such as Fe^{2+}.

2.4.6 Radical–Radical Annihilation

- In this process of radical–radical annihilation, an antioxidant, after reacting with a free radical, continues to respond with another radical, neutralizing both radicals.
- In this way, certain antioxidants undergo radical–radical annihilation by reacting with free radicals.
- Vitamin C gets converted into an ascorbyl radical after being given an electron. Further, it combines with another ascorbyl radical, leading to the formation of a stable molecule [28].

- For example, Vitamin C oxidizes into ascorbyl radicals after reacting with hydroxyl and superoxide radicals by radical–radical annihilation.

2.4.7 Antioxidant Enzyme Activation

- Enzymes including catalase (CAT), Superoxide Dismutase (SOD), and Glutathione peroxidase (GPx) transform reactive oxygen species such as Hydrogen Peroxide (H_2O_2) and superoxide anion (O_2^-) into not so harmful molecules such as water (H_2O) and oxygen (O_2).
- Antioxidants, such as phenols, stimulate the transcription of antioxidant response elements (AREs). This leads to a rise in the production of these protective enzymes.
- For example, in the Nrf2 pathway, a compound that is available in cruciferous vegetables, that is, sulforaphane, improves the antioxidant activity of the cell by increasing the expression of Superoxide Dismutase (SOD) as well as catalase (CAT) [29].

2.4.8 Quenching of Singlet Oxygen ($^1O_2$)

- Singlet oxygen, which is known as the highly reactive form of oxygen, is inactivated by some antioxidants, such as carotenoids, which have powerful quenching properties that function by exploiting the energy of singlet oxygen by reducing it to a lower energy stage without becoming reactive themselves [30].
- Carotenoids having conjugated double-bond structures prevent singlet oxygen from reacting with other molecules by effectively exploiting its energy.
- For example, lycopene in tomatoes guards' cells from oxidative damage by absorbing the energy of singlet oxygen species (*quenching*) [31].

2.5 Categories of Antioxidants

Antioxidants are primarily classified into two types:

i. Enzymatic antioxidants
ii. Nonenzymatic antioxidants.

2.5.1 Enzymatic Antioxidant

Enzymatic antioxidants are vital molecules that guard cells from oxidative damage, scavenge free radicals, and play a role in cellular defense mechanisms. These antioxidants break down free radicals and peroxide species into less reactive molecules. Some of the enzymatic antioxidants include: (a) Superoxide Dismutase (SOD)—the dissociation of superoxide anion (O_2^-) into oxygen (O_2) and Hydrogen Peroxide

(H_2O_2) is catalyzed by the enzyme Superoxide Dismutase (SOD) [27, 32]. Hydrogen peroxide is further detoxified by catalase and glutathione peroxidase. (b) Catalase (CAT)—hydrogen peroxide is dissociated into oxygen (O_2) and water (H_2O) by the enzyme catalase. The enzyme catalase prevents oxidative damage in tissues. (c) Glutathione Peroxidase (GPx)—By using glutathione (GSH) as a coenzyme, glutathione peroxidase breaks down organic and hydrogen peroxide into alcohol and water. In this process, glutathione (GSH) is oxidized into glutathione disulfide (GSSG), but it is regenerated by Glutathione Reductase (GR). (d) Glutathione Reductase (GR)—glutathione reductase maintains overall redox homeostasis and the activity of GPx. The enzyme glutathione reductase regenerates glutathione (GSH) from its oxidative form. (e) Thioredoxin (Trx) and Thioredoxin Reductase (TrxR)—the redox balance of the cell is maintained by Thioredoxin (Trx) by combining with Thioredoxin Reductase (TrxR), which allows the transfer of electrons, which facilitates protein disulfide reduction. (f) Peroxiredoxins (Prx)—peroxidase enzymes such as Peroxiredoxins (Prx) dissociate lipid hydroperoxides, peroxynitrite, and hydrogen peroxides. This enzyme has unique properties, because it regenerates through the redox cycle, making it resistant to oxidative stress.

2.5.2 Nonenzymatic Antioxidants

Nonenzymatic antioxidants are those substances that do not have any enzymatic activity. They are always present in external sources such as fruits, vegetables, etc. These antioxidants do not catalyze radicals; instead, they donate a hydrogen atom or electron, suppressing the activity of highly reactive molecules. Vitamins such as vitamin C, which is water-soluble, neutralize ROS settled in extracellular spaces and generate other nonenzymatic antioxidants like vitamin E. Vitamin E, which is lipid-soluble and known for its protective function, guards the cell membrane from lipid peroxidation. Glutathione is a substrate for the glutathione peroxidase enzyme, a tripeptide that directly neutralizes ROS. Glutathione also plays a key role in certain processes, like regulating cell death and proliferation, detoxification, and immune response. Melatonin improves the function of enzymatic antioxidants like catalase, glutathione peroxidase, and superoxide dismutase. They also instantly scavenge reactive oxygen species (ROS) and reactive nitrogen species (RNS). Polyphenols inhibit the Fenton reaction that generates ROS. Polyphenols like flavonoids chelate metal ions or donate hydrogen atoms to neutralize radicals. Certain well-known polyphenols are resveratrol, catechins, and quercetin. Carotenoids (e.g., beta-carotene, lycopene, and lutein) protect the cell membrane and cellular lipids by transferring electrons or quenching the singlet oxygen. Oxidative damage is reduced by absorbing the energy of singlet oxygen. Carotenoids protect the skin from the sun's UV radiation, possess anticancer properties, and provide eye health. Ubiquinone or coenzyme Q10, is a lipid-soluble coenzyme, aids in the electron transport chain inside the mitochondria, and can recycle vitamin E. This antioxidant is involved in energy production and guards mitochondria from oxidative stress.

3 In Vitro Evaluation of Antioxidant Activity

Under in vitro conditions, the evaluation of antioxidant activity is considered a vital tool in research that facilitates scientists to evaluate both the synthetic and organic compounds, allowing them to determine the capacity of the antioxidants to maintain oxidative stress. Several assays for evaluating antioxidant capacity under in vitro conditions have been developed recently. The below-mentioned assays are some of the most popularly used in vitro assays for antioxidant evaluation.

3.1 DPPH(2,2-Diphenyl-1-Picrylhydrazyl) Assay

DPPH technique is the most used assay under in vitro conditions. It is one of the effective methods for evaluating antioxidant activity (Fig. 3). Sample preparation: the plant material containing antioxidants from any plant source is made into fine powder. Then, the powdered plant material is dissolved in either water or methanol, and a 1 mg/mL solution is prepared. The prepared samples are diluted to stop the decolorization upon reacting with DPPH. Preparation of DPPH solution: DPPH powder (*approximately* 3.7 mg) is dissolved in 25 mL of ethanol or methanol, and a 375 μM solution is obtained. The absorbance rate of the DPPH solution must be around 1.0 at 515–518 nm to confirm the accuracy rate. As DPPH solution has a short shelf life, the solution must be frequently prepared, and the storage area must be dark. Incubation: 1 mL of DPPH solution and 1 mL of sample solution are mixed thoroughly. The mixture is incubated in the dark at room temperature for about 30–40 min so that the reaction occurs effectively under certain conditions. Measurement of absorbance rate using spectroscopy: by using spectroscopy, the absorbance rate of the mixture is evaluated at 517 nm [33]. By comparing it to DPPH (*control solution*) free of antioxidants, the percentage of DPPH scavenged is calculated. The formula is given below:

$$\text{DPPH scavenged}\,(\%) = \left[\frac{\text{Absorbance of Control} - \text{Absorbance of Sample}}{\text{Absorbance of Control}} \times 100\right]$$

Calculation of IC50—the DPPH scavenging percentage is plotted against the concentration of antioxidants to determine the IC50 (the required concentration to reduce DPPH by 50%). A water-soluble vitamin E known as Trolox is frequently used as a standard for comparison.

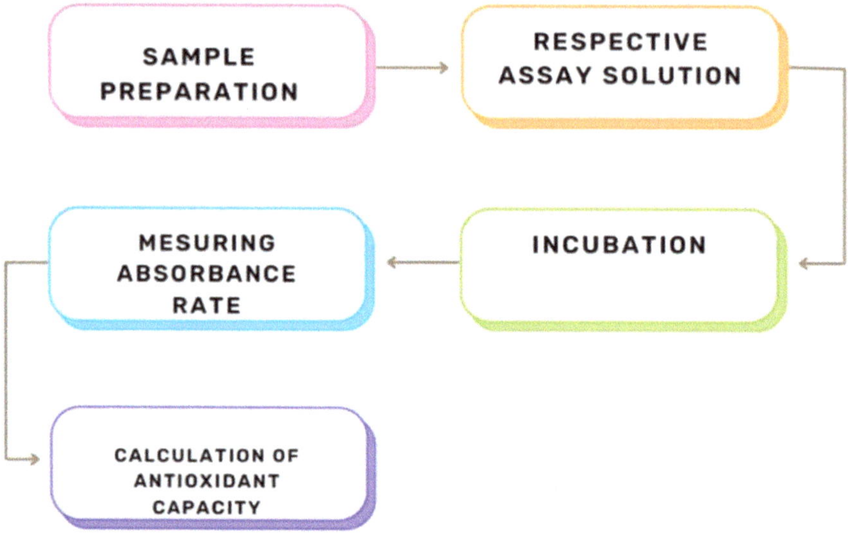

Fig. 3 Steps involved in measuring antioxidant capacity in assays

3.2 ABTS (2,2′-Azinobis(3-Ethylbenzothiazoline-6-Sulfonic Acid)) Assay

In this technique, antioxidants react with the ABTS radical cation ($ABTS^{+}$), a stable and water-soluble radical. As a result, ABTS is reduced by antioxidants, which leads to the loss of blue-green color, and the reaction is measured by using spectrophotometry. Both the changes in absorbance and standard antioxidant (Trolox) are compared, and the results of this experiment are expressed as Trolox Equivalent Antioxidant Capacity (TEAC) values (Fig. 3).

Principle of the ABTS assay ABTS radical cation ($ABTS^{\cdot+}$), that is, a blue-green chromophore, is generated in the ABTS assay, which absorbs light at a wavelength of 734 nm. Potassium persulfate or hydrogen peroxide is an oxidizing agent that oxidizes ABTS(2,2′-azino-bis(3-ethylbenzothiazoline-6-sulfonic acid) to generate positive radical ion. When an antioxidant is added, it donates its hydrogen atom or electron to reduce $ABTS^{\cdot+}$, which eventually reduces the intensity of the blue-green color. The intensity of the color is measured using spectrophotometry, and antioxidant capacity corresponds to the reduction of color. Methodology: specific crucial steps are involved in preparing and measuring antioxidant capacity in an ABTS assay. Precise sample handling and reagent preparation is important (Fig. 3). Preparation of $ABTS{\cdot}^{+}$ Solution: To prepare the solution, an oxidizing agent, mostly ammonium persulfate (APS), is added to ABTS (2,2′-azino-bis(3-ethylbenzothiazoline-6-sulfonic acid)) and left for about 12–16 h in the dark. A stable ABTS radical is produced during the reaction at a characteristic absorbance of 734 nm. Setting up of assay: the radical is diluted for assay consistency at an absorbance rate of 0.700 at 734 nm. A reference standard is taken. Usually, Trolox, an analogy of vitamin E, is used. To create a calibration curve, different concentrations of Trolox solution are prepared, and the capacity of antioxidants in test samples is evaluated by comparing color absorbance at 734 nm. Procedure: mixing of samples diluted with $ABTS^{\cdot+}$ solution and sample solution or Trolox (approximately 1 mL) are added in equal volumes in a cuvette or microplate. For example, 20 μL of sample or standard is added to 200 μL of $ABTS^{\cdot+}$ solution. Incubation: the mixture is incubated at room temperature for about 5–6 min to facilitate the reaction between $ABTS^{\cdot+}$ and antioxidants. Based on the reactivity of the antioxidants, the reaction time is changed. Absorbance is measured using a microplate reader or spectrophotometer at 734 nm after incubation. Upon comparing the incubated solution with the control (solution of $ABTS^{\cdot+}$ having no antioxidants), there is a decrease in absorbance rate in the incubated solution, which shows antioxidant activity. The percentage of ABTS inhibition is calculated based on the reduction of absorbance rate. By using the below formula, antioxidant capacity in the samples can be calculated:

$$\text{Inhibition } (\%) = \left[\frac{\text{Absorbance of Control} - \text{Absorbance of Sample}}{\text{Absorbance of Control}} \times 100\right]$$

Convert this inhibition percentage to Trolox equivalents (TEAC) by plotting against the Trolox standard curve for comparative purposes. This comparison provides a standardized measure of antioxidant capacity in Trolox equivalent units (μM TE/mg).

3.3 FRAP (Ferric Reducing Antioxidant Power) Assay

This method measures the antioxidant capacity of the sample in which ferric (Fe^{3+}) is reduced to ferrous (Fe^{2+}) ions [34]. This reduction results in forming a colored complex known as ferrous-tripyridyltriazine (Fe-TPTZ), which is absorbed at 593 nm. The antioxidant power of the sample is in proportion to the color intensity. This method involves evaluating the sample's antioxidant activity and comparing the sample's standard and absorbance. A standard curve is also made using the known antioxidant concentrations (*usually ascorbic acid or trolox*). FRAP assay works on the principle of transfer of electrons, which is a simple and cost-effective method. The main principle involved is the reduction of ferric (Fe^{3+}) to ferrous (Fe^{2+}) under acidic conditions in the presence of 2,4,6-tripyridyl-s-triazine (TPTZ) at 593 nm, where the colorless complex is turned blue, which is measured spectrophotometrically. The number of antioxidants in the sample is proportional to the amount of a blue complex formed. The FRAP assay is less effective for lipid-soluble antioxidants, whereas it can be very effective for antioxidants that are soluble in water. Pairing FRAP with other assays, such as DPPH, can give more accurate results. The FRAP (ferric reducing antioxidant power) assay is used to determine the capability of the antioxidants in the sample to reduce Ferric ions (Fe^{3+}) to Ferrous ions

(Fe^{2+}) in the presence of TPTZ (2,4,6-tripyridyl-s-triazine), a colorimetric agent (Fig. 3). Preparation of FRAP reagent: 20 mM of $FeCl_3$, 0.3 M of acetate buffer with pH 3.6, and 10 mM of TPTZ in 40 mM of HCl are mixed at a ratio of 10:1:1. The prepared reagent is light-sensitive, so it should be kept in the dark and freshly prepared. Preparation of sample: A solvent is used for sample preparation depending on the solubility of antioxidants, which are present in the sample. Most antioxidants in food, biological samples, and plants dissolve in solvents like water and ethanol. Preparation of standard: at a concentration of around 1 mM, a stock solution of Trolox (*standard antioxidant*) is prepared. An analogy of Vitamin E, Trolox, is a water-soluble compound used to measure antioxidant capacity. A standard curve is created by preparing serial dilutions of Trolox solution, ranging from 10, 20, 40, 80, and 100 μM. Procedure: 2 mL of the FRAP reagent is added to 100 μL standard or sample. The mixture is incubated at room temperature for about 4–10 min, facilitating the reduction reaction and forming a blue-colored Fe^{2+}-TPTZ complex. The antioxidant capacity in the sample and the intensity of the blue color are correlated. Estimation: absorbance is measured at 593 nm by using a spectrophotometer. For calculating the antioxidant activity, a standard curve created by using Trolox or other standard antioxidants is compared with the absorbance readings of the sample. The result is expressed in terms of micromolar Trolox equivalents (μmol TE/g) per gram of sample.

$$\text{Slope of the Trolox standard curve} = \left[\frac{\text{Absorbance of sample} - \text{absorbance of blank}}{\text{Antioxidant Activity } (\mu\text{mol Trolox Equivalents})}\right]$$

3.4 ORAC (Oxygen Radical Absorbance Capacity) Assay

This method is developed to measure the antioxidant capacity of substances such as nutrients, foods, and various chemical compounds. It is a widely used method, and this assay can quantify the substance's ability to neutralize oxygen radicals (reactive species), which damage cells, DNA, and proteins and cause oxidative stress. This assay is used in food science, nutrition, and cosmetics, and it helps identify antioxidant-rich compounds and foods. Principle: This assay uses the principle of hydrogen atom transfer (HAT). This measures the antioxidant's capacity to prevent the oxidative degeneration of fluorescent probes (mostly fluorescein), which is initiated by a kind of radical generator, that is, AAPH (2,2′-azobis(2-amidinopropane) dihydrochloride). The antioxidants protect the fluorescein (fluorescent probe) by delaying or inhibiting its degradation by scavenging the free radicals. Peroxyl radicals are produced by the thermal decomposition of compounds such as AAPH (2,2′-azobis(2-amidinopropane) dihydrochloride). The ORAC assay helps to estimate the sample's antioxidant ability to scavenge peroxyl radicals. Oxidative stress is initiated by this radical source (peroxyl radical), which is then assessed by quantifying the decay of the fluorescence probe as it reacts with radicals. The fluorescence decay is delayed by the antioxidants present in the sample, facilitating the estimation of antioxidant capacity. Preparation of reagents: physiological pH is maintained at 7.4 using phosphate-buffered saline (PBS). Fluorescein is made at a concentration of around 80 nM and is mixed in phosphate buffer. The AAPH (2,2′-azobis(2-amidinopropane) dihydrochloride), a radical generator, is often prepared at around 153 mM. At 37 °C, AAPH (water-soluble azo), upon decomposition, produces peroxyl radicals, which eventually initiate the oxidation of fluorescein. To create a standard curve, Trolox, an analog of vitamin E, is dissolved in phosphate buffer with concentrations of around 6.25 to 100 μM (Fig. 3). Phosphate buffer dilutes the samples (biological fluids, food extracts, etc.) to confirm that the antioxidant capacity falls within the expected range. Samples are often filtered or centrifuged with plant or food extract to remove excess particles: 25 μL of the sample solution, buffer (for blanks), or Trolox solution, and 150 μL of the fluorescein solution are added to each well in the microplate. Further, 2,2′-azobis(2-amidinopropane) dihydrochloride (AAPH) is added to each well, which, under incubation conditions, produces peroxyl radicals, which eventually leads to the degradation of the fluorescein probe. Fluorescence intensity is reduced by the oxidation of fluorescein over time. Fluorescence decay is measured at the absorbance rate of around 485 nm at an interval of 1 min in a microplate reader. Antioxidants in the sample delay the decay, facilitating comparison to the Trolox standard. For each sample, the area under the curve (AUC) is calculated by plotting fluorescence decay. The net AUC is determined based on the difference between the blank AUC and the sample AUC, which gives antioxidant capacity. Results are expressed in terms of Trolox equivalents (TE) per gram of the sample.

$$\text{ORAC Value (TE)} = \text{Net AUC} \times \frac{\text{Trolox concentration}}{\text{AUCTrolox4}}$$

Net AUC is calculated by using the below formula,

$$\text{Net AUC} = \text{AUC sample} - \text{AUC blank}$$

The AUC is calculated using the following formula:

$$\text{AUC} = 1 + \frac{F0}{F1} + \frac{F0}{F2} + \ldots \frac{F0}{Fn}$$

3.5 TBARS (Thiobarbituric Acid Reactive Substances) Assay

When lipids break down, a by-product known as malondialdehyde (MDA) is formed. Lipid peroxidation is measured by detecting malondialdehyde using the Thiobarbituric Acid Reactive Substances (TBARS) assay. Lipid peroxidation shows oxidative stress in the biological system. This assay is specifically designed to detect Malondialdehyde (MDA) formed during lipid oxidation. The free radical formed causes cell damage and inflammation by breaking the cell membrane. A red MDA-TBA adduct is produced when malondialdehyde reacts with thiobarbituric acid (TBA) at 532 nm under heat and acidic conditions. *Principle*: TBARS works on detecting Malondialdehyde (MDA) (a by-product of lipid breakdown), which shows oxidative stress in the cell. Under acidic conditions, Thiobarbituric Acid (TBA) is heated with biological samples, which gives a red-coloured complex at 532 nm. The lipid peroxidation levels present in the samples are quantified by subjecting the MDA-TBA complex to colorimetric or fluorometric measurement. This is often used to study diseases connected to oxidative stress, such as neurodegenerative and cardiovascular diseases. This assay has limitations as it reacts with substances other than MDA, which may affect the specificity of the results. *Methodology*: By estimating Malondialdehyde (MDA) and other such compounds, lipid peroxidation levels are measured by using TBARS (Thiobarbituric Acid Reactive Substances) assay (Fig. 3). This assay can be used for many biological sources, such as cell lysate, serum, plasma, and tissue homogenates. Samples are homogenized in Tris–HCl buffer (pH 7.4) or phosphate-buffered saline (PBS). This mixture is then centrifuged to remove excess debris. Properly storing samples on ice or in a freezer is essential to maintain their shelf life or prevent lipid peroxidation. After centrifugation, the supernatant of the sample is added to the HCl solution, (TCA) trichloroacetic acid, and (TBA) thiobarbituric acid. Adding all these solutions to the supernatant precipitates the protein, and an acidic condition is created so that MDA reacts with TBA. 15% TCA, 0.375% TBA, and 0.25 M HCl are added to provide desired reaction conditions. The prepared mixture is heated for 15–30 min at 95 °C. Under this condition, MDA reacts with TBA, which eventually leads to the formation of a colored MDA-TBA complex. A glycol bath is often used for more specific detection. The solution is cooled after heating, and 1-butanol or n-butanol is added to the solution to extract the MDA-TBA complex. The sample was centrifuged to separate the aqueous and organic phases, and the colored organic phase was extracted for further analysis. By using a microplate reader or spectrophotometer, the MDA-TBA complex present in the organic phase is measured at 532 nm [35]. A standard calibration with known MDA concentrations is used to determine the MDA levels in the sample. The result is often expressed in terms of nmol/mg for protein samples and nmol/mL for plasma/serum samples. The inhibition rate of lipid peroxidation by antioxidants in samples can be determined by comparing control and treated samples.

Table 1 Comparative analysis of in vitro antioxidant assays

Assay	Mechanism	Merits	Demerits	Applications
DPPH	Utilizes electron transfer to neutralize free radicals, resulting in measurable color change	Rapid, simple, and cost-effective	Limited efficacy for lipophilic antioxidants	Screening antioxidant potential in plant extracts
ABTS	Employs ABTS radical cation reduction, producing a distinct color change	Effective for both hydrophilic and lipophilic antioxidants	Susceptible to variability due to solvent interactions	Evaluation of antioxidants in food products and beverages
FRAP	Measures the reduction of Fe^{3+} to Fe^{2+} in acidic conditions, forming a colored complex	Highly effective for hydrophilic antioxidants	Limited sensitivity for fat-soluble antioxidants	Assessment of antioxidant capacity in pharmaceuticals and nanoparticles
ORAC	Evaluates the antioxidant's ability to inhibit fluorescence decay of fluorescent probes	Highly sensitive and biologically relevant	Require specialized equipment's; time-intensive	Assessment of antioxidants in nutraceuticals and dietary supplements
TBARS	Quantifies malondialdehyde (MDA) levels as an indicator of lipid peroxidation	Effective for evaluating lipid oxidation in biological samples	Possible interference with non-MDA aldehydes	Used to assess oxidative stress in clinical and aging studies

The Formula for Calculating Antioxidant Capacity in TBARS Assay

See Table 1.

$$\text{Antioxidant Activity (\%inhibition)} = \left[\frac{\text{Absorbance of Control} - \text{Absorbance of Sample}}{\text{Absorbance of Control}} \times 100\right]$$

3.6 Advantages of In Vitro Antioxidant Assays

Compared to in vivo methods, in vitro antioxidant assays have the following advantages, listed point by point (Fig. 4). (i) In vitro assays reduce the use of animals for testing and support the 3Rs principle (Reduction, Refinement, and Replacement). (ii) Faster results are obtained than in vivo, which is very helpful in drug development. (iii) In vitro assays are cost-efficient and require less workforce and fewer resources. (iv) Organ-on-a-chip technology, which replicates the function of human organs, is supported by in vitro assays. (v) Rapid screening of molecular interactions in samples using high-throughput mass spectrometry is also supported by in vitro assays. (vi)

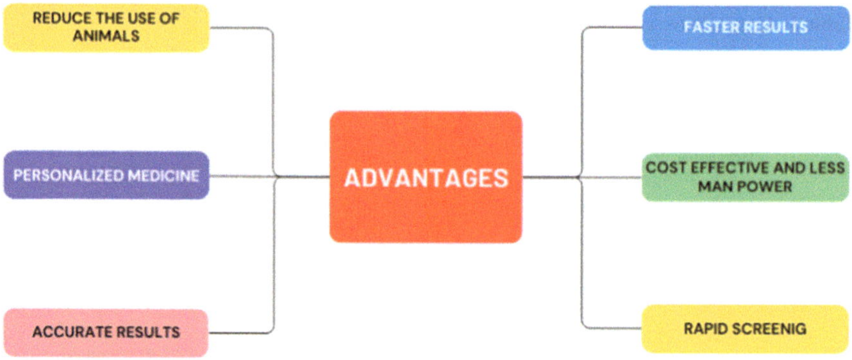

Fig. 4 Advantages of in vitro antioxidant assays

Recent developments in vitro assays are helpful for personalized medicine and for understanding complex biological mechanisms. (vii) By using this assay, the assay performer can obtain accurate results.

3.7 Limitations of In Vitro Antioxidant Assays

Compared to in vivo methods, in vitro antioxidant assays have the following limitations, listed point by point (Fig. 5). (i) They cannot mimic the complex biological process occurring in whole organisms. (ii) Organ-on-a-chip technologies, despite showing accuracy, cannot mimic the functions of the whole organ of the human. (iii) The results may vary because of the differences between assay conditions and cell lines. (iv) These assays usually require specific conditions which might not be available in the laboratories. (v) In toxicology, detecting kidney or hepatic toxicity might be difficult, as these may require a whole organ for toxic responses.

4 In Vivo Evaluation of Antioxidant Activity

4.1 Necessity of In Vivo Testing

In vivo assessment of antioxidant activity is essential for elucidating these compounds' physiological relevance and therapeutic potential. While in vitro studies provide valuable preliminary insights into the antioxidant capacity of various substances, they often cannot replicate the complexities inherent in living organisms. The utilization of appropriate animal models enables assessing the efficacy of antioxidants, their pharmacokinetics, and their safety profiles. Ultimately, integrating both in vitro and in vivo approaches will enhance the understanding of mechanisms by

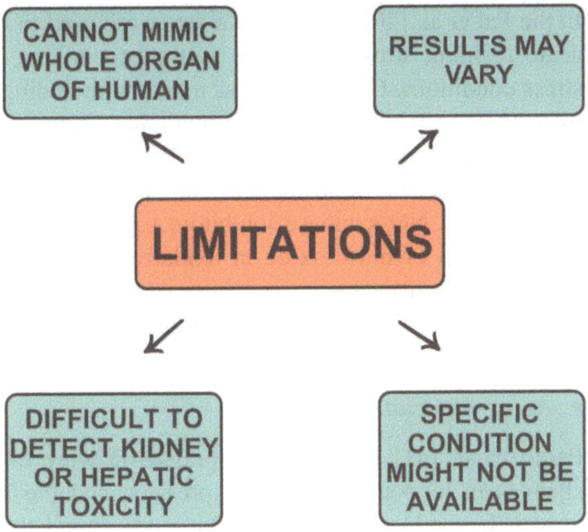

Fig. 5 Limitations of in vitro antioxidant assays

which antioxidants operate within biological systems and facilitate the development of effective interventions for oxidative stress-related diseases.

4.2 Commonly Used Animal Models

Several animal models are used to study the antioxidant effects on disease prevention, which include rodent models (mice & rats), the nematode *Caenorhabditis elegans*, *Drosophila melanogaster* (fruit flies), and *Danio rerio* (Zebrafish).

4.2.1 Rodent Models

Rodent models, particularly mice and rats, are highly effective for in vivo evaluations of antioxidants due to their physiological and genetic similarities to humans. Their well-characterized antioxidant systems and the ability to undergo genetic manipulation facilitate the detailed assessment of antioxidant activity in oxidative stress [36]. These models have been extensively employed in studies of various conditions, such as diabetes, cancer, aging, cardiovascular diseases, and neurodegenerative diseases including Alzheimer's and Parkinson's [37, 38]. The relatively short lifespan of rodents, along with the ability to control environmental variables, enhances the reproducibility of research on chronic diseases and the long-term effects of antioxidant treatments [39].

- *Oxidative-Stress-Related Diseases*

In the Pal et al. [40] study, Wistar rats were used to model oxidative stress through acute restraint stress (RSx1), a method to induce oxidative damage like human stress conditions. Oxidative stress markers, including glutathione and malondialdehyde (MDA) (showing lipid peroxidation) were significantly altered in stressed rats. Administering antioxidants like L-ascorbic acid (L-AA) and L-arginine (a NO precursor) restored these markers, showing their potential to reduce oxidative stress. Additionally, NO synthase inhibitors had mixed effects, demonstrating the intricate function of nitric oxide in oxidative pathways. This highlights the value of rodent models in evaluating oxidative stress and antioxidant therapies in vivo, which are essential for exploring therapeutic options for human oxidative stress-related diseases.

- *Aging-Associated Research*

Pardo et al. [41] used female Sprague–Dawley rats to investigate aging-related changes in hippocampal gene expression and the impacts of IGF-I (Insulin-like growth factor-1) gene therapy. The study found that IGF-I therapy in aging rats upregulated key genes involved in synaptic plasticity, neurogenesis, and memory function, such as Sypl2, NNAT, and Itga-8. It also increased miR-186, a microRNA that declines with age and is linked to Alzheimer's disease. This finding demonstrates that rodent models are crucial for aging studies, as they help identify aging-associated molecular targets and test potential therapies like IGF-I to mitigate cognitive decline.

- *Neurodegenerative Disease*

Yang et al. [42] demonstrated that long-term photo biomodulation (PBM) treatment significantly improved cognitive deficits and reduced Alzheimer's disease (AD) pathology in transgenic AD rats. PBM decreased memory loss, amyloid plaques, and tau hyperphosphorylation while preserving neuronal integrity. The mechanisms involved included regulating glial polarization, inhibiting neuroinflammation, and enhancing microglial recruitment for amyloid clearance. This study emphasizes the importance of rodent models in exploring potential therapeutic interventions for neurodegenerative diseases like AD.

4.2.2 *C. elegans* Model

The nonparasitic nematode *Caenorhabditis elegans (C.elegans)*, a free-living worm, has become an invaluable model organism for biomedical research because of its simplicity, genetic tractability, and remarkable homology with humans. *C. elegans* exhibits 60–80% genome homology with humans, sharing many physiological, anatomical, and metabolic traits [43]. It is small (around 1 mm in length and 80 μm in diameter) and has a short lifecycle of 3.5 days at 20 °C and a typical lifespan of 2–3 weeks. *C. elegans* can be easily cultivated in large numbers in the laboratory on agar plates, where it feeds on a standard strain of nonpathogenic *Escherichia coli (E. coli)* [44]. Its rapid reproduction rate allows it to work with large populations, facilitating genetic analyses and high-throughput screening for identifying hyper-virulent and attenuated strains and preliminary testing of novel pharmaceutical compounds [45]. One of the most significant attributes of *C. elegans* as a model organism is its transparency, which facilitates direct visualization of cellular structures, including neurons, through fluorescent protein tagging, especially green fluorescent protein (GFP) [46]. This characteristic is advantageous for research on neurodegenerative diseases, including Parkinson's disease (PD), as it permits in vivo observation and quantification of neuronal cell death, thereby providing a real-time indicator of neurodegeneration. Moreover, *C. elegans* is widely used in nanotoxicology research because of its multicellular structure, which enables the detection of organism-level endpoints and interactions between nanoparticles and various biological targets, allowing for a comprehensive assessment of nanoparticle biosafety [47]. The organism is highly suited for genetic analysis because of its ability to reproduce rapidly, generating approximately 300 progenies within a single reproductive cycle and using both self-fertilization and cross-fertilization [48]. Its rapid life cycle, prolific reproduction, and compatibility with molecular analyses at individual and population levels establish *C. elegans* as an invaluable platform for toxicological and genetic investigations, with applications extending to genetic toxicology, biomedicine, and beyond.

- *Studies on Lifespan and Stress Resistance*

Recent investigations have highlighted the role of antioxidants in enhancing lifespan and stress resistance in *C. elegans*. For instance, treatments with antioxidant compounds such as resveratrol and N-acetylcysteine (NAC) have been shown to extend the lifespan of *Caenorhabditis elegans* by reducing oxidative damage and activating cellular defense mechanisms. According to a study, supplementing *C. elegans* with NAC greatly extended both its mean and maximal lifespan, with a mean lifespan extension of up to 30.5% at optimal concentrations (5 mM) [49]. Similarly, resveratrol had been demonstrated to exert beneficial effects on longevity in *C. elegans* through its antioxidant properties, promoting health span by activating key longevity pathways. Activating these pathways helps mitigate oxidative stress, enhancing the organism's ability to cope with environmental challenges [50]. Research indicates that mutations in genes associated with oxidative stress response can significantly affect both lifespan and stress resistance in this nematode model. Studies have shown that *C. elegans* responds to oxidative stress through various mechanisms, including activating specific transcription factors such as SKN-1, which is vital in regulating antioxidant defense pathways and promoting longevity [51] (Table 2).

4.3 Induction of Oxidative Stress in Animal Models

Inducing oxidative stress in animal models suggests significant insights into the pathophysiological mechanisms that contribute to diverse conditions, including cancer, neurodegeneration, and cardiovascular disorders. Several agents are commonly used to induce oxidative stress, with hydrogen peroxide, being one of

Table 2 Overview of in vivo models for antioxidant assessment

Model	Advantages	Limitations	Research applications
Rodent models (mice, rats)	Share physiological and genetic similarities with humans, well-established protocols for controlled experimentation	Ethical concerns; resource-intensive	Used in aging, neurodegeneration, and cardiovascular research
C. elegans (nematodes)	Transparent body enables direct observation; rapid reproduction	Limited immune system resemblance to humans	Investigating antioxidant effects, lifespan extension, and stress-resistance studies
Zebrafish (*Danio rerio*)	Genetic similarity to humans; suitable for high-throughput screening	Requires specialized aquatic maintenance systems	Effective in oxidative stress evaluation, neurotoxicity assessments, and drug discovery

the most prevalent, because it rapidly generates ROS. Besides H_2O_2, animal models also use other agents, such as paraquat and carbon tetrachloride (CCl_4), to induce oxidative stress.

4.3.1 *Hydrogen Peroxide (H_2O_2)*

Hydrogen Peroxide, a reactive oxygen species (ROS), is frequently used to induce oxidative stress in various biological models. It is a simple peroxide that can readily penetrate cellular membranes, generating hydroxyl radicals through Fenton reactions upon its interaction with transition metals like iron [52]. H_2O_2 induces oxidative stress by promoting the production of highly reactive hydroxyl radicals, which are capable of damaging cellular components, including lipids, proteins, and nucleic acids. This process disrupts cellular homeostasis and can lead to apoptosis or necrosis if the oxidative damage is severe [53]. H_2O_2 induces oxidative stress in mammalian cells by generating aldehydic DNA lesions (ADLs), which contribute to the formation of mutagenic DNA adducts at sub-millimolar concentrations, demonstrating a more efficient induction of damage [54]. A recent study showed that H_2O_2 exposure resulted in weight loss, morphological changes in the intestine, and increased oxidative stress markers, including malondialdehyde and protein carbonyls. Additionally, H_2O_2 triggered autophagy and apoptosis in intestinal epithelial cells, emphasizing its role in disrupting cellular homeostasis and contributing to intestinal health deterioration [55]. Studies have shown that high concentrations of H_2O_2 are often required to observe significant oxidative damage because of its rapid conversion into oxygen by catalase, complicating the interpretation of experimental results [56].

4.3.2 Paraquat

Paraquat, a bipyridyl herbicide, is frequently used in oxidative stress research because of its ability to produce superoxide radicals via redox cycling. When administered in animal models, paraquat undergoes reduction by cellular NADPH, creating a cycle where it generates superoxide anions ($O_2^{\cdot -}$) continuously, which are later dismutated to hydrogen peroxide [57]. This results in systemic oxidative damage, beneficial for studying Parkinsonian models, as paraquat targets dopaminergic neurons similarly to neurodegenerative patterns seen in Parkinson's disease (PD) [58]. Paraquat's accumulation in the lungs makes it especially suitable for modeling respiratory oxidative stress, offering insight into idiopathic pulmonary fibrosis and chronic obstructive pulmonary disease (COPD) [59]. Animal studies have shown that paraquat exposure results in lipid peroxidation and mitochondrial damage in lung tissues, highlighting antioxidant responses against respiratory oxidative stress [60]. Using antioxidants in models of paraquat exposure has been shown to reduce oxidative biomarkers, which may guide the development of respiratory treatments [61]. Furthermore, paraquat-induced oxidative stress helps to test neuroprotective antioxidants. In rodent models, paraquat's neurotoxicity is correlated with dopaminergic degeneration, enabling the evaluation of neuroprotective agents, such as resveratrol. Studies have shown that antioxidants can mitigate paraquat-induced neuroinflammation and apoptosis, providing insights into potential therapies for neurodegenerative diseases [62].

4.3.3 Carbon Tetrachloride (CCl_4)

CCl_4 is widely used to induce hepatic oxidative stress because of its metabolism with hepatic cytochrome P450 enzymes, resulting in trichloromethyl radicals ($CCl_3^{\cdot}$). These radicals initiate lipid peroxidation, causing cellular damage, necrosis, and fibrosis, resembling liver injury in human conditions such as cirrhosis and hepatitis [63]. The CCl_4 model enables the study of the hepatoprotective effects of antioxidants, making it valuable for evaluating therapeutic agents in liver injury models [64]. The CCl_4 model also contributes to understanding the anti-fibrotic effects of antioxidants. Animal studies have shown that antioxidants like silymarin and curcumin reduce collagen deposition and inflammation in CCl_4-induced liver fibrosis [65]. This model is vital for evaluating the systemic efficacy of antioxidants in reducing CCl_4-induced oxidative damage across multiple organ systems.

4.4 Biomarkers for Antioxidant Activity

The assessment of antioxidant activity is essential for understanding the physiological responses to oxidative stress and the potential therapeutic effects of antioxidants. Various biomarkers are used to evaluate oxidative stress and the efficacy of antioxidant interventions. These biomarkers are often classified as lipid peroxidation markers, DNA damage indicators, protein oxidation markers, and enzymatic and nonenzymatic antioxidants.

4.4.1 Enzymatic Biomarkers

The measurement of enzymatic antioxidants, which includes catalase, superoxide dismutase (SOD), and glutathione peroxidase (GPx), is essential for evaluating the body's defense mechanisms against oxidative stress. Such enzymes are critical in neutralizing reactive oxygen species (ROS) and maintaining cellular homeostasis. Their levels in tissues or serum can serve as biomarkers for oxidative stress and overall health status.

- *Superoxide Dismutase*

Superoxide dismutase (SOD) is a crucial antioxidant enzyme that catalyzes the dismutation of superoxide radicals ($O_2^{\cdot-}$) into hydrogen peroxide and molecular oxygen (O_2). This enzymatic reaction is vital for mitigating oxidative stress, as superoxide radicals are highly reactive species generated during various metabolic processes, including mitochondrial respiration [66]. SOD exists in multiple isoforms, such as mitochondrial Mn-SOD, cytosolic Cu/Zn SOD, and extracellular SOD. Each isoform is localized to specific cellular compartments to manage oxidative stress effectively. This compartmentalization allows for localized control of superoxide levels, essential for cellular function and signaling [67]. Elevated levels of superoxide result in significant cell destruction, contributing to the pathogenesis of many health conditions, including cardiovascular disorders, neurodegenerative condition, and cancer [68]. The activity of SOD can be evaluated using xanthine oxidase to generate superoxide in vitro and measuring the reduction in superoxide levels or the increase in H_2O_2 using specific substrates [69].

- *Catalase*

Catalase is a pivotal enzymatic antioxidant that catalyzes the decomposition of H_2O_2 into water and oxygen, thereby inhibiting the accumulation of this potentially harmful substance within cells. This enzymatic reaction is crucial for cellular protection against oxidative damage, as H_2O_2 can form highly reactive hydroxyl radicals if not adequately detoxified [70]. By efficiently converting H_2O_2 into harmless products, catalase is critical in maintaining cellular redox balance and preventing oxidative stress [71]. Measurement of catalase activity involves adding a known concentration of H_2O_2 to a sample containing catalase and measuring the rate at which H_2O_2 is consumed or oxygen is released. It uses colorimetric methods that assess the decrease in absorbance of H_2O_2, typically around 240 nm, or by monitoring oxygen production during the enzymatic reaction [72].

- *Glutathione Peroxidase*

Glutathione peroxidase (GPx) is an essential antioxidant enzyme that reduces hydrogen peroxide and lipid peroxides using glutathione as a substrate, protecting cells from oxidative damage. GPx exists in several isoforms, with GPx1 being the most prevalent in various tissues [73]. The activity of GPx can be evaluated using spectrophotometric assays at 412 nm, which measure the oxidation of NADPH or the reduction of hydrogen peroxide over time, typically involving coupled enzyme reactions to enhance sensitivity and accuracy [74]. It regulates redox homeostasis and inhibits ferroptosis, playing significant roles in the pathophysiology of diseases like cardiovascular disorders, neurodegenerative conditions, and cancer, presenting potential therapeutic targets for oxidative-stress-related interventions [75].

4.4.2 Nonenzymatic Biomarkers

Nonenzymatic antioxidants are crucial biomarkers in assessing cellular oxidative stress and antioxidant responses. Among these, glutathione (GSH), ascorbate, and tocopherol are widely studied for their roles in modulating oxidative stress at the cellular level.

- *Glutathione*

Glutathione (GSH) is a tripeptide composed of glutamine, cysteine, and glycine and is a major intracellular antioxidant. It protects cells from oxidative stress by directly scavenging free radicals and participating in various enzymatic reactions that detoxify harmful compounds [76]. The levels of GSH can be evaluated using spectrophotometric assays that measure the reduction of 5,5′-dithiobis (2-nitrobenzoic acid) (DTNB) to measure total glutathione or through high-performance liquid chromatography (HPLC) for more precise quantification [77]. Low levels of GSH indicate oxidative stress and have been shown to contribute to several diseases, such as cancer and neurodegenerative disorders [78].

- *Ascorbate*

Ascorbate, commonly known as vitamin C, is a vital water-soluble antioxidant widely used as a biomarker in antioxidant assays due to its reliable reflection of oxidative status within cells and tissues [79]. As an essential antioxidant, ascorbate can directly scavenge reactive oxygen species (ROS) and regenerate other antioxidants, making it a sensitive marker for redox balance [80]. It is often measured in its reduced form (ascorbate) and oxidized form (dehydroascorbic acid, DHA) to assess oxidative stress [81]. The stability of ascorbate as a biomarker is enhanced by acidic deproteinization, which prevents oxidation during storage and handling, providing reproducible results across various assay protocols [82]. It is an essential micronutrient with roles in multiple physiological processes, including immune function, collagen synthesis, and iron absorption [83].

- *Tocopherol*

α-Tocopherol is a primary lipid-soluble antioxidant in the human body and an essential form of vitamin E. It is critical in protecting cell membranes from oxidative damage by inhibiting lipid peroxidation [84]. Mechanistically, α-tocopherol donates a hydrogen atom to free radicals, stabilizing them and preventing chain reactions that could damage lipids, proteins, and DNA [85]. It can halt the propagation of lipid peroxyl radicals, thus preserving cellular integrity and functionality [86]. Plasma and tissue α-tocopherol levels are indicators of antioxidant status and are commonly measured using HPLC [87]. High doses of α-tocopherol can reduce oxidative biomarkers in patients with atherosclerosis and improve endothelial function [88]. Furthermore, its role in neuroprotection has been explored in models of Alzheimer's disease, where it may help mitigate oxidative neuronal damage [89].

4.4.3 Lipid Peroxidation Biomarkers

Lipid peroxidation biomarkers are critical for assessing oxidative damage in tissues and fluids. They include malondialdehyde (MDA) and F2-isoprostanes, which serve as lipid peroxidation indicators.

- *Malondialdehyde (MDA)*

Malondialdehyde, a by-product of polyunsaturated fatty acid peroxidation, is widely used to measure lipid damage in tissues and fluids, such as plasma or urine [90]. It reacts with thiobarbituric acid (TBA) to form a red, pink coloured complex, known as thiobarbituric acid—reactive substances (TBARS), which is quantifiable (at 532 nm) and often used as a lipid peroxidation marker [91]. Modern techniques, such as HPLC and gas chromatography, improve the specificity and reliability of MDA detection compared to traditional TBA assays, which can sometimes detect other aldehydes [92]. MDA levels correlate strongly with the extent of lipid peroxidation and are often elevated in various disease states involving oxidative stress, including ischemic stroke and cardiovascular diseases [93].

- *F2-Isoprostanes*

F2-isoprostanes, produced by free radical-catalyzed peroxidation of arachidonic acid, are considered one of the most reliable and specific markers for oxidative stress [94]. Unlike MDA, F2-isoprostanes are chemically stable and are not produced through enzymatic pathways, making them ideal for oxidative stress studies [95]. Analytical methods, such as mass spectrometry, allow for accurate quantification of F2-isoprostanes in biological samples like plasma, urine, and tissue [96]. Elevated levels have been associated with health conditions such as Alzheimer's disease, atherosclerosis, and pulmonary disorders [97].

4.4.4 Protein Oxidation Biomarkers

Protein oxidation biomarkers are valuable indicators of oxidative stress. Key biomarkers include protein carbonyl groups and nitrotyrosine.

- **Protein Carbonyls**

Carbonyl groups are formed on proteins through direct oxidation of amino acid side chains or indirectly by reactions with lipid peroxidation products like malondialdehyde (MDA). These oxidized protein carbonyls are stable and thus serve as reliable indicators of oxidative damage [98]. Detection methods typically involve derivatization with 2,4-dinitrophenylhydrazine (DNPH), forming a hydrazone that can be quantified spectrophotometrically or by Western blotting [99]. Elevated levels of protein carbonyls have been linked to aging, neurodegenerative disorders, and cardiovascular diseases [100].

- **Nitrotyrosine**

Nitrotyrosine is a specific marker formed by the nitration of tyrosine residues, typically induced by reactive nitrogen species, such as peroxynitrite ($ONOO^-$) [101]. Detection of nitrotyrosine is commonly performed using immunohistochemistry, enzyme-linked immunosorbent assays (ELISA), or mass spectrometry [102]. Elevated nitrotyrosine levels are often associated with conditions of increased inflammation and oxidative stress, including atherosclerosis, sepsis, and neurodegeneration [103].

4.4.5 DNA Damage Markers

8-Hydroxy-2′-deoxyguanosine (8-OHdG) is a crucial biomarker for oxidative DNA damage, formed when reactive oxygen species (ROS) interact with DNA, specifically guanine residues [104]. Measurement of 8-OHdG is commonly performed through urine samples, utilizing methods such as high-performance liquid chromatography (HPLC) and enzyme-linked immunosorbent assay (ELISA) [105]. Elevated levels of 8-OHdG are indicative of oxidative stress and have been implicated in several pathophysiological conditions, including cancer, CVD and neurodegenerative diseases, diabetes, and periodontal diseases [106].

4.5 In Vivo Pharmacokinetics of Antioxidants

The in vivo pharmacokinetics of antioxidants encompass their absorption, distribution, metabolism, and excretion (ADME) properties, which are critical for understanding their efficacy and safety in biological systems. Antioxidants must be adequately absorbed and distributed to exert their effects. Factors such as

Table 3 Biomarkers for antioxidant evaluation

Biomarker	Classification	Function	Detection method
Superoxide dismutase (SOD)	Enzyme	Neutralizes superoxide radicals	Spectrophotometric assays
Malondialdehyde (MDA)	Lipid peroxidation marker	Reflects lipid oxidative degradation and cellular damage	TBARS assay or HPLC
Glutathione (GSH)	Nonenzymatic antioxidant	Maintains cellular redox balance and detoxifies ROS	DTNB assay or HPLC
Hydroxy-2′-deoxyguanosine (8-OHdG)	DNA damage marker	Indicates oxidative-stress-induced DNA damage	ELISA or HPLC
Catalase	Enzyme	Decomposes H_2O_2 into water and oxygen, protecting cells from oxidative damage	Colorimetric methods

lipophilicity, molecular weight, and formulation significantly influence the absorption rates of antioxidant compounds [107]. The metabolic pathways of antioxidants can vary considerably between in vitro and in vivo settings. Studies have shown that some antioxidants undergo extensive first-pass metabolism, which can alter their bioavailability and pharmacological activity [108]. The excretion routes of antioxidants also play a crucial role in determining their therapeutic potential. Compounds that are efficiently excreted can lead to reduced systemic exposure and diminished efficacy [109]. Antioxidants that are more polar and have an inclination to bind to pro-oxidants are likely to have enhanced renal excretion [110] (Table 3).

5 Comparative Analysis of In Vitro and In Vivo Methods

5.1 Correlation Between In Vitro and In Vivo Study

Translating in vitro antioxidant activity to in vivo outcomes is a multifaceted challenge crucial for developing effective antioxidant therapies.

- *Physiological Conditions*

In vitro studies are often conducted under highly controlled conditions that fail to mimic the physiological complexity of living organisms [111]. The chemical milieu within biological systems, including various macromolecules, fluctuating pH levels, and different ionic strengths, can significantly alter the behavior and effectiveness of antioxidants [112]. For example, proteins and lipids in biological fluids can enhance or diminish the antioxidant capacity, leading to variable outcomes when transitioning from a laboratory to a physiological environment.

- *Bioavailability Constraints*

Bioavailability is a critical factor affecting the efficacy of antioxidants in vivo. Various compounds that exhibit potent antioxidant properties in vitro may demonstrate limited bioavailability because of poor solubility, rapid degradation, or extensive first-pass metabolism [113]. Curcumin, despite its recognized antioxidant capacity, often fails to achieve sufficient plasma levels post oral administration, limiting its therapeutic application [114]. Thus, while in vitro assays can establish potential efficacy, they may not accurately predict clinical outcomes.

- *Metabolic Transformations*

Antioxidants often undergo significant metabolic transformations in vivo that can alter their pharmacological profiles. Metabolic processes, particularly those mediated by liver enzymes, may convert active antioxidants into less effective or toxic metabolites [115]. For instance, resveratrol is metabolized into various conjugates, some of which may lack the antioxidant activity observed in vitro, complicating the correlation between laboratory findings and therapeutic effects [116].

- *Interindividual Variability*

Individual variability in genetics, health status, and microbiota composition can further complicate the translation of antioxidant activity from in vitro to in vivo [117]. Differences in metabolic rates and absorption profiles among individuals can lead to inconsistent therapeutic responses, challenging the generalizability of findings derived from standardized in vitro assays [118].

- *Dosing Discrepancies*

The dosages under in vitro studies often do not reflect clinically relevant levels achievable in vivo. In vitro experiments may utilize substantially higher concentrations than those that can be safely administered in human subjects [119]. The formulation and delivery methods of antioxidants can significantly influence their efficacy, causing careful consideration of these factors during the translation process.

5.2 Factors Influencing In Vivo Efficacy

Understanding the factors that impact the efficacy of antioxidants in vivo is crucial for optimizing therapeutic outcomes (Fig. 6). Key considerations include the following:

- *Bioavailability*

Bioavailability refers to the proportion of an administered antioxidant that reaches the systemic circulation and is available for therapeutic effect [120]. Solubility, stability,

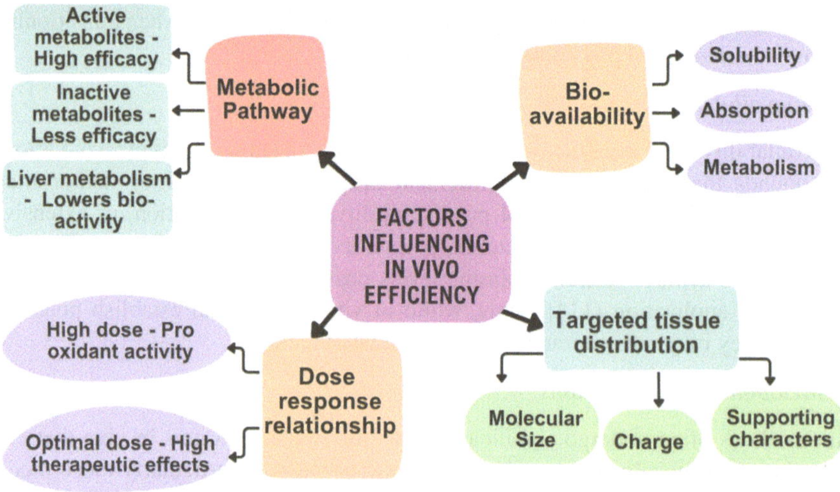

Fig. 6 Factors influencing in vivo efficiency

and formulation influence bioavailability [121]. For example, lipophilic antioxidants may have enhanced absorption when delivered in lipid-based formulations, whereas hydrophilic compounds might require modifications to improve their absorption rates in the GI tract [122].

- *Metabolic Pathways*

The metabolic pathways of antioxidants can lead to their conversion into active or inactive metabolites, impacting their overall efficacy. Various antioxidants undergo phase I and phase II metabolic processes in the liver, which can alter their bioactivity [123]. The variability in metabolic rates among individuals can cause significant differences in antioxidant effectiveness [124].

- *Targeted Tissue Distribution*

The therapeutic efficacy of antioxidants depends on their targeted distribution to specific tissues. Molecular size and charge influence an antioxidant's ability to traverse cellular barriers and reach target tissues [125]. Innovative drug delivery systems, such as nanoparticle formulations, have been developed to improve tissue targeting and enhance therapeutic effects [126].

- *Dose–Response Relationship*

The relationship between the dose of an antioxidant and its biological effect is not always linear. Low doses may provide beneficial effects, while high doses can lead to pro-oxidant activity or adverse effects [127]. Understanding the optimal dosing is critical for maximizing therapeutic benefits while minimizing potential toxicity [128].

6 Emerging Technologies in Antioxidant Evaluation

The evaluation of the antioxidant activity helps in the emergence of upgraded technologies to overcome the drawbacks faced by traditional antioxidant evaluation methods. The significant areas of these technological advancements include high-throughput screening (HTS), omics approaches, and nanotechnology-based delivery systems.

6.1 High-Throughput Screening

High-throughput screening is an efficient method used in biotechnology to test many biological samples for biological or chemical activity, such as antioxidant activity. This method has brought significant revolutionary changes by efficiently screening large compound libraries of antioxidant capacity [129]. These automated high-throughput screening platforms use miniaturized assays, highly efficient detection systems, and various computational tools to simultaneously process a large number of samples to identify the potent antioxidants. The recent advancement in high-throughput screening is the integration of luminescence-based and fluorescence-based assays, which provides enhanced specificity and sensitivity in evaluating the antioxidant activity [130].

6.2 Omics Approaches

The omics approaches include genomics, transcriptomics, proteomics, and metabolomics, which play a significant role in antioxidant activity at the molecular level (Fig. 7). The omics approaches help in the analysis of the expression of genes, the interaction of proteins, and the metabolic reactions that occur in response to the antioxidant reactions, which provide insights to study the mechanism of the antioxidant action [129].

6.2.1 Genomics and Transcriptomics

These technologies help identify genes that respond to antioxidants. Transcriptomics reveals the changes in gene expression patterns in response to the antioxidant mechanisms [131].

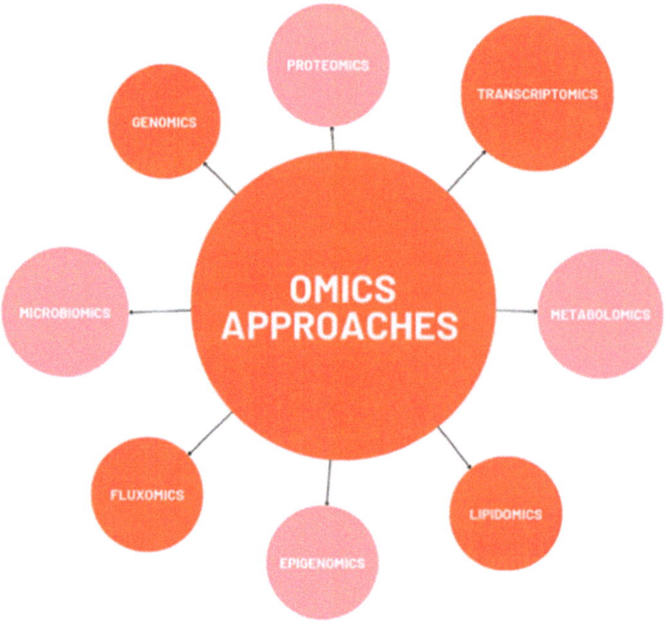

Fig. 7 Schematic representation of omics approaches plays a vital role in the antioxidant activity at the molecular level

6.2.2 Proteomics

Proteomics quantitatively identifies and analyzes the proteins involved in the antioxidant activity and explains the posttranslational changes and protein interactions in the antioxidant activity.

6.2.3 Metabolomics

Metabolomics helps assess the metabolite concentrations and the reactions in the antioxidant pathways. It also helps evaluate the biochemical changes in the antioxidant reactions [132].

6.3 Nanotechnology in Antioxidant Delivery

Nanotechnology has evolved into a transformative tool for improving the delivery and efficiency of antioxidants. Nanocarriers are used for the targeted delivery of the antioxidants, enhancing the bioavailability and maintaining stability in antioxidant reactions. Nanocarriers used in antioxidant studies are liposomes, polymeric nanoparticles, nanocrystals, and dendrimers. These nanocarriers prevent the premature degradation of antioxidants, improve solubility, and perform controlled release of antioxidants, increasing the therapeutic efficiency [133]. Integrating all these technologies together can help pave the way for future discoveries in antioxidant evaluation. The high-throughput and omics technologies help in the rapid analysis of complex data, and nanotechnology helps efficiently deliver antioxidants to target tissues.

7 Applications of Antioxidant Evaluation

The application of antioxidants is visualized in various fields because of their ability to eradicate free radicals and prevent the oxidative stress that is the primary cause of a wide range of diseases. In the pharmaceutical industry, antioxidants are increasingly being used as therapeutic agents, mainly in treating and preventing cancer, neurodegenerative disorders, and cardiovascular diseases. The antioxidant application is also explored in the nutraceutical and cosmetic industries.

7.1 Pharmaceutical Applications

The applications of antioxidants focus on preventing oxidative stress, which is being made applicable in the therapeutic measures of various diseases. Oxidative stress occurs due to an imbalance between the production of reactive oxidative species and the body's antioxidant defense. This imbalance can damage lipids, proteins, and DNA, leading to cell dysfunction and facilitating disease progression [134].

7.1.1 Cardiovascular Diseases

Oxidative stress is studied as the most significant factor in the cause of various cardiovascular diseases such as atherosclerosis, heart failure, and ischemia–reperfusion injuries. The reactive oxidative species promote oxidation of lipids, which acts as the cause of plaque in the arteries. The oxidation of lipids also triggers inflammation and apoptosis in myocardial cells [135]. Thus, antioxidant studies can help in the therapeutic prevention and treatment of heart diseases. Antioxidants that are used as therapeutic approaches in heart diseases are (i) Coenzyme Q10 (CoQ10), which is popularly known for its role in cellular energy metabolism, has also shown various benefits by reducing cardiovascular risks by reducing the oxidative damage in the myocardial cells and lipids, thus preventing the blood pressure and endothelial damage [136]. (ii) Curcumin: Usually derived from turmeric, its antioxidant properties help to reduce lipid oxidation and inflammation. It also activates various pathways and protects the heart from ischemia–reperfusion injury, oxidative stress, and damage [137]. (iii) Resveratrol: Present in grapes and red wine. It is highly cardio-protective, as it can activate the antioxidant response element. This element helps in preventing inflammation, improving endothelial function, and preventing oxidative stress, therefore reducing the risk of atherosclerosis and heart failure [138]. (iv) Carotenoids: Carotenoids like lycopene and beta-carotene are potent antioxidants that can lower the risk of cardiovascular diseases by eradicating free radicals and protecting low-density cholesterol from oxidation [139]. (v) Flavonoids: Flavonoids are substances found in various fruits and teas, such as quercetin, that play an essential role in maintaining cardiovascular health by improving blood vessel function and reducing blood pressure levels through antioxidant usage.

7.1.2 Neurodegenerative Diseases

Neurodegenerative conditions like Alzheimer's disease and Parkinson's disease are highly influenced by the oxidative stress caused by oxidation in the neurons. The accumulation of the reactive oxidative species leads to the degradation of the nerve cells, which contributes to the poor signaling of neural conduction, which leads to motor dysfunction [140]. (i) Vitamin E: A lipid-soluble antioxidant plays a vital role in preventing the oxidative damage of the cell membranes. Vitamin E can help prevent Alzheimer's and other neurodegenerative diseases [141]. (ii) Melatonin: This sleep-regulating hormone also helps in preventing mitochondrial dysfunction, which acts as a standard feature in neurodegenerative disorders. (iii) Glutathione is an endogenous antioxidant that plays a critical role in detoxifying the reactive oxygen species in the brain. Glutathione supplements have helped in the treatment of oxidative stress in neurodegenerative disorders. (iv) Resveratrol: This polyphenol is known for its neuroprotective effects, showing promising effects in the models of Alzheimer's and Parkinson's disease [138]. (v) Coenzyme Q10 (CoQ10): Plays a vital role in mitochondrial function. It also shows benefits in the clinical traits of Parkinson's disease by providing cellular energy and offering neural protection [136].

7.1.3 Cancer

Several antioxidants are being used in the treatment and prevention and have also shown effective results, even with different factors such as intensity of the condition, dosage of the drug, and form of the antioxidant. Common antioxidants being used in therapeutic approaches to cancer and their effects are (i) Vitamin E: This vitamin is adequate in establishing protection against prostate cancer and colorectal cancers. The therapeutic effectiveness of Vitamin E in the prevention of cancer depends on the dosage and the health condition of the patient. (ii) Selenium: Selenium, along with Vitamin E, has been studied as having the potential to reduce the risk of cancer, mainly gastrointestinal cancer [142]. (iii) Lycopene and Beta-Carotene: These carotenoids, found in colorful vegetables and fruits, are beneficial in reducing the risk of lung and prostate cancer [143].

7.2 Nutraceuticals and Functional Foods

In recent years, nutraceuticals and functional foods rich in antioxidants have reduced the effects of oxidative stress, which contributes to major chronic diseases, including cancer. Nutraceuticals and supplements are fortified with antioxidants such as vitamins C and E, polyphenols, flavonoids, and carotenoids. These antioxidants help to reduce cellular damage because of oxidative stress and, thus, reduce the risk of chronic health progression [144]. Functional foods are known for their ability to acquire health benefits because of their abundance of antioxidant properties. Antioxidants in fruits and vegetables are more effective in preventing and combating oxidative stress than synthetic supplements. Thus, the antioxidants present in nutraceuticals and functional foods can prevent cellular damage because of oxidative stress, thus preventing the progression of chronic disease conditions.

7.3 Cosmetics and Antiageing

Antioxidants play a significant role in the cosmetics industry and in antiaging products, which aim to maintain the skin barrier and cells by preventing them from oxidative damage. When used on the skin, antioxidants also promote rejuvenating damaged skin cells and tissues. Antioxidants neutralize the reactive oxidative species (ROS). The ROS species contribute to the skin cells' aging process by accelerating collagen breakdown, an essential protein responsible for skin elasticity. Common antioxidants such as vitamin C, E and plant extracts have been shown to reduce oxidative damage, promoting collagen synthesis and skin elasticity [145]. A study of the antioxidant activity using assays has demonstrated how botanical extracts and essential oils show strong antiaging effects. Indian sandalwood oil has been studied as a tremendous antioxidant-rich compound that neutralizes ROS and reduces collagenase activity, an enzyme that breaks down collagen in reaction to UV radiation and environmental pollution. Also, *Jasminum sambac* and *Artocarpus altilis* plant extracts have excellent antioxidant and anti-inflammatory effects, protecting the skin against environmental changes.

8 Challenges and Future Directions

The challenges faced in evaluating antioxidant capacity using in vitro and in vivo assays and future directory measures to enhance the methodologies and discoveries of new compounds and techniques to overcome the current challenges are discussed below.

8.1 Standardization of Methods in Antioxidant Evaluation

The inconsistent application of antioxidant evaluation indicates a significant backlash in antioxidant research. All the common assays, such as DPPH, ABTS, ORAC, and FRAP, vary significantly in their chemical properties and capacity to measure different antioxidant properties. For example, the FRAP assays help to assess the reduction of the power of the free radicals, while the DPPH and ABTS are more focused on eradicating the free radicals. Recent studies and articles denote a standardized assay and method for calibrating the data acquired from the assays. This improvement would enhance the research across different contexts [146].

8.2 Improving Bioavailability

The major challenge in antioxidant research is translating the antioxidant potential observed in the assays to therapeutic applications. Antioxidants from natural sources demonstrate poor solubility, high metabolism, and low absorption in vivo conditions [147]. This limits the efficiency in therapeutic conditions despite showing excellent antioxidant activity in laboratory conditions. Advanced delivery systems such as nanocarriers, nanoencapsulation, and liposomal delivery are used to make these naturally derived antioxidants more effective in therapeutic situations. These delivery systems help protect the antioxidants from degradation, facilitating the targeted delivery, thus increasing the bioavailability of the antioxidants in the body [148]. Also, new research has shown structural modifications of the antioxidants by conjugating them with molecules that enhance their solubility and cell uptake. The conjugated molecules are known as bio-enhancers, which enhance the antioxidant properties of the antioxidants.

8.3 Personalized Antioxidant Therapy

The emerging personalized medicine has started to extend in antioxidant research. Genetic modifications of an individual can also affect the antioxidant action in an individual's body and influence the body's response against oxidative stress [149]. Gene variations lead to changes in the genes encoding vital oxidative enzymes such as glutathione peroxidase and superoxide dismutase, which are the primary prevention line from oxidative stress and help regulate antioxidant intake. Hence, personalized antioxidant therapy can help increase the antioxidants' effectiveness and reduce oxidative damage [150]. This approach especially benefits in managing chronic conditions such as cardiovascular disorders, neurodegenerative disorders, diabetes, and cancer.

9 Conclusion

To comprehensively understand the evaluation of antioxidant capacity, it is necessary to integrate and compare the insights from both in vitro and in vivo studies. Assays used under in vitro conditions are an essential tool in drug development and research and are also effective in studying toxicity, drug efficiency, and cell response. They are cost-effective, require minimal human resources, and do not consume time. Even though it has advantages, it also possesses certain limitations. They lack physiological reactions/interactions seen in living organisms, eventually leading to inaccurate results in the in vitro conditions. As technology evolves, these systems can be further enhanced, which might reduce the need for animal testing. In vivo, evaluation of antioxidant activity is essential for elucidating these compounds' physiological relevance and therapeutic potential. While in vitro studies provide valuable preliminary insights into the antioxidant capacity of various substances, they often cannot replicate the complexities inherent in living organisms. The utilization of appropriate animal models enables assessing the efficacy of antioxidants, their pharmacokinetics, and their safety profiles. Ultimately, integrating both in vitro and in vivo approaches will enhance the understanding of mechanisms by which antioxidants operate within biological systems and facilitate the development of effective interventions for oxidative-stress-related diseases.

References

1. Andrés, C. M. C., de la Lastra, J. M. P., Juan, C. A., Plou, F. J., & Pérez-Lebeña, E. (2024). Antioxidant metabolism pathways in vitamins, polyphenols, and selenium: Parallels and divergences. *International Journal of Molecular Sciences, 25*(5), 2600.
2. Bibi Sadeer, N., Montesano, D., Albrizio, S., Zengin, G., & Mahomoodally, M. F. (2020). The versatility of antioxidant assays in food science and safety—Chemistry, applications, strengths, and limitations. *Antioxidants, 9*(8), 709.
3. Chaudhary, P., Janmeda, P., Docea, A. O., Yeskaliyeva, B., Razis, A. F. A., Modu, B., Calina, D., & Sharifi-Rad, J. (2023). Oxidative stress, free radicals and antioxidants: Potential crosstalk in the pathophysiology of human diseases. *Frontiers in chemistry, 11*, 1158198.
4. Halliwell, B., & Gutteridge, J. M. C. (2015). *Free radicals in biology and medicine*. Oxford University Press.
5. Nakai, K., & Tsuruta, D. (2021). What are reactive oxygen species, free radicals, and oxidative stress in skin diseases? *International Journal of Molecular Sciences, 22*(19), 10799.
6. Irato, P., & Santovito, G. (2021). Enzymatic and non-enzymatic molecules with antioxidant function. *Antioxidants, 10*, 579. https://doi.org/10.3390/antiox10040579
7. Vanova, A., Gerasimova, E., & Gazizullina, E. (2020). Study of antioxidant properties of agents from the perspective of their action mechanisms. *Molecules, 25*(18), 4251.
8. Carvalho, J. R. B., Meireles, A. N., Marques, S. S., Gregório, B. J. R., Ramos, I. I., Silva, E. M. P., Barreiros, L., & Segundo, M. A. (2023). Exploiting kinetic features of ORAC assay for evaluation of radical scavenging capacity. *Antioxidants, 12*(2), 505.
9. Kotha, R. R., Tareq, F. S., Yildiz, E., & Luthria, D. L. (2022). Oxidative stress and antioxidants—A critical review on in vitro antioxidant assays. *Antioxidants, 11*(12), 2388.
10. Munteanu, I. G., & Apetrei, C. (2021). Analytical methods used in determining antioxidant activity: A review. *International Journal of Molecular Sciences, 22*(7), 3380.

11. Danet, A. F. (2021). *Recent advances in antioxidant capacity assays*. IntechOpen.
12. Masenga, S. K., Kabwe, L. S., Chakulya, M., & Kirabo, A. (2023). Mechanisms of oxidative stress in metabolic syndrome. *International Journal of Molecular Sciences, 24*(9), 7898. https://doi.org/10.3390/ijms24097898
13. Kaur, D., Yousuf, B., & Qadri, O. S. (2024). *Syzygium cumini* anthocyanins: Recent advances in biological activities, extraction, stability, characterisation and utilisation in food systems. *Food Production, Processing and Nutrition, 6*(1), 34.
14. George, J., Edwards, D., Pun, S., & Williams, D. (2022). Evaluation of antioxidant capacity (ABTS and CUPRAC) and total phenolic content (Folin-Ciocalteu) assays of selected fruit, vegetables, and spices. *International Journal of Food Science, 2022*(1), 2581470.
15. Calderon Moreno, R., Navas-Acien, A., Escolar, E., Nathan, D. M., Newman, J., Schmedtje Jr, J. F., Diaz, D., Lamas, G. A., & Fonseca, V. (2019). Potential role of metal chelation to prevent the cardiovascular complications of diabetes. *The Journal of Clinical Endocrinology & Metabolism, 104*(7), 2931–2941.
16. Zhu, L., Luo, M., Zhang, Y., Fang, F., Li, M., An, F., Zhao, D., & Zhang, J. (2023). Free radical as a double-edged sword in disease: Deriving strategic opportunities for nanotherapeutics. *Coordination chemistry reviews, 475*, Article 214875.
17. Martínez-Cabanas, M., López-García, M., Rodríguez-Barro, P., Vilariño, T., Lodeiro, P., Herrero, R., Barriada, J. L., & de Vicente, M. E. S. (2021). Antioxidant capacity assessment of plant extracts for green synthesis of nanoparticles. *Nanomaterials,* 11(7), 1679.
18. Christodoulou, M. C., Orellana Palacios, J. C., Hesami, G., Jafarzadeh, S., Lorenzo, J. M., Domínguez, R., Moreno, A., & Hadidi, M. (2022). Spectrophotometric methods for measurement of antioxidant activity in food and pharmaceuticals. *Antioxidants, 11*(11), 2213.
19. Stobiecka, M., Król, J., & Brodziak, A. (2022). Antioxidant activity of milk and dairy products. *Animals, 12*(3), 245.
20. Spiegel, M., Kapusta, K., Kołodziejczyk, W., Saloni, J., Żbikowska, B., Hill, G. A., & Sroka, Z. (2020). Antioxidant activity of selected phenolic acids–ferric reducing antioxidant power assay and QSAR analysis of the structural features. *Molecules, 25*(13), 3088.
21. Asma, U., Bertotti, M. L., Zamai, S., Arnold, M., Amorati, R., & Scampicchio, M. (2024). A kinetic approach to oxygen radical absorbance capacity (ORAC): Restoring order to the antioxidant activity of hydroxycinnamic acids and fruit juices. *Antioxidants, 13*(2), 222.
22. Wołosiak, R., Drużyńska, B., Derewiaka, D., Piecyk, M., Majewska, E., Ciecierska, M., Worobiej, E., & Pakosz, P. (2021). Verification of the conditions for determination of antioxidant activity by ABTS and DPPH assays—A practical approach. *Molecules, 27*(1), 50.
23. Kim, H., & Chin, K. B. (2020). Evaluation of antioxidant activity of *Cudrania tricuspidata* (CT) leaves, fruit powder and CT fruit in pork patties during storage. *Food Science of Animal Resources, 40*(6), 881.
24. Rudrapal, M., Khairnar, S. J., Khan, J., Dukhyil, A. B., Ansari, M. A., Alomary, M. N., Alshabrmi, F. M., Palai, S., Deb, P. K., & Devi, R. (2022). Dietary polyphenols and their role in oxidative stress-induced human diseases: Insights into protective effects, antioxidant potentials and mechanism (s) of action. *Frontiers in pharmacology, 13*, 806470.
25. Sun, S., Liu, Z., Lin, M., Gao, N., & Wang, X. (2024). Polyphenols in health and food processing: Antibacterial, anti-inflammatory, and antioxidant insights. *Frontiers in Nutrition, 11*, 1456730.
26. Moreno, R. C., Navas-Acien, A., Escolar, E., Nathan, D. M., Newman, J., Schmedtje, J. F., Diaz, D., Lamas, G. A., & Fonseca, V. (2019). Potential role of metal chelation to prevent the cardiovascular complications of diabetes. *The Journal of Clinical Endocrinology & Metabolism, 104*(7), 2931–2941. https://doi.org/10.1210/jc.2018-01484
27. Sies, H., (Ed.) (2019). *Oxidative stress: Eustress and distress*. Academic Press.
28. Varesi, A., Campagnoli, L. I. M., Carrara, A., Pola, I., Floris, E., Ricevuti, G., Chirumbolo, S., & Pascale, A. (2023). Non-enzymatic antioxidants against Alzheimer's disease: Prevention, diagnosis and therapy. *Antioxidants, 12*, 180. https://doi.org/10.3390/antiox12010180

29. Guerreiro, B. M., Carvalho Silva, J., Lima, J. C., Reis, M. A. M., & Freitas, F. (2020). Antioxidant potential of the bio-based fucose-rich polysaccharide FucoPol supports its use in oxidative stress-inducing systems. *Polymers, 13*(18), 3020.
30. Mendonça, J. S., Guimarães, R. C. A., Zorgetto-Pinheiro, V. A., Fernandes, C. D. P., Marcelino, G., Bogo, D., Freitas, K. C., et al. (2022). Natural antioxidant evaluation: A review of detection methods. *Molecules, 27*(11), 3563.
31. Santos-Sánchez, N. F., Salas-Coronado, R., Villanueva-Cañongo, C., & Hernández-Carlos, B. (2019). Antioxidant compounds and their antioxidant mechanism. *Antioxidants, 10,* 1–29.
32. Rajput, V. D., Harish, Singh, R. K., Verma, K. K., Sharma, L., Quiroz-Figueroa, F. R., Meena, M., Gour, V. S., Minkina, T., Sushkova, S., et al. (2021). Recent developments in enzymatic antioxidant defence mechanism in plants with special reference to abiotic stress. *Biology, 10,* 267. https://doi.org/10.3390/biology10040267
33. Rumpf, J., Burger, R., & Schulze, M. (2023). Statistical evaluation of DPPH, ABTS, FRAP, and Folin-Ciocalteu assays to assess the antioxidant capacity of lignins. *International Journal of Biological Macromolecules, 233,* Article 123470.
34. Rothe, J., Fischer, R., Cotterchio, C., Gastl, M., & Becker, T. (2023). Analytical determination of antioxidant capacity of hop-derived compounds in beer using specific rapid assays (ORAC, FRAP) and ESR-spectroscopy. *European Food Research and Technology, 249*(1), 81–93.
35. Abeyrathne, E. D. N. S., Nam, K., & Ahn, D. U. (2021). Analytical methods for lipid oxidation and antioxidant capacity in food systems. *Antioxidants, 10*(10), 1587.
36. Vandamme, T. F. (2014). Use of rodents as models of human diseases. *Journal of Pharmacy and Bioallied Sciences, 6*(1), 2–9.
37. Jankowsky, J. L., Savonenko, A., Schilling, G., Wang, J., Xu, G., & Borchelt, D. R. (2002). Transgenic mouse models of neurodegenerative disease: Opportunities for therapeutic development. *Current Neurology and Neuroscience Reports, 2*(5), 457–464.
38. Renugadevi, J., & Prabu, S. M. (2010). Quercetin protects against oxidative stress-related renal dysfunction by cadmium in rats. *Experimental and Toxicologic Pathology, 62*(5), 471–481.
39. Bryda, E. C. (2013). The mighty mouse: The impact of rodents on advances in biomedical research. *Missouri Medicine, 110*(3), 207–211.
40. Pal, G., Mishra, H. P., Suvvari, T. K., Tanwar, A., Ghosh, T., Verma, P., Pal, A., Patial, K., Mahapatra, C., Amanullah, N. A., & Shukoor, S. A. (2023). Oxidative stress in wistar rats under acute restraint stress and its modulation by antioxidants and nitric oxide modulators. *Cureus, 15*(8).
41. Pardo, J., Abba, M. C., Lacunza, E., Ogundele, O. M., Paiva, I., Morel, G. R., Outeiro, T. F., & Goya, R. G. (2018). IGF-I gene therapy in aging rats modulates hippocampal genes relevant to memory function. *Journals of Gerontology. Series A, Biological Sciences and Medical Sciences, 73*(4), 459–467.
42. Yang, L., Wu, C., Parker, E., Li, Y., Dong, Y., Tucker, L., Brann, D. W., Lin, H. W., & Zhang, Q. (2022). Non-invasive photobiomodulation treatment in an Alzheimer disease-like transgenic rat model. *Theranostics, 12*(5), 2205–2231.
43. Gonzalez-Moragas, L., Berto, P., Vilches, C., Quidant, R., Kolovou, A., Santarella-Mellwig, R., Schwab, Y., Stürzenbaum, S., Roig, A., & Laromaine, A. (2017). In vivo testing of gold nanoparticles using the *Caenorhabditis elegans* model organism. *Acta biomaterialia., 15*(53), 598–609.
44. Markaki, M., & Tavernarakis, N. (2010). Modeling human diseases in *Caenorhabditis elegans. Biotechnology Journal, 5*(12), 1261–1276.
45. Marsh, E. K., & May, R. C. (2012). *Caenorhabditis elegans*, a model organism for investigating immunity. *Applied and Environmental Microbiology, 78*(7), 2075–2081.
46. Harrington, A. J., Hamamichi, S., Caldwell, G. A., & Caldwell, K. A. C. (2010). Elegans as a model organism to investigate molecular pathways involved with Parkinson's disease. *Developmental dynamics: An official publication of the American Association of Anatomists, 239*(5), 1282–1295.
47. Wu, T., Xu, H., Liang, X., & Tang, M. (2019). *Caenorhabditis elegans* as a complete model organism for biosafety assessments of nanoparticles. *Chemosphere, 221*, 708–726.

48. Meneely, P. M., Dahlberg, C. L., & Rose, J. K. (2019). Working with worms: *Caenorhabditis elegans* as a model organism. *Current Protocols Essential Laboratory Techniques, 19*(1), Article e35.
49. Oh, S. I., Park, J. K., & Park, S. K. (2015). Lifespan extension and increased resistance to environmental stressors by *N*-acetyl-L-cysteine in *Caenorhabditis elegans. Clinics, 70*(5), 380–386.
50. Zhou, K. I., Pincus, Z., & Slack, F. J. (2011). Longevity and stress in *Caenorhabditis elegans. Aging (Albany NY), 3*(8), 733–753. https://doi.org/10.18632/aging.100367
51. Park, S. K., Tedesco, P. M., & Johnson, T. E. (2009). Oxidative stress and longevity in *Caenorhabditis elegans* as mediated by SKN-1. *Aging Cell, 8*(3), 258–269. https://doi.org/10.1111/j.1474-9726.2009.00473
52. Ransy, C., Vaz, C., Lombès, A., & Bouillaud, F. (2020). Use of H_2O_2 to cause oxidative stress, the catalase issue. *International Journal of Molecular Sciences, 21*(23), 9149. https://doi.org/10.3390/ijms21239149
53. Bucher, J. R. (2019). Oxidative stress and radical-induced signalling. In: R. A. Baan, B. W. Stewart, & K. Straif (Eds.), *Tumour site concordance and mechanisms of carcinogenesis.* Lyon (FR): International Agency for Research on Cancer. (IARC Scientific Publications, No. 165).
54. Nakamura, J., Purvis, E. R., & Swenberg, J. A. (2003). Micromolar concentrations of hydrogen peroxide induce oxidative DNA lesions more efficiently than millimolar concentrations in mammalian cells. *Nucleic Acids Research, 31*(6), 1790–1795. https://doi.org/10.1093/nar/gkg263
55. Zhong, Y., Ma, T., Fu, Z., Chen, A., Yu, J., Huang, Y., & Fu, J. (2023). Effects of hydrogen peroxide-induced oxidative stress on intestinal morphology, redox status, and related molecules in squabs. *Animals, 13*(4), 749. https://doi.org/10.3390/ani13040749
56. Heo, S., Kim, S., & Kang, D. (2020). The role of hydrogen peroxide and peroxiredoxins throughout the cell cycle. *Antioxidants (Basel), 9*(4), 280. https://doi.org/10.3390/antiox9040280.PMID:32224940;PMCID:PMC7222192
57. Mohammadi Bardbori, A., & Ghazi-Khansari, M. (2008). Alternative electron acceptors: Proposed mechanism of paraquat mitochondrial toxicity. *Environmental Toxicology and Pharmacology, 26*, 1–5. https://doi.org/10.1016/j.etap.2008.02.009
58. da Silva, S., da Costa, C. D., Naime, A. A., Santos, D. B., Farina, M., & Colle, D. (2024). Mechanisms mediating the combined toxicity of paraquat and maneb in SH-SY5Y neuroblastoma cells. *Chemical Research in Toxicology., 37*(8), 1269–1282.
59. Chen, J., Su, Y., Lin, F., Iqbal, M., Mehmood, K., Zhang, H., & Shi, D. (2021). Effect of paraquat on cytotoxicity involved in oxidative stress and inflammatory reaction: A review of mechanisms and ecological implications. *Ecotoxicology and Environmental Safety, 224*, 112711. ISSN 0147–6513. https://doi.org/10.1016/j.ecoenv.2021.112711
60. Tyagi, N., & Singh, R. (2020). Paraquat-induced oxidative stress and lung inflammation. In: Chakraborti, S., Parinandi, N., Ghosh, R., Ganguly, N., & Chakraborti, T. (Eds.), *Oxidative stress in lung diseases*. Springer. https://doi.org/10.1007/978-981-32-9366-3_11
61. Li, W., Li, M., Chen, K., Tang, Y., Yin, R., Lan, L., & Hong, G. (2022). Oxaloacetate acid ameliorates paraquat-induced acute lung injury by alleviating oxidative stress and mitochondrial dysfunction. *Frontiers in Pharmacology, 13*(13), 1029775. https://doi.org/10.3389/fphar.2022.1029775
62. Srivastava, G., Dixit, A., Yadav, S., Patel, D. K., Prakash, O., & Singh, M. P. (2012). Resveratrol potentiates cytochrome P450 2d22-mediated neuroprotection in maneb- and paraquat-induced Parkinsonism in the mouse. *Free Radical Biology and Medicine, 52*(8), 1294–1306. ISSN 0891–5849. https://doi.org/10.1016/j.freeradbiomed.2012.02.005
63. Fareed, M. M., Khalid, H., Khalid, S., & Shityakov, S. (2024). Deciphering molecular mechanisms of carbon tetrachloride-induced hepatotoxicity: A brief systematic review. *Current Molecular Medicine, 24*(9), 1124–1134.
64. Meng, X., Tang, G. Y., Liu, P. H., Zhao, C. J., Liu, Q., & Li, H. B. (2020). Antioxidant activity and hepatoprotective effect of 10 medicinal herbs on CCl4-induced liver injury in mice. *World Journal of Gastroenterology, 26*(37), 5629–5645. https://doi.org/10.3748/wjg.v26.i37.5629

65. Mata-Santos, H. A., Dutra, F. F., Rocha, C. C., Lino, F. G., Xavier, F. R., Chinalia, L. A., Hossy, B. H., Castelo-Branco, M. T., Teodoro, A. J., Paiva, C. N., & dos Santos, P. A. (2014). Silymarin reduces profibrogenic cytokines and reverses hepatic fibrosis in chronic murine schistosomiasis. *Antimicrobial Agents and Chemotherapy, 58*(4), 2076–2083.
66. Wang, Y., Branicky, R., Noë, A., & Hekimi, S. (2018). Superoxide dismutases: Dual roles in controlling ROS damage and regulating ROS signaling. *Journal of Cell Biology, 217*(6), 1915–1928. https://doi.org/10.1083/jcb.201708007
67. Fukai, T., & Ushio-Fukai, M. (2011). Superoxide dismutases: Role in redox signaling, vascular function, and diseases. *Antioxidants & Redox Signaling, 15*(6), 1583–1606. https://doi.org/10.1089/ars.2011.3999
68. Chidambaram, S. B., Anand, N., Varma, S. R., Ramamurthy, S., Vichitra, C., Sharma, A., Mahalakshmi, A. M., & Essa, M. M. (2024). Superoxide dismutase and neurological disorders. *IBRO Neuroscience Reports, 16*. ISSN 2667–2421. https://doi.org/10.1016/j.ibneur.2023.11.007
69. Zhang, J., Chen, R., Yu, Z., & Xue, L. (2017). Superoxide dismutase (SOD) and catalase (CAT) activity assay protocols for *Caenorhabditis elegans*. *Bio Protocol, 7*(16), Article e2505. https://doi.org/10.21769/BioProtoc.2505
70. Rasheed, Z. (2024). Therapeutic potentials of catalase: Mechanisms, applications, and future perspectives. *International Journal of Health Sciences (Qassim), 18*(2), 1–6.
71. Anwar, S., Alrumaihi, F., Sarwar, T., Babiker, A. Y., Khan, A. A., Prabhu, S. V., & Rahmani, A. H. (2024). Exploring therapeutic potential of catalase: strategies in disease prevention and management. *Biomolecules, 14*(6), 697. https://doi.org/10.3390/biom14060697
72. Hadwan, M. H., & Ali, S. K. (2018). New spectrophotometric assay for assessments of catalase activity in biological samples. *Analytical Biochemistry, 542*, 29–33. ISSN 0003–2697, https://doi.org/10.1016/j.ab.2017.11.013
73. Handy, D. E., & Loscalzo, J. (2022). The role of glutathione peroxidase-1 in health and disease. *Free Radical Biology Medicine, 188*, 146–161. https://doi.org/10.1016/j.freeradbiomed.2022.06.004
74. Sattar, A., Matin, A., Hadwan, M., Muntadhar, M. H., & Mohammed, R. (2024). Rapid and effective protocol to measure glutathione peroxidase activity. *Bulletin of the National Research Centre, 48*. https://doi.org/10.1186/s42269-024-01250-x
75. Zhang, W., Liu, Y., Liao, Y., Zhu, C., & Zou, Z. (2024). GPX4, ferroptosis, and diseases. *Biomedicine and Pharmacotherapy, 174*, 116512. ISSN 0753–3322. https://doi.org/10.1016/j.biopha.2024.116512
76. Cankurtaran, M., & Ceylan, S. (2023). Chapter 14—Vitamins, minerals, and antiaging drugs in older adults. In: İ. Vargel, & F. F. Özgür (Eds.), *Beauty, aging, and antiaging* (pp. 135–142). Academic Press. ISBN 9780323988049. https://doi.org/10.1016/B978-0-323-98804-9.00010-4
77. Tipple, T. E., & Rogers, L. K. (2012). Methods for the determination of plasma or tissue glutathione levels. *Methods in Molecular Biology, 889*, 315–324. https://doi.org/10.1007/978-1-61779-867-2_20
78. Ballatori, N., Krance, S. M., Notenboom, S., Shi, S., Tieu, K., & Hammond, C. L. (2009). Glutathione dysregulation and the etiology and progression of human diseases. *Biological Chemistry, 390*(3), 191–214. https://doi.org/10.1515/BC.2009.033
79. Salem, M. A., Zayed, A., Merghany, R. M., & Ezzat, S. M. (2022). Chapter 5—Roles of antioxidants in the prevention and management of coronavirus disease. In: C. Egbuna (Ed.), *In drug discovery update, coronavirus drug discovery* (pp. 85–104). Elsevier. ISBN 9780323955744. https://doi.org/10.1016/B978-0-323-95574-4.00020-2
80. Pehlivan, F. E. (2017). Vitamin C: An antioxidant agent. *Vitamin C, 2*(2), 23–35.
81. Lykkesfeldt, J. (2020). On the effect of vitamin C intake on human health: How to (mis) interprete the clinical evidence. *Redox Biology, 1*(34), Article 101532.
82. Pullar, J. M., Bayer, S., & Carr, A. C. (2018). Appropriate handling, processing and analysis of blood samples is essential to avoid oxidation of vitamin C to dehydroascorbic acid. *Antioxidants, 7*(2), 29. https://doi.org/10.3390/antiox7020029

83. Abdullah, M., Jamil, R. T., & Attia, F. N. (2023). *Vitamin C (ascorbic acid).* InStatPearls [Internet]. StatPearls Publishing.
84. Rizvi, S., Raza, S. T., Ahmed, F., Ahmad, A., Abbas, S., & Mahdi, F. (2014). The role of vitamin e in human health and some diseases. *Sultan Qaboos University Medical Journal, 14*(2), e157–e165.
85. Tucker, J. M., & Townsend, D. M. (2005). Alpha-tocopherol: Roles in prevention and therapy of human disease. *Biomedicine & Pharmacotherapy, 59*(7), 380–387. https://doi.org/10.1016/j.biopha.2005.06.005
86. Balakrishnan, R., Satish Kumar, C. S., Rani, M. U., Srikanth, M. K., Boobalan, G., & Reddy. A. G. (2013). An evaluation of the protective role of α-tocopherol on free radical induced hepatotoxicity and nephrotoxicity due to chromium in rats. *Indian Journal of Pharmacology, 45*(5), 490–495. https://doi.org/10.4103/0253-7613.117778
87. Baxter, L. L., Marugan, J. J., Xiao, J., Incao, A., McKew, J. C., Zheng, W., & Pavan, W. J. (2012). Plasma and tissue concentrations of α-tocopherol and δ-tocopherol following high dose dietary supplementation in mice. *Nutrients, 4*(6), 467–490. https://doi.org/10.3390/nu4060467
88. Coornaert, I., Breynaert, A., Hermans, N., De Meyer, G. R., & Martinet, W. (2024). α-Tocopherol inhibits atherogenesis and improves cardiac function in mice independently of its antioxidant properties. *Vascular Biology, 6*(1).
89. Plascencia-Villa, G., & Perry, G. (2021). Preventive and therapeutic strategies in Alzheimer's disease: Focus on oxidative stress, redox metals, and ferroptosis. *Antioxidative Redox Signaling, 34*(8), 591–610. https://doi.org/10.1089/ars.2020.8134
90. Cordiano, R., Di Gioacchino, M., Mangifesta, R., Panzera, C., Gangemi, S., & Minciullo, P. L. (2023). Malondialdehyde as a potential oxidative stress marker for allergy-oriented diseases: An update. *Molecules, 28*(16), 5979. https://doi.org/10.3390/molecules28165979
91. De Leon, J. A. D., & Borges, C. R. (2020). Evaluation of oxidative stress in biological samples using the thiobarbituric acid reactive substances assay. *Journal of Visualized Experiments, 159.* https://doi.org/10.3791/61122
92. Wang, W., Zhang, Z., Liu, X., Cao, X., Wang, L., Ding, Y., & Zhou, X. (2022). An improved GC-MS method for malondialdehyde (MDA) detection: Avoiding the effects of nitrite in foods. *Foods, 11*(9), 1176. https://doi.org/10.3390/foods11091176
93. Lee, R., Margaritis, M., Channon, K. M., & Antoniades, C. (2012). Evaluating oxidative stress in human cardiovascular disease: Methodological aspects and considerations. *Current Medicinal Chemistry, 19*(16), 2504–2520. https://doi.org/10.2174/092986712800493057
94. Holder, C., Adams, A., McGahee, E., Xia, B., Blount, B. C., & Wang, L. (2020). High-throughput and sensitive analysis of free and total 8-isoprostane in urine with isotope-dilution liquid chromatography-tandem mass spectrometry. *ACS Omega, 5*(19), 10919–10926.
95. Miller, E., Morel, A., Saso, L., & Saluk, J. (2024). Isoprostanes and neuroprostanes as biomarkers of oxidative stress in neurodegenerative diseases. *Oxidative Medicine and Cell Longevity, 2014*, 572491. https://doi.org/10.1155/2014/572491
96. Milne, G. L., Gao, B., Terry, E. S., Zackert, W. E., & Sanchez, S. C. (2013). Measurement of F2-isoprostanes and isofurans using gas chromatography-mass spectrometry. *Free Radical Biology and Medicine, 59*, 36–44. https://doi.org/10.1016/j.freeradbiomed.2012.09.030
97. Bono, R., & Romanazzi, V. (2015). Isoprostanes as biomarkers of disease and early biological effect. In: *General methods in biomarker research and their applications* (pp. 383–404). Springer Science+ Business Media Dordrecht.
98. Rodríguez-García, A., García-Vicente, R., Morales, M. L., Ortiz-Ruiz, A., Martínez-López, J., & Linares, M. (2020). Protein carbonylation and lipid peroxidation in hematological malignancies. *Antioxidants (Basel), 9*(12), 1212. https://doi.org/10.3390/antiox9121212
99. Bruno, S., Ronda, L., Paredi, G., Bettati, S., & Mozzarelli, A. (2018). Protein carbonylation detection methods: A comparison. *Data in brief, 1*(19), 2215–2220.
100. Gonos, E. S., Kapetanou, M., Sereikaite, J., Bartosz, G., Naparło, K., Grzesik, M., & Sadowska-Bartosz, I. (2018). Origin and pathophysiology of protein carbonylation, nitration and chlorination in age-related brain diseases and aging. *Aging (Albany NY), 10*(5), 868.

101. Franco, M. C., & Estévez, A. G. (2014). Tyrosine nitration as mediator of cell death. *Cellular and Molecular Life Science, 71*(20), 3939–3950. https://doi.org/10.1007/s00018-014-1662-8
102. Teixeira, D., Fernandes, R., Prudêncio, C., & Vieira, M. (2016). 3-Nitrotyrosine quantification methods: Current concepts and future challenges. *Biochimie, 1*(125), 1–1.
103. Bandookwala, M., & Sengupta, P. (2020). 3-Nitrotyrosine: A versatile oxidative stress biomarker for major neurodegenerative diseases. *International Journal of Neuroscience, 130*(10), 1047–1062. https://doi.org/10.1080/00207454.2020.1713776
104. Omari Shekaftik, S., & Nasirzadeh, N. (2021). 8-Hydroxy-2′-deoxyguanosine (8-OHdG) as a biomarker of oxidative DNA damage induced by occupational exposure to nanomaterials: A systematic review. *Nanotoxicology, 15*(6), 850–864. https://doi.org/10.1080/17435390.2021.1936254
105. Zhang, C., Nestorova, G., Rissman, R. A., & Feng, J. (2013). Detection and quantification of 8-hydroxy-2′-deoxyguanosine in Alzheimer's transgenic mouse urine using capillary electrophoresis. *Electrophoresis, 34*(15), 2268–2274. https://doi.org/10.1002/elps.201300036
106. Goriuc, A., Cojocaru, K. A., Luchian, I., Ursu, R. G., Butnaru, O., & Foia, L. (2024). Using 8-Hydroxy-2′-Deoxiguanosine (8-OHdG) as a reliable biomarker for assessing periodontal disease associated with diabetes. *International Journal of Molecular Sciences, 25*(3), 1425. https://doi.org/10.3390/ijms25031425
107. Nicolescu, A., Babotă, M., Barros, L., Rocchetti, G., Lucini, L., Tanase, C., Mocan, A., Bunea, C. I., & Crişan, G. (2023). Bioaccessibility and bioactive potential of different phytochemical classes from nutraceuticals and functional foods. *Frontiers in Nutrition, 27*(10), 1184535. https://doi.org/10.3389/fnut.2023.1184535
108. Herman, T. F., & Santos, C. (2024). First-pass effect. In: *StatPearls* [Internet]. StatPearls Publishing.
109. Garza, A. Z., Park, S. B., & Kocz, R. (2024). Drug elimination. In: *StatPearls* [Internet]. StatPearls Publishing.
110. Roumeliotis, S., Roumeliotis, A., Dounousi, E., Eleftheriadis, T., & Liakopoulos, V. (2019). Dietary antioxidant supplements and uric acid in chronic kidney disease: A review. *Nutrients, 11*(8), 1911. https://doi.org/10.3390/nu11081911
111. Shi, D., Mi, G., Wang, M., & Webster, T. J. (2019). In vitro and ex vivo systems at the forefront of infection modeling and drug discovery. *Biomaterials, 198*, 228–249. https://doi.org/10.1016/j.biomaterials.2018.10.030
112. Cristea, E., Ghendov-Mosanu, A., Patras, A., Socaciu, C., Pintea, A., Tudor, C., & Sturza, R. (2021). The influence of temperature, storage conditions, pH, and ionic strength on the antioxidant activity and color parameters of rowan berry extracts. *Molecules, 26*(13), 3786. https://doi.org/10.3390/molecules26133786
113. Aqil, F., Munagala, R., Jeyabalan, J., & Vadhanam, M. V. (2013). Bioavailability of phytochemicals and its enhancement by drug delivery systems. *Cancer Letters, 334*(1), 133–141. https://doi.org/10.1016/j.canlet.2013.02.032
114. Dei Cas, M., & Ghidoni, R. (2019). Dietary curcumin: Correlation between bioavailability and health potential. *Nutrients, 11*(9), 2147. https://doi.org/10.3390/nu11092147
115. Li, S., Tan, H. Y., Wang, N., Zhang, Z. J., Lao, L., Wong, C. W., & Feng, Y. (2015). The role of oxidative stress and antioxidants in liver diseases. *International Journal of Molecular Sciences, 16*(11), 26087–26124. https://doi.org/10.3390/ijms161125942
116. Gambini, J., Inglés, M., Olaso, G., Lopez-Grueso, R., Bonet-Costa, V., Gimeno-Mallench, L., Mas-Bargues, C., Abdelaziz, K. M., Gomez-Cabrera, M. C., Vina, J., & Borras, C. (2015). Properties of resveratrol: In vitro and in vivo studies about metabolism, bioavailability, and biological effects in animal models and humans. *Oxidative Medicine and Cellular Longevity, 2015*(1), Article 837042.
117. Favari, C., Rinaldi de Alvarenga, J. F., Sánchez-Martínez, L., Tosi, N., Mignogna, C., Cremonini, E., Manach, C., Bresciani, L., Del Rio, D., & Mena, P. (2024). Factors driving the inter-individual variability in the metabolism and bioavailability of (poly)phenolic metabolites: A systematic review of human studies. *Redox Biology, 71*, 103095. https://doi.org/10.1016/j.redox.2024.103095

118. Auer, S. K., Salin, K., Anderson, G. J., & Metcalfe, N. B. (2018). Individuals exhibit consistent differences in their metabolic rates across changing thermal conditions. *Comparative Biochemistry and Physiology Part A: Molecular & Integrative Physiology*, 217, 1–6. https://doi.org/10.1016/j.cbpa.2017.11.021
119. Jansson-Löfmark, R., Hjorth, S., & Gabrielsson, J. (2020). Does in vitro potency predict clinically efficacious concentrations? *Clinical Pharmacology and Therapeutics, 108*(2), 298–305. https://doi.org/10.1002/cpt.1846
120. Price, G., & Patelm D. A. (2024). Drug bioavailability. In: *StatPearls* [Internet]. StatPearls Publishing.
121. Murakami, T., Bodor, E., & Bodor, N. (2021). Factors and dosage formulations affecting the solubility and bioavailability of P-glycoprotein substrate drugs. *Expert Opinion on Drug Metabolism & Toxicology, 17*(5), 555–580. https://doi.org/10.1080/17425255.2021.1902986
122. Preeti, S. S., Saharan, R., Narwal, S., Malik, R., Gahlot, V., Khalid, A., Najmi, A., Zoghebi, K., Halawi, M. A., Albratty, M., & Mohan, S. (2023). Exploring LIPIDs for their potential to improves bioavailability of lipophilic drugs candidates: A review. *Saudi Pharmaceutical Journal, 31*(12), 101870. https://doi.org/10.1016/j.jsps.2023.101870
123. Bešlo, D., Golubić, N., Rastija, V., Agić, D., Karnaš, M., Šubarić, D., & Lučić, B. (2023). Antioxidant activity, metabolism, and bioavailability of polyphenols in the diet of animals. *Antioxidants (Basel), 12*(6), 1141. https://doi.org/10.3390/antiox12061141
124. Salin, K., Auer, S. K., Rudolf, A. M., Anderson, G. J., Cairns, A. G., Mullen, W., Hartley, R. C., Selman, C., & Metcalfe, N. B. (2015). Individuals with higher metabolic rates have lower levels of reactive oxygen species in vivo. *Biology Letters, 11*(9), 20150538. https://doi.org/10.1098/rsbl.2015.0538
125. Losada-Barreiro, S., Sezgin-Bayindir, Z., Paiva-Martins, F., & Bravo-Díaz, C. (2022). Biochemistry of antioxidants: Mechanisms and pharmaceutical applications. *Biomedicines, 10*(12), 3051. https://doi.org/10.3390/biomedicines10123051
126. Dang, Y., & Guan, J. (2020). Nanoparticle-based drug delivery systems for cancer therapy. *Smart Materials in Medicine, 1*(1), 10–19.
127. Guérard, M., Baum, M., Bitsch, A., Eisenbrand, G., Elhajouji, A., Epe, B., Habermeyer, M., Kaina, B., Martus, H. J., Pfuhler, S., Schmitz, C., Sutter, A., Thomas, A. D., Ziemann, C., & Froetschl, R. (2015). Assessment of mechanisms driving non-linear dose-response relationships in genotoxicity testing. *Mutation Research/Reviews in Mutation Research, 763*, 181–201. https://doi.org/10.1016/j.mrrev.2014.11.001
128. Borgert, C. J., Fuentes, C., & Burgoon, L. D. (2021). Principles of dose-setting in toxicology studies: The importance of kinetics for ensuring human safety. *Archives of Toxicology, 95*(12), 3651–3664. https://doi.org/10.1007/s00204-021-03155-4
129. Sari, B., Isik, M., Eylem, C. C., Bektas, C., Okesola, B. O., Karakaya, E., Emregul, E., Nemutlu, E., & Derkus, B. (2022). Omics technologies for high-throughput-screening of cell–biomaterial interactions. *Molecular Omics, 18*(7), 591–615. https://doi.org/10.1039/d2mo00060a
130. Wu, Q., Zheng, J., Sui, X., Fu, C., Cui, X., Liao, B., Ji, H., Luo, Y., He, A., Lu, X., Xue, X., Tan, C. S. H., & Tian, R. (2024). High-throughput drug target discovery by fully automated proteomics sample preparation platform. *Chemical Science, 15*(8), 2833–2847. https://doi.org/10.1039/d3sc05937e
131. Bordbar, A., Monk, J. M., King, Z. A., & Palsson, B. O. (2014). Constraint-based models predict metabolic and associated cellular functions. *Nature Reviews Genetics, 15*(2), 107–120. https://doi.org/10.1038/nrg3643
132. Valenzuela-Riffo, F., & Morales-Quintana, L. (2020). Study of the structure and binding site features of FaEXPA2, an α-expansin protein involved in strawberry fruit softening. *Computational Biology and Chemistry, 87*, Article 107279. https://doi.org/10.1016/j.compbiolchem.2020.107279
133. Rantanen, J., & Fatouros, D. G. (2021). Preface: Additive manufacturing in pharmaceutical product design. *Advanced Drug Delivery Reviews, 178*, Article 113991. https://doi.org/10.1016/j.addr.2021.113991

134. Nikhade, N., Telrandhe, R., & Ansari, M. (2019). A review of natural an-tioxidants in medicinal plants.
135. Duthie, G. G., Wahle, K. W. J., & James, W. P. T. (1989). Oxidants, antioxidants and cardiovascular disease. *Nutrition Research Reviews, 2*(1), 51–62.
136. Rabanal-Ruiz, Y., Llanos-González, E., & Alcain, F. J. (2021). The use of coenzyme Q10 in cardiovascular diseases. *Antioxidants, 10*(5), 755.
137. Wongcharoen, W., & Phrommintikul, A. (2009). The protective role of curcumin in cardiovascular diseases. *International Journal of Cardiology, 133*(2), 145–151.
138. Wang, H., Yang, Y. J., Qian, H. Y., Zhang, Q., Xu, H., & Li, J. J. (2012). Resveratrol in cardiovascular disease: What is known from current research? *Heart Failure Reviews, 17*, 437–448.
139. Eghbaliferiz, S., & Iranshahi, M. (2016). Prooxidant activity of polyphenols, flavonoids, anthocyanins and carotenoids: Updated review of mechanisms and catalyzing metals. *Phytotherapy Research, 30*(9), 1379–1391.
140. Casetta, I., Govoni, V., & Granieri, E. (2005). Oxidative stress, antioxidants and neurodegenerative diseases. *Current Pharmaceutical Design, 11*(16), 2033–2052.
141. Ulatowski, L. M., & Manor, D. (2015). Vitamin E and neurodegeneration. *Neurobiology of Disease, 84*, 78–83.
142. Hecht, F., Zocchi, M., Alimohammadi, F., & Harris, I. S. (2024). Regulation of antioxidants in cancer. *Molecular Cell, 84*(1), 23–33.
143. Druesne-Pecollo, N., Latino-Martel, P., Norat, T., Barrandon, E., Bertrais, S., Galan, P., & Hercberg, S. (2010). Beta-carotene supplementation and cancer risk: A systematic review and metaanalysis of randomized controlled trials. *International Journal of Cancer, 127*(1), 172–184.
144. Lobo, V., Patil, A., Phatak, A., & Chandra, N. (2010). Free radicals, antioxidants and functional foods: Impact on human health. *Pharmacognosy Reviews/Bioinformatics Trends/Pharmacognosy Review, 4*(8), 118. https://doi.org/10.4103/0973-7847.70902
145. Valko, M., Leibfritz, D., Moncol, J., Cronin, M. T., Mazur, M., & Telser, J. (2006). Free radicals and antioxidants in normal physiological functions and human disease. *The International Journal of Biochemistry & Cell Biology, 39*(1), 44–84. https://doi.org/10.1016/j.biocel.2006.07.001
146. de Juan-Franco, E., Pérez, A., Ribas, V., Sánchez-Hernández, J. A., Blanco-Vaca, F., Ordóñez-Llanos, J., & Sánchez-Quesada, J. L. (2009). Standardization of a method to evaluate the antioxidant capacity of high-density lipoproteins. *International Journal of Biomedical Science: IJBS, 5*(4), 402.
147. Porrini, M., & Riso, P. (2008). Factors influencing the bioavailability of antioxidants in foods: A critical appraisal. *NMCD. Nutrition Metabolism and Cardiovascular Diseases, 18*(10), 647–650.
148. Liao, Z., Wong, S. W., Yeo, H. L., & Zhao, Y. (2020). Smart nanocarriers for cancer treatment: Clinical impact and safety. *NanoImpact, 20*, Article 100253.
149. Peneş, N. O., Weber, B., & Păun, S. D. (2017). Role of genetic polymorphism in nutritional supplementation therapy in personalized medicine. *Romanian Journal of Morphology and Embryology, 58*(1), 53–58.
150. Yang, F., Wendusubilige, Kong, J., Zong, Y., Wang, M., Jing, C., Ma, Z., Li, W., Cao, R., Jing, S., Gao, J., Li, W., & Wang, J. (2023). Identifying oxidative stress-related biomarkers in idiopathic pulmonary fibrosis in the context of predictive, preventive, and personalized medicine using integrative omics approaches and machine-learning strategies. *EPMA Journal, 14*(3), 417–442.

Mechanisms of Antioxidant and Prooxidant Interactions: Redox Dynamics in Plants

Renuka Suresh, Sadananda Mal, Shweta Panchal, and Ramanathan Karuppasamy

Abstract Redox reactions support processes such as maintaining homeostasis and metabolic functions in cells, making it possible for the plants to perfuse oxygenated conditions. The balance, however, could be interfered with by exogenous prooxidants, especially herbicides, which produce excess reactive oxygen species (ROS), leading to oxidative stress. Herbicide use, which is normally meant for weeds, also causes lipid peroxidation followed by cell death, which hinders the development and productivity of a plant. In terms of herbicide tolerance, crop plants possess certain enzymatic systems like cytochrome P450. On the contrary, excessive application of these chemicals into and onto the environment may negatively affect it and the non-target organisms, including farmers, through indirect effects. Studies have proven that prooxidant herbicides like glyphosate induce oxidative stress among agricultural workers; consequently, there are health-related issues, including respiratory disorders and other oxidative stress-associated conditions. This chapter calls for urgent attention toward alternative, eco-friendly weed management, like bioherbicides, to reduce health and environmental risks from synthetic herbicides.

Keywords Redox reaction · Prooxidants · Herbicides · Oxidative stress · Reactive oxygen species · Bioherbicides

1 Introduction to Redox Reactions in Biological Organisms

Redox state is one of those terms that have changed in meaning with the seasons. The term originally refers to the relative amounts of the oxidized and reduced forms of a given redox pair. It has become very broad in recent years and could, for instance, refer to the overlying redox state in complex biological systems, such as cells [1]. Furthermore, redox (reduction–oxidation) reactions are fundamental to life processes across all biological organisms. These reactions, which involve the

R. Suresh · S. Mal · S. Panchal · R. Karuppasamy (✉)
Department of Biotechnology, School of Bio Sciences and Technology, Vellore Institute of Technology, Vellore, India
e-mail: kramanathan@vit.ac.in

K. Kasinathan et al. (eds.), *Recent Advances in Oxidative Stress Associated Chronic Diseases Volume 1*, https://doi.org/10.1007/978-981-96-9670-3_8

transfer of electrons between molecules, are essential for various functions such as energy production, growth regulation, and stress responses. These cellular functions are very complicated to control, being reliant on a wide range of components. Compounds based on oxygen are an instrumental part of the small, relatively reactive molecules that control cellular activities. These molecules have historically been referred to as reactive oxygen species (ROS) [2]. Plants, like animals, are obligate aerobic organisms, requiring oxygen for mitochondrial energy production. Interestingly, plants possess a unique ability to tolerate much higher oxygen concentrations than animals. This is mostly because, during the day, their green tissues continuously produce oxygen through photosynthesis. As a result, in comparison to non-green components like roots, green leaves have oxygen concentrations that are 2.5 times greater. Furthermore, compared to animal cells, the oxygen levels in both green and non-green plant tissues remain noticeably greater [3]. Plant redox processes are especially important because of this oxygen-rich environment. They control growth and development, as well as essential functions, including respiration, photosynthesis, and cellular signaling. Thus, a delicate balance between oxidants and antioxidants must be preserved for cellular integrity and appropriate responses to environmental stimuli. Therefore, the balance between oxidants and antioxidants has to be maintained delicately in terms of maintaining cellular integrity and proper responses to environmental stimuli. Any change in this redox homeostasis could lead to the generation of ROS in a process referred to as oxidative stress. Remarkably, ROS, which comprise superoxide radicals (O_2^-), hydrogen peroxide (H_2O_2), and hydroxyl radicals (OH^-), play a dual role in plant biology. On one hand, these molecules may harm cellular structures and functions when present in excessive quantities. On the other hand, when present in low concentrations, ROS serve as important signaling molecules. Therefore, the relationship between plants and ROS is complex and subtle. Such a delicate balancing act indicates the sophistication of plant biochemistry and how important redox homeostasis is to plant health and function.

Considering normal physiological conditions, various enzyme systems work diligently to maintain intracellular reactive oxygen species at low, manageable levels. However, this delicate balance can be disrupted by prooxidants, which are endobiotic or xenobiotic substances that induce oxidative stress, either by generating ROS or by inhibiting antioxidant systems. There are many types of prooxidants classified under exogenous and endogenous types [4]. To counteract the potential cellular damage caused by these prooxidants, organisms have evolved several strategies. Indeed, in light of the oxidative stress-related free radical theory, antioxidants are considered the first line of defense against cellular stress. The endogenous antioxidant defense system comprises a complex network of both enzymatic and non-enzymatic molecules, strategically distributed throughout the cytoplasm and various cell organelles, which orchestrate a complex cascade of reactions to convert harmful ROS into more stable molecules, such as water and oxygen, commonly referred to as ROS scavengers [5]. Consequently, we can refine our understanding of oxidative stress as not merely the presence of ROS, but more accurately as an imbalance between prooxidants and antioxidants. This perspective underscores the importance of maintaining equilibrium in cellular redox states for optimal organism health and functions [4]. In the context of agriculture, understanding redox dynamics in plants is crucial when addressing agricultural challenges, especially the widespread use of herbicides that could further complicate plant redox homeostasis. In modern farming practices, herbicides have supplanted traditional weeding methods. Specific target-dependent herbicides, however, are prooxidants that promote oxidative stress by generating excessive reactive oxygen species (ROS), leading to lipid peroxidation and damage to cellular structures, ultimately affecting the weed's growth and yield. Additionally, studies by [6] have identified important detoxification enzymes in crop plants, particularly cytochrome P450s, which have shed light on the selective effectiveness of prooxidant herbicides like PPO. Weeds lack these essential detoxifying factors, which causes them to die from an excessive buildup of ROS, whereas crop plants gain from this enzymatic defense. Most interestingly, cytochrome P450s (CYPs) not only function in herbicide detoxification but also provide critical multifunctional catalytic activity for the synthesis of important secondary metabolites and antioxidants. Therefore, they also augment the general resilience and sustainability of crop plants [7].

While effective in weed control, excessive chemical herbicide application can have detrimental effects on the environment and farmers. In a study conducted by [8], it was reported that prooxidant herbicides like glyphosate can enter the body through the digestive, respiratory, and cutaneous systems while mixing and spraying the herbicides, and cleaning the equipment. Previous studies have also found that farmers who were found to use glyphosate were linked to various illnesses, including respiratory disorders [9, 10]. Furthermore, laboratory findings added weight to the findings that exposure to glyphosate causes adverse biological effects, such as oxidative stress in farmers [11]. These documented health risks to farmers, including respiratory disorders and oxidative stress from herbicide exposure, underscore the urgent need for safer alternatives. New sustainable methods of weed management should involve exploring bioherbicides synthesized from natural plant compounds. In a nutshell, this chapter examines the complex interrelationships among oxidative stress caused by herbicides, redox biology, and the health of both plants and agricultural laborers. As prooxidant herbicides interfere with redox homeostasis in plants and expose serious health risks from exposure to farmers, knowledge of these mechanisms is important for the development of less harmful and more sustainable agricultural practices. By studying the redox balance in plants in the light of resulting exposure to herbicides, this chapter aims to illuminate the urgent need for weed management paradigms that focus on both environmental and human health. To achieve the research objective as stated above, the remainder of this chapter is organized as follows: The next section will provide a brief on the fundamentals of redox biology in plants. The second section will provide details on herbicides as prooxidants. While crop plants demonstrate remarkable tolerance mechanisms, it is equally important to understand how plants respond to herbicide-induced oxidative stress, which we examine in the third section. Moving beyond plant responses, we will also focus on examining the crucial aspect of how these prooxidant herbicides affect the health and safety of agricultural workers. Finally, considering these challenges, we conclude by exploring sustainable alternatives and future directions in weed management strategies.

2 Fundamentals of Redox Biology in Plants

Redox biology is an integral component of plant metabolism, with the cellular redox state being governed by the oxidation or reduction of various redox-active molecules that participate in numerous metabolic processes [12]. However, they must continuously manage the challenge of maintaining a balanced redox state to ensure proper physiological functions and development in an oxidizing environment. It is now clear that plants coordinate a diverse array of reactions involving reactive molecular species, redox modifications, and redox-regulating proteins, all of which play a crucial role in maintaining cellular homeostasis [13–15]. Plants have evolved a sophisticated redox signaling network, with chloroplasts, peroxisomes, and mitochondria functioning as critical redox signaling centers that mediate intracellular communication and regulate responses to stress. In chloroplasts, reductants like ferredoxin (Fdx) and NADPH are generated through the photosynthetic electron transport chain (ETC) and, together with ATP, are utilized for the synthesis of sugar phosphates, amino acids, and various other metabolites, which are distributed to the rest of the cell. Additionally, NAD(P)H metabolism plays a central role in key cellular processes such as glycolysis, fermentation, and the oxidative pentose phosphate (OPP) pathway in the cytosol, respiratory electron transport, the tricarboxylic acid (TCA) cycle, and biosynthetic activities in mitochondria, as well as photorespiration in mitochondria, plastids, and peroxisomes [16–19].

2.1 *Chloroplastidial Redox Biology*

Chloroplasts conduct oxygenic photosynthesis through light-dependent photosystems (PSII and PSI), where electrons derived from water molecules are transferred via the ferredoxin–thioredoxin (FDX–FTR–TRX) pathway to $NADP^+$, producing NADPH as reducing power. This process also generates a proton gradient across the thylakoid membrane, which drives ATP synthesis through the CF0F1 ATP synthase complex [20, 21]. These ATP and NADPH are subsequently utilized in the dark reactions to convert inorganic CO_2 into organic sugars, forming the primary mechanism of carbon fixation in higher plants [22]. During the light reactions of photosynthesis, electron transfer occurs in the presence of oxygen, leading to the generation of reactive oxygen species (ROS) in both photosystems. At PSII, singlet oxygen is produced, while PSI generates superoxide anions, which can cause oxidative damage to cells [4, 23–25]. To mitigate this, photosynthetic processes are closely linked to the antioxidant defense system. Chloroplasts contain both enzymatic and non-enzymatic ROS scavenging mechanisms. For instance, Cu/Zn and Fe-superoxide dismutases (SODs) are enzymatic systems that catalyze the conversion of superoxide anions into H_2O_2. Although both ROS production and scavenging can contribute to oxidative stress, ROS signaling is essential for modulating gene expression and post-translational modifications, particularly for scavenging H_2O_2 and enabling the plant to adapt to environmental conditions [26, 27]. In this context, alongside the FDX–FTR–TRX pathway, the NADPH-dependent thioredoxin reductase C (NTRC) system plays a pivotal role in the efficient redox regulation of 2-Cys peroxiredoxins (PRXs), ensuring the reductive activity of chloroplast TRXs in the light. The NTRC–2-Cys PRX system finely regulates the redox state of chloroplast enzymes, optimizing photosynthesis under varying light conditions. During darkness, 2-Cys PRXs enable the rapid oxidative inactivation of chloroplast enzymes by facilitating the transfer of reducing equivalents from these enzymes, through TRXs, to hydrogen peroxide [17, 28–32].

2.2 *Mitochondrial Redox Biology*

Plant mitochondria structure closely resembles that of other eukaryotes and plays a vital role in their function as efficient energy-producing organelles [33]. In addition to the conventional components of the mitochondrial electron transport chain (mtETC), such as Complexes I–IV and ATP synthase (Complex V), plant mitochondria possess several unique elements, including alternative NADH dehydrogenases (NDs) and alternative oxidases (AOXs) [34–36]. Mitochondrial redox biology and metabolism are primarily governed by its key flux pathways, particularly respiration and the photorespiration pathway, which is shared with the chloroplast, peroxisome, and cytosol [37, 38]. Oxidative reactions in glycolysis and the TCA cycle result in the production of NADH and $FADH_2$ molecules. The mETC oxidizes these reduced coenzymes, releasing free energy that drives proton translocation across the inner mitochondrial membrane, establishing an electrochemical proton gradient. This energy is harnessed as protons flow back through the F0 channel of ATP synthase. The F1 catalytic site then uses this energy to synthesize ATP from ADP and inorganic phosphate (Pi) in the mitochondrial matrix [29, 38–40]. Like chloroplasts, mitochondria are organelles where thioredoxins (TRXs play a central role in connecting redox signaling to the regulation of key organelle processes. TRXs act as key regulators of the mitochondrial TCA cycle and the citrate shunt pathway, which exports cycle intermediates to the cytosol for biosynthesis. TRXs also implicated in the regulation of mitochondrial electron transport, both indirectly through fumarase (FUM, linked to NADH-derived electron accumulation, and directly via succinate dehydrogenase (SDH, which feeds electrons into the mETC at complex II [41–43]. During the interaction of oxygen with complex IV and AOX, water is produced by transferring four electrons. However, interactions between oxygen and the reduced forms of other electron transport components, such as flavins and ubiquinone, lead to the generation of ROS. Following ROS generation, manganese superoxide dismutases (MnSODs) convert superoxide into hydrogen peroxide (H_2O_2). To prevent H_2O_2 accumulation, a range of antioxidant defense systems is active [44]. Ascorbate and tripeptide glutathione, glutathione peroxidase (GPX)-like enzymes, and peroxiredoxins (PRXs) play key roles in these systems, effectively scavenging H_2O_2 [13, 45, 46].

2.3 Peroxisomal Redox Biology

The peroxisome plays a crucial role in cellular redox regulation. On one hand, it is involved in the beta-oxidation of fatty acids, jasmonic acid biosynthesis, and auxin metabolism. On the other hand, it participates in photorespiration [47]. A wide array of reductase and dehydrogenase enzymes within the peroxisome contribute to regulating cellular redox balance. Peroxisomes harbor the highest concentrations of H_2O_2 among cellular organelles. In green tissues, the primary source of peroxisomal H_2O_2 is glycolate oxidase (GOX) during the photorespiration cycle [48]. Additionally, fatty-acid β-oxidation, polyamine catabolism, and sarcosine oxidase significantly contribute to H_2O_2 production in peroxisomes. The regulation of H_2O_2 levels within peroxisomes is achieved through a balance between its generation and scavenging, primarily mediated by the enzyme catalase (CAT) [29, 49]. In addition to reactive oxygen species (ROS), peroxisomes also generate nitrogen oxide (NO) and reactive nitrogen species (RNS). Peroxisomal NO production has been associated with NO synthase-like (NOS^{-1}) activity. Other nitrogen-derived species, including peroxynitrite ($ONOO^-$) and nitrosoglutathione (GSNO), are also present. Peroxisomal superoxide dismutase (SOD) regulates $ONOO^-$ levels by controlling the availability of $O_2{}^-$, while catalase (CAT) degrades peroxynitrite. Together, these enzymes play a crucial role in modulating the interplay between H_2O_2 and NO/$ONOO^-$ signaling pathways. Additionally, urate, a well-known peroxynitrite scavenger, may contribute to regulating $ONOO^-$ in peroxisomes. S-nitrosoglutathione reductase (GSNOR), which modulates NO and S-nitrosothiol levels, has also been identified in plant peroxisomes [49–53]. Ultimately, maintaining redox balance is critical for plant health, influencing growth, stress responses, and essential metabolic functions. Key organelles such as chloroplasts, mitochondria, and peroxisomes play vital roles in generating redox signals that support cellular homeostasis. Furthermore, the plant's antioxidant defenses counteract ROS and RNS, safeguarding cellular integrity. Advances in sensor technology hold promise for deeper insights into plant redox dynamics, enhancing our understanding of plant resilience and adaptation.

3 Herbicides as Prooxidants

The agricultural sector is dependent on herbicides to control weed growth and improve crop yields. These substances confer benefits and risks; while they are supposed to be selective in their actions, they can just as well harm the crops they protect. Furthermore, when herbicides exhibit prooxidant activity in plant cells, they induce overproduction of ROS, which then disrupts the physiological redox balance of the cell. Manifestations of oxidative stress may vary and include decreased leaf area index, reduced SDW, stunted growth of the whole plant, and altered metabolism. For stressed plants, it can generally be related to the generation of ROS and the consequent oxidative damage to plant cells [4]. ROS can be introduced by herbicides through multiple pathways. When these pesticides, including herbicides, enter the environment as pollutants and are absorbed by plants, they function as xenobiotics, triggering widespread cellular and tissue responses that often result in damage [54]. Furthermore, herbicide exposure typically causes ROS levels to spike within plant cells, destabilizing the crucial equilibrium between oxidative and antioxidative processes. This oxidative stress cascade leads to a series of damaging effects: Membranes become compromised, lipids undergo peroxidation, and both proteins and DNA suffer structural damage. These oxidative impacts are fundamental to understanding the toxic effects herbicides have on plants [55].

3.1 Mechanism of Herbicide-Induced ROS Generation

Interestingly, herbicides vary in their mode of action: Some exert their effect by generating ROS directly as part of their primary function, while others produce ROS indirectly due to the stress imposed on plants by these herbicides (Table 1). Regardless of the mechanism, the results can be catastrophic, leading to severe biochemical disruptions, cellular damage, and often death of the plant.

3.1.1 Direct Generation of Reactive Oxygen Species (ROS)

Some herbicides directly generate ROS by interacting with specific enzymes or redox-active components in plant cells. For example, photosystem II inhibitors like atrazine disrupt the photosynthetic electron transport system (ETS), leading to the generation of singlet oxygen. In addition, protoporphyrinogen oxidase (PPO) inhibiting herbicides, such as forseen, inhibit the PPO enzyme, which catalyzes chlorophyll biosynthesis. Under such conditions, protoporphyrinogen IX accumulates in the cell and leaks out of chloroplasts, where it spontaneously oxidizes in the presence of light and oxygen, generating singlet oxygen and other ROS. This ROS drives lipid peroxidation, damaging cell membranes, oxidizing proteins, and ultimately causing pigment loss, especially chlorophyll and carotenoids. This degradation of cellular structures leads to plant collapse and death [56, 57].

3.1.2 Disruption of Antioxidant and Redox Systems

Herbicides can disturb the antioxidative systems of the plants as they identify them, thus diminishing cellular defense capabilities against detoxifying ROS. Some herbicides interfere with redox-sensitive signaling pathways, hence altering the already delicate balance between ROS production and scavenging. This breaks the control of ROS production upon protein cellular communication, energy transfer, and defense

Table 1 Classes of herbicides with prooxidant activity

S.no.	Class of herbicide	Biological process	Active ingredient	Chemical family
1	ALS (acetohydroxyacid synthase) inhibitor	Amino acid biosynthesis	Metsulfuron, bensulfuron, chlorimuron, halosulfuron	Imidazolinone, pyrimidinyl benzoates, sulfonylurea, triazolinones, triazolopyrimidine type 1, triazolopyrimidine type 2
2	Photosystem II inhibitors	Photosynthesis (electron transfer)	Propanil, atrazine, prometon, hexazinone, diuron	Amide, triazine, triazinone, uracil, urea
3	PPO (protoporphyrinogen oxidase) inhibitors	Cell membrane disruptors	Fomesafen, lactofen, saflufenacil, carfentrazone	Diphenyl ether, N-phenyl imide, N-phenyl triazolinone
4	Photosystem I electron diverter	Cell membrane disruptors	Diquat, paraquat	Pyridinium
5	HPPD (4-hydroxyphenylpyruvate dioxygenase) inhibitors	Pigment inhibitors	Isoxaflutole, pyrasulfotole, mesotrione, tembotrione	Isoxazole, pyrazol, triketone
6	Glutamine synthetase inhibitors	Nitrogen metabolism inhibitor	Glufosinate	Phosphinic acid
7	EPSP (5-enolpyruvyl-shikimate-3-phosphate) synthase inhibitor	Amino acid biosynthesis	Glyphosate	Glycine
8	Synthetic auxins (TIR1, AFB1-5, and unknown auxin receptor)	Growth regulators	2,4-D, dicamba, aminopyralid, halauxifen, quinclorac	Benzoate, phenoxy carboxylate, pyridine carboxylate, pyridyloxy carboxylate, quinoline carboxylate

mechanisms. Inhibitors of the glutamine synthetase (GS) enzyme—glufosinate—withhold the ROS accumulation through the indirect inhibition of critically central antioxidant enzyme activity that allows the cell to maintain proper redox homeostasis. According to Takano et al. [58], glufosinate inhibits GS, disrupting key photosynthetic processes like photorespiration, carbon fixation, and electron flow that regulate ROS levels. This disruption overwhelms the antioxidant defense, forcing oxygen from PSII to accept excess electrons and generate ROS. Thus, inhibition of GS indirectly reinforces the burden on the antioxidant system by surging ROS, which it cannot rectify. Conversely, glyphosate-based herbicides like Roundup inhibit 5-enolpyruvylshikimate-3-phosphate synthase (EPSPS) within the shikimic acid pathway, preventing the plant from completing aromatic amino acid synthesis, which is critical for antioxidant functions as well as electron transport, hence reducing the capability of the plant to scavenge ROS [54].

3.1.3 Interaction of Herbicides with Cellular Redox System

Typically, in most cases, herbicides are involved in the interaction of cellular redox systems and result in oxidative damage to key biomolecules such as lipids, proteins, and nucleic acids. Lipid peroxidation is an outcome of herbicide activity, like diclofop-methyl, in generating reactive lipid species that would propagate the oxidative damage for further oxidation within the cell. In simpler terms, they induce a lipid peroxidation chain reaction in which ROS strikes polyunsaturated fatty acids in the cell membrane. Formed lipid radicals and attack from oxygen give birth to lipid peroxyl radicals, from which they are even able to keep on reacting with any of their nearby lipids, spreading injury all over the membranes and enhancing cell permeability. The result of this chain reaction is reactive lipid species, partly represented as malondialdehyde (MDA) and 4-hydroxynonenal (HNE), which can incur damage to proteins and DNA and the whole metabolite within the cells, intensifying oxidative stress and cell damage within weed species [59, 60].

3.2 Types of ROS Generated by Herbicides

Herbicide-induced ROS generation primarily includes the production of specific types of ROS, each with unique reactivity and roles in cellular damage. ROS are mainly generated in the chloroplasts, mitochondria, and peroxisomes.

3.2.1 Superoxide Radicals ($O_2^{\bullet -}$)

Superoxide radicals are often the initial ROS produced in herbicide-stressed cells, primarily in chloroplasts, where electron transport is disrupted with a half-life less than 1ms. Superoxide radicals are more reactive than O_2 and are produced either through the partial reduction of oxygen in electron transport systems or through enzymatic reactions triggered by the herbicide, which could ultimately cause oxidative damage in plants [55].

3.2.2 Hydrogen Peroxide (H_2O_2)

Hydrogen peroxide is the most abundant nonradical ROS that reacts fast with other molecules with a half-life greater than 1 ms [18]. H_2O_2 is a likely candidate for stress signaling because it can diffuse across membranes and initiate oxidative damage throughout the cell. It could also be beneficial to the plants, but in excess amounts, it could cause irreversible oxidative damage to various molecules. They likely target bound and unbound iron, disrupting iron metabolism and generating •OH [55].

3.2.3 Hydroxyl Radicals (•OH)

Hydroxyl radicals are highly reactive and capable of causing extensive cellular damage by initiating lipid peroxidation and oxidizing DNA, proteins, and other macromolecules. Herbicides that induce redox cycling, such as protoporphyrinogen oxidase inhibitors, lead to the formation of hydroxyl radicals [55].

3.2.4 Singlet Oxygen ($^1O_2$)

Singlet oxygen is a highly reactive, excited form of oxygen with a half-life of 1–4 μs generated primarily through photosystem II inhibition in chloroplasts. According to Dumanović et al. [5], $^1O_2$ contributes to oxidative stress, damages cell membranes and proteins, and attacks Trp, Tyr, and Met residues on DNA. In short, herbicides with prooxidant activity pose significant oxidative challenges to plant cells, disrupting redox homeostasis and generating harmful ROS. To counteract these effects, plants have evolved sophisticated antioxidant defense systems that enhance their herbicide tolerance. These defense systems, composed of enzymatic and non-enzymatic antioxidants, play a crucial role in neutralizing ROS and stabilizing cellular structures under herbicide stress.

4 Antioxidant Defense Systems and Herbicide Tolerance in Plants

To counteract these damaging oxidative effects of herbicides, plants have evolved complex antioxidant defense systems that include both enzymatic and non-enzymatic antioxidants.

4.1 *Enzymatic Antioxidants*

Enzymatic antioxidants are the first line of defense against ROS in plants, neutralizing harmful species and mitigating cellular damage.

4.1.1 Superoxide Dismutase (SOD)

Superoxide dismutase catalyzes the conversion of superoxide radicals (O_2^-) into hydrogen peroxide (H_2O_2), a less reactive molecule. SOD plays a pivotal role in controlling ROS levels within cellular compartments, including chloroplasts, mitochondria, and peroxisomes, where oxidative stress is often heightened. By converting superoxide radicals into H_2O_2, SOD prevents the formation of more reactive and harmful species like hydroxyl radicals, which can damage lipids, proteins, and DNA [61].

4.1.2 Catalase (CAT)

CAT is essential for decomposing hydrogen peroxide into water and oxygen, which is a crucial step in maintaining cellular redox balance. They are active in peroxisomes, where they remove H_2O_2 generated during photorespiration and fatty-acid oxidation. Catalase activity has been closely associated with herbicide tolerance, as herbicide-induced ROS often leads to excessive H_2O_2 accumulation [56].

4.1.3 Glutathione Peroxidase (GPx)

GPx uses glutathione (GSH) as an electron donor to reduce hydrogen peroxide and organic hydroperoxides. By preventing lipid peroxidation and protecting membrane structures, GPx is essential for preserving cellular integrity during stress conditions. It operates synergistically with catalase and other peroxidases, enhancing the plant's resilience to herbicide-induced oxidative stress [62].

4.1.4 Peroxidases (POD)

Peroxidases catalyze the reduction of hydrogen peroxide by various electron donors, including phenolics, ascorbate, and glutathione (GSH). These enzymes are widely distributed in plant tissues, where they are involved in multiple physiological processes, including lignification and cross-linking of cell wall components [61].

4.2 Non-enzymatic Antioxidants

By directly scavenging ROS and halting oxidative damage brought on by herbicide stress, non-enzymatic antioxidants serve as an additional line of defense in addition to enzymatic antioxidants [56].

4.2.1 Ascorbic Acid (AsA)

Vitamin C, often referred to as ascorbic acid, is a water-soluble non-enzymatic antioxidant [56]. Its structure facilitates easy electron donation, which lowers reactive oxygen species (ROS) and mitigates oxidative stress. AsA can donate one or two electrons, forming mono-dehydroascorbate (MDHA), which could ultimately form dehydroascorbate (DHA). AsA is regenerated to maintain its antioxidant capacity through recycling pathways involving MDHA and DHA reductases. This regeneration ensures a high pool of active AsA levels in plants that contribute to herbicide tolerance by stabilizing cellular structures and preventing ROS accumulation [63].

4.2.2 Glutathione

Glutathione (GSH), a low molecular weight tripeptide containing cysteine, glutamate, and glycine, plays several roles in cellular defense against oxidative stress. In addition to directly detoxifying ROS and participating in the ascorbate–glutathione cycle, GSH is a cofactor for glutathione peroxidase and glutathione-S-transferases (GSTs), involved in herbicide detoxification. The redox potential of GSH/GSSG (oxidized form) is a crucial indicator of cellular redox status and resilience to herbicides [61, 63].

4.2.3 Tocopherols

Tocopherols and tocotrienols, known as vitamin E, are strong non-enzymatic antioxidants. By neutralizing singlet oxygen and lipid radicals, these compounds prevent lipid peroxidation and preserve membrane integrity. Because of its methylation pattern, α-tocopherol is the most active antioxidant, while α-tocotrienol shows higher effectiveness in the membrane environment. Through interactions with other antioxidants such as ascorbate, vitamin E can regenerate itself, maintaining its protective role against oxidative stress in photosynthetic tissues [52, 61].

4.2.4 Carotenoids

Carotenoids, including β-carotene, lycopene, lutein, and zeaxanthin, act as non-enzymatic antioxidants in plants, capturing ROS and protecting membrane lipids from peroxidation. They quench singlet oxygen and scavenge other ROS, protecting chlorophyll and other photosynthetic components from oxidative damage. Carotenoids also help dissipate excess light energy as heat through a process known as non-photochemical quenching, which is critical in mitigating herbicide-induced photooxidative stress [61, 64].

4.2.5 Phenolic Compounds

Phenolic compounds, such as flavonoids and phenolic acids, are secondary metabolites that serve as potent antioxidants. They directly scavenge ROS and modulate signaling pathways involved in stress responses. Phenolics also contribute to the structural reinforcement of cell walls, which can limit herbicide penetration and reduce cellular damage. Elevated phenolic content has been linked to improved herbicide tolerance in crop plants [65]. Hence, it is comprehensible that the intricate balance between oxidative stress and antioxidant defense is central to herbicide tolerance. Both enzymatic and non-enzymatic antioxidants work synergistically to protect plant cells from herbicide-induced damage. By limiting oxidative damage, these systems enable plants to maintain cellular homeostasis and recover from stress.

5 Role of CYP450 in Herbicide Tolerance, Detoxification, and Involvement in Antioxidant Production

A common mode of tolerance to herbicides in agronomic crops is by metabolism brought about by the enzyme system, Cytochrome P450 monooxygenases (CYP450s), which are a family of enzymes involved in the metabolism of various endogenous and xenobiotic compounds, including herbicides [66, 67]. In plants, especially monocots, CYP450s contribute to herbicide tolerance through the detoxification of herbicide molecules, transforming them into less harmful metabolites that can be further conjugated or compartmentalized. This enzyme is activated by certain chemicals called safeners. When this safener is applied in combination with herbicides, it provides tolerance in monocot crops such as *Sorghum bicolor* (L.) Moench., *Oryza sativa* L., Zea mays L., *Triticum aestivum* L. against herbicides, there is no formation of ROS [67]. Additionally, a study conducted by Pandian et al. [7] mentioned that CYP450s play an indirect role in modulating antioxidant levels by influencing secondary metabolic pathways, including the biosynthesis of phenolic compounds and other metabolites that act as non-enzymatic antioxidants. This enhances the plant's antioxidant defense system, further boosting herbicide

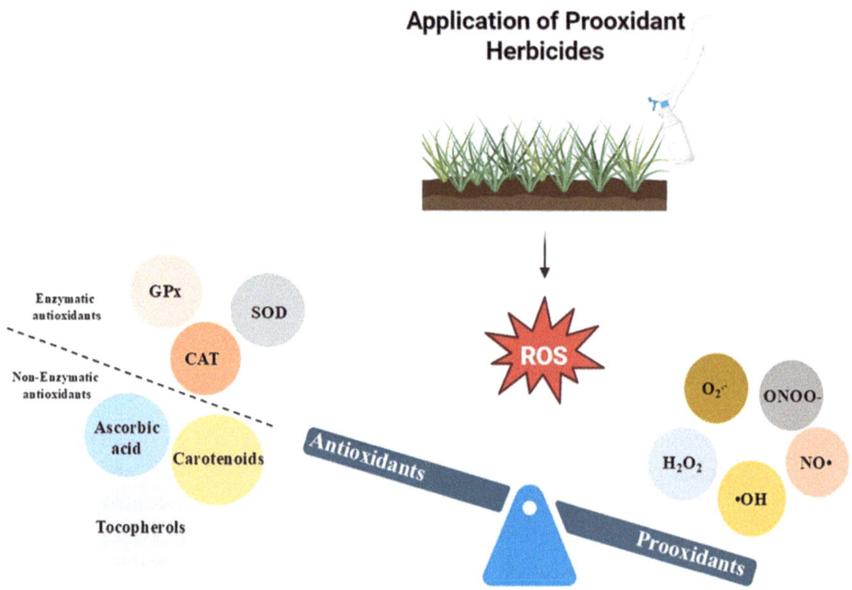

Fig. 1 Imbalance of prooxidant–antioxidant on application of prooxidant herbicides results in oxidative damage

tolerance by reducing oxidative damage. Therefore, CYP450 enzymes are crucial in protecting plants from both biotic and abiotic stresses, including the oxidative damage induced by herbicides. On the other hand, certain weeds lack CYP450 enzymes, hindering their ability to metabolize herbicides effectively. As a result, these weeds must rely heavily on antioxidant compounds to combat the oxidative stress induced by herbicide exposure. However, the absence of CYP450 enzymes and the inability to metabolize the herbicide can lead to an imbalance between antioxidants and reactive oxygen species (ROS), potentially causing a buildup of oxidative stress that becomes lethal to the weeds (Fig. 1) [6, 66].

6 Impact of Prooxidant Herbicides on Farmers

Farmers are frequently exposed to prooxidant herbicides, particularly through inhalation, dermal contact, and accidental ingestion during herbicide application. Many herbicides are designed to induce oxidative damage in plant tissues; however, their prooxidant properties can also initiate oxidative stress in humans. The effects are often compounded by inadequate protective equipment, extended exposure periods, and improper handling. Commonly used herbicides such as paraquat, atrazine, and Roundup are known to produce reactive oxygen species in biological systems, posing health risks to farmers. These risks are amplified by repetitive application, potentially leading to chronic health issues [68, 69].

6.1 Occupational Herbicide Exposure: Health Impact on Agricultural Workers

Agricultural workers face high occupational health risks due to regular exposure to herbicides, particularly through dermal absorption and inhalation during mixing, application, and post-application activities. Recent studies have shown worrying links between herbicide exposure and various health conditions. Research relevant to glyphosate, among the most widely used herbicides in the world, has documented its potential to induce oxidative stress across multiple organ systems [11]. Studies have shown elevated markers of oxidative stress and inflammation in farmers exposed to pesticides, including herbicides, which correlate with an increased risk of depression and neurological disorders [70]. Among the exposed farmers, clinical studies have documented cases of hepatotoxicity, nephrotoxicity, and adverse effects on cardiovascular and reproductive systems among exposed workers [69, 71]. Concerns are findings showing glyphosate's presence in human sperm and its correlation with oxidative stress parameters in agricultural workers [72]. Quite often, chronic exposure and poor protective measures during herbicide application aggravate these health impacts, providing an urgent need for the development and implementation of safety measures and alternative weed management approaches. Beyond direct occupational exposure, herbicide residues persist in the environment and enter the food chain. These compounds accumulate in agricultural products, potentially exposing consumers to low-level, chronic exposure through their diet. For that reason, this creates a concerning cycle where herbicides not only affect farmers but also reach consumers' plates through residual contamination in food products [73]. These concerns state the vital need to develop and implement safer alternatives, particularly bioherbicides derived from natural products, which could offer effective weed control while minimizing health and environmental risks [66].

7 Alternative Approaches on Hazardous Chemical Herbicides and Future Directions

The search for sustainable weed management solutions has spotlighted bioherbicides as a promising alternative to synthetic herbicides, which often pose health and environmental risks. Bioherbicides, derived from natural organisms or compounds, offer an eco-friendlier approach by targeting specific weeds with minimal impact on surrounding flora and fauna [74]. For instance, seaweeds have shown potential as a source of bioherbicidal compounds, with their natural phytochemicals demonstrating effective weed control while being biodegradable and nontoxic to humans

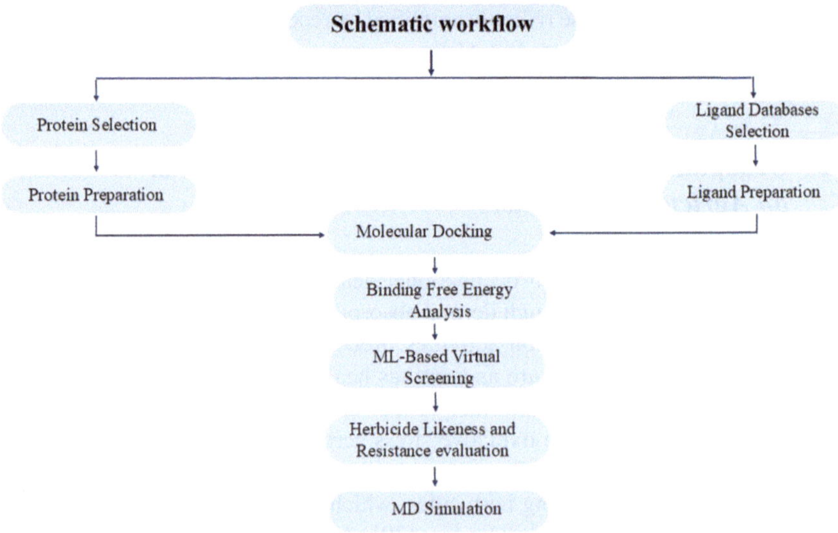

Fig. 2 Outline of the in silico screening cascade applied to discern the novel bioherbicides

and the environment [75, 76]. Another area under research revolves around mechanically approaching enzymes from specific weed types, like acetyl-CoA carboxylase, possibly lessening the need for broad-spectrum chemicals. The ACCase is a biotin-dependent enzyme and is key to the duty of fatty-acid biosynthesis. Additionally, this group A herbicide inhibiting the homomeric ACCase from plastids of grassy weeds has gained wide acceptance in the weed management of grassy weeds in the paddy without affecting crops. The targeted herbicides offer selective weed control and provide some economic advantages to the farmer by reducing their exposure to hazardous chemicals. With the developments in computational biology, in silico approaches are now becoming effective in the identification of future bioherbicides. Virtual screening and molecular docking techniques with machine learning-based tools allow researchers to predict, assess, and optimize future bioherbicidal molecules with enhanced efficacy [76, 77]. Database nontoxic bioherbicides, like toxic plant phytotoxins, assist in an assessment of potential micro-pollution, aiding in the identification of environmentally friendly bioherbicidal candidate selections (Figs. 2 and 3) [66, 77]. Overall, these alternatives not only protect farmers from chemical-related health risks but also promote sustainable agricultural practices by reducing soil and water contamination, supporting long-term environmental health. Bioherbicides offer a pathway to safer weed management and improved farmer well-being, laying the groundwork for a healthier, more resilient farming ecosystem.

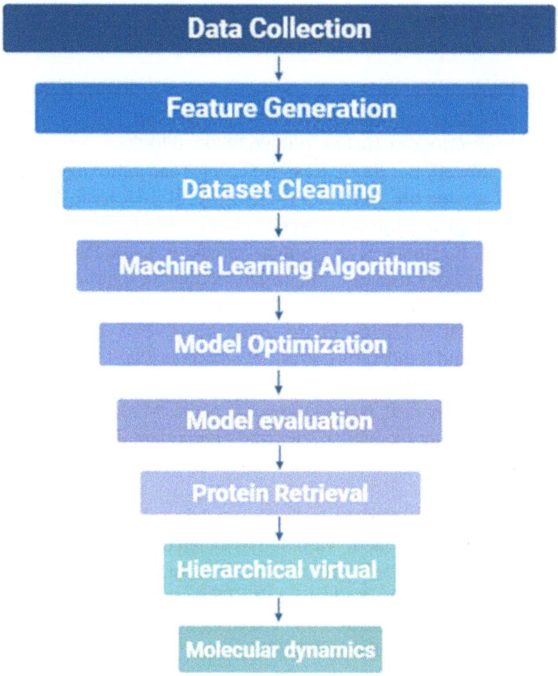

Fig. 3 Outline of the in silico screening cascade applied in the machine learning-assisted virtual screening for bioherbicide discovery

8 Conclusion

In conclusion, maintaining redox homeostasis is crucial for plant health, which is often challenged by the application of prooxidant herbicides. While these chemicals are effective in weed control, they introduce significant oxidative stress in both plants and humans, particularly farmers who handle these chemicals. The balance between prooxidant herbicides and plant antioxidant defenses is key to understanding herbicide tolerance and resistance mechanisms in plants. However, the adverse health impacts on humans necessitate a shift toward safer, sustainable weed management practices. Bioherbicides and other natural alternatives present promising solutions for reducing environmental contamination and health risks, paving the way for sustainable agriculture that prioritizes the well-being of both plants and humans.

Acknowledgements The authors thank the management of Vellore Institute of Technology for providing the necessary resources and facilities to carry out this work. Additionally, the authors would like to acknowledge the assistance of Biorender for its valuable tools in creating quality diagrams used in this study.

Author Contributions R. K. designed the chapter framework and supervised the work. R. S. prepared the table and figures. S. M. and S. P. contributed to the writing of Sect. 2 and were involved in proofreading. R. S. and R. K. contributed to writing the remaining sections in this chapter. All authors reviewed and approved the manuscript.

References

1. Oktyabrsky, O. N., & Smirnova, G. V. (2007). Redox regulation of cellular functions. In *Biochemistry (Moscow)* (Vol. 72, Issue 2, pp. 132–145). Maik Nauka-Interperiodica Publishing. https://doi.org/10.1134/S0006297907020022.
2. Halliwell, B. (2006). Reactive species and antioxidants. Redox biology is a fundamental theme of aerobic life. In *Plant Physiology* (Vol. 141, Issue 2, pp. 312–322). American Society of Plant Biologists. https://doi.org/10.1104/pp.106.077073.
3. Das, P., Nutan, K. K., Singla-Pareek, S. L., & Pareek, A. (2015). Oxidative environment and redox homeostasis in plants: Dissecting out significant contribution of major cellular organelles. *Frontiers in Environmental Science, 2*(JAN). https://doi.org/10.3389/fenvs.2014.00070.
4. Pospíšil, P. (2016). Production of reactive oxygen species by photosystem II as a response to light and temperature stress. *Frontiers in Plant Science, 7*(December). https://doi.org/10.3389/fpls.2016.01950.
5. Dumanović, J., Nepovimova, E., Natić, M., Kuča, K., & Jaćević, V. (2021a). The significance of reactive oxygen species and antioxidant defense system in plants: A concise overview. In *Frontiers in Plant Science* (Vol. 11). Frontiers Media S.A. https://doi.org/10.3389/fpls.2020.552969.
6. Adhikari, P., Goodrich, E., Fernandes, S. B., Lipka, A. E., Tranel, P., Brown, P., & Jamann, T. M. (2020). Genetic variation associated with PPO-inhibiting herbicide tolerance in sorghum. *PLoS ONE, 15*(10 October). https://doi.org/10.1371/journal.pone.0233254.
7. Pandian, B. A., Sathishraj, R., Djanaguiraman, M., Prasad, P. V. V., & Jugulam, M. (2020). Role of cytochrome P450 enzymes in plant stress response. In *Antioxidants* (Vol. 9, Issue 5). MDPI. https://doi.org/10.3390/antiox9050454.
8. Sidthilaw, S., Sapbamrer, R., Pothirat, C., Wunnapuk, K., & Khacha-ananda, S. (2022). Effects of exposure to glyphosate on oxidative stress, inflammation, and lung function in maize farmers, Northern Thailand. *BMC Public Health, 22*(1). https://doi.org/10.1186/s12889-022-13696-7.
9. Damalas, C. A., & Koutroubas, S. D. (2016). Farmers' exposure to pesticides: Toxicity types and ways of prevention. *Toxics, 4*(1), 1.
10. Tudi, M., Li, H., Li, H., Wang, L., Lyu, J., Yang, L., & Connell, D. (2022). Exposure routes and health risks associated with pesticide application. *Toxics, 10*(6), 335.
11. Chang, V. C., Andreotti, G., Ospina, M., Parks, C. G., Liu, D., Shearer, J. J., Rothman, N., Silverman, D. T., Sandler, D. P., Calafat, A. M., Beane Freeman, L. E., & Hofmann, J. N. (2023). Glyphosate exposure and urinary oxidative stress biomarkers in the agricultural health study. *JNCI: Journal of the National Cancer Institute, 115*(4), 394–404. https://doi.org/10.1093/jnci/djac242.
12. Geigenberger, P., Smirnoff, N., Van Breusegem, F., Dietz, K.-J., & Noctor, G. (2021). Plant redox biology—on the move. *Plant Physiology, 186*(1), 1–3. https://doi.org/10.1093/plphys/kiab103
13. Foyer, C. H., & Noctor, G. (2009). Redox regulation in photosynthetic organisms: Signaling, acclimation, and practical implications. *Antioxidants & Redox Signaling, 11*(4), 861–905. https://doi.org/10.1089/ars.2008.2177
14. Noctor, G., Reichheld, J.-P., & Foyer, C. H. (2018). ROS-related redox regulation and signaling in plants. *Seminars in Cell & Developmental Biology, 80*(August), 3–12. https://doi.org/10.1016/j.semcdb.2017.07.013
15. Zaffagnini, M., Fermani, S., Marchand, C. H., Costa, A., Sparla, F., Rouhier, N., Geigenberger, P., Lemaire, S. D., & Trost, P. (2019). Redox homeostasis in photosynthetic organisms: Novel and established thiol-based molecular mechanisms. *Antioxidants & Redox Signaling, 31*(3), 155–210. https://doi.org/10.1089/ars.2018.7617
16. Peláez-Vico, M. Á., Fichman, Y., Zandalinas, S. I., Van Breusegem, F., Karpiński, S. M., & Mittler, R. (2022). ROS and redox regulation of cell-to-cell and systemic signaling in plants during stress. *Free Radical Biology and Medicine, 193*(November), 354–362. https://doi.org/10.1016/j.freeradbiomed.2022.10.305

17. Serrato, A. J., Juan Fernández-Trijueque, J. B., Chueca, A. & Sahrawy, M. (2013). Plastid Thioredoxins: A 'One-for-All' redox-signaling system in plants. *Frontiers in Plant Science, 4*(November). https://doi.org/10.3389/fpls.2013.00463.
18. Shigeoka, S., & Maruta, T. (2014). Cellular redox regulation, signaling, and stress response in plants. *Bioscience, Biotechnology, and Biochemistry, 78*(9), 1457–1470. https://doi.org/10.1080/09168451.2014.942254
19. Suzuki, N., Koussevitzky, S., Mittler, R., & Miller, G. (2012). ROS and redox signalling in the response of plants to abiotic stress. *Plant, Cell & Environment, 35*(2), 259–270. https://doi.org/10.1111/j.1365-3040.2011.02336.x
20. Hanke, G., & Mulo, P. (2013). Plant type ferredoxins and ferredoxin-dependent metabolism. *Plant, Cell & Environment, 36*(6), 1071–1084. https://doi.org/10.1111/pce.12046
21. Rochaix, J.-D. (2011). Regulation of photosynthetic electron transport. *Biochimica et Biophysica Acta (BBA)—Bioenergetics, 1807*(3), 375–83. https://doi.org/10.1016/j.bbabio.2010.11.010.
22. Buchanan, B. B. (2016). The carbon (Formerly Dark) reactions of photosynthesis. *Photosynthesis Research, 128*(2), 215–217. https://doi.org/10.1007/s11120-015-0212-z
23. Hoh, D., Froehlich, J. E., & Kramer, D. M. (2024). Redox regulation in chloroplast thylakoid lumen: The Pmf changes everything, again. *Plant, Cell & Environment, 47*(8), 2749–2765. https://doi.org/10.1111/pce.14789
24. Khorobrykh, S., Havurinne, V., Mattila, H., & Tyystjärvi, E. (2020). Oxygen and ROS in photosynthesis. *Plants, 9*(1), 91. https://doi.org/10.3390/plants9010091
25. Triantaphylidès, C., & Havaux, M. (2009). Singlet oxygen in plants: production, detoxification and signaling. *Trends in Plant Science, 14*(4), 219–228. https://doi.org/10.1016/j.tplants.2009.01.008
26. Pilon, M., Karl, R., & Wiebke, T. (2011). The biogenesis and physiological function of chloroplast superoxide dismutases. *Biochimica et Biophysica Acta (BBA) – Bioenergetics, 1807*(8), 989–98. https://doi.org/10.1016/j.bbabio.2010.11.002.
27. Pinnola, A., & Bassi, R. (2018). Molecular mechanisms involved in plant photoprotection. *Biochemical Society Transactions, 46*(2), 467–482. https://doi.org/10.1042/BST20170307
28. Cejudo, F. J., González, M.-C., & Pérez-Ruiz, J. M. (2021). Redox regulation of chloroplast metabolism. *Plant Physiology, 186*(1), 9–21. https://doi.org/10.1093/plphys/kiaa062
29. Geigenberger, P., & Fernie, A. R. (2014). Metabolic control of redox and redox control of metabolism in plants. *Antioxidants & Redox Signaling, 21*(9), 1389–1421. https://doi.org/10.1089/ars.2014.6018
30. Nikkanen, L., & Rintamäki, E. (2019). Chloroplast thioredoxin systems dynamically regulate photosynthesis in plants. *Biochemical Journal, 476*(7), 1159–1172. https://doi.org/10.1042/BCJ20180707
31. Schürmann, P., & Buchanan, B. B. (2008). The ferredoxin/thioredoxin system of oxygenic photosynthesis. *Antioxidants & Redox Signaling, 10*(7), 1235–1274. https://doi.org/10.1089/ars.2007.1931
32. Serrato, A. J., Pérez-Ruiz, J. M., Spínola, M. C., & Cejudo, F. J. (2004). A novel NADPH Thioredoxin reductase, localized in the chloroplast, which deficiency causes hypersensitivity to abiotic stress in Arabidopsis thaliana*. *Journal of Biological Chemistry, 279*(42), 43821–43827. https://doi.org/10.1074/jbc.M404696200
33. Møller, I. M., Rasmusson, A. G., & Van Aken, O. (2021). Plant mitochondria – past, present and future. *The Plant Journal, 108*(4), 912–959. https://doi.org/10.1111/tpj.15495
34. Elhafez, D., Murcha, M. W., Clifton, R., Soole, K. L., Day, D. A., & Whelan, J. (2006). Characterization of mitochondrial alternative NAD(P)H dehydrogenases in arabidopsis: Intraorganelle location and expression. *Plant and Cell Physiology, 47*(1), 43–54. https://doi.org/10.1093/pcp/pci221
35. Rasmusson, A. G., Geisler, D. A., & Møller, I. M. (2008). The multiplicity of dehydrogenases in the electron transport chain of plant mitochondria. *Mitochondrion, 8*(1), 47–60. https://doi.org/10.1016/j.mito.2007.10.004

36. Vanlerberghe, G. (2013). Alternative oxidase: A mitochondrial respiratory pathway to maintain metabolic and signaling homeostasis during abiotic and biotic stress in plants. *International Journal of Molecular Sciences, 14*(4), 6805–6847. https://doi.org/10.3390/ijms14046805
37. Bauwe, H., Hagemann, M., & Fernie, A. R. (2010). Photorespiration: Players, partners and origin. *Trends in Plant Science, 15*(6), 330–336. https://doi.org/10.1016/j.tplants.2010.03.006
38. Noctor, G., De Paepe, R., & Foyer, C. H. (2007). Mitochondrial redox biology and homeostasis in plants. *Trends in Plant Science, 12*(3), 125–134. https://doi.org/10.1016/j.tplants.2007.01.005
39. Schwarzländer, M., & Finkemeier, I. (2013). Mitochondrial energy and redox signaling in plants. *Antioxidants & Redox Signaling, 18*(16), 2122–2144. https://doi.org/10.1089/ars.2012.5104
40. Van Aken, O. (2021). Mitochondrial redox systems as central hubs in plant metabolism and signaling. *Plant Physiology, 186*(1), 36–52. https://doi.org/10.1093/plphys/kiab101.
41. Banze, M., & Follmann, H. (2000). Organelle-specific NADPH Thioredoxin reductase in plant mitochondria. *Journal of Plant Physiology, 156*(1), 126–129. https://doi.org/10.1016/S0176-1617(00)80283-7
42. Daloso, D.M., Karolin, M., Toshihiro, O., Alexandra, F., Takayuki, T., Alexandra, B., & Christophe, R., et al. (2015). Thioredoxin, a master regulator of the tricarboxylic acid cycle in plant mitochondria. *Proceedings of the National Academy of Sciences, 112*(11). https://doi.org/10.1073/pnas.1424840112.
43. Laloi, C., Rayapuram, N., Chartier, Y., Grienenberger, J.-M., Bonnard, G., & Meyer, Y. (2001). Identification and characterization of a mitochondrial thioredoxin system in plants. *Proceedings of the National Academy of Sciences, 98*(24), 14144–14149. https://doi.org/10.1073/pnas.241340898
44. Møller, I. M. (2001). Plant mitochondria and oxidative stress: Electron transport, NADPH turnover, and metabolism of reactive oxygen species. *Annual Review of Plant Biology, 52*(Volume 52, 2001), 561–91. https://doi.org/10.1146/annurev.arplant.52.1.561.
45. Liebthal, M., Maynard, D., & Dietz, K.-J. (2018). Peroxiredoxins and redox signaling in plants. *Antioxidants & Redox Signaling, 28*(7), 609–624. https://doi.org/10.1089/ars.2017.7164
46. Navrot, N., Collin, V., Gualberto, J., Gelhaye, E., Hirasawa, M., Rey, P., Knaff, D. B., Issakidis, E., Jacquot, J.-P., & Rouhier, N. (2006). Plant glutathione peroxidases are functional peroxiredoxins distributed in several subcellular compartments and regulated during biotic and abiotic stresses. *Plant Physiology, 142*(4), 1364–1379. https://doi.org/10.1104/pp.106.089458
47. Pan, R., Liu, J., Wang, S., & Jianping, Hu. (2020). Peroxisomes: Versatile organelles with diverse roles in plants. *New Phytologist, 225*(4), 1410–1427. https://doi.org/10.1111/nph.16134
48. Walker, C. L., Pomatto, L. C. D., Tripathi, D. N., & Davies, K. J. A. (2018). Redox regulation of homeostasis and proteostasis in peroxisomes. *Physiological Reviews, 98*(1), 89–115. https://doi.org/10.1152/physrev.00033.2016
49. Sandalio, L. M., Peláez-Vico, M. A., Molina-Moya, E., & Romero-Puertas, M. C. (2021). Peroxisomes as redox-signaling nodes in intracellular communication and stress responses. *Plant Physiology, 186*(1), 22–35. https://doi.org/10.1093/plphys/kiab060
50. Antonenkov, V. D., Grunau, S., Ohlmeier, S., & Kalervo Hiltunen, J. (2010). Peroxisomes are oxidative organelles. *Antioxidants & Redox Signaling, 13*(4), 525–537. https://doi.org/10.1089/ars.2009.2996
51. Barroso, J. B., Corpas, F. J, Carreras, A., Sandalio, L. M., Valderrama R., Palma, J. M., Lupiáñez, J. A., & del Río, L. A. (1999). Localization of nitric-oxide synthase in plant peroxisomes*. *Journal of Biological Chemistry, 274*(51), 36729–33. https://doi.org/10.1074/jbc.274.51.36729.
52. Corpas, F. J., & Barroso, J. B. (2014). Peroxynitrite (ONOO−) is endogenously produced in Arabidopsis peroxisomes and is overproduced under cadmium stress. *Annals of Botany, 113*(1), 87–96. https://doi.org/10.1093/aob/mct260
53. Gebicka, L., & Joanna, D. (2009). Catalytic scavenging of peroxynitrite by catalase. *Journal of Inorganic Biochemistry, Special Issue Containing Contributions from the First Latin American Meeting on Biological Inorganic Chemistry – LABIC 2008, 103*(10), 1375–79. https://doi.org/10.1016/j.jinorgbio.2009.07.011.

54. Reshi, Z. A., Ahmad, W., Lukatkin, A. S., & Javed, S. Bin. (2023). From nature to lab: A review of secondary metabolite biosynthetic pathways, environmental influences, and in vitro approaches. In *Metabolites* (Vol. 13, Issue 8). Multidisciplinary Digital Publishing Institute (MDPI). https://doi.org/10.3390/metabo13080895.
55. Traxler, C., Gaines, T. A., Küpper, A., Luemmen, P., & Dayan, F. E. (2023). The nexus between reactive oxygen species and the mechanism of action of herbicides. *Journal of Biological Chemistry, 299*(11), Article 105267. https://doi.org/10.1016/j.jbc.2023.105267
56. Caverzan, A., Piasecki, C., Chavarria, G., Stewart, C. N., & Vargas, L. (2019). Defenses against ROS in crops and weeds: The effects of interference and herbicides. In *International Journal of Molecular Sciences* (Vol. 20, Issue 5). MDPI AG. https://doi.org/10.3390/ijms20051086.
57. Sherwani, S. I., Arif, I. A., & Khan, H. A. (2015). Modes of action of different classes of herbicides. In *Herbicides, physiology of action, and safety*. InTech. https://doi.org/10.5772/61779
58. Takano, H. K., Beffa, R., Preston, C., Westra, P., & Dayan, F. E. (2020). A novel insight into the mode of action of glufosinate: How reactive oxygen species are formed. *Photosynthesis Research, 144*(3), 361–372. https://doi.org/10.1007/s11120-020-00749-4
59. Barrera, G., Pizzimenti, S., Daga, M., Dianzani, C., Arcaro, A., Cetrangolo, G. P., Giordano, G., Cucci, M. A., Graf, M., & Gentile, F. (2018). Lipid peroxidation-derived aldehydes, 4-hydroxynonenal and malondialdehyde in aging-related disorders. In *Antioxidants* (Vol. 7, Issue 8). MDPI. https://doi.org/10.3390/antiox7080102.
60. Endale, H. T., Tesfaye, W., & Mengstie, T. A. (2023). ROS induced lipid peroxidation and their role in ferroptosis. In *Frontiers in cell and developmental biology* (Vol. 11). Frontiers Media SA. https://doi.org/10.3389/fcell.2023.1226044.
61. Dumanović, J., Nepovimova, E., Natić, M., Kuča, K., & Jaćević, V. (2021b). The significance of reactive oxygen species and antioxidant defense system in plants: A concise overview. In *Frontiers in Plant Science* (Vol. 11). Frontiers Media S.A. https://doi.org/10.3389/fpls.2020.552969.
62. Pei, J., Pan, X., Wei, G., & Hua, Y. (2023). Research progress of glutathione peroxidase family (GPX) in redoxidation. *Frontiers in Pharmacology, 14*. https://doi.org/10.3389/fphar.2023.1147414.
63. Hasanuzzaman, M., Bhuyan, M. H. M. B., Anee, T. I., Parvin, K., Nahar, K., Mahmud, J. A., & Fujita, M. (2019). Regulation of ascorbate-glutathione pathway in mitigating oxidative damage in plants under abiotic stress. *Antioxidants, 8*(9), 384. https://doi.org/10.3390/antiox8090384
64. Swapnil, P., Meena, M., Singh, S. K., Dhuldhaj, U. P., Harish, & Marwal, A. (2021). Vital roles of carotenoids in plants and humans to deteriorate stress with its structure, biosynthesis, metabolic engineering and functional aspects. *Current Plant Biology, 26*, 100203. https://doi.org/10.1016/j.cpb.2021.100203.
65. Mutha, R. E., Tatiya, A. U., & Surana, S. J. (2021). Flavonoids as natural phenolic compounds and their role in therapeutics: An overview. *Future Journal of Pharmaceutical Sciences, 7*(1), 25. https://doi.org/10.1186/s43094-020-00161-8
66. Antony, A., & Karuppasamy, R. (2022). Searching of novel herbicides for paddy field weed management—A case study with Acetyl-CoA carboxylase. *Agronomy, 12*(7). https://doi.org/10.3390/agronomy12071635.
67. Nandula, V. K., Riechers, D. E., Ferhatoglu, Y., Barrett, M., Duke, S. O., Dayan, F. E., Goldberg-Cavalleri, A., Tétard-Jones, C., Wortley, D. J., Onkokesung, N., Brazier-Hicks, M., Edwards, R., Gaines, T., Iwakami, S., Jugulam, M., & Ma, R. (2019). Herbicide metabolism: Crop selectivity, bioactivation, weed resistance, and regulation. *Weed Science, 67*(2), 149–175. https://doi.org/10.1017/wsc.2018.88
68. Rajak, P., Roy, S., Ganguly, A., Mandi, M., Dutta, A., Das, K., Nanda, S., Ghanty, S., & Biswas, G. (2023). Agricultural pesticides—friends or foes to biosphere? *Journal of Hazardous Materials Advances, 10*, Article 100264. https://doi.org/10.1016/j.hazadv.2023.100264
69. Zanchi, M. M., Marafon, F., Marins, K., Bagatini, M. D., & Zamoner, A. (2024). Redox imbalance and inflammation: A link to depression risk in Brazilian pesticide-exposed farmers. *Toxicology, 501*, Article 153706. https://doi.org/10.1016/j.tox.2023.153706

70. Mazuryk, J., Klepacka, K., Kutner, W., & Sharma, P. S. (2024). Glyphosate: Hepatotoxicity, nephrotoxicity, hemotoxicity, carcinogenicity, and clinical cases of endocrine, reproductive, cardiovascular, and pulmonary system intoxication. *ACS Pharmacology & Translational Science, 7*(5), 1205–1236. https://doi.org/10.1021/acsptsci.4c00046
71. Kori, R. K., Singh, M. K., Jain, A. K., & Yadav, R. S. (2018). Neurochemical and behavioral dysfunctions in pesticide exposed farm workers: A clinical outcome. *Indian Journal of Clinical Biochemistry, 33*(4), 372–381. https://doi.org/10.1007/s12291-018-0791-5
72. Vasseur, C., Serra, L., El Balkhi, S., Lefort, G., Ramé, C., Froment, P., & Dupont, J. (2024). Glyphosate presence in human sperm: First report and positive correlation with oxidative stress in an infertile French population. *Ecotoxicology and Environmental Safety, 278*, Article 116410. https://doi.org/10.1016/j.ecoenv.2024.116410
73. Mohd Ghazi, R., Nik Yusoff, N. R., Abdul Halim, N. S., Wahab, I. R. A., Ab Latif, N., Hasmoni, S. H., Ahmad Zaini, M. A., & Zakaria, Z. A. (2023). Health effects of herbicides and its current removal strategies. *Bioengineered*, *14*(1). https://doi.org/10.1080/21655979.2023.2259526.
74. Islam, A. K. M. M., Karim, S. M. R., Kheya, S. A., & Yeasmin, S. (2024). Unlocking the potential of bioherbicides for sustainable and environment friendly weed management. *Heliyon, 10*(16), Article e36088. https://doi.org/10.1016/j.heliyon.2024.e36088
75. Chukwuma, O. C., Tan, S. P., Hughes, H., McLoughlin, P., O'Toole, N., & McCarthy, N. (2024). *The potential of seaweeds as a rich natural source for novel Bioherbicide formulation/development*. Cambridge University Press. https://doi.org/10.1017/wsc.2024.1
76. Suresh, R., & Karuppasamy, R. (2025). Seaweed-based PPO inhibitors as a new frontier in biological weed control for sorghum cultivation: From ocean to field. *Protoplasma*. https://doi.org/10.1007/s00709-025-02049-x
77. Antony, A., & Karuppasamy, R. (2023). An integrated machine learning-based virtual screening strategy for biological weeding in maize field: A case study with HPPD. *Journal of Plant Diseases and Protection, 130*, 1433–1449. https://doi.org/10.1007/s41348-023-00784-y

Basic Understanding of Oxidative Stress and the Pathophysiology of Chronic Diseases in Human

Varshni Premnath and Shanthi Veerappapillai

Abstract Oxidative stress, a condition characterized by an excess of harmful reactive oxygen and nitrogen species overwhelming the body's protective antioxidant defenses, is a key factor in the development of many chronic diseases. Endogenous sources, primarily mitochondrial electron transport, are key contributors to reactive oxygen and nitrogen species production, which, when uncontrolled, lead to cellular and molecular damage. Environmental factors, such as toxins and radiation, further exacerbate this imbalance. Chronic oxidative stress impacts vital biomolecules, disrupting cellular function and contributing to aging and various chronic diseases, which include cardiovascular disease (CVD), diabetes mellitus (DM), liver disorders, cancer, and neurodegenerative conditions. In CVD, oxidative stress promotes endothelial dysfunction and plaque formation, while in diabetes, it impairs insulin sensitivity and pancreatic β-cell function, aggravating hyperglycemia. Furthermore, excessive ROS in liver cells trigger inflammation, fibrosis, and increased risk of hepatocellular carcinoma, while neurodegenerative diseases and age-related dysfunctions are similarly driven by progressive oxidative damage. This chapter offers a comprehensive overview of oxidative stress mechanisms underlying these diseases, exploring therapeutic interventions aimed at reducing ROS, enhancing antioxidant defenses, and addressing the challenges and future directions of research.

Keywords Oxidative stress · Chronic diseases · Reactive oxygen species · Liver diseases · Cardiovascular diseases · Diabetes mellitus

---

V. Premnath · S. Veerappapillai (✉)
Department of Biotechnology, School of Bio Sciences and Technology, Vellore Institute of Technology, Vellore, Tamil Nadu, India
e-mail: shanthi.v@vit.ac.in

K. Kasinathan et al. (eds.), *Recent Advances in Oxidative Stress Associated Chronic Diseases Volume 1*, https://doi.org/10.1007/978-981-96-9670-3_9

1 Introduction

Oxidative stress (OxS) occurs when there is an imbalance between the production of reactive oxygen species (ROS) and reactive nitrogen species (RNS) and the body's ability to neutralize them [1]. This leads to an accumulation of these highly reactive molecules, including free radicals like superoxide anions, hydroxyl radicals, hydrogen peroxide, and nitric oxide (NO). The primary sources of ROS are endogenous, with the mitochondrial electron transport chain being a major contributor. Environmental factors such as toxins, ultraviolet radiation, tobacco smoke, and certain drugs contribute to the overproduction of ROS [2]. When the level of ROS exceeds the body's antioxidant capacity, cellular damage can occur, affecting vital molecules like lipids, proteins, and DNA. The body's defense system against ROS consists of two primary types of antioxidants: enzymatic and non-enzymatic. ROS can be immediately neutralized by enzymatic antioxidants such as glutathione peroxidase, catalase, and superoxide dismutase, which break them down into less dangerous molecules [3]. According to [4], vitamins C and E are examples of non-enzymatic antioxidants that scavenge free radicals to stop cell damage. An insufficient defense can compromise cellular function, impairing the cell's viability and potentially leading to cell death [5]. The impact of prolonged OxS is significant, it not only contributes to cellular aging and tissue damage but also plays a crucial role in the development and progression of several chronic diseases. As OxS has a wide range of consequences and is linked to several medical problems, it is a prominent area of study interest. Conditions closely linked to prolonged OxS include cardiovascular disease (CVD, diabetes mellitus (DM, liver disorders, cancer, and neurodegenerative diseases [6].

For instance, in cardiovascular disease, OxS plays a major role in endothelial dysfunction, which damages blood vessel linings, leading to the deposition of atherosclerotic plaques and increasing the risk of heart failure [7]. In diabetes, OxS exacerbates insulin resistance by causing dysfunction of pancreatic β-cells, contributing to persistent hyperglycaemia [8]. Similarly, in liver diseases, excessive ROS contribute to liver cell damage, leading to inflammation, fibrosis, and cirrhosis, ultimately increasing the risk of hepatocellular carcinoma [9]. ROS play a significant role in the aging process by contributing to mitochondrial dysfunction, causing progressive cellular damage, triggering chronic inflammation, and increasing the susceptibility to age-related diseases [10]. Age-related diseases, including many neurodegenerative conditions, are strongly associated with chronic OxS [11]. Understanding the connection between OxS and aging could reveal therapeutic targets to potentially slow aging and mitigate age-related diseases. Given its central role in chronic diseases, targeting OxS is a promising area for developing new therapies. Emerging therapeutic approaches include the use of synthetic antioxidants, increasing synthesis of antioxidant enzymes, and natural antioxidant supplements to boost the body's defense against ROS [12]. The impact of oxidative stress extends across various chronic diseases, as summarized in Table 1. This chapter offers a thorough understanding of the role of OxS in the progression of chronic diseases, highlighting the challenges and exploring future research directions in the field.

Table 1 Impact of oxidative stress on chronic diseases

S. no	Chronic disease	Oxidative stress impact
1	Cardiovascular diseases	Endothelial dysfunction, plaque formation, atherosclerosis
2	Diabetes mellitus	β-cell dysfunction, insulin resistance, mitochondrial stress
3	Liver disorders	Lipid accumulation, fibrosis, ROS-induced DNA mutations
4	Neurodegenerative diseases	Mitochondrial dysfunction, protein aggregation, apoptosis
5	Cancer	DNA mutations, increased proliferation, immune evasion

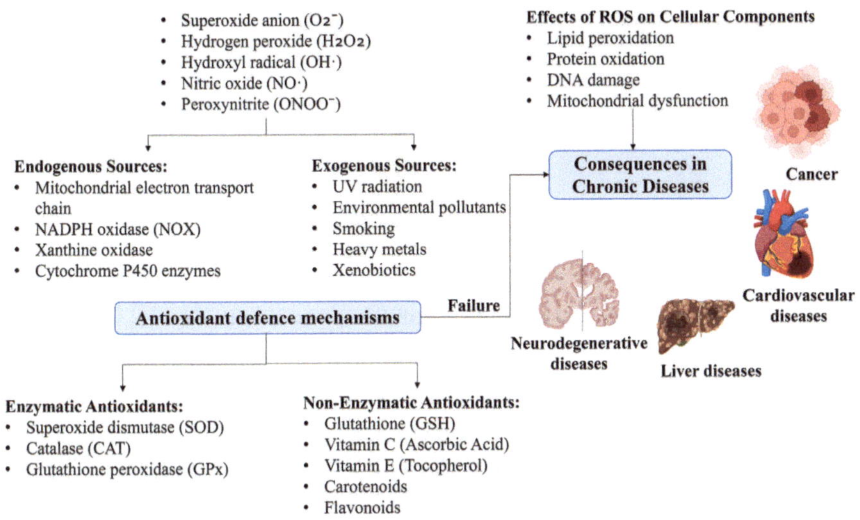

Fig. 1 Oxidative stress and its role in chronic diseases

Oxidative stress results from an imbalance between ROS production and antioxidant defense, as illustrated in Fig. 1, which outlines its role in chronic disease pathology. Uncovering the complex interplay between OxS and chronic disease may pave the way for innovative diagnostic tools, novel therapeutic approaches, and improved disease management strategies.

2 Mechanisms of Oxidative Stress in Chronic Diseases

The ways in which OxS contributes to chronic diseases are intricate and multi-faceted. ROS and RNS not only damage cellular components, leading to impaired cellular function, but also trigger inflammatory responses, further exacerbating tissue damage [13]. Additionally, OxS can impair mitochondrial function and disrupt redox signaling pathways, which are essential for regulating cell growth, differentiation,

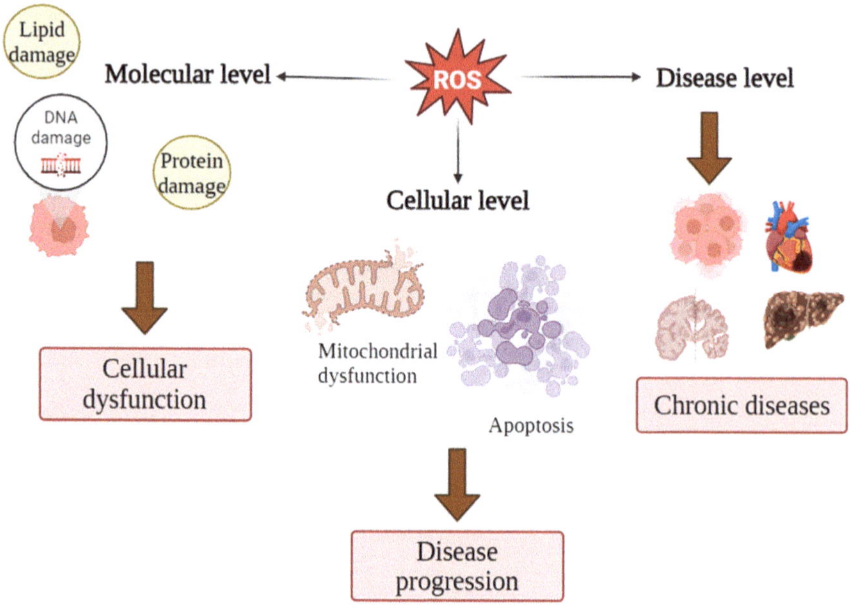

Fig. 2 Impact of ROS on molecular, cellular, and disease progression levels

and programmed cell death. This can lead to aberrant cell proliferation, contributing to cancer development. Understanding these mechanisms enables the development of targeted therapeutic approaches to alleviate the detrimental effects of OxS and enhance chronic disease management [2]. The impact of ROS is evident at molecular, cellular, and systemic levels, leading to disease progression, as depicted in Fig. 2.

2.1 *Cardiovascular Diseases (CVDs)*

OxS significantly influences the development and progression of CVDs, including atherosclerosis, hypertension, myocardial infarction, and heart failure. In the cardiovascular system, it primarily originates from mitochondrial activity, particularly during energy production through oxidative phosphorylation. Here, mitochondria play a dual role: They are essential for cellular energy but also generate ROS as by-products, which, if uncontrolled, can be harmful. Enzymes, such as NADPH oxidase (NOX), xanthine oxidase (Xo), and lipoxygenases, also contribute significantly to ROS production [14]. This can harm cells and poses specific risks to cardiovascular health by disrupting cellular functions and promoting inflammation, ultimately impacting heart and vessel health.

2.1.1 Atherosclerosis

Atherosclerosis is a core driver of ischemic heart disease, stroke, and peripheral artery disease, with OxS playing a critical role in its progression. The disease begins with damage to endothelial cells, leading to the accumulation of low-density lipoprotein (LDL) cholesterol in the vessel walls, which subsequently results in plaque formation [15]. The enzymes such as NOX, Xo, and other mitochondrial enzymes are central to ROS production, plaque formation, and inflammation in atherosclerosis, specifically, NOX1, NOX2, and Xo [16, 17]. The early stages of atherosclerosis involve LDL infiltration into the arterial wall, triggering inflammation. OxS modifies it to form oxidized LDL, which macrophages consume, creating foam cells that contribute to plaque buildup [18].

2.1.2 Hypertension

OxS also exacerbates hypertension by generating oxygen-free radicals like superoxide anions, hydroxyl radicals, and nitric oxide (NO) radicals. Elevated ROS levels contribute to increased vascular resistance by altering signaling pathways that regulate vascular contraction and relaxation. Moreover, oxidative damage to endothelial cells leads to increased vascular stiffness, further elevating blood pressure and contributing to the hypertensive state [19]. Although OxS is not the only cause of hypertension, it increases the blood pressure when combined with other risk factors, which include angiotensin II, endothelin-1, aldosterone, and high salt intake [20].

2.1.3 Myocardial Infarction (MI)

During MI, increased ROS production in cardiomyocytes is linked to a rapid rise in intracellular calcium (Ca^{2+}), which promotes cell death through apoptosis and necrosis. Under hypoxic conditions, Ca^{2+} levels rise mainly due to release from the sarcoplasmic reticulum and plasma membrane depolarization, facilitating further Ca^{2+} influx. Both ROS and Ca^{2+} can activate the mitochondrial permeability transition pore, causing mitochondrial dysfunction, which is characterized by swelling, loss of membrane potential, and content leakage. This cascade increases cell death during reperfusion, expanding the damaged, necrotic area of the heart [21].

2.2 Diabetes Mellitus (DM)

Diabetes is a metabolic disorder characterized by high blood sugar levels due to insulin resistance or insulin deficiency [22]. Persistent hyperglycemia in type 2 diabetes mellitus (T2DM) induces OxS by increasing ROS and RNS production, leading to insulin resistance and β-cell dysfunction [8]. This oxidative imbalance plays a significant role in T2DM's pathophysiology, exacerbating hyperglycemia and causing cellular damage. Mitochondria, the main source of ROS during cellular respiration, produce superoxide anions and hydrogen peroxide, which under normal conditions account for about 1–2% of oxygen consumption. However, in hyperglycemic states, the percentage can rise sharply, with nearly 10% of oxygen consumption resulting in free radicals [23]. Excessive ROS damages mitochondria, leading to insulin resistance, β-cell death, and worsened hyperglycemia.

2.2.1 Polyol Pathway Activation

The polyol pathway, which is responsible for metabolizing glucose to sorbitol and fructose, is a major contributor to OxS in diabetes. In hyperglycemic conditions, this pathway's flux increases, using up NADPH, thus diminishing cellular antioxidant capacity. The pathway also disrupts the NADH/NAD + balance, increasing ROS production. This leads to protein glycation and lipid peroxidation, which further propagate inflammation and oxidative damage within tissues [24].

2.2.2 Protein Kinase C (PKC) Activation Pathway

Hyperglycemia can activate PKC, particularly through diacylglycerol. This activation leads to an increase in ROS production via NADPH oxidase and mitochondrial pathways. PKC activation is linked to endothelial dysfunction, inflammation, and vascular complications commonly observed in diabetes [23].

2.2.3 Hexosamine Pathway

In hyperglycemia, excess glucose fuels the hexosamine pathway, leading to increased conversion of fructose-6-phosphate into UDP-N-acetylglucosamine. This UDP-GlcNAc modifies transcription factors, altering gene expression and potentially causing pathological changes in cells, such as glomerular cells and endothelial cells, leading to complications like carotid artery plaque formation in diabetes. Hyperglycemia heightens ROS production via mitochondrial electron transport chain activity, inducing DNA damage and activating poly ADP-ribose polymerase, which subsequently inhibits glyceraldehyde-3-phosphate dehydrogenase. This inhibition causes glycolytic intermediates to accumulate, promoting advanced glycation end-product (AGE) formation and PKC activation, while fructose-6-phosphate accumulation further drives the hexosamine pathway. Concurrently, aldose reductase reduces glucose, consuming NADPH and amplifying OxS. In insulin-resistant cells, elevated free fatty acids are oxidized in the mitochondria, increasing FADH and NADH, which also enhance ROS production, creating oxidative damage like that seen in hyperglycemia. Together, these mechanisms converge to increase ROS and cellular dysfunction, underpinning chronic complications in diabetes through a unified pathway [25].

2.2.4 Advanced Glycation End Products (AGEs) and Inflammation

Chronic hyperglycemia results in the formation of AGEs, which not only directly damage biomolecules but also stimulate pro-inflammatory cytokines. AGEs enhance ROS generation, perpetuating OxS and promoting vascular inflammation, a contributor to diabetic complications such as cardiovascular disease. AGEs interact with their receptor on cell surfaces, activating redox-sensitive transcription factors that upregulate inflammatory cytokines and contribute to insulin resistance and vascular damage [26]. The interplay between ROS overproduction, mitochondrial dysfunction, inflammatory pathways, and glycation processes emphasizes the need for strategies that can mitigate oxidative damage to manage diabetes and prevent its long-term consequences.

2.3 Liver Diseases

OxS is a key factor driving various liver diseases, including non-alcoholic fatty liver disease (NAFLD), alcoholic liver disease (ALD), chronic hepatitis B and C infections, and even advanced conditions like cirrhosis and hepatocellular carcinoma (HCC), as outlined in Fig. 3.

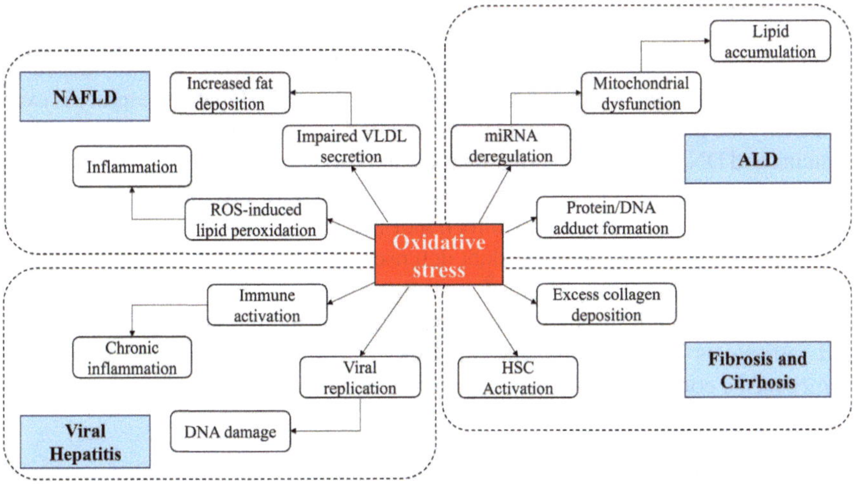

Fig. 3 The role of oxidative stress in liver diseases

2.3.1 Non-alcoholic Fatty Liver Disease (NAFLD)

NAFLD is characterized by the buildup of fat in the liver in individuals without excessive alcohol intake or the presence of liver conditions like autoimmune disorders, drug-induced liver injury, or viral hepatitis. NAFLD can progress from simple fat accumulation (NAFL) to a more serious condition called non-alcoholic steatohepatitis (NASH), which involves inflammation, liver cell damage, and potential scarring (fibrosis). In severe cases, NASH can lead to cirrhosis and HCC [27]. OxS plays an important role in NAFLD pathogenesis. Excess ROS arise mainly from mitochondria, where fatty acid oxidation and oxidative phosphorylation naturally generate ROS. According to the *"two-hits"* and *"multiple hits"* theories, OS is crucial in the transition from simple steatosis to NASH, inducing lipid peroxidation, which leads to inflammation and fibrosis via hepatic stellate cell (HSC) activation. Additionally, ROS inhibit the ability of hepatocytes to release very low-density lipoprotein (VLDL), leading to a buildup of fat within the liver. ROS also contribute to hepatic insulin resistance, inflammation, and apoptosis by activating various intracellular signaling pathways [28]. Moreover, the transcription factor Nrf2 plays a vital role in antioxidant defense and lipid metabolism in NAFLD. In response to high ROS, Nrf2 moves to the nucleus to activate antioxidant genes, supporting cellular redox balance, detoxifying enzymes, and lipid regulation. Studies suggest Nrf2 may reduce lipid accumulation and improve mitochondrial function, yet some research indicates it could also increase lipids under certain conditions [29]. This dual role highlights the complexity of OS regulation in NAFLD, suggesting Nrf2 as both a therapeutic target and a factor requiring careful modulation to balance benefits in NAFLD management.

2.3.2 Alcoholic Liver Disease (ALD)

ALD is characterized by chronic ethanol metabolism, which generates substantial ROS, mainly through cytochrome P450 enzyme CYP2E1. Hepatocyte injury is exacerbated by the generation of acetaldehyde and ROS, which form adducts with cellular proteins and DNA, promoting inflammation and hepatocyte damage. Chronic alcohol consumption induces CYP2E1 expression, which increases ROS production, particularly in mitochondria, leading to mitochondrial damage. This, in turn, disrupts mitochondrial integrity, lowers antioxidant defenses, and drives OxS [30]. Additionally, NOX enzymes further amplify ROS generation, contributing to lipid accumulation and hepatocyte death through various pathways, including apoptosis, necroptosis, and ferroptosis [31]. OxS-induced miRNAs also impair mitochondrial and lipid homeostasis, contributing to disease progression [32].

2.3.3 Viral Hepatitis

Hepatitis B (HBV) and Hepatitis C (HCV) infections promote OxS through a combination of direct viral actions and immune-mediated mechanisms. Chronic hepatitis infections generate OxS both by altering cellular redox states and inducing immune responses that heighten ROS production. HCV infection leads to a substantial increase in ROS levels, rising to five orders of magnitude in liver tissue [33]. HBV also exploits OxS to enhance viral replication and promote chronic infection. Excessive ROS levels damage cellular components, which can lead to persistent inflammation, hepatocyte death, and liver fibrosis. With chronic HBV infection, redox state fluctuations influence viral activity, with higher OxS often corresponding to heightened viral replication [34]. Over time, the chronic oxidative and inflammatory environment in both HBV and HCV infections can induce DNA mutations, promoting the development of HCC. Mitochondria, a major site of ROS generation through oxidative phosphorylation, are particularly vulnerable to viral manipulation, as viruses exploit these organelles to support their replication cycle. As viruses like HBV and HCV stimulate ROS production, they also modulate redox conditions to sustain favorable environments for their life cycles while evading ROS-induced cell death. This persistent redox imbalance not only favors viral survival but also sets the stage for long-term cellular damage, fibrosis, and ultimately, the progression to HCC [35].

2.3.4 Fibrosis

Liver fibrosis is a disease characterized by the excessive buildup of scar tissue, primarily composed of collagen, within the liver. This excessive scarring disrupts the liver's normal structure and function. Hepatic stellate cells (HSCs) are key players in this process, as they are responsible for producing the excess collagen that leads to fibrosis [36]. Under normal conditions, HSCs remain in a quiescent state, storing vitamin A and performing various support functions for liver cells. During liver injury, OxS stimulates the activation of HSCs, transforming them into myofibroblast-like cells that synthesize excessive ECM proteins, leading to liver fibrosis [37]. ROS not only stimulate HSCs to produce ECM proteins but also induce their differentiation to a fibrogenic phenotype. Redox imbalance, resulting from excessive reactive oxygen and nitrogen species, further promotes HSC activation. Damaged hepatocytes, Kupffer cells, and infiltrating lymphocytes contribute to ROS/RNS production, which drives HSC proliferation and ECM deposition, linking immune response and OxS to fibrosis progression.

2.4 Cancer

OxS is increasingly recognized as a key contributor to both tumor initiation and progression. ROS, produced at elevated levels due to redox imbalance, induce DNA damage, including mutations, strand breaks, and chromosomal alterations, which can compromise genome integrity and promote carcinogenesis. High levels of ROS induce DNA damage, including mutations, double-strand breaks, and base modifications, which disrupt cellular homeostasis. This DNA damage promotes genomic instability, a hallmark of cancer [38]. Oxidative stress not only contributes to cellular damage but also induces mutations in key metabolic genes, such as IDH, leading to altered redox homeostasis and tumor progression [39]. ROS can activate signaling pathways like MAPK and PI3K/Akt, which promote cell growth, survival, and the metastasis of cancer. By upregulating genes involved in growth and angiogenesis, ROS-mediated signaling helps tumors expand and invade nearby tissues [40]. Cancer cells with high ROS levels often develop mechanisms to avoid apoptosis, supporting continued growth even under OxS. This includes activation of prosurvival pathways and downregulation of apoptotic signals, making cancer cells more resilient [27]. ROS influence the tumor microenvironment, contributing to inflammation that fosters tumor growth. Inflammatory cells, such as macrophages and neutrophils, produce additional ROS/RNS, further enhancing the oxidative environment and promoting immune evasion by cancer cells.

2.4.1 Colorectal Cancer (CRC)

CRC is a major global health concern linked to lifestyle factors, such as diet, obesity, and inactivity, which increase OxS and genomic instability. OxS with excessive ROS/RNS plays a dual role: Under normal conditions, it regulates cell proliferation and survival, but in CRC, metabolic and oncogenic changes overwhelm antioxidant defenses. This imbalance leads to redox adaptation, which contributes to drug resistance. However, dependence of CRC cells on elevated ROS makes them susceptible to further oxidative insults, suggesting potential for therapies targeting redox perturbations [41]. Inflammation and chronic OS are crucial in CRC development, as OS can oxidize biomolecules and activate inflammatory pathways, supporting tumor initiation and survival. Notably, chronic conditions like inflammatory bowel disease elevate CRC risk due to their link with OS [42]. The Nrf2/KEAP1 pathway, which normally protects cells from OS and inflammation, is vital in CRC. KEAP1 degrades Nrf2, a transcription factor regulating antioxidant responses. Phytochemicals have shown promise in modulating this pathway, potentially inhibiting cancer progression. Hence, phytochemicals targeting Nrf2/KEAP1 could offer a preventative or therapeutic approach for CRC. This understanding highlights the role of OS in CRC pathogenesis and potential redox-based interventions [43].

2.4.2 Leukemia

OxS is key to leukemia pathogenesis, influencing cell growth, proliferation, and survival. Leukemia cells produce high ROS levels through pathways like mitochondrial DNA mutation, oncogene activation, and increased NADPH and cytochrome P450 activities [44]. These pathways promote genomic instability, drug resistance, and relapse. Current chemotherapy employs ROS to induce apoptosis but has significant drawbacks, impacting quality of life. Emerging research suggests that antioxidants, when used alongside chemotherapy, can reduce toxicity, enhance efficacy, and improve patient outcomes in acute myeloid leukemia [45].

2.4.3 Hepatocellular Carcinoma (HCC)

HCC, the most common form of primary liver cancer, often arises in individuals with cirrhosis and is driven by metabolic and OxS, which contribute to tumor initiation, progression, and therapeutic resistance. Although curative treatments are available for early-stage HCC, advanced disease limits these options, highlighting the urgent need for improved biomarkers and therapies [46]. As an inflammation-driven cancer, HCC is linked to chronic inflammation, a condition that induces oxidative and nitrosative stress, leading to lipid peroxidation. This process amplifies OxS and generates reactive aldehydes, which can form DNA adducts, promoting genomic instability and driving tumorigenesis [47]. Moreover, elevated ROS levels in HCC create a prooxidative environment, causing cellular senescence and metabolic reprogramming. These interconnected factors establish a cellular environment that supports tumor growth and survival [48]. Cancer cells undergo a well-documented shift in metabolism known as the Warburg effect, wherein cells preferentially utilize glycolysis over oxidative phosphorylation, even under aerobic conditions. This metabolic reprogramming in HCC serves to provide biosynthetic precursors for rapid cell proliferation, supporting the synthesis of proteins, lipids, and nucleotides [49]. Chronic liver diseases like HCV and NASH induce OxS, contributing to HCC development [50]. HCV is known to cause high levels of OxS, contributing to a markedly high incidence of HCC in patients with cirrhosis. HCV weakens the immune system, particularly by targeting antigen-presenting cells such as Kupffer cells and macrophages. These cells, when disrupted by HCV, release both pro-inflammatory and immunosuppressive cytokines, leading to chronic inflammation. Kupffer cells upregulate PD-L1 in response to HCV, and prolonged inflammation leads to hepatocyte apoptosis and DNA damage, facilitating HCC development [51]. The *"multiple hit"* theory explains NASH progression, where the first hit is triglyceride accumulation in hepatocytes, and subsequent hits include oxidative and ER stress, immune responses, and inflammation. Visceral obesity in NAFLD patients induces inflammatory cytokines, such as IL-17, which promotes neutrophil infiltration and NASH progression, ultimately aiding in HCC development. Elevated IL-17 levels in HCC patients correlate with recurrence risk after surgery [52]. In HCC associated with metabolic dysfunction-associated fatty liver disease, OxS also contributes to cellular senescence, altering hepatic cellular function and promoting an inflammatory environment. This senescence-driven alteration in hepatocyte function and metabolism can initiate a transition from simple steatosis to inflammation and fibrosis, ultimately setting the stage for HCC development [48]. Obesity, a rising global health concern, is a significant risk factor for HCC development due to its association with chronic inflammation, metabolic dysregulation, and OxS. Excessive fat accumulation in hepatocytes leads to an overload of fatty acids, resulting in increased production of harmful ROS. High levels of ROS damage mitochondria, DNA, and further activate inflammatory pathways. This includes dysregulation of BCL-2 proteins and protein tyrosine phosphatases (PTPs), which modulate ROS-mediated signaling [53].

The tumor microenvironment (TME) in HCC, shaped by both OxS and metabolic reprogramming, plays a crucial role in supporting cancer progression. Elevated lactate production, a byproduct of the Warburg effect, acidifies the TME. This acidic environment suppresses the immune system by hindering the function of cytotoxic T-cells and promoting the activity of regulatory T-cells. Additionally, ATP catabolism produces immunosuppressive adenosine, further dampening anti-tumor immune responses. This immunosuppressive TME allows HCC cells to evade immune surveillance, fostering a permissive environment for cancer progression [49].

2.5 *Neurodegenerative Diseases*

OxS is a major factor in the development and progression of various neurodegenerative disorders, including Alzheimer's disease (AD), Parkinson's disease (PD), Huntington's disease (HD), and amyotrophic lateral sclerosis (ALS) [11]. Figure 4 explains the role of oxidative stress in neurodegenerative diseases.

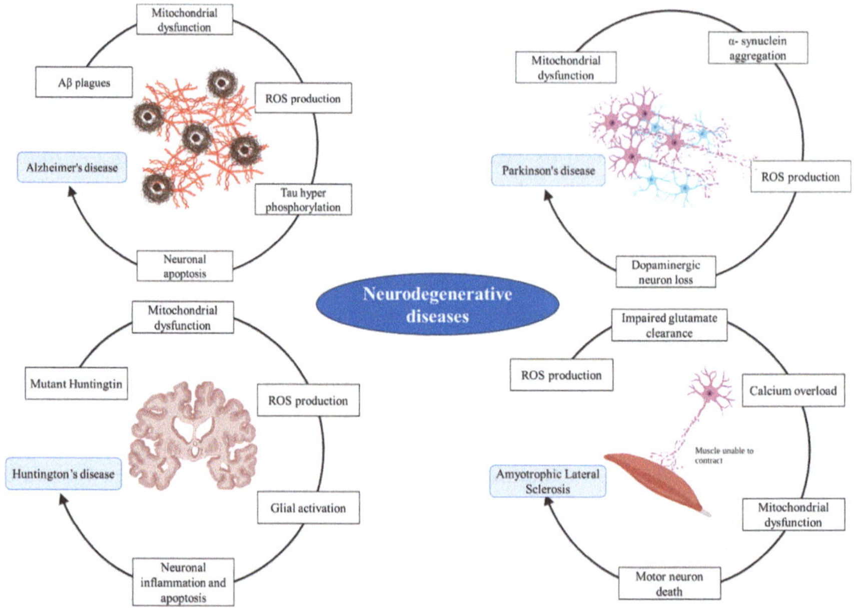

Fig. 4 Oxidative stress in neurodegenerative diseases

2.5.1 Alzheimer's Disease (AD)

The amyloid cascade hypothesis proposes that the accumulation of amyloid-beta (Aβ) peptides in the brain constitutes an initiating event in the pathogenesis of AD. Aβ can increase ROS production through interactions with mitochondrial membranes, disrupting electron transport and exacerbating mitochondrial dysfunction, which leads to additional ROS generation [54]. Elevated ROS levels can activate kinases that drive tau hyperphosphorylation, resulting in tau aggregation and impaired neuronal transport. Damaged tau protein accumulates within neurons, disrupting cellular structures and leading to apoptosis [55].

2.5.2 Parkinson's Disease (PD)

Parkinson's disease involves the loss of dopamine neurons in the substantia nigra, leading to motor deficits. A major contributing factor to PD pathology is mitochondrial dysfunction in these neurons, leading to excessive ROS generation and subsequent OS. ROS-induced oxidative damage to dopamine neurons is exacerbated by the accumulation of α-synuclein, a protein that aggregates abnormally in PD and enhances mitochondrial dysfunction [56]. Dopaminergic neurons are particularly susceptible to OS due to dopamine's inherent propensity to oxidize. Dopamine undergoes enzymatic and non-enzymatic oxidation, producing ROS and reactive quinones that damage cellular components. This damage impairs neuron function and triggers apoptotic pathways, accelerating neuronal loss. Additionally, oxidized dopamine is neurotoxic and can further elevate ROS levels, creating a self-amplifying cycle of oxidative damage that contributes significantly to PD pathogenesis [57].

2.5.3 Huntington's Disease (HD)

HD is caused by a mutation in the HTT gene that leads to the production of misfolded mutant huntingtin protein (mHtt), which accumulates in neurons, especially in regions like the striatum and cortex [58]. Normally, neurons maintain a dynamic balance between the production of ROS and their neutralization by antioxidants. This balance safeguards cells from oxidative damage. However, in HD, this balance is disrupted, leading to excessive ROS accumulation and subsequent damage to crucial cellular components. This OxS contributes to neuronal dysfunction and cell death [59]. The misfolded huntingtin protein disrupts normal protein quality control mechanisms and impairs the cell's oxidative defenses. Additionally, mHtt activates glial cells, microglia, and astrocytes, which release ROS and pro-inflammatory cytokines, further increasing OxS and triggering chronic neuroinflammation in the brain [60]. In response to the high energy demands of neurons, mitochondria rely on coordinated processes of fission, fusion, and trafficking. However, mHtt disrupts these mitochondrial dynamics, worsening mitochondrial dysfunction, ROS production, and calcium imbalance, leading to apoptotic pathways and further neuronal degeneration [61].

2.5.4 Amyotrophic Lateral Sclerosis (ALS)

ALS involves the degeneration of motor neurons in the spinal cord, brainstem, and motor cortex [62]. Individuals with ALS often exhibit elevated levels of inflammation and OxS. This is evident in increased markers such as 3-nitrotyrosine, 4-HNE, 8-OHdG, and MDA [50]. Elevated OxS exacerbates neurodegeneration through several mechanisms. Excessive ROS at synapses reduce glutamate transport, increasing synaptic glutamate and overstimulating receptors, leading to harmful calcium influx into motor neurons. This calcium enters mitochondria, impairing their function and increasing ROS production. ROS from damaged neurons affect glial synapses, similarly, enhancing glutamate concentration, calcium influx, and ROS production, impairing glial function and increase neuroinflammation [63].

2.6 Oxidative Stress and Aging

Aging is a natural process characterized by a gradual decline in physical function and an increased susceptibility to age-related diseases like heart disease, cancer, and neurodegenerative disorders [64]. Human mitochondrial DNA (mtDNA) is vulnerable to damage from reactive oxygen and nitrogen species due to its proximity to ROS production sites. Over time, ROS accumulation and oxidative damage drive age-associated functional decline by impairing cellular and mitochondrial homeostasis. OxS accelerates aging and contributes to age-related conditions, including cardiovascular diseases, neurodegenerative disorders, sarcopenia, and cancer [65].

3 Therapeutic Interventions Targeting Oxidative Stress

Targeting OxS has emerged as a promising strategy for managing chronic diseases, where excessive ROS contribute to cellular damage, inflammation, and disease progression across conditions such as CVD, neurodegeneration, diabetes, and cancer. In Alzheimer's disease, OxS plays a significant role in the neuronal degeneration observed alongside amyloid-β and tau accumulation. Promising therapeutic interventions, targeting OxS and related pathways, are under preclinical or clinical trials for AD and mild cognitive impairment. These compounds focus on modulating redox metals and preventing neuronal death [66]. In diabetic cardiomyopathy (DbCM), OxS arises from heightened ROS production by sources like mitochondria, NOX, and Xo. Excessive ROS disrupt cardiac cell function by damaging mitochondrial DNA, causing lipid peroxidation, and modifying proteins through post-translational modifications, ultimately leading to cell death and cardiac dysfunction. Current therapeutic strategies aim to enhance endogenous antioxidant defenses, such as catalase and superoxide dismutase, which are upregulated in response to ROS but are often insufficient in halting damage progression. Targeting ROS production and enhancing

Table 2 Antioxidant defense mechanisms

S. no	Antioxidant	Mechanism of action	Key role
1	Superoxide dismutase	Converts superoxide radicals into hydrogen peroxide	First-line defense against ROS
2	Catalase	Breaks down hydrogen peroxide into water and oxygen	Prevents accumulation of H2O2
3	Glutathione peroxidase	Reduces lipid hydroperoxides	Protects against oxidative damage
4	Glutathione reductase	Reduces lipid hydroperoxides	Protects against oxidative damage
5	Thioredoxin	Reduces oxidized proteins and regulates redox signaling	Protects against oxidative stress in cells
6	Peroxiredoxins	Detoxifies peroxides, including hydrogen peroxide	Regulates intracellular ROS levels
7	Vitamin C (ascorbic acid)	Donates electrons to neutralize ROS	Protects biomolecules from oxidative damage
8	Vitamin E (tocopherol)	Inhibits lipid peroxidation	Protects cell membranes from oxidative damage

antioxidant defenses pharmacologically have shown potential in protecting cardiac cells from ROS-mediated damage, slowing DbCM progression, and reducing heart failure risk associated with diabetes [67]. Despite significant preclinical success, many antioxidant-based therapies have yet to demonstrate efficacy in clinical trials, highlighting a gap in understanding the mechanisms and timing of antioxidant intervention in diseases like atherosclerosis and cancer. The efficacy of antioxidants depends on the specific pathways through which they operate and the stage of disease progression [12]. Table 2 lists the key antioxidant defense mechanisms that help control oxidative stress.

4 Challenges and Future Directions

Targeting OxS as a therapeutic strategy presents unique challenges, particularly in complex, multifactorial diseases like type 2 diabetes mellitus, Alzheimer's disease, and cardiovascular conditions. While OxS is a well-established contributor to the pathogenesis of these diseases, translating this knowledge into effective therapeutic interventions remains a significant challenge. These include understanding the intricate balance of ROS within cellular signaling, determining the most beneficial timing for antioxidant interventions, and addressing the variability in patient response to antioxidant therapies. For instance, in T2DM, excessive ROS production contributes to insulin resistance, impaired beta-cell function, and complications like neuropathy and retinopathy. While antioxidant-based treatments have been explored, their clinical outcomes have been inconsistent. This is likely due to the intricate nature of ROS-related signaling pathways involved in glucose regulation and insulin sensitivity [68]. In neurodegenerative conditions, ROS disrupt mitochondrial function, resulting in cellular damage and progression of neurodegeneration. Despite strong preclinical evidence linking mitochondrial OxS to these pathologies, antioxidant therapies have not consistently demonstrated clinical success. This inconsistency is partly due to the difficulty in targeting mitochondrial ROS specifically and efficiently across different tissues and disease stages [69]. Advancements in selective antioxidant delivery, such as mitochondria-targeted antioxidants, could allow for a more localized and effective reduction in OxS without disrupting essential redox signaling in healthy cells. Combining antioxidants with other therapeutic strategies, such as anti-inflammatory agents or metabolic modulators, may also enhance efficacy. Ultimately, overcoming these challenges will require a detailed understanding of ROS as both signaling molecules and contributors to disease, enabling the development of more tailored and efficacious therapies to mitigate OxS in chronic diseases.

5 Conclusion

OxS is a central mechanism in the progression and pathology of numerous chronic diseases, impacting cardiovascular, metabolic, hepatic, neurodegenerative, and oncological conditions. The intricate interplay between the production of ROS and antioxidant defenses underscores the challenge in designing targeted therapies, as ROS not only drive disease progression but are also integral to normal cellular signaling. While preclinical studies have shown promising results, the translation to clinical success remains limited, emphasizing the need for nuanced approaches that consider the specific disease context, timing, and patient variability in ROS-related damage. Future research should prioritize the development of targeted therapies that specifically modulate ROS, mitochondrial-focused antioxidants, and personalized treatment plans tailored to individual OxS levels. By deepening our understanding of the molecular pathways involving ROS and refining antioxidant therapies, we can develop effective disease management strategies that improve the quality of life for individuals with chronic conditions. Unraveling the role of OxS in chronic disease progression and aging remains crucial, as it offers valuable insights for developing innovative diagnostic, preventive, and therapeutic approaches to combat OxS and its far-reaching impact on human health.

Acknowledgements The authors thank the management of Vellore Institute of Technology for providing the necessary resources and facilities to carry out this work.

Author Contributions All authors have contributed equally to this chapter.

References

1. Reddy, V. P. (2023). Oxidative stress in health and disease. *Biomedicines, 11*(11), 2925.
2. Sharifi-Rad, M., Anil Kumar, N. V., Zucca, P., Varoni, E. M., Dini, L., Panzarini, E., & Sharifi-Rad, J. (2020). Lifestyle, oxidative stress, and antioxidants: Back and forth in the pathophysiology of chronic diseases. *Frontiers in Physiology, 11*, 694.
3. Kıran, T. R., Otlu, O., & Karabulut, A. B. (2023). Oxidative stress and antioxidants in health and disease. *Journal of Laboratory Medicine, 47*(1), 1–11.
4. Liu, Z., Ren, Z., Zhang, J., Chuang, C. C., Kandaswamy, E., Zhou, T., & Zuo, L. (2018). Role of ROS and nutritional antioxidants in human diseases. *Frontiers in Physiology, 9*, Article 360203.
5. Aramouni, K., Assaf, R., Shaito, A., Fardoun, M., Al-Asmakh, M., Sahebkar, A., & Eid, A. H. (2023). Biochemical and cellular basis of oxidative stress: Implications for disease onset. *Journal of Cellular Physiology, 238*(9), 1951–1963.
6. Manzoor, M. F., Arif, Z., Kabir, A., Mehmood, I., Munir, D., Razzaq, A., & Rusu, A. (2022). Oxidative stress and metabolic diseases: Relevance and therapeutic strategies. *Frontiers in Nutrition, 9*, Article 994309.
7. Dubois-Deruy, E., Peugnet, V., Turkieh, A., & Pinet, F. (2020). Oxidative stress in cardiovascular diseases. *Antioxidants, 9*(9), 864.
8. Yaribeygi, H., Sathyapalan, T., Atkin, S. L., & Sahebkar, A. (2020). Molecular mechanisms linking oxidative stress and diabetes mellitus. *Oxidative Medicine and Cellular Longevity, 2020*(1), 8609213.
9. Gabbia, D., Cannella, L., & De Martin, S. (2021). The role of oxidative stress in NAFLD–NASH–HCC transition—focus on NADPH oxidases. *Biomedicines, 9*(6), 687.
10. Maldonado, E., Morales-Pison, S., Urbina, F., & Solari, A. (2023). Aging hallmarks and the role of oxidative stress. *Antioxidants, 12*(3), 651.
11. Teleanu, D. M., Niculescu, A. G., Lungu, I. I., Radu, C. I., Vladâcenco, O., Roza, E., & Teleanu, R. I. (2022). An overview of oxidative stress, neuroinflammation, and neurodegenerative diseases. *International Journal of Molecular Sciences, 23*(11), 5938.
12. Forman, H. J., & Zhang, H. (2021). Targeting oxidative stress in disease: Promise and limitations of antioxidant therapy. *Nature Reviews Drug Discovery, 20*(9), 689–709.
13. Jomova, K., Raptova, R., Alomar, S. Y., Alwasel, S. H., Nepovimova, E., Kuca, K., & Valko, M. (2023). Reactive oxygen species, toxicity, oxidative stress, and antioxidants: Chronic diseases and aging. *Archives of toxicology, 97*(10), 2499–2574.
14. Jin, S., & Kang, P. M. (2024). A systematic review on advances in management of oxidative stress-associated cardiovascular diseases. *Antioxidants, 13*(8), 923.
15. Jebari-Benslaiman, S., Galicia-García, U., Larrea-Sebal, A., Olaetxea, J. R., Alloza, I., Vandenbroeck, K., & Martín, C. (2022). Pathophysiology of atherosclerosis. *International Journal of Molecular Sciences, 23*(6), 3346.
16. Batty, M., Bennett, M. R., & Yu, E. (2022). The role of oxidative stress in atherosclerosis. *Cells, 11*(23), 3843.
17. Poznyak, A. V., Grechko, A. V., Orekhova, V. A., Chegodaev, Y. S., Wu, W. K., & Orekhov, A. N. (2020). Oxidative stress and antioxidants in atherosclerosis development and treatment. *Biology, 9*(3), 60.
18. Jiang, H., Zhou, Y., Nabavi, S. M., Sahebkar, A., Little, P. J., Xu, S., & Ge, J. (2022). Mechanisms of oxidized LDL-mediated endothelial dysfunction and its consequences for the development of atherosclerosis. *Frontiers in Cardiovascular Medicine, 9*, Article 925923.
19. Zhang, Z., Zhao, L., Zhou, X., Meng, X., & Zhou, X. (2023). Role of inflammation, immunity, and oxidative stress in hypertension: New insights and potential therapeutic targets. *Frontiers in Immunology, 13*, 1098725.
20. Touyz, R. M., Rios, F. J., Alves-Lopes, R., Neves, K. B., Camargo, L. L., & Montezano, A. C. (2020). Oxidative stress: A unifying paradigm in hypertension. *Canadian Journal of Cardiology, 36*(5), 659–670.

21. de Almeida, A. J. P. O., de Almeida Rezende, M. S., Dantas, S. H., de Lima Silva, S., de Oliveira, J. C. P. L., de Lourdes Assunção Araújo de Azevedo, F., & de Medeiros, I. A. (2020). Unveiling the role of inflammation and oxidative stress on age-related cardiovascular diseases. *Oxidative Medicine and Cellular Longevity, 2020*(1), 1954398.
22. Cole, J. B., & Florez, J. C. (2020). Genetics of diabetes mellitus and diabetes complications. *Nature Reviews Nephrology, 16*(7), 377–390.
23. Black, H. S. (2022). A synopsis of the associations of oxidative stress, ROS, and antioxidants with diabetes mellitus. *Antioxidants, 11*(10), 2003.
24. Darenskaya, M. A., Kolesnikova, L. A., & Kolesnikov, S. I. (2021). Oxidative stress: Pathogenetic role in diabetes mellitus and its complications and therapeutic approaches to correction. *Bulletin of Experimental Biology and Medicine, 171*(2), 179–189.
25. Ghasemi-Dehnoo, M., Amini-Khoei, H., Lorigooini, Z., & Rafieian-Kopaei, M. (2020). Oxidative stress and antioxidants in diabetes mellitus. *Asian Pacific Journal of Tropical Medicine, 13*(10), 431–438.
26. Khalid, M., Petroianu, G., & Adem, A. (2022). Advanced glycation end products and diabetes mellitus: Mechanisms and perspectives. *Biomolecules, 12*(4), 542.
27. Zahra, K. F., Lefter, R., Ali, A., Abdellah, E. C., Trus, C., Ciobica, A., & Timofte, D. (2021). The involvement of the oxidative stress status in cancer pathology: A double view on the role of the antioxidants. *Oxidative Medicine and Cellular Longevity, 2021*(1), 9965916.
28. Smirne, C., Croce, E., Di Benedetto, D., Cantaluppi, V., Comi, C., Sainaghi, P. P., & Pirisi, M. (2022). Oxidative stress in non-alcoholic fatty liver disease. *Livers, 2*(1), 30–76.
29. Arroyave-Ospina, J. C., Wu, Z., Geng, Y., & Moshage, H. (2021). Role of oxidative stress in the pathogenesis of non-alcoholic fatty liver disease: Implications for prevention and therapy. *Antioxidants, 10*(2), 174.
30. Tan, H. K., Yates, E., Lilly, K., & Dhanda, A. D. (2020). Oxidative stress in alcohol-related liver disease. *World Journal of Hepatology, 12*(7), 332.
31. Michalak, A., Lach, T., & Cichoż-Lach, H. (2021). Oxidative stress—A key player in the course of alcohol-related liver disease. *Journal of Clinical Medicine, 10*(14), 3011.
32. Yang, Y. M., Cho, Y. E., & Hwang, S. (2022). Crosstalk between oxidative stress and inflammatory liver injury in the pathogenesis of alcoholic liver disease. *International Journal of Molecular Sciences, 23*(2), 774.
33. Paracha, U. Z., Fatima, K., Alqahtani, M., Chaudhary, A., Abuzenadah, A., Damanhouri, G., & Qadri, I. (2013). Oxidative stress and hepatitis C virus. *Virology Journal, 10*, 1–9.
34. Popa, G. L., & Popa, M. I. (2022). Oxidative stress in chronic hepatitis B—An update. *Microorganisms, 10*(7), 1265.
35. Foo, J., Bellot, G., Pervaiz, S., & Alonso, S. (2022). Mitochondria-mediated oxidative stress during viral infection. *Trends in Microbiology, 30*(7), 679–692.
36. Acharya, P., Chouhan, K., Weiskirchen, S., & Weiskirchen, R. (2021). Cellular mechanisms of liver fibrosis. *Frontiers in Pharmacology, 12*, Article 671640.
37. Berumen, J., Baglieri, J., Kisseleva, T., & Mekeel, K. (2021). Liver fibrosis: Pathophysiology and clinical implications. *WIREs Mechanisms of Disease, 13*(1), Article e1499.
38. Hayes, J. D., Dinkova-Kostova, A. T., & Tew, K. D. (2020). Oxidative stress in cancer. *Cancer Cell, 38*(2), 167–197.
39. Murali, P., & Karuppasamy, R. (2023). Discovery of novel and potent inhibitors against mutational variants of idh1 protein for glioma therapy: A fragment-based approach. *Journal of Computational Biophysics and Chemistry, 22*(01), 43–61.
40. Arfin, S., Jha, N. K., Jha, S. K., Kesari, K. K., Ruokolainen, J., Roychoudhury, S., & Kumar, D. (2021). Oxidative stress in cancer cell metabolism. *Antioxidants, 10*(5), 642.
41. Basak, D., Uddin, M. N., & Hancock, J. (2020). The role of oxidative stress and its counteractive utility in colorectal cancer (CRC). *Cancers, 12*(11), 3336.
42. Bardelčíková, A., Šoltys, J., & Mojžiš, J. (2023). Oxidative stress, inflammation and colorectal cancer: An overview. *Antioxidants, 12*(4), 901.
43. Lee, D. Y., Song, M. Y., & Kim, E. H. (2021). Role of oxidative stress and Nrf2/KEAP1 signaling in colorectal cancer: Mechanisms and therapeutic perspectives with phytochemicals. *Antioxidants, 10*(5), 743.

44. Kaweme, N. M., Zhou, S., Changwe, G. J., & Zhou, F. (2020). The significant role of redox system in myeloid leukemia: From pathogenesis to therapeutic applications. *Biomarker Research, 8*, 1–12.
45. Dong, C., Zhang, N. J., & Zhang, L. J. (2021). Oxidative stress in leukemia and antioxidant treatment. *Chinese Medical Journal, 134*(16), 1897–1907.
46. Premnath, V., & Veerappapillai, S. (2024). Unveiling miRNA–gene regulatory axes as promising biomarkers for liver cirrhosis and hepatocellular carcinoma. *ACS Omega, 9*(44), 44507–44521.
47. Fu, Y., & Chung, F. L. (2018). Oxidative stress and hepatocarcinogenesis. *Hepatoma research, 4.*
48. Anastasopoulos, N. A., Charchanti, A. V., Barbouti, A., Mastoridou, E. M., Goussia, A. C., Karampa, A. D., & Glantzounis, G. K. (2023). The role of oxidative stress and cellular senescence in the pathogenesis of metabolic associated fatty liver disease and related hepatocellular carcinoma. *Antioxidants, 12*(6), 1269.
49. Mossenta, M., Busato, D., Dal Bo, M., & Toffoli, G. (2020). Glucose metabolism and oxidative stress in hepatocellular carcinoma: Role and possible implications in novel therapeutic strategies. *Cancers, 12*(6), 1668.
50. Park, H. R., & Yang, E. J. (2021). Oxidative stress as a therapeutic target in amyotrophic lateral sclerosis: Opportunities and limitations. *Diagnostics, 11*(9), 1546.
51. Ivanov, A. V., Bartosch, B., Smirnova, O. A., Isaguliants, M. G., & Kochetkov, S. N. (2013). HCV and oxidative stress in the liver. *Viruses, 5*(2), 439–469.
52. Wang, Z., Li, Z., Ye, Y., Xie, L., & Li, W. (2016). Oxidative stress and liver cancer: Etiology and therapeutic targets. *Oxidative Medicine and Cellular Longevity, 2016*(1), 7891574.
53. Brahma, M. K., Gilglioni, E. H., Zhou, L., Trépo, E., Chen, P., & Gurzov, E. N. (2021). Oxidative stress in obesity-associated hepatocellular carcinoma: Sources, signaling and therapeutic challenges. *Oncogene, 40*(33), 5155–5167.
54. Misrani, A., Tabassum, S., & Yang, L. (2021). Mitochondrial dysfunction and oxidative stress in Alzheimer's disease. *Frontiers in Aging Neuroscience, 13*, Article 617588.
55. Bai, R., Guo, J., Ye, X. Y., Xie, Y., & Xie, T. (2022). Oxidative stress: The core pathogenesis and mechanism of Alzheimer's disease. *Ageing Research Reviews, 77*, Article 101619.
56. Chang, K. H., & Chen, C. M. (2020). The role of oxidative stress in Parkinson's disease. *Antioxidants, 9*(7), 597.
57. Dorszewska, J., Kowalska, M., Prendecki, M., Piekut, T., Kozłowska, J., & Kozubski, W. (2021). Oxidative stress factors in Parkinson's disease. *Neural Regeneration Research, 16*(7), 1383–1391.
58. Bono-Yagüe, J., Gómez-Escribano, A. P., Millán, J. M., & Vázquez-Manrique, R. P. (2020). Reactive species in Huntington disease: Are they really the radicals you want to catch? *Antioxidants, 9*(7), 577.
59. Alqahtani, T., Deore, S. L., Kide, A. A., Shende, B. A., Sharma, R., Chakole, R. D., & Ghosh, A. (2023). Mitochondrial dysfunction and oxidative stress in Alzheimer's disease, and Parkinson's disease, Huntington's disease and amyotrophic lateral sclerosis-an updated review. *Mitochondrion, 71*, 83–92.
60. D'Egidio, F., Castelli, V., Cimini, A., & d'Angelo, M. (2023). Cell rearrangement and oxidant/antioxidant imbalance in Huntington's disease. *Antioxidants, 12*(3), 571.
61. Jurcau, A. (2022). Molecular pathophysiological mechanisms in Huntington's disease. *Biomedicines, 10*(6), 1432.
62. Cunha-Oliveira, T., Montezinho, L., Mendes, C., Firuzi, O., Saso, L., Oliveira, P. J., & Silva, F. S. (2020). Oxidative stress in amyotrophic lateral sclerosis: Pathophysiology and opportunities for pharmacological intervention. *Oxidative Medicine and Cellular Longevity, 2020*(1), 5021694.
63. Hemerková, P., & Vališ, M. (2021). Role of oxidative stress in the pathogenesis of amyotrophic lateral sclerosis: Antioxidant metalloenzymes and therapeutic strategies. *Biomolecules, 11*(3), 437.
64. Hajam, Y. A., Rani, R., Ganie, S. Y., Sheikh, T. A., Javaid, D., Qadri, S. S., & Reshi, M. S. (2022). Oxidative stress in human pathology and aging: Molecular mechanisms and perspectives. *Cells, 11*(3), 552.

65. Warraich, U.-A., Hussain, F., & Kayani, H. U. R. (2020). Aging—oxidative stress, antioxidants and computational modeling. *Heliyon, 6*(5), e04107.
66. Plascencia-Villa, G., & Perry, G. (2021). Preventive and therapeutic strategies in Alzheimer's disease: Focus on oxidative stress, redox metals, and ferroptosis. *Antioxidants & Redox Signaling, 34*(8), 591–610.
67. Byrne, N. J., Rajasekaran, N. S., Abel, E. D., & Bugger, H. (2021). Therapeutic potential of targeting oxidative stress in diabetic cardiomyopathy. *Free Radical Biology and Medicine, 169*, 317–342.
68. Bhatti, J. S., Sehrawat, A., Mishra, J., Sidhu, I. S., Navik, U., Khullar, N., & Reddy, P. H. (2022). Oxidative stress in the pathophysiology of type 2 diabetes and related complications: Current therapeutics strategies and future perspectives. *Free Radical Biology and Medicine, 184*, 114–134.
69. Kowalczyk, P., Sulejczak, D., Kleczkowska, P., Bukowska-Ośko, I., Kucia, M., Popiel, M., & Kaczyńska, K. (2021). Mitochondrial oxidative stress—A causative factor and therapeutic target in many diseases. *International Journal of Molecular Sciences, 22*(24), 13384.
70. Karkucinska-Wieckowska, A., Simoes, I. C., Kalinowski, P., Lebiedzinska-Arciszewska, M., Zieniewicz, K., Milkiewicz, P., & Wieckowski, M. R. (2022). Mitochondria, oxidative stress and nonalcoholic fatty liver disease: A complex relationship. *European Journal of Clinical Investigation, 52*(3), Article e13622.
71. Takaki, A., & Yamamoto, K. (2015). Control of oxidative stress in hepatocellular carcinoma: Helpful or harmful? *World Journal of Hepatology, 7*(7), 968.

Oxidative Stress as a Dominant Mechanism in the Pathophysiology of Chronic Diseases: Insights into Molecular Pathways, Clinical Implications, and Advancing Therapeutic Interventions

Nivethitha Chandran, Mageshwari Kannan, Subashini Chandrasekaran, and Johnson Retnaraj Samuel Selvan Christyraj

Abstract An imbalance between ROS and antioxidant defenses gives rise to oxidative stress and it is a major factor in the development and progression of chronic diseases due to oxidative stress. We are going to explore the molecular and cellular pathways through which oxidative stress promotes the breakthrough and advancement of chronic illness, including cardiovascular disorders, diabetes, Alzheimer's and Parkinson's disease, chronic inflammatory conditions, and cancer. It triggers a series of cellular disturbances, resulting in DNA damage, mitochondrial dysfunction, lipid oxidation, and protein oxidation. These disruptions weaken cellular structure and activate inflammatory signaling pathways, which worsen tissue damage and drive disease progression. Additionally, chronic inflammation, metabolic disturbances, and the immune system create a harmful cycle that accelerates disease progression. This chapter also examines how factors such as age, lifestyle habits (like diet, smoking, and environmental exposures), and genetic factors influence oxidative stress levels, increasing the risk of chronic diseases in specific populations. It also analyzes the potential of both pharmacological and dietary antioxidant therapies to reduce oxidative stress and restore redox balance, focusing on innovative therapeutic methods, including targeted ROS-scavenging compounds and gene therapies aimed at enhancing antioxidant capacity. It also explains the significance of how the pathogenesis of chronic diseases is related to oxidative stress. It emphasizes the importance of developing novel therapies that reduce oxidative stress, slow the progression of illness, and improve patients' lives.

Nivethitha Chandran, Mageshwari Kannan and Subashini Chandrasekaran are the authors contributed equally.

N. Chandran · M. Kannan · S. Chandrasekaran · J. R. S. Selvan Christyraj (✉)
Regeneration and Stem Cell Biology Lab, Centre for Molecular and Nanomedical Sciences, International Research Centre, Sathyabama Institute of Science and Technology, Chennai, Tamil Nadu, India
e-mail: johnnbt@sathyabama.ac.in

K. Kasinathan et al. (eds.), *Recent Advances in Oxidative Stress Associated Chronic Diseases Volume 1*, https://doi.org/10.1007/978-981-96-9670-3_10

Keywords Oxidative stress · Biomarkers · Antioxidant · Key signaling pathways · Chronic disease · Therapeutic implications · Free radicals

1 Introduction

An imbalance between oxidants and antioxidants, favoring oxidants, causes disturbances in redox signaling and control, which may cause molecular damage. Its main concept is sustaining a stable redox balance at a particular setpoint in an open metabolic system, providing a fundamental redox tone. When there is a deviation from this stable redox balance, it is viewed as stress, triggering a stress response. Oxidative stress is viewed as harmful due to the attack of oxygen-free radicals, such as proteins and DNA. However, it is also crucial for physiological adaptation and control of intracellular signaling. Consequently, an accurate definition of oxidative stress might be an imbalance caused by oxidative forces exceeding antioxidant systems capacity [1]. Here, we list chronic illnesses commonly linked to oxidative stress. (i) *Cardiovascular diseases*: The generation of reactive oxygen species (ROS) triggers high blood pressure and high cholesterol, which causes oxidative stress [2]. Reactive oxygen species (ROS) play a dual function in cardiovascular health. In small amounts, they offer benefits such as preventing atherosclerosis, encouraging new blood vessel formation, and activating heart-protective mechanisms. However, in large quantities, ROS cause oxidative stress, harming cells and contributing to cardiovascular diseases such as cardiac failure, arrhythmias, ischemia/reperfusion damage, endothelial dysfunction, and atherosclerosis [3]. (ii) *Neurodegenerative disease*: Neuronal sensitivity to oxidative stress varies across different groups in the brain. For example, neurons with more synapses and longer axons need more energy for functions like long-term plasticity and axonal transport. This increased energy requirement makes them more vulnerable to oxidative damage. The mechanisms involved in neurodegeneration surround several processes, such as protein misfolding and aggregation, aberrant signaling pathways involving kinases, dysregulation of calcium in neurons, and compromised synaptic transmissions [4]. These mechanisms contribute to dysfunction and death in the development of neurodegenerative disease [5]. (iii) *Diabetes*: Diabetes, a primary cause of oxidative stress, leads to atherogenic effects. Diabetes promotes oxidative damage in the development of atherosclerosis and other cardiovascular issues [6]. This leads to a higher generation of superoxide anions and a reduction in efficient ATP synthesis. This imbalance affects energy metabolism and is attributed to oxidative stress, which is similar to diabetic complications [7]. (iv) *Cancer*: The evolution of cancer in humans is a complicated process involving a range of cellular and molecular modifications that are influenced by various internal and external factors [8]. Both endogenous stimuli, such as genetic alterations and metabolic changes, as well as exogenous factors, including environmental pollutants and dietary habits, play significant roles in this complex transformation [9]. In cancer cells, a persistent condition of oxidative stress occurs due to dysfunction in their mitochondria and changes in their metabolic activities. This plays a vital role in cancer improvement and progression, influencing various pathways that boost tumor development and endurance [10].

In chronic disease, oxidative stress is a major factor that causes an imbalance between reactive oxygen species (ROS) and antioxidants, leading to oxidative stress. It is provoked by numerous internal and external aspects, such as immune responses, infections, pollution, poor nutrition, or inactivity, which increase ROS beyond the body's ability to control [1]. This imbalance results in cellular malfunction by harming DNA, lipids, and proteins that speed up aging and contribute to chronic diseases. For example, oxidative stress is linked to cardiovascular disease, diabetes, neurodegenerative disorders, and certain cancers by inducing inflammation, DNA mutations, and lipid damage. All this causes disease development. Additionally, oxidative stress and chronic inflammation feed into one another, creating a cycle that aggravates cases such as type 2 diabetes, Alzheimer's disease, and atherosclerosis. Tackling oxidative stress through antioxidant-rich diets and lifestyle adjustments shows promise. Antioxidants like fruits, vegetables, and vitamin supplements are crucial in neutralizing ROS, which can help slow disease progression. Research suggests that natural antioxidants in phytochemicals—such as polyphenols in berries and carotenoids in carrots—can provide specific protective effects by reducing oxidative damage and enhancing immune responses, thereby preventing or postponing related diseases. It shows the importance of preventive strategies like a proper diet and regular physical activity, to boost the body's defense against oxidative stress and lower the danger of chronic illnesses [11].

2 Mechanisms of Oxidative Stress

2.1 Historical Background

The word free radical was first coined by Moses Gomberg in 1950. He was the first researcher to discover that the first stable free radical consists of a central triplet carbon atom named triphenylmethyl. Understanding the free radical chain reaction took 30 years. In 1968, the first enzyme activity against free radicals was discovered by McCord and Fridovich, termed superoxide dismutase. Loschen found the reactive oxygen species formed in cellular metabolic respiration in 1971. Halliwell and Gutteridge observed that reactive oxygen species contained oxygen-free radicals in 1989 [12].

2.2 Free Radicals

Every atom, molecule, or fragment containing one or more unpaired electrons in its outer orbits is called a free radical. They function as a reductant or oxidant because they donate or accept electrons from other molecules [13]. They are electrophilic, so they quickly target the high-level electron density mostly seen in nitrogen and carbon–carbon double bonds. Free radicals, which are highly reactive and unstable, are created by unpaired electrons. Free radicals play crucial roles in our body in fighting against diseases. They harm the cells and tissues in our body when present at high levels, targeting DNA, RNA, proteins, and carbohydrates [14]. The free radicals are further classified as follows.

2.2.1 Superoxide Dismutase Radicals

It is generated when molecular oxygen receives an electron. It is not reactive. It cannot easily cross the lipid membrane, so it stops the production from the compartment. Superoxide is especially produced in an electron-rich area surrounding nothing but the thinner mitochondrial membrane associated with the respiratory chain [15]. Xanthine oxidase, lipoxygenase, cyclooxygenase, and NADPH-dependent oxidase are enzymes that produce superoxide. Superoxide reactivity is pH-dependent with non-radical molecules. The radical serves as a stronger reducing agent than as an oxidizing agent when $O_2^{\bullet-}$ interacts with iron complexes, converting Fe^{3+} into Fe^{2+}.

$$O_2 + e^- \rightarrow O_2^{\bullet-}$$

$$O_2 + Fe^{+2} \rightarrow Fe^{3+} + O_2^{\bullet-}(\text{auto} - \text{oxidation})$$

$O_2^{\bullet-}$ acts as an oxidizing agent and has the ability to oxidize Cu^{3+}, Fe^{4+}, and Mn^{3+}, as well as dopamine, adrenaline, and ascorbate, resulting in their transformation [16].

2.2.2 Hydroxyl Radicals

It forms when the hydrogen peroxide and metal ion interact with one another. This is called the Fenton reaction [17]. Hydroxyl radical is a highly reactive free radical. Free radicals vigorously interact with molecules, leading to serious cellular destruction compared to free radicals [18]. The hydroxyl is formed based on the Fenton reaction, where the hydrogen peroxide H_2O_2 interacts with the metal ion Fe^{2+} or Cu^{+}, which are typically complexed with proteins such as ferritin, a protein that stores ions, and ceruloplasmin, a plasma copper that carries the protein. The discharge ions from ferritin occur because of excess $O_2^{\bullet-}$ production, and the free ions released from ferritin undergo a Fenton reaction to form $OH^{\bullet}$ while in stress conditions [19].

$$Fe^{2+} + H_2O_2 \rightarrow Fe^{3+} + HO^{\bullet} + {}^{-}OH$$

$$Fe^{3+} + H_2O_2 \rightarrow Fe^{2+} + HOO^{\bullet} + H^{+}$$

2.2.3 Hydrogen Peroxide

Two electrons are used to reduce oxygen. It is not regarded as a radical because of the lack of an unpaired electron. It also produces additional enzymes such as amino oxidase, xanthine, and glucose monoamine [16]. Although it lacks unpaired electrons and is not considered a radical, it can still damage cells at low concentrations [13]. H_2O_2 is involved in cell proliferation, differentiation, and signal transduction. H_2O_2 is removed by enzymes such as glutathione peroxidase, peroxiredoxins, and catalase.

2.2.4 Singlet Oxygen

A type of non-radical reactive oxygen species changes from a ground state to an excited state by utilizing two unpaired electrons. O_2 exists in two forms initially. It reaches the delta state ($1\Delta g$), which is further excited to a higher energy level known as the sigma state [15]. The delta state is extremely reactive,it is usually generated when neutrophils and electrophiles are activated [20]. O_2 is a strong oxidizer that harms DNA and tissues, mainly in the eyes and skin. It is produced by several enzymatic reactions involving lipoxygenases, dioxygenases, and lactoperoxidase. High concentration of O_2 exposure to light can cause porphyrins, resulting in blistering, eruption, edema, and scarring [21].

2.2.5 Peroxyl Radicals

Peroxyl radicals originate from oxygen in biological systems. They are produced during lipid oxidation chain reactions, including the oxidation of polyunsaturated fats, which spoil lipid-containing foods. The perhydroxyl radical (HOO•) is the most basic type of peroxyl radical. It is made by adding a proton to superoxide. Peroxyl radicals are generated by LPO after the breakdown of lipoperoxidation in the presence of transition metal ions by UV or heat [22].

$$ROOH + Fe^{2+}RO^{\cdot} \rightarrow Fe^{3+}RO\cdot \rightarrow RO\cdot + OH^{-}$$

$$ROOH + Fe^{3+}RO\cdot \rightarrow ROO\cdot + H^{+} + Fe^{2+}RO\cdot$$

2.3 *ROS in Oxidative Stress*

Reactive oxygen species are the byproducts of cellular metabolism. Mainly found inside the mitochondria, they are the primary source of ROS. The electron transport chain is where most ROS in mitochondria are found [23]. The cells maintain the

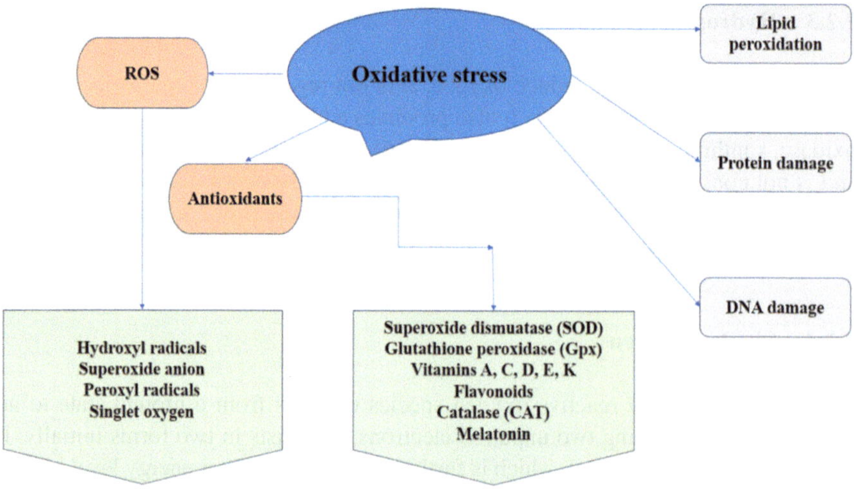

Fig. 1 Schematic showing the role of ROS in oxidative stress and its consequences for the human body. Enhanced generation of free radicals causes an imbalance in the body's capacity. ROS are also produced in the body by external factors, such as pollution

ratio of ROS and antioxidant defenses,enzymes involved are catalase, glutathione peroxidase, and superoxide dismutase [24, 25]. If excess ROS are produced, this results in damage to DNA, proteins, and lipids and can lead to cell death. ROS are also produced in response to environmental stress factors like ionizing radiation, growth factors, inflammatory cytokines, UV light, chemical oxidants, chemotherapy drugs, toxins, transition metals, and hyperoxia, which increase ROS production. Antioxidants like vitamins, flavonoids, and polyphenols decrease oxidative stress [26] as shown in Fig. 1.

2.4 *Endogenous Sources of Oxidative Stress*

More than 50 enzymes produced in human cells are responsible for producing $O_2^{\bullet-}$ and H_2O_2, with NADPH oxidase and mitochondria serving as the primary producers of these molecules.

2.4.1 Mitochondria in the Electron Transport Chain

Mitochondrial dysfunction happens if mitochondria cannot generate enough energy for cells, which changes cellular health and impacts the functions of different organs. This malfunction leads to extreme oxidative stress, leading to cell destruction and causing health problems such as neurodegenerative diseases like Alzheimer's disease, cardiovascular diseases like heart failure, and diabetes. The respiratory chain of mitochondria, found in the inner membrane of mitochondria, is a critical structural and functional component, and it consists of V complexes, which transfer electrons across these interconnected complexes to activate the phosphorylation of ADP into ATP. Mitochondria generate 1–3% reactive oxygen species (ROS) while synthesizing ATP by utilizing ATP synthase, which uses mitochondrial membrane potential. It usually reaches equilibrium at a low level due to an antioxidant system in mitochondria. Mitochondrial dysfunction severely affects cellular energy conversion, impacting tissue that depends on chemical energy produced by mitochondria over time. Thus, it shows that mitochondria produce a high level of ROS. The mitochondrial superoxide radical is formed at two specific sites: the electron transport chain-complex I (NADH) and complex III (ubiquinone-cytochrome c reductase). When metabolism is normal, it serves as a primary location for generating ROS in complex III, resulting in frequent attacks by free radicals in the mitochondrial respiratory chain. Superoxide is referred to as a location for complex I and complex III (Fig. 2). Because of their high reactivity, hydroxyl radicals, such as proteins, DNA, and lipids, can harm various macromolecules inside the mitochondria. When DNA damage in mitochondria remains unsolved, it leads to complex I and complex III dysfunction, increasing O_2 reduction to form superoxide. Due to the damage to mitochondria, DNA becomes more vulnerable to oxidative damage, which increases the oxidative stress by lowering the production of vital proteins essential for the electron transport chain, which forms a continuous cycle for the formation of reactive oxygen species (ROS), and damage to organelles results in apoptosis (Fig. 3). Both nitric oxide (NO) and peroxynitrite (ONOO) can harm the mitochondrial respiratory chain. Nitration can change the mitochondrial protein, causing dysfunction in various enzymes in electron transport chains, such as cytochrome c oxidase and nicotinamide adenine dinucleotide dehydrogenase. Continuous encounters with reactive oxygen species damage proteins, nucleic acids, and cellular lipids, also causing oxidative destruction of mitochondria. Acute encounters with ROS result in the dysfunction of aconitase in the tricarboxylic acid cycle and Fe-S in electron transport chain complexes I, II, and III, leading to the termination of mitochondrial energy generation. Numerous studies indicate that increasing oxidative stress and changes in apoptosis are linked to different age-related neurodegenerative disorders. Numerous studies show that mitochondria are essential for various neurodegenerative conditions, including Parkinson's disease, amyotrophic lateral sclerosis, Huntington's disease, and Alzheimer's disease (AD). Malfunction of mitochondria is a primary cause of pathogenesis [27].

2.4.2 Endoplasmic Reticulum

It is an organelle consisting of specialized proteins and enzymes responsible for protein folding. This process is complicated; it only allows properly folded proteins that pass through the Golgi apparatus to be transferred to their designated sites within the cell. To enable protein folding, elevated calcium levels optimize chaperone protein function and control the strongly oxidizing environment to promote the formation of

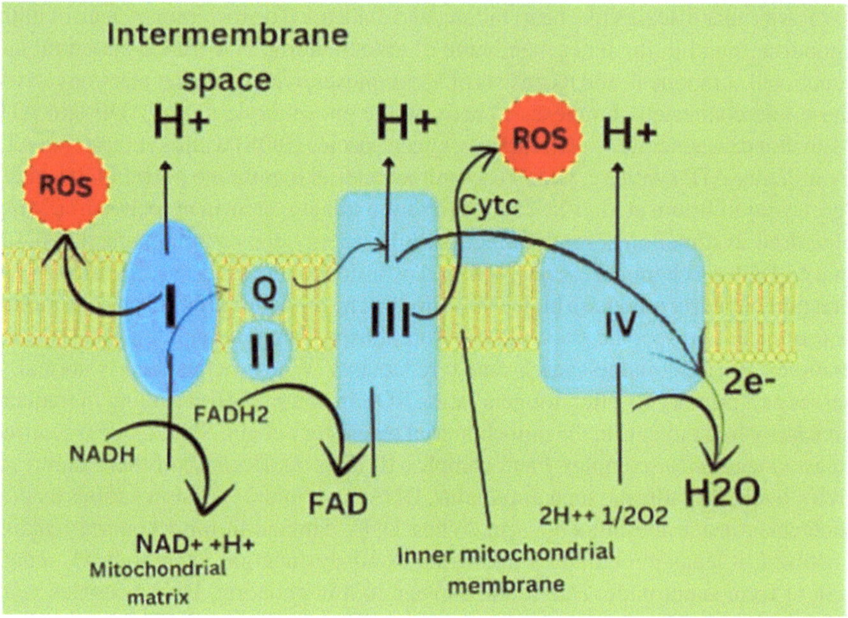

Fig. 2 Electron transport chain (ETC), responsible for mitochondrial ATP and ROS generation. ROS: reactive oxygen species, TCA: tricarboxylic acid, NADH & NAD^+: reduced and oxidized forms of nicotinamide adenine dinucleotide, FADH2 and FAD: reduced and oxidized forms of flavin adenine dinucleotide; ATP: adenosine triphosphate; ADP: adenosine diphosphate; UCP: uncoupling protein, Cyst c: cytochrome c, and Co Q: coenzyme Q

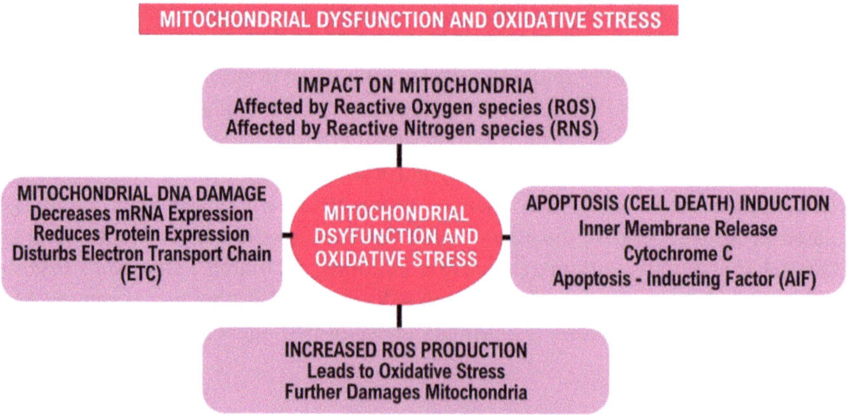

Fig. 3 Dysfunction of mitochondria and oxidative stress. Numerous factors, including reactive oxygen species and reactive nitrogen species, impact mitochondria. Mitochondrial DNA damage decreases mRNA expression and reduces protein expression, disturbing the electron transport chain. This increases the production of reactive oxygen species, which triggers the release of cytochrome c and apoptosis-inducing factor from the inner membrane, ultimately leading to apoptosis

disulfide bonds between proteins [28]. Misfolded proteins can reveal hydrophobic regions in their amino acids, which support the aggregation of proteins. This aggregation causes ER stress. This condition is linked to the pathophysiology of chronic disease, including diabetes mellitus and neurodegenerative disease, triggering the "ER stress response" or "unfolded protein response" when reacting to ER stress. Protein kinase R-like endoplasmic reticulum, inositol-requiring enzyme-1 (IRE1), and activating transcription factor 6 (ATF6) are the three ER sensors that trigger and regulate the UPR [29]. At the beginning of UPR, three primary signaling cascades are initiated by the chaperone protein Bip, which attaches to unfolded proteins. This leads to a shift in the binding stability away from protein by PKR-like endoplasmic reticulum kinase (PERK), activating transcription factor 6 (ATF-6), and inositol-requiring enzyme 1 (IRE1). IRE1 is a transmembrane protein and possesses both ribonuclease activities and protein kinase,it forms dimers or oligomers by activating IRE1, which triggers some regulatory sites in its protein kinase domains to become transphosphorylated. This process converts IRE1 into its active phosphorylated form, known as IRE1p. To activate nuclease activity, it requires adenosine nucleotides (ATP/ADP) as a cofactor [30]. After IRE1 activation, transforming the unspliced variant (XBP1u) of the mRNA changes it into its spliced form (XBP1s) by removing 26 nucleotides of intron known as XBP1 (X-box binding protein 1) from the mRNA. This conversion leads to the production of a more effective transcriptional activator. The 90-kDa protein known as ATF6 is continuously present in the cells. A vast domain facing the ER lumen is a single-pass type 2 transmembrane protein structure. ATF6 can form monomers, dimers, and oligomers. ATF6 moves to the Golgi apparatus in response to ER stress, split by Site-1 and Site-2 proteases, forming an active transcription factor ATF6N. XBP transcription is directly impacted by a cytosolic fragment (ATF6f) that is produced by ATF6. Recent research indicates that the redox status of ATF6 impacts its capacity to translocate and process. ATF6 dimers joined by a C467-467 disulfide bond mainly translocate to the Golgi apparatus. When ERp18, an ER resident oxidoreductase, amplified the formation of ATF6 dimers (through reduction) and the conversion of ATF6 to ATF6N, the absence of ERp18 increased the rate of ATF6 dimerization. After cleavage, the N-terminal fragment of ATF6 (ATF6N) is released and transferred to the nucleus, where it interacts with cis-regulatory elements such as ERSE I, ERSE II, and UPRE promoter sequences, activating transcription of UPR target genes [31–33].

2.4.3 Peroxisomes

Peroxisomes break down fatty acids, generating hydrogen peroxide (H_2O_2) as a primary byproduct. This occurs through enzymes like D-amino acid oxidase and fatty acyl-CoA oxidase, which oxidize fatty acids and amino acids. Due to the high concentrations of peroxiredoxin 5 and catalase, hydrogen peroxide breaks down into water (H_2O). An essential function of hydrogen peroxide (H_2O_2) generated in peroxisomes is to trigger the transcription factor FOXO3 and inhibit phosphatidylinositol-3,4,5-trisphosphate 3-phosphatase (PTEN). Additionally, it noted that reactive oxygen species generated by peroxisomes may regulate autophagy, a cellular mechanism in charge of breaking down and recycling various components. This control is linked to the tuberous sclerosis complex (TSC), which maintains the peroxisome mTORC1 response to reactive oxygen species. This interaction results in the termination of mTORC1, thus triggering the early phases of the autophagy mechanism [34].

2.5 *Exogenous Sources of Oxidative Stress*

Exogenous sources of oxidative stress are external factors that promote the formation of reactive oxygen species and free radicals in the body. These sources may result in environmental factors, lifestyle habits, and dietary consumption.

2.5.1 Cigarette Smoke

Among the numerous oxidants, free radicals, and organic compounds found in cigarette smoke are superoxide and nitric oxide. Moreover, intake of smoke into the lungs triggers an internal mechanism, resulting in the collection of neutrophils and macrophages, which enhances the oxidative damage [35].

2.5.2 UV Light

Ultraviolet light is a section of the electromagnetic radiation spectrum that lies between X-rays and visible light. UV light is classified into three categories: UVA (longest), UVB (medium), and UVC (shortest) [36]. Synthesizing vitamin D depends on ultraviolet A and ultraviolet B. Too much exposure to rays can cause skin harm and sunburn and may even trigger cancer development [37]. UV radiation produces various free radicals through miscellaneous pathways that directly alter the cellular components or through the process of photosensitization. It also damages the enzyme CAT's structure, enhancing the generation of nitric oxide synthase (NOS). UV radiation lowers the protein kinase C (PKC), resulting in a boost in the generation of reactive oxygen species [38].

2.5.3 Ionizing Radiation

Ionizing radiation (IR) is produced by essential radioisotopes in the soil and aurora particles that strike the Earth's surface [39]. Electromagnetic waves carry this energy, such as X-rays, gamma rays, and UV radiation. Ionizing radiation has diverse applications in medicine and other fields. It also triggers the radiolysis of water molecules, a significant source of ROS formation. IR affects mitochondria, which results in mitochondrial dysfunction, which raises ROS production. Szumiel reports that mutations in mitochondrial DNA alter the epigenetic regulation of nuclear DNA by reducing methyltransferase activity, resulting in widespread DNA hypomethylation [40, 41].

2.5.4 Heavy Metals

Heavy metals are key contaminants related to cell damage. It is defined by having densities and large atomic weights that are five times that of water. Arsenic (As), cadmium (Cd), lead (Pb), mercury (Hg), zinc (Zn), and nickel (Ni) are found in various organisms like plants, animals, and even microorganisms. It plays a significant role in boosting the production of reactive oxygen species and breaking the cellular antioxidant defenses in oxidative stress [42]. Heavy metals generate free radicals, resulting in altered DNA bases in various ways that enhance lipid peroxidation and disturb sulfhydryl groups and calcium balance. For example, chromium, arsenic, lead, cadmium, and mercury have been found to raise oxidative stress levels by reducing vital antioxidants like glutathione and affecting mitochondrial function [43]. Figure 4 shows the overall source of oxidative stress.

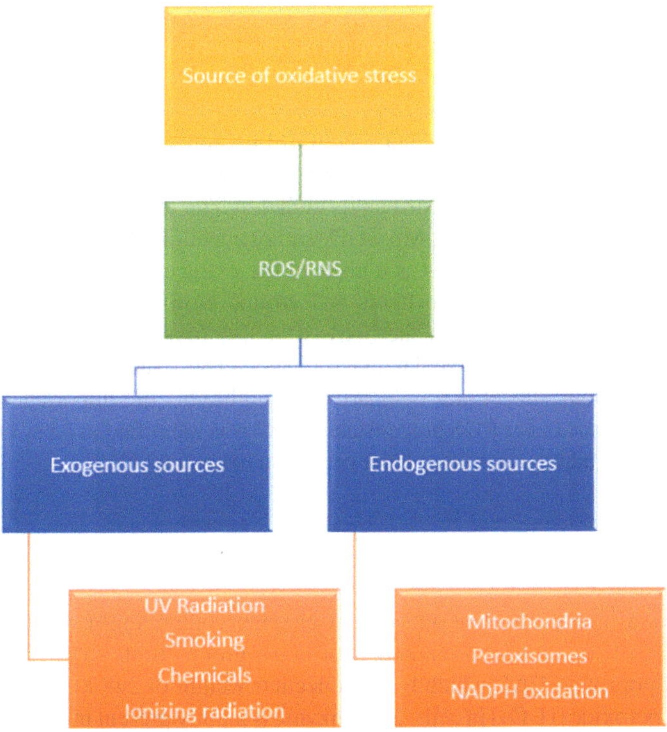

Fig. 4 Diagrammatic representation of endogenous and exogenous sources of oxidative stress

2.6 Antioxidant

The human body contains different antioxidants that work to counteract the effects of oxidants. These antioxidants can be classified into enzymatic and non-enzymatic antioxidants.

2.6.1 Enzymatic Antioxidants

Enzymatic antioxidants, known as proteins, promote metabolic activities to counteract reactive oxygen species and shield cells from oxidative damage. Typically produced within the body, these antioxidants are crucial for sustaining cellular health.

Superoxide Dismutase

The primary source of reactive oxygen species (ROS) is superoxide, which transforms into superoxide dismutase (SOD), essential to maintaining cellular health. SODs are metalloproteins that facilitate the transition of superoxide anions into hydrogen peroxide [44]. There are three forms of superoxide dismutase (SOD), each situated in specific areas of the cell close to reactive oxygen species (ROS) sources: SOD1 is found in the cytosol, and SOD2 and SOD3 are found in mitochondria and the external space. The SOD1 dimer and SOD3 tetramer utilize copper and zinc as cofactors, so they are referred to as Cu–Zn SOD, while SOD2, a tetramer known as Mn-SOD, uses manganese (Mn) as its cofactor. The three forms of SOD found in human lungs are CuZn-SOD, Mn-SOD, and extracellular EC-SOD. While extracellular SOD (EC-SOD) is most abundant in the extracellular matrix, particularly in areas rich in type I collagen fibers and around both pulmonary and systemic blood vessels found throughout the body, it is also found in alveolar epithelium, bronchial epithelium, and alveolar macrophages. The existence of distinct SOD isoforms in various subcellular compartments highlights the importance of maintaining strict control over ROS homeostasis. SOD was identified as a key component in the oxidative damage resulting from cadmium exposure [45].

Catalase

Catalase is a highly effective antioxidant enzyme comprising a protein structure. Catalase is a hemoprotein containing four heme groups in its structure. The main role of hemoprotein CAT, which has a molecular weight of 248 kDa, is to detoxify hydrogen peroxide (H_2O_2) in cells. This enzyme is mainly present in aerobic bacteria, plants, and animals and is primarily in the peroxisomes of cells [46]. Except for erythrocytes and human vascular cells, it is widely expressed in peroxisomes of all mammalian cell types. Many mammalian cells transport into peroxisomes using a modified PTS1 sequence [47].

Glutathione Peroxidase (GSH-PX)

The unique amino acid selenocysteine is present in the active site of GSH-PX, a tetrameric enzyme. They convert lipid peroxides into their respective alcohols using low molecular weight thiols. Living things' antioxidant systems depend on the GSH-Px enzyme to function effectively under normal and oxidative stress circumstances [48]. It is crucial in protecting against carcinogens, free radicals, and peroxides. Studies have shown that eight GSH-Px enzymes, known as selenoproteins (Se GPxs), are present in the human body in various cell types. The only monomeric enzymes are GSH-Px-4, GSH-Px-7, and GSH-Px-8, while the GSH-Px enzymes are tetrameric. A cytosolic enzyme, GSH-Px1, also known as cellular GSH-Px (cGSH-Px), has a tetrameric structure in all cells. It functions against organic hydroperoxides and hydrogen peroxide (H_2O_2) [49]. The human liver and gastrointestinal tract contain SH-Px 2, known as gastrointestinal GSH-Px (GSH-Px-GI), and it is absent in the kidneys, heart, and lungs [50]. GSH-Px-3 is the only member of the GSH-Px family found in the extracellular compartment. It is also called extracellular GSH-Px and, in mammals, is one of the essential extracellular antioxidant enzymes. GSH-Px4, also known as phospholipid GSH-Px (PH-GSH-Px), is found in the cell membrane, cytosol, and mitochondria [51].

2.6.2 Non-Enzymatic Antioxidants

Neutralizing reactive oxygen species (ROS) without using enzymes shields tissues and cells from oxidative stress and damage. Glutathione, β-carotene, flavonoids, and others are some examples of non-enzymatic antioxidants.

Glutathione

GSH has significant antioxidant properties and is a tripeptide with a nucleophilic structure. In most cells, the glutathione system functions as a primary redox buffer, mainly distributed in the cytosol, nuclei, peroxisomes, and mitochondria [52]. Although it is found in various tissues, it acts as a primary site of production in the liver. Glutathione protects against xenobiotics and heavy metals such as cadmium [50].

Melatonin

The neurohormone melatonin (N-acetyl-5-methoxytryptamine) is produced by the pineal gland and secreted by the bone marrow, gastrointestinal tract, retina, and ovary. It is made from the amino acid tryptophan. It regulates circadian rhythms and is a powerful antioxidant, safeguarding cell membranes from lipid peroxidation [53]. In the skin, it protects deeper tissues from harmful sunlight and supports the exchange of pigment granules. It also helps defend against oxidative damage to cholesterol derivatives, bile acids, and mucous membranes. Melatonin can readily penetrate all intracellular components and protect the organelles, cell membrane, and nucleus from damage caused by free radicals because it dissolves in both water and lipid phases [54].

Vitamin E

Vitamin E is characterized by its strong antioxidant properties and oil solubility. It contains eight isoforms, including α-, β-, γ-, and δ-tocopherol, as well as stereoisomers α-, β-, γ-, and δ-tocotrienol [55]. A-tocopherol is considered the most biologically active form, especially in cell membranes. The primary characteristic of alpha-tocopherol is its capacity to protect the cell membrane from the damaging effects of free radicals and protect against lipid peroxidation. According to research, cataracts, arthritis, schemia, certain cardiovascular illnesses, colon and prostate cancer, and neurological disorders are protected by utilizing vitamin E [56].

Vitamin C

There are two redox forms of vitamin C, or ascorbic acid (AA), including the reduced form, which becomes deprotonated at physiological pH. Because of its strong electron-donating ability, AA can be oxidized by two electrons to form dehydroascorbic acid (DHA). It can also undergo one-electron oxidation, forming a semi-dehydroascorbyl radical. Furthermore, vitamin C can detoxify endogenous and exogenous reactive oxygen species (ROS), as well as proteins and lipids that ROS have altered [57]. A meta-analysis including 400,000 patients showed that the use of vitamins C and E decreased the incidence of coronary heart disease. Research suggests that vitamin C can control smooth muscle-mediated vasodilation, endothelial cell proliferation and apoptosis, and other processes that are essential to the progression of cardiovascular disorders [58, 59].

Flavonoids

A diverse class of polyphenolic chemicals is commonly present in plant-based meals like vegetables, fruits, and olive oil, as well as in beverages like wine and tea. It possesses significant antioxidants and has strong chelating properties, which are naturally present in plants, but the human body is unable to produce them. They are categorized into types such as flavanols (found in kale, lettuce, cocoa, green tea), flavones (parsley, celery, thyme), flavanones (citrus fruits, such as oranges, petitgrain, grapes, and lemons), isoflavones (soybeans and other legumes), and anthocyanins (cyanidin, pelargonidin, delphinidin, peonidin, and malvidin). Flavanones' ability to scavenge free radicals has been associated with a number of health advantages. Some isoflavones, like genistein and daidzein, have estrogenic properties in some animals and are considered phytoestrogens. Studies indicate that flavonoids possess various antioxidant capabilities, including chelating metals, LDL oxidation prevention, free radical scavenging, and vasodilating action. These characteristics help flavonoids protect against damage caused by cadmium (Cd) [60].

2.7 Imbalance of ROS Production and Antioxidant Defense Leads to Oxidative Damage

An imbalance between the body's antioxidant defense systems and the production of reactive oxygen species is key to causing oxidative damage. This imbalance mainly arises from higher ROS production, lower antioxidant defenses, or both.

2.7.1 ROS-Antioxidant Imbalance Happens Due to Increasing ROS Production

Mitochondrial dysfunction: Mitochondria serve as a leading source of ROS, produced as byproducts during ATP synthesis. When mitochondrial function is disrupted by factors like aging, genetic mutations, or toxin damage, increased ROS leakage and generation decrease the efficiency of the electron transport chain [61]. Environmental stress, including UV radiation, pollution, smoking, and contact with heavy metals (such as mercury and cadmium), can directly trigger ROS generation. For example, UV light promotes ROS production in skin cells, results in photodamage, and raises the risk of skin cancer [62]. Inflammation: During the immune response, cells like macrophages and neutrophils generate ROS to eliminate pathogens. However, in chronic inflammation, such as rheumatoid arthritis or long-term infections like Parkinson's and Alzheimer's disorder, OS production becomes excessive, surpassing the antioxidant defense system [63].

2.7.2 Decreased Antioxidant Defense

(i) Age-Related Decline: The body's natural antioxidant defenses typically weaken as we age. The levels of enzymatic and non-enzymatic antioxidants frequently

decrease, making it more difficult to neutralize ROS [64]. Nutritional *(ii) Deficiencies:* Glutathione and other antioxidants like vitamins C and E are vital in neutralizing ROS. Diets deficient in these nutrients weaken the fight against ROS in our body, elevating the sensitivity of cells to oxidative stress. *(iii) Genetic factors:* Genetic polymorphisms change the expression function of antioxidant enzymes, resulting in differences in antioxidant defense mechanisms between individuals. For example, oxidative stress can result from mutations in the gene producing superoxide dismutase (SOD), leading to a decreased ability to neutralize superoxide radicals [65].

2.7.3 Imbalance Leads to Oxidative Damage

When ROS levels increase or antioxidant defenses are weakened, oxidative stress damages cellular structures. *Membrane Damage*: ROS damage polyunsaturated fatty acids in cell membranes, triggering a chain reaction of lipid peroxidation. This process weakens the structure and the cell membrane integrity, resulting in enhanced permeability, impaired cell function, and, ultimately, cell death [66]. *(i) Protein:* ROS can alter proteins by oxidizing amino acid residues, resulting in structural changes, enzyme inactivation, and damaged cellular signaling. This affects cellular functions, mainly when it involves enzymes or transport proteins that are crucial for cell survival [67]. (ii) *DNA Damage:* ROS can directly create breaks in DNA strands and alter bases, which can lead to mutations. If these mutations go unrepaired, they may result in genomic instability, a factor that contributes to cancer and various degenerative diseases [68].

3 Pathophysiological Impact of Oxidative Stress

3.1 Cellular and Molecular Effects

3.1.1 Damage to DNA, Proteins, and Lipids

Oxidative damage can affect ribonucleic acids and deoxyribonucleic acids. mRNA and tRNA are also affected, and rRNA is a relatively straightforward biopolymer consisting of four distinct nucleic acids. Its integrity is crucial for cell division and survival. Oxidative modifications to nucleic acid polymers, such as transcription and DNA replication, are examples of oxidative modifications to nucleic acid polymers that might result in cell senescence and mutations to cause death. Given the valuable nature of the genetic code, it seems to be a prime target for oxidative harm. Even under normal physiological circumstances, the extent of oxidative damage can be significant, with estimates suggesting base modifications occur once per 13,000 bases in nuclear DNA. Damage to mitochondrial DNA has been linked to cancer aging. Oxidants can produce diverse DNA damage products, many of which have been thoroughly studied. Single and double-strand breaks, sister chromatid swaps, DNA–DNA and DNA–protein cross-links, and base alterations are the different categories of DNA damage. All four bases of DNA can undergo oxidative modification. These changes may include hydroxylation, ring contraction, ring opening, and ring saturation. Some changes typically lead to a loss of aromaticity and planarity, which results in local distortions within the double helix. Depending on the type of DNA damage, the altered bases may either remain attached or detach from the DNA molecule, creating apurinic/apyrimidinic (AP) sites. There is increasing evidence that redox endonucleases are active in higher eukaryotes, addressing DNA damage that arises in cells due to oxidative challenges, including single-strand breaks and base damage. Apart from excision repair mechanisms, there may also be direct repair of oxidatively modified DNA bases [69]. Reactive oxygen species (ROS) cause higher post-translational alteration in proteins than in DNA or lipids. Due to protein backbone oxidation, the protein fragmentation process starts with hydroxy radicals, facilitating the removal of hydrogen atoms from residues of amino acids. This results in a form of carbon-centered radical. This radical reacts with O_2 and produces an alkyl peroxyl radical, which, through intermediates such as alkyl peroxide and alkoxyl radical, leads to the formation of hydroxylated protein derivatives. All amino acid residues are susceptible to modification by ROS attacks, resulting in chain oxidation and the creation of cross-links between proteins. The sulfur concentration in the structure makes methionine and cysteine residues especially vulnerable to oxidation. The production of disulfides results from the oxidation of the cysteine residues, which can be reverted to cysteine under mild conditions, such as the presence of reducing agents like NADH, without enzymatic involvement. Methionine is produced by oxidizing methionine residues, which can also be restored by the methionine sulfoxide reductase that facilitates the reduction back to methionine. The reversibility in the oxidation of methionine highlights its potential role in the cellular regulatory processes. In the brain, tyrosine, a significant α-amino acid, serves as a precursor of neurotransmitters like dopamine and is easily nitrated, with nitrotyrosine often serving as an indicator of nitrosative stress. Tryptophan, an aromatic amino acid, is highly susceptible to ROS modification. Direct oxidation of lysine, arginine, proline, and threonine results in the formation of carbonyl derivatives, which are frequently used as indicators of oxidative stress on proteins due to their apparent connection to aging and oxidative challenges. An excessive buildup of oxidative modifications on amino acid residues can significantly impair cellular functions. The ineffective degradation of Aβ and tau proteins is one contributing factor to the increased accumulation of these proteins in the brains of Alzheimer's disease patients. Under oxidative stress, oxidative protein damage initiates various chemical reactions, including oxidation of amino acid side chains, polypeptide chain fragmentation, cross-link creation, and conformational changes. Generally, these oxidative modifications are non-reversible and can serve several protein functions, which are notable for exceptions, being the reversible oxidation of methionine and cysteine side chains. The capability allows for the restoration of protein function and the removal of ROS, which are crucial in defending cells from oxidative damage. There are two primary difficulties in analyzing protein damage. Firstly, numerous proteins are altered by reactive intermediates in cellular and tissue proteomes. This phenomenon was initially highlighted in electrophoretic studies of liver proteins modified by yellow pigment in rats treated with azo dye and has been corroborated through similar methodologies. Secondly, modified proteins constitute only a minor portion of the proteins found in the cellular and tissue extracts [70].

The first oxidative damage in the form of lipid peroxidation was examined. Phospholipids in membranes are continuously subjected to oxidative stress. The lipid peroxidation process consists of chain reactions initiated by removing a carbon-centered lipid radical that will react with oxygen to remove a hydrogen atom from carbon in an unsaturated fatty acyl chain and will cause a lipid peroxidation chain reaction by removing a hydrogen atom from a neighboring unsaturated fatty acid. The resulting lipid hydroxy peroxide (LOOH) readily decomposes into different reactive species, such as alkoxyl radicals (LO), aldehydes, alkanes, lipid epoxides, and alcohols. Peroxidized membranes become rigid, lose their selective permeability, and may lose their structural integrity under severe conditions. Dialdehydes can function as cross-linking agents and are believed to contribute to aggregation, forming the age pigment lipofuscin. Oxidized lipid bilayers become more favorable phospholipase enzyme substrates. Phospholipase A2 (PLA2) targets the sn-2 position of the phospholipid glycerol backbone, producing a free fatty acid and a lysophospholipid. Lipid oxidation products and peroxidized membranes are a constant threat to aerobic cells. It is now widely acknowledged that cells have evolved a number of defense mechanisms for integrity and homeostasis by repairing oxidation-damaged components, in addition to blocking the start of peroxidation (with substances like vitamin E) [69].

3.1.2 Mitochondrial Dysfunction

Five linked complexes form the mitochondrial respiration chain within the inner membrane. They are regulated by the phosphorylation of ADP and ATP, including proteins, DNA, and lipids. Damaged mitochondria can cause DNA in complex I or III. These effects may lead to a reduction of O_2 by electrons. It results in the form of superoxide. Mitochondria are double-membrane cytoplasmic organelles. Mitochondrion reproduces itself with its genome. There are 800–1000 copies of mtDNA in each mitochondrion. They are inherited maternally and arranged in a compactly packed nucleoprotein structure known as the nucleoid. Some mechanisms are used to reduce the oxidative stress caused by the overproduction of free radicals in the cells. The enzymatic protection system comprises the valuable effects of various antioxidant enzymes. It includes superoxide dismutase (SOD), catalase (CAT), glutathione reductase (GR), and glutathione peroxidase (GPx). There are also non-enzymatic defenses consisting of antioxidant composites. It protects cells from oxidative stress, which is vital for maintaining the growth of cells, avoiding cell death pathways, and removing all the damaged mitochondria. Mitochondrial morphology adapts dynamically in response to external stimuli and the availability of nutrients, varying significantly across different cell and tissue types. The mechanistic connection between diabetes and oxidative stress/mitochondrial dysfunction has received increased attention in recent years. In the past 20 years, type 2 diabetes mellitus has surged, becoming a pressing public health issue all over the world. Currently, there are 382 million individuals diagnosed with type 2 diabetes, and it is expected to increase to 438 million in 2030 in India, China, and the United States [27, 71].

3.1.3 Inflammatory Response Activation

Inflammation is a disease condition characterized by the infiltration of immune cells such as monocytes, macrophages, lymphocytes, and PMN plasma cells within the walls of blood vessels, the movement of these immune cells surrounding tissue, and the release of oxygen species by these cells, resulting in tissue damage. PMN initially produces the ROS through the enzyme NADPH oxidase 2, the superoxide anion (O_2^-) generated either endogenously or through the enzymatic action (through SOD) converts into hydrogen peroxide (H_2O_2). H_2O_2 can form other higher reactive radicals, including the hydroxyl radical (–OH), when catalyzed by the enzyme myeloperoxidase (MPO). These radicals contribute to tissue damage and oxidize various proteins and other molecules. ROS produced by inflammatory cells produce oxidative stress to eliminate pathogens and activate pathways that enhance inflammation. The production of pro-inflammatory cytokines and chemokines is initiated when ROS activate kinases, including PKC, JNK, and p38 MAPK; they stimulate transcription factors. The process of recruitment and adherence is intricate, involving cytokine chemokine gradients to regulate and cell adhesion molecules (CAMs) to manage the interaction between the immune cells and the vessel walls. Following the disruption of adherens junctions induced by MMIF, MMPS play a crucial role in enabling immune cells to enter the surrounding tissues. Cell damage and apoptosis caused by inflammasomes play a role in enhanced inflammatory disease progression. Inflammasomes have been associated with numerous inflammatory diseases, such as multiple sclerosis (MS), Alzheimer's disease (AD), and Parkinson's disease (PD), as well as metabolic conditions like atherosclerosis, type 2 diabetes, and obesity. ROS-induced cell death can happen through direct damage to cellular membranes or structures or by triggering enzymes that facilitate cell death through lipid peroxidation byproducts like 4-hydroxy-2-nonenol (HNE) [72].

3.2 *Key Signaling Pathways*

Pathway models elucidate the complex interplay between cellular components. The signaling is interconnected and complex to separate; cellular responses become clearer through pathway-specific research. Key signaling pathways regulate cellular functions, including WNT, SHH, Notch, MAPK, mTOR, JAK-STAT, and NF-kB, which are critical for cell function.

3.2.1 NF-KB Pathways

The transcription factor, nuclear factor NF-kB, modulates expression in different elements of the immune system. These elements include chemokines, pro-inflammatory cytokines, and enzymes like cyclooxygenase-2 and nitric oxide synthase. These proteins contribute to the immediate immune response and influence the targeted adaptive response, such as major histocompatibility complexes that are essential for the initial period of specific immunity, in addition to cytokines such as interleukin (IL), IL-12, and interferon, which regulate differentiation and lymphocyte growth. It is also associated with cancer development. NF-kB refers to different dimeric complexes formed by components of the Rel protein family, which include cRel, Rel A (p65), Rel B, NF-kB1 (p50 and its precursor p105), and NF-kB2 (p52 and its precursor p100). Each protein has a roughly 300-amino-acid N-terminal Rel homology domain responsible for translocation, dimerization to the nucleus, and binding to DNA. The dimeric combination of p50-p65 is the most frequently observed. The NF-kB complex includes precursor proteins p105 and p100 under inactivation. Other IkB can inactivate transcriptional complexes at specific promoters. A closely connected protein, Bcl-3, interacts with p50 and p52 dimers to form transcription initiation complexes. The classical activation of NF-kB relies on the breakdown of IkB. A prerequisite for this process is the prior phosphorylation of IkB on two critical serines located at the N-terminus by IkB kinases (IKKs). Following the phosphorylation of IkB proteins, the E3 IkB ubiquitin ligase complex β-TrCP-SCF binds to IkB, polyubiquitinating it and marking it for elimination by the 265 proteasomes. The IKK complex is activated by different mitogen-activated protein kinases, such as TAK1, MEKK1, and MEKK3, in response to different stimuli, including those from toll-like receptors (TLRs). TLRs recognize pathogen-associated molecules and stimulate the production of pro-inflammatory proteins and co-stimulatory molecules, which initiate adaptive immunity. TLRs utilize the same signaling components of the IL-1 and IL-18 receptors. These receptors have a specific Toll/IL-1R (TIR) domain, which allows them to pair with other proteins containing the same domain, forming a complex that recruits adaptor proteins inside the cell. The interaction between IRAK and TRAF-6 results in the activation of TAK1. IRAK is pivotal in this activation pathway since it facilitates the movement of TAK1-binding protein 2 (TAB2) from the membrane into the cytosol, where TAB2 interacts with TRAF-6 and connects TRAF-6 to TAK1. The latter, aided by TAB1, becomes activated and activates the IKK complex. The activation of TAK1 by TRAF-6 is contingent upon nonclassical polyubiquitination. TRAF-6 can also activate other MAP3Ks that stimulate the IKK complex. It is associated with a novel adaptor protein, ECSIT, which sequentially activates MEKK-1 and IKKs. TRAF-6 interacts with MEKK3, which is essential for activating IKKs by TLR4 and IL-1R. TRAF-6 also connects with another adaptor, p62, to activate PKC, resulting in the phosphorylation of p65 in Fig. 5 [73].

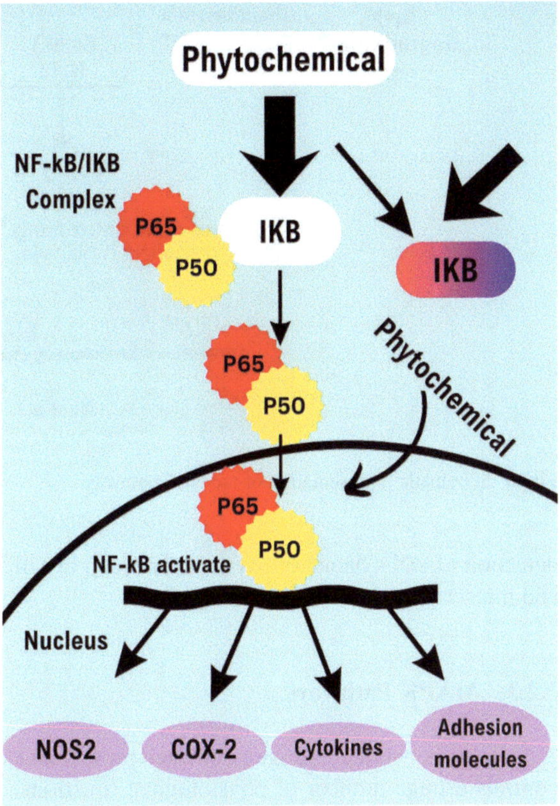

Fig. 5 NF-kB pathway in oxidative stress

3.2.2 Nrf2 Pathways

Nrf2 (nuclear factor erythroid 2-related factor 2) is a transcription factor involved in the cellular reaction to oxidative stress. It is connected to Parkinson's disease and regulates the cellular shield against the toxic and oxidative insults by promoting gene expression involved in the oxidative stress response and drug detoxification [74]. NRF2 plays a crucial role in coordinating the cells against harmful substances and oxidative damage. NRF2, produced by the gene of nuclear factor, erythroid 2-like 2 (NFE2L2), is a member of the Cap'n'Collar (CNC) subfamily of basic leucine zipper (bZIP) transcription factors. Nrf2 activation has gained significant research attention as a potential therapeutic target for numerous chronic diseases [75]. Its biological effects can be assessed mainly by its transcriptional activities and intracellular localization within the cells. It includes antioxidant proteins that act directly, enzymes for phase I and II detoxification of electrophiles, transport mechanisms for toxic substances, metabolism of free radicals, suppression of inflammation, maintenance of glutathione levels, functioning of the proteasome,

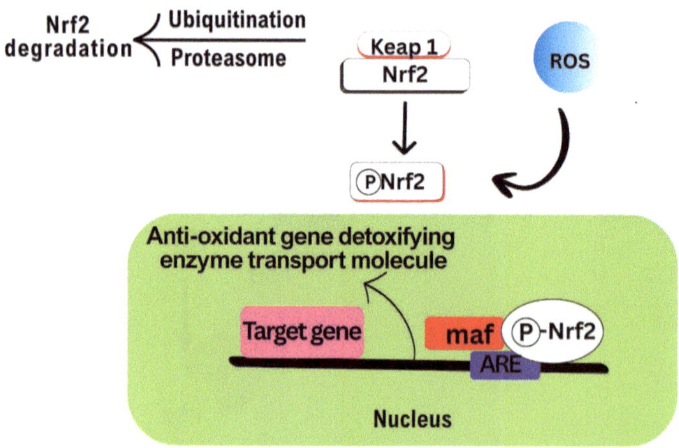

Fig. 6 Schematic representation of the Nrf2 pathway

detection of DNA damage, and the expression of different associated growth factors and transcription in Fig. 6 factors [74].

3.2.3 MAPK Pathways

A complex network of mitogen-activated protein kinase (MAPK) signaling pathways controls a huge number of physiological functions, such as cell division, growth, and apoptosis. One consequence of oxidative energy metabolism is that it is an important physiological regulator of a number of signaling pathways, such as the MAPK pathway. ERK, JNK, and p38 kinases are activated by oxidative stress, and they also activate the MAPK signaling pathway. ROS are produced in response to oxidative stress, causing this activation, which can have both apoptotic and survival-promoting effects. ROS activate PLC-gamma and Src, leading to the phosphorylation of Ras and Raf, which ultimately activates ERK both directly and indirectly. Once activated, ERK regulates ROS levels positively by inducing p22phox to increase ROS production and negatively by activating Nrf2 to upregulate antioxidants. ROS also stimulates ASK1 and transports a few pathways, such as releasing thioredoxin (Trx) and PP2C epsilon-induced ASK1 positive regulator. The activation of JNK and p38 is initiated by activated ASK1. Like ERK, JNK displays both proapoptotic and prosurvival processes in response to oxidative stress. By stimulating SIRT1, which inhibits p53-dependent transcription, and FoXO, they increase antioxidant synthesis, and JNK can improve cellular lifespan. However, active JNK can trigger apoptosis in the nucleus and cytoplasm via several pathways. JNK inhibits proapoptotic proteins such as p665HC and BIM/BMF, while positively influencing antiapoptotic proteins such as Bcl-xL. Proapoptotic gene expression is triggered by transcription factors that JNK activates in the nucleus. When p38 is active, it simultaneously moves to the

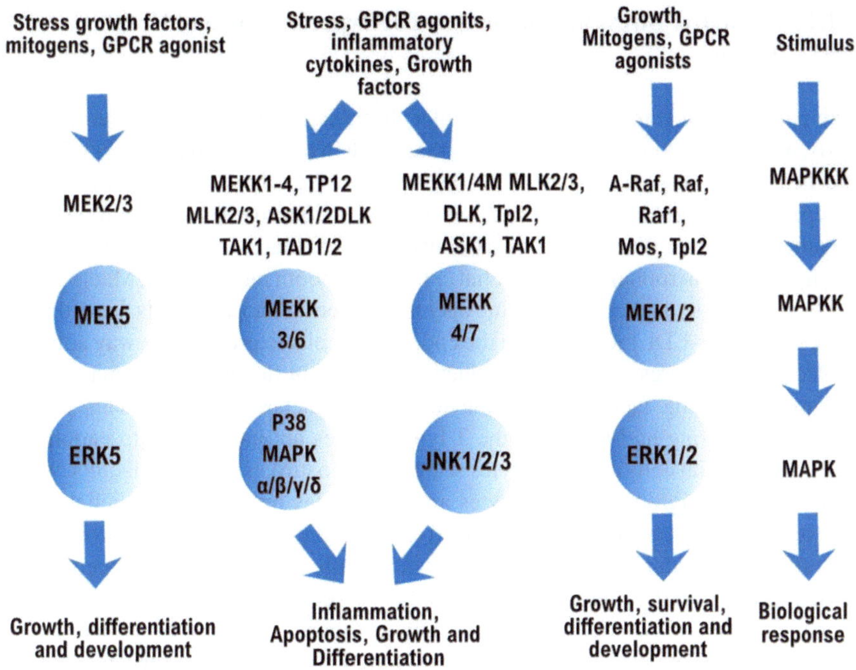

Fig. 7 Schematic representation of the MAPK pathway

nucleus and uses MSK1/2 to activate the transcription factors ATF2 and CREB. This then stimulates the activation of many proapoptotic genes in Fig. 7; therefore, JNK and p38 are essential for modulating cellular reactions to stress and coordinating signals for survival and apoptosis [76].

4 Oxidative Stress in Chronic Diseases

4.1 Cardiovascular Diseases

Cardiovascular diseases are complex disorders that are the leading cause worldwide, according to the WHO (World Health Organization). Reactive oxidative species (ROS) are produced by different oxidase enzymes, which include nicotinamide adenine dinucleotide phosphate (NADPH) oxidase, xanthine oxidase, endothelial nitric oxide synthase (eNOS), cyclooxygenase, glucose oxidase, lipooxygenase, and the mitochondrial electron transport chain. Different pieces of evidence suggest that oxidative stress significantly contributes to the onset and progression of cardiovascular diseases, such as hypertension, dyslipidemia, diabetes mellitus, atherosclerosis, myocardial infarction, angina pectoris, and heart failure [77].

4.1.1 Oxidative Stress in Atherosclerosis, Hypertension, and Heart Failure

The pathological impact of cardiovascular diseases, mainly caused by atherosclerosis, involves remodeling blood vessels that result in blood flow parameters affecting the heart and nervous system. The occurrence of cardiovascular disease led to atherosclerosis problems. Oxidized lower-density lipoproteins may be transported by atherosclerosis. The sources of oxidants for the oxidative modifications of phospholipids are macrophages, endothelial cells, and smooth muscle cells. Macrophages serve as a primary source of oxidizing agents for altering phospholipids. Low-density lipoprotein (Ox-LDL) will damage the endothelial cells to produce the way adhesion molecules are expressed, like chemotactic and selectin factors such as macrophage colony-stimulating factor and monocyte chemoattractant protein-1. This process leads to activating and attaching the T-lymphocytes and monocytes to endothelial cells. Endothelial cells and leukocytes secrete chemoattractant growth factors, which can cause the migration of monocytes and leukocytes in the subendothelial space. The huge cardiovascular disorder results from the difficulties of atherosclerosis. Over 50 million individuals in the United States are affected by systolic hypertension, which contributes to different vascular conditions such as stroke and atherosclerosis. Genetic factors of hypertension are more than 50 genes that have been identified as playing a main role in regulating blood pressure. Its involvement of receptor AT1 in hypertension regulation is shown through in vivo studies and animal experiments. Angiotensin II influences hypertension primarily through interaction between the renin–angiotensin system. The activation of receptors AT1 in blood vessels triggers NADPH/NAD(P)H oxidase activation in vascular cells. The resulting oxidative damage is thought to be a common link between atherosclerosis and hypertension. The stimulation of NADPH oxidase by receptor AT1 activation, Ang II, also directly affects NADP activation by relocating the small GTPase Rac1 to the cell membrane or phosphorylating it to relocate it to the cell membrane on p47phox. The activation of NADPH oxidase, simulated by stretching, is not noted in p47phox, the hypertensive response to Ang II infusion. Mice with vascular smooth muscle cells transgenic for active Rac1 showed signs of hypertension. Antioxidant treatment returned the hypertension to normal, and ROS generation was seen. Furthermore, in mouse studies, hypertension has been linked to oxidative stress brought on by increased 12-LO [78]. The pathogenesis of numerous chronic illnesses, including heart failure, is influenced by oxidative stress. It results from the mismatch between the body's antioxidant defenses and the production of reactive oxygen species (ROS). Excess ROS production can harm cells and perhaps result in cell death. Those with chronic heart failure also experience this case. Myocardial antioxidants include nicotinamide adenine dinucleotide (NAD^+), catalase, and superoxide dismutase. Reduction in SOD, catalase, and GPx activity has been observed in heart failure models [79]. In 2012, heart failure (HF) cost more than $108 billion annually, making it the world's largest cause of morbidity and mortality. It tends to mimic rising HF hospitalizations globally as incidence rises with age. Exercise capacity is severely reduced when heart failure (HF) occurs because it usually manifests as fluid retention leading to peripheral edema and pulmonary congestion (congestive symptoms) and a decrease in cardiac output (low-output symptoms). Heart reduced ejection fraction (HF-rEF) is mostly caused by myocardial infarction (MI) and coronary artery disease. MI causes compensatory alterations in the heart's unaffected regions. The extracellular matrix and cardiac myocytes exhibit structural and functional changes that are referred to as cardiac remodeling. Several indicators, such as uric acid (UA), malondialdehyde (MDA), C-reactive protein (CRP), and the superoxide dismutase enzymes Mn and CuZn, can be used to evaluate the relationship between oxidative stress and heart failure [80].

4.2 Neurogenerative Diseases

Neurodegenerative diseases are weak conditions determined by progressive damage to the central nervous system cells, often degenerating over time and currently lacking a cure.

4.2.1 Alzheimer's Disease

The more typical kind of dementia is AD. Cognitive and memory impairments are features of AD. Age is a risk factor for AD. The pathological symptoms of AD include neurofibrillary tangles and senile plaques, which contain tau proteins and β-amyloid peptides. Aggregation processes and protein misfolding may be linked to cellular alterations brought on by aging. Structural magnetic resonance imaging and histological studies in AD revealed reduced metabolic activity and neuronal loss in the entorhinal cortex and the CA1 area of the hippocampus. Myelinated fibers that link neurons in various cortical regions may be impacted, which could explain the other connections of higher-order brain systems and loss of the memories they had stored in AD. The basic cause of age-dependent memory deficiencies is the inability to access stored memories [81]. Presenilin PSEN1 and PSEN2, the genes encoding beta amyloid precursor protein, were found to have missense mutations in familial forms. Senile plaques and neurofibrillary tangles are the causes of protein aggregation. Numerous indicators of oxidative damage in AD are elevated, which may be due to glial recruitment and activation, mitochondrial dysfunction from peptide Aβ42, or a combination. In neurons, it turns out to be excitotoxicity and calcium excess [82]. Nearly all people with Down syndrome experience AD in their middle years as opposed to their later years. The production and secretion of β-APP are increased when glutamate traumatizes neurons in mixed cell cultures of neurons and glia. As a result, microglia produce more IL-1, which stimulates AChE production and activity in neurons. An essential part of the pathophysiology of a unique AD feature is IL-1. Alzheimer's disease is a deadly neurological condition marked by memory loss and gradual cognitive and functional impairment. The development of neurofibrillary tangles and abnormal cerebral Aβ synthesis or accumulation are pathological markers of AD. Amyloid synthesis protein in neurons initiates a proteolytic process to produce Aβ, which is then present in the extracellular environment in various aggregation states in AD. By releasing proteases, microglial cells also contribute to the phagocytosis of Aβ fibrils and the breakdown of extracellular Aβ. A single intracerebroventricular injection of Aβ oligomers was enough to cause a memory establishment defect in an acute mouse model of AD. Pro-inflammatory cytokine production and microglial cell activation were linked to this. There are currently no biological markers to identify the disease as early pathogenic indicators. Therefore, developing disease-modifying strategies that focus on the neuroinflammatory phase is difficult. When managing AD clinically, innate immunity-focused treatments should consider that the central nervous system innate immune cells are most likely experiencing a period of unresponsiveness rather than pro-inflammatory activation [83]. The most common neurodegenerative disease, AD, severely interferes with day-to-day functioning. It is possible that over-activation of N-methyl-D-aspartate-type glutamate receptors causes the severe OS seen in AD patients. By increasing cell permeability and the accompanying production of neurotoxic amounts of ROS (reactive nitrogen species), it has been demonstrated to result in an excessive influx of Ca^{2+}. Environmental elements like radiation, toxins, and pollution can cause OS. Compounds with neuroprotective, anti-inflammatory, and antioxidant properties are primarily used in AD therapies. While adenoviral Nrf2 gene transfer has shown protective effects against the toxicity of Aβ deposition, transgenic mice with AD symptoms showed weaker Nrf2-ARE pathways in their brains. Thus, the transcriptional regulation of endogenous antioxidants may be of great potential for reducing the symptoms of AD. Moreover, metal ion chelators, histone deacetylase inhibitors, and traditional Chinese and Ayurvedic therapies have shown promise in the treatment of AD [84].

4.2.2 Parkinson's Disease

Parkinson's disease is another common age-related neurodegenerative disease that is caused by misfolded proteins. Synuclein is the misfolded protein that gathers and forms Lewy bodies. It begins in the lower brainstem and olfactory areas. The substantia nigra pars compacta dopamine neurons that extend to the striatum are lost in Parkinson's disease (PD). According to a longitudinal cohort study, older PD patients experienced dementia earlier and had significantly shorter disease courses than patients with younger onset. In addition to other age-related plaque pathology, these patients with older onsets had significantly more synuclein-containing Lewy bodies throughout their brains. Age may also affect responsiveness to medications. Older PD patients are more liable to suffer disruptive actions of anticholinergic drugs yet have fewer levodopa-induced dyskinesia. The extrapyramidal signs (EPS) in patients with PD might also occur in elderly people. For instance, in non-PD-aged people, rest tremor is lacking. In contrast to the standard unilateral onset of Parkinson's disease (PD), these age-associated extrapyramidal signs are symmetrical and do not interact with dopaminergic therapy [85]. The familiar sorts of PD are quite infrequent. Still, investigations of families with earlier-onset, Lewy body-positive, autosomal dominant PD contribute to the recognition of two mutations in the gene encoding alpha-synuclein, a smaller synaptic protein. The pathological hallmark of the disease is huge cytoplasmic inclusions referred to as Lewy bodies. The ubiquitin ligase family is Parkin, which is accountable for ubiquitin, a process that tags proteins for degradation by means of the proteasomal pathway. A growing body of evidence endorses the role of impaired mitochondrial function in PD. Inhibition of complex I mitochondrial electron transfer by MPTP can cause human Parkinsonism, and unidentified PD patients have minimized activity of complexes I and II/III in mitochondria isolated from the platelets. There is plenty of evidence corresponding with oxidative stress in PD. For example, the increased iron content in zona compacta and nitrotyrosine in Lewy bodies, apoptosis, and caspase chain activation were also clearly identified in PD [82].

4.2.3 Huntington's Disease

Huntington's disease is an autosomal dominant disorder with higher penetrance. The increased GAG recurs in the mutant gene and has been expressed as an elongated huntingtin protein. The following abnormal protein is divided into fragments conjugated with ubiquitin that aggregate, becoming neuronal intranuclear inclusions. These intranuclear aggregates are toxic, carrying transcriptional dysfunction, proteasome disability, mitochondrial impairment, free radical damage, and excitotoxicity [82]. Chemical shift magnetic resonance spectroscopy was recently used on 16 HD patients to directly demonstrate a deficiency in oxidative phosphorylation in HD. The phosphocreatine-to-inorganic phosphate proportion decreased, and inorganic phosphate increased in the resting muscles of HD patients. In addition, compared to normal controls, HD patients had a higher lactate-to-pyruvate ratio in their CSF fluid. Our initial results show high levels of 8-hydroxy-2-deoxyguanosine in both nuclear and mitochondrial DNA from the cerebral cortex, with evidence of oxidative damage in HD.

4.2.4 ROS Impact on Neurons and Glial Cells

In brain aging or various inflammatory, vascular, and neurodegenerative diseases in the central system, damage to the mitochondria and oxidative injury are important. A comparison of gene expression in active lesions revealed that when progressive multiple sclerosis is progressing, these pathways of tissue injury are a lot more prominent compared to lesions of other chronic inflammatory diseases or Alzheimer's disease. The prevalent oxidative injury observed in MS lesions provides an opportunity to examine the response of individual cell populations in the central nervous system. Neurons require a more significant amount of energy due to their electrical activity. They appear to have limited ability to control oxidative stress. The limited ability may be because they have a different regulator of antioxidant defense, including Nrf2, even when there is inflammation. Energy deficiency and subsequent functional damage in neuronal impulse conditions occur due to mitochondrial injury. It can cause axonal and neuronal degeneration by disrupting ionic balance and accumulating Ca++ [86]. Glial cells are essential in maintaining homeostasis within the central nervous system (CNS), affecting both healthy and diseased states. Each glial cell type can impact the functioning of others, and dysfunction in any one can trigger self-reinforcing harmful processes, such as communication between glial and impaired neuronal function. However, glial cells could be significant cellular therapeutic targets for CNS-related issues. In this context, these cells' cellular, molecular, and neurochemical alterations, which can have toxic effects or compromise their protective capabilities toward neurons and glial cells, are called glucotoxicity. Glyo protection relates to the specific actions of glial cells, whether under normal or pathological circumstances, that enable them to safeguard both glial and neuronal cells, leading to an overall enhancement of CNS function.

4.2.5 Types and Functions of Glial Cells

Astrocytes

Astrocytes are the most prevalent type of glial cells, known for their different homeostatic roles within the CNS. They possess a complex structure with various star-shaped processes that enable them to interact with neurons, blood vessels, and other cell types. These structural characteristics of astrocytes render them crucial components in linking peripheral and central systems, and they are also associated with several functional responsibilities, including the development and maintenance of the blood–brain barrier (BBB), provision of oxygen or nutrients to the brain, control of synaptic transmission, and facilitation of adaptability at the tripartite synapse. Interaction between astrocytes, synapses, and blood vessels provides metabolic support to neurons by linking synaptic activity to glucose utilization. Additionally, astrocytes can react to abnormal conditions in the CNS through changes in their morphology, physiology, metabolism, and gene expression, collectively referred to as reactive astrogliosis, which can disrupt their homeostatic functions.

Oligodendrocytes

The primary role of oligodendrocytes is to create the myelin sheath, a specialized and lipid-rich structure that wraps around the axons of numerous neurons in a spiral manner. In the adult CNS, oligodendrocyte progenitor cells (OPCs) allow for ongoing proliferation and differentiation into mature, myelinating oligodendrocytes. The process of differentiation and myelination involves sequential steps regulated by different molecular and cell signals, including those from neurons and other glial cells. However, oligodendrocytes have been found to carry out immune functions by expressing a range of immune-related receptors and immunomodulatory molecules, which play a vital role in both oligodendrocytes and glial cells.

Microglia

Microglia serve as a primary immune cell within the CNS, distributed throughout the brain tissue, and play a main role in maintaining brain stability. In brain injury or infection, they are typically the primary cells that become activated and carry out various well-established functions. In addition to their traditional immune role, microglia play a role in numerous homeostatic functions, including the secretion of trophic factors and the generation and maintenance of other neuronal cell types. They also contribute to the maturation, regulation, and adaptability of synapses and neural circuits, facilitate the removal of cell debris, oversee myelination, and play a role in memory formation, as well as learning, essential activities like pathogen detection, inflammatory responses, and phagocytosis [87].

4.3 Diabetes and Metabolic Disorders

4.3.1 Hyperglycaemia-Induced Oxidative Stress, Insulin Resistance, and β-Cell Dysfunction

In vivo oxidative stress brought on by hyperglycemia occurs before the clinical presentation of problems from diabetes. According to Wolff and Dean, non-enzymatic protein glycation was first proposed as a mechanism for glucose-induced toxicity, playing a significant role in the development of diabetic complications through the generation of reactive oxygen species (ROS), such as hydroxyl radicals and superoxide, via glucose autoxidation catalyzed by transition metals. These pathways consist of the polyol route, protein kinase C (PKC), and advanced glycation end products (AGEs) and their receptors (RAGE); other biological pathways have also been experimentally linked to hyperglycemia. The activation of these pathways involved the emergence of problems linked with diabetes, β-cell dysfunction, and insulin resistance. Both decreased insulin secretion and insulin resistance are key characteristics. Both decreased insulin secretion and insulin resistance are key characteristics related to the pathogenesis of type 2 diabetes. Numerous factors trigger insulin resistance, which typically develops several years before type 2 diabetes manifests. A significant proportion of the general population is affected by insulin resistance, which has multiple contributing factors. Its results in a genetic basis are mostly responsible for insulin resistance. Insulin resistance also arises from environmental factors such as inactivity, pregnancy, obesity, and elevated hormone levels. During the initial phase, the body compensates for insulin resistance with hyperinsulinemia, maintaining normal glucose tolerance. Elevating insulin, free fatty acids, and/or glucose can trigger stress-sensitive pathways and activate them, increasing the generation of oxidative stress and reactive oxygen species. This can exacerbate insulin secretion and action, thus hastening the transition to type 2 diabetes. The β-cell is another target of oxidative stress. It is the charge of β-cells to secrete an adequate amount of insulin in response to glucose activation. Numerous studies have indicated that the integrity of β-cells arises from prolonged exposure to elevated glucose levels, increased free fatty acid (FFA) concentrations, or both. There is substantial indication that people with type 2 diabetes who have high blood sugar levels over time have impaired β-cell function. Low amounts of antioxidant enzymes demonstrate that β-cells have a limited capacity for free radicals, thus rendering them susceptible to reactive oxygen species (ROS) and reactive nitrogen species (RNS). Transgenic mice or pancreatic islets that express antioxidant enzymes aid in a number of these unfavorable adverse effects. Glucose oxidation and insulin secretion are suppressed. Antioxidants like zinc, aminoguanidine, the spin-trapping chemical α-phenyl-tret-butyl nitrone, and N-acetyl-L-cysteine (NAC) can protect islets from lipid peroxidation products, maintain β-cell toxicity, decrease the production of glycation end products, and prevent NF-kB activation. Lately, β-cell function has been examined in islets after glutamine, fructose-6-phosphate, and aminotransferase—the rate-limiting enzyme in the hexosamine biosynthesis pathway—were elevated. The hexosamine pathway's activation led to a marked decline in glucose-stimulated insulin release or other indicators of β-cell function and increased hydrogen peroxide. These obstructive effects were moderated through the treatment with the antioxidant NAC.

4.4 Cancer

4.4.1 ROS Involvement in Carcinogenesis, Tumor Progression, and Metastasis

Carcinogenesis is a complex process that involves multiple steps, including the mutation of key genes and an increase in cell proliferation. ROS (reactive oxygen species) are produced in excess by both internal and external sources. Internal sources comprise various intracellular organelles as well as sources related to inflammation; external sources include xenobiotics and radiation. Crucial to the carcinogenic process, the resulting oxidative stress can lead to mutations in essential cellular genes when antioxidant defense mechanisms and DNA repair processes are impaired. The rise in cancer cases over the past 50–60 years can largely be attributed to two main reasons: the aging population and the spread of cancer-causing substances, which are found beyond workplaces as well as in the broader environment. There is supporting evidence that life span exposure to carcinogenic agents and carcinogenic processes was increased during the developmental life [88]. The rising threat of developing conditions related to oxidative damage challenges the community of scientists regarding their molecular origins or characteristics. The environmental changes continue to occur, and the buildup of carcinogens is on the rise. Lifestyle factors, while not classified as direct causes of cancer, are risk factors closely linked to cancer development. The critical timing and mechanisms through which these factors contribute to cancer development remain only partially understood. It is widely recognized that the reactive oxygen species (ROS) ultimately result in DNA damage, where inadequate cellular repair processes may contribute to premature aging and programmed cell death. On the other hand, disruptions in the signaling pathways for the metabolic protein turnover caused by ROS might lead to opposing outcomes, attracting malfunctioning or abnormal proteins and substances that can initiate tumor development. As a result, DNA damage is involved in the process of carcinogenesis,it is also linked to the aging process in living organisms, including cells. The extra actions that ROS are likely to take are potentially significant, potentially impacting p53, cellular proliferation, and metastasis. Chronic inflammation increases the possibility of developing cancer. The function of ROS is likely intricate since ROS sometimes act as anti-inflammatory agents [89]. Cancer is a leading cause of death, and its overall incidence is expected to rise further. Reactive oxygen species (ROS) play a significant role in the development and progression of various tumor types, acting as a critical factor in carcinogenesis and inflammation. Reactive oxygen species (ROS) play a complex and context-dependent role in malignancy. Antioxidant defenses are specifically meant for cancer prevention, upholding the notion that maintaining a healthy lifestyle and consuming a diet rich in fruits and vegetables is linked to reduced risks of tumors, morbidity, and cardiovascular diseases. Reactive oxygen species (ROS) can stem from both internal and external sources. Mitochondria are recognized as the main producers of internal ROS, contributing to increased intracellular oxidative stress. The production of ROS especially occurs in the electron transport chain on the inner mitochondrial membrane during oxidative phosphorylation, a process that generates adenosine triphosphate (ATP) from oxygen and simple sugars. Reactive oxygen species (ROS) play an important role in physiological processes, but excessive ROS results in conditions that depend on a proper balance between deficient and excessive ROS levels. Notably, ROS also strongly influence the cancer microenvironment and are thought to promote tumor angiogenesis, transformation, and progression in various contexts. At moderate collection, ROS activates the tumor cell survival signaling cascade, including mitogen-activated protein kinase/extracellular signal-regulated kinases 1/2 (MAPK/ERK1/2), p38, c-Jun N-terminal kinase (JNK), and phosphoinositide-3-kinase/protein kinase B (PI3K/Akt), which in sequence activate nuclear factor kappa-light-chain-enhancer of activated B cells (NF-κB), matrix metalloproteinases (MMPs), and vascular endothelial growth factor (VEGF). In extreme concentrations, ROS can cause tumor cell apoptosis. As a result, it depends on the ROS levels to either raise tumorigenesis or bring about apoptosis. The main problem is the two-fold conduct of ROS, efficiently having to do with the collection bias that needs to be carefully considered to deter tumor angiogenesis and transformation, for ROS to present potential curative goals both exogenously and endogenously. In general, additional research is required to comprehend the potential of ROS as an active anticancer approach and therapy-relying malignancies [90]. Metastasis is involved in the transmission of cancer cells to distant organs from the primary tumor, and it is the primary reason for health disorders or death. Reactive oxygen species (ROS) play a significant role in the movement and invasion of cancer cells. ROS have a variety of adverse effects and consequences. ROS are first produced as the byproducts of mitochondrial electron transport during aerobic respiration. The mesenchymal transition (EMT) is lost by the epithelial cells as the key factor in metastasis,the protein p53 is important in cell migration through the action of the cytokine TGF-β1. Mitochondrial dysfunction can lead to an increase in ROS production, which subsequently raises the expression of C-X-C motif chemokine 14 (CXC14) through the AP-1 signaling pathway and increases cytosolic Ca^{2+} levels to facilitate cell motility. ROS activate ERK1\2 and increase ROS production in cancer cells. Thus, the growth of premalignant cells may be moderated by applying topical antioxidants that target KIf9. Mitochondrial Ca^{2+} also plays an important role in cancer metastasis. The vimentin protein is significantly engaged in the early stage and development of cancer, which includes EMT and metastasis. Oxidative damage driven by HIF-1 regulates the translation of the vimentin gene, enhancing the creation of invadopodia during cancer cell activity and invasion. Inhibiting vimentin levels through RNA interference can diminish cell metastasis and lower tumor volume. Additionally, ROS cause epigenetic alterations within the promoter regions of E-cadherin and other tumor suppressor genes, promoting tumor growth and metastasis. It may lead to hypermethylation of the promoter regions by enhancing snail-facilitated DNA methylation with the assistance of DNA methyltransferase 1 (DNMT1) and histone deacetylase 1 (HDAC1) [91].

4.5 *Chronic Kidney Diseases*

4.5.1 ROS and Kidney Tissue Damage, Fibrosis, and Renal Failure

In chronic kidney disease patients, oxidative stress is connected to the pathogenesis of atherosclerosis, the occurrence of renal injury, and worsening in disease burden. An imbalance between the generation of free radical production and antioxidant defenses leads to oxidative stress. As a result, certain biomolecules tend to get oxidized, leading to structural and functional modification in the biomolecules. The production of the oxidants generally takes place in the mitochondria, using the enzymes known as mitochondrial cytochrome oxidase. The ROS products thus produced may contribute to the progression of renal injury and pathogenesis of atherosclerosis in CKD patients. Recently, the pathophysiology of oxidative stress in people with CKD has been well established: (i) increased concentrations of circulating biomarkers, (ii) reduced antioxidant defenses due to an inability to remove ROS, and (iii) the number of oxidative stress markers in patients with atherosclerotic CKD [92].

4.6 Pulmonary Diseases

4.6.1 Oxidative Stress in COPD, Asthma, and Lung Fibrosis

Patients with chronic obstructive pulmonary disease (COPD) have direct exogenous oxidative stress due to cigarette smoking and air pollution, as well as endogenous oxidative stress is due to reactive oxygen species (ROS) released by activated inflammatory cells, mostly neutrophils and macrophages. These are the major factors driving the pathophysiology of COPD and also further progress, including amplifying acute exacerbations. Oxidative stress causes a reduction in antioxidant defenses. It is obvious that oxidative stress is a key contributing factor in COPD and may be responsible for a number of the pathophysiological changes that occur. Oxidative stress stimulates intracellular signaling pathways that direct the synthesis and release of inflammatory mediators [93]. Overproduction of ROS leads to tissue damage and deregulation of cell signaling, which results in the pathogenesis of different diseases, including asthma. Compared to other tissues, the lung is subjected to increased oxygen tension. Exogenous oxidants and pollutants boost the synthesis of oxidant production and stimulate inflammatory cells to produce many free radicals. The risk of oxidative stress may be reduced by increasing endogenous antioxidants and ensuring proper intake of dietary antioxidants [94].

5 Therapeutic Implications

5.1 Natural and Synthetic Antioxidants

Natural antioxidants are substances that can help to prevent the oxidative stress, reducing the effects caused by free radicals. Some of the important compounds were discovered in plants with antioxidant capabilities, such as (a) polyphenols, including lignans, curcuminoids, phenolic acids, stilbenes, flavonoids, and chalcones; (b) carotenoids; (c) capsacinoids; (d) isothiocyanates; and (e) catechins. Synthetic antioxidants are chemical substances that delay or prevent oxidation, which helps to balance the dangerous reaction of reactive oxygen species (ROS).

5.1.1 Cardiovascular Disease and Antioxidants

Coronary heart disease can be represented as a large single cause of mortality, approximately 28% of deaths in the United States in 1983. Many investigations have not maintained the ongoing interest of exploring the possible connection between vitamin C and cardiovascular disease causes, as indicated by a review of the scientific literature. Since 1961, almost 12 review articles have dealt with this question. Similarly, because of this opinion of the surgeon general report, some of them are impressed by this data on vitamin C to propose cardiovascular disease; it is the end stage of vitamin C deficiency [95]. The first origin of this superoxide was referred to as mitochondria, xanthine oxidase, nicotinamide dinucleotide phosphate (NADPH) oxidase, and a specific circumstance of NOS in the endothelium.

NADPH Oxidase

It can be produced by the superoxide enzyme, described first in neutrophils, and also demonstrates that the adventitia enzyme contains at least five isoforms in endothelium and smooth muscle cells (NOXI 5).

Xanthine Oxidase

Xanthine oxidase reductase facilitates the step-by-step alteration of hypoxanthine into uric acid and xanthine through hydroxylation. This enzyme has two forms that vary in substrate specificity in oxidation. Dehydrogenation primarily uses the electron acceptor, NAD^+. Still, it can distribute molecular oxygen electrons. Through xanthine dehydrogenase, thiol oxidation, and proteolytic degradation, it was sourced in mammals; then it cannot oxidize in the form of an oxidase. Thereafter, it can be ready to transfer the electrons to molecular oxygen [96].

Flavonoids

A group of plant compounds was found in plants and food, which are based on plants. Flavonoids can be rich in diets and have been associated with nutritional recommendations, a healthy way of life, and prevention against chronic diseases. In brief, some flavonoids have to demonstrate potent anti-inflammatory activity and also reduce inflammation, which is a result of various causes, including antibody-resistant bacteria [97]. Flavonoids are phenolic phytochemicals that are important to the human diet [98]. In plants, aromatic amino acids (such as phenylalanine, tyrosine, and malonate) were formed. Flavonoids' structure is a flavan nucleus, which has three rings containing 15 carbon atoms (C_6–C_3–C_6) [99].

5.1.2 Neurodegenerative Disease with Natural Antioxidants

Neurodegenerative disease is a group of diseases that have clinical significance and causes [100]. The central nervous system (CNS) needs significant energy, which is produced by mitochondria due to the increase in metabolic rate. The major effects of oxidative stress may lead to a wide range of CNS disorders, such as neurodegenerative conditions [101]. The interaction between microglial cells and astrocytes can be done in both directions, suggesting that microglia can stimulate astrocytes by using cytokine-based communication [102], promoting astrogliosis and the spread of neurodegenerative diseases. Activation of microglial cells involved in the progression and mitigation of pathophysiological mechanisms, which is associated with a number of neurodegenerative disorders [103]. Here, we consider the effects of natural antioxidants like vitamins E and C, flavonoids, and polyphenols, which help to prevent neurodegenerative diseases.

Vitamins E and C

Vitamins E and C are popular substances and are mostly suggested for patients' treatment. In animal and clinical studies, this vitamin E deficiency has been commonly used to investigate its involvement in nigral neurodegeneration and in the growth or advancement of PD. It was found that ingestion of high-dose vitamin E (1000 mg/kg) or carotene (200 mg/kg) before exposure prevented glutathione (GSH) depletion [104] and provided partial protection to dopaminergic nigrostriatal neurons in mice that received a single injection of 40 mg/kg MPTP (N-methyl-1,2,3,6 tetrahydropyridine) [105]. In different aspects of the study, 91 individuals aged over 65 years without any disease were chosen. After the follow-up time duration of 4.3 years, it was found that none of the 27 individuals who needed vitamin E developed AD, compared with 2.5 expected cases based on sex, age, and years of education ($p = 0.03$). Similarly, none of the 23 individuals who used vitamin C developed AD, compared with the expected 3.2 cases ($p = 0.04$). These data will help prescribe high amounts of vitamins E and C, which can decrease the risk of AD [106].

Green Tea Polyphenols

The polyphenols from green tea and their main components, (-)-epigallocatechin-3-gallate, have various pharmaceutical actions, including anticarcinogenic and antimutagenic effects. It is believed that potent antioxidative properties are responsible for these beneficial effects. In fact, polyphenols act as vigorous antioxidants against free radicals, like DPPH (1,1-diphenyl-2-picrylhydrazyl) radicals [107], superoxide anion, hydroxyl radicals, and lipid free radicals [108].

Genistein

Genistein, a primary compound of soy isoflavone, has been studied for its capability to bind with the receptor for estrogen. It has antioxidant properties and promotes the increase of glutathione reduction in cells [109]. It inhibits protein tyrosine kinase (PTK) and has various physiological functions [110]. The neuroprotective impact of genistein against Aβ25-35-induced apoptosis in cultured hippocampal neurons was examined [111]. Exposure to Aβ25-35 for 24 h increased DCF fluorescence intensity compared to the control. These findings indicate that genistein provides neuronal protection against Aβ25-35 at nanomolar concentrations via an ER-mediated pathway. The antioxidative properties are mainly attributed to the micromolar level and the neuroprotective impact of genistein.

5.1.3 Diabetes

Diabetes is a chronic disease identified by a high blood sugar level because of a defect in insulin production, action, or both. This condition, defined by insufficient insulin secretion or insulin action, can be related to chronic hyperglycemia and disturbances of carbohydrate, lipid, and protein metabolism [112].

Melatonin

Melatonin, also called N-acetyl-5-methoxytryptamine, is a type of steroid that can be created by the pineal gland and plays a role in regulating reproductive processes, modulating the activity of the immune system, controlling the formation of tumors, and efficiently inhibiting oxidative stress. The most important function of melatonin is to multiply the formation of radicals in oxygen metabolism, therefore potentially saving these membranes, proteins, and DNA from free radicals. Melatonin needs to play a role in naturally occurring free radical-related diseases like diabetes [113].

Lipoic Acid

Lipoic acid is a dithiol substance that can be obtained from octanoic acid and used as a powerful antioxidant, as a heavy metal poisoning detoxificant, and has been suggested as a potential way to enhance cognitive impairment along with aging [114]. There is evidence to prove that lipoic acid is a useful agent for the problems related to diabetes mellitus [115].

Catalase (CAT)

Catalase is located in the peroxisomes and plays a primary role in regulating hydrogen peroxide and metabolism. In chromosome 11p13, this catalase gene is present. Metabolism, specifically of exon 2 nearer to the gene, has been found to be an important reason for the development of type 2 diabetes [116]. Next, the oxidative stress modulates the structure of the tyrosine residue in catalase to reduce its activity. A highly reactive small molecule is hydrogen peroxide,it can be formed as a natural result of an energy source. Excess amounts of absorption of hydrogen peroxide may cause serious effects, including no proteins, DNA, RNA, and lipids [117].

Vitamin C (Ascorbic Acid)

Vitamin C (ascorbic acid) is an important hydrophilic antioxidant in plasma and the cytosol of many cells. It is involved in counteracting various water-soluble oxidants and works with vitamin E to stop lipid oxidation [118]. It is also able to increase the production of nitric oxide by stabilizing the NOS cofactor BH4 in endothelial cells.

5.1.4 Liver Disease

The liver is a crucial organ targeted by ROS. The primary parenchymal cells are affected in the liver due to injury from oxidative stress [119]. ROS can be produced from the mitochondria, microsomes, and peroxisomes in parenchymal cells, which is related to the gene expression for liver fatty acid oxidation [120]. In addition to natural products, different compounds have been studied for their potential in reducing oxidative stress, including L-thiamine, vitamin E, N-acetyl cysteine, raxofelast, and betaine [121].

5.1.5 Limitations and Challenges of Antioxidant Therapies

Antioxidant therapies have been widely researched for their potential to reduce oxidative stress, a factor in numerous diseases like cardiovascular disease, cancer, neurodegenerative disorders, and aging. However, several challenges and limitations have limited their success in clinical settings.

(a) *Poor Targeting and Low Bioavailability*: Many antioxidants lack precision in targeting specific sites of oxidative damage within cells or tissues, and their absorption and stability can be low. Often, they are poorly absorbed in the gastrointestinal tract or fail to enter cells effectively. For example, polyphenols in foods like green tea and berries, though beneficial, have low bioavailability due to rapid breakdown and limited intestinal absorption.
(b) *Prooxidant Effects at High Levels*: At higher doses, certain antioxidants can act as prooxidants, generating free radicals rather than neutralizing them. This can worsen oxidative stress instead of relieving it, potentially harming cells. For example, vitamin C and beta-carotene, while protective at low levels, can sometimes produce reactive oxygen species (ROS) in the presence of metals such as iron and copper, shifting from antioxidant to prooxidant behavior.
(c) *Limited Success in Clinical Trials*: Many clinical trials on antioxidant supplements, such as vitamins C and E, have shown disappointing results. Some studies have shown little or no impact on disease prevention, and a few have even linked supplements to higher mortality. For example, In vitamin e, selenium and vitamin E cancer prevention trial (SELECT) at high doses did not prevent prostate cancer and was even along with slight improvement in risk.

(d) *Complexity of Oxidative Stress Pathways*: Oxidative stress involves complex interactions of ROS, antioxidants, and signaling pathways that control cell survival, inflammation, and programmed cell death. Non-specific antioxidant therapies can interfere with these processes and may disrupt normal cell functions, such as immune responses, where ROS play essential roles. For example, ROS are crucial in immune cell signaling, so broadly suppressing ROS can negatively affect immune function.
(e) *Interaction with Cellular Signaling Pathways*: ROS are not only damaging molecules but also play an essential role in signaling pathways that regulate cell proliferation, apoptosis (programmed cell death), and other crucial functions. Excessive antioxidant use can disrupt these pathways, potentially interfering with natural processes of cell turnover and immune responses. For example, ROS function as signaling molecules that modulate the pathways of NF-kB and MAPK, which are involved in inflammation and stress responses. Suppressing ROS levels indiscriminately with antioxidants can impair these signaling functions, potentially reducing the body's ability to respond to infections or inflammation effectively.
(f) *Limitations of Conventional Antioxidant Therapies*: Traditional antioxidants like vitamins C and E lack specificity, often failing to reach the specific cellular locations where oxidative stress is most harmful. There is a growing interest in targeted therapies that include enzyme-based treatments or peptides that mimic antioxidant enzymes to improve efficacy. For example, synthetic compounds that mimic superoxide dismutase (SOD) and catalase are being studied for their ability to target oxidative stress more precisely than conventional antioxidants, especially in diseases like ALS and Parkinson's, where specific neuronal damage is a concern.

5.2 Novel Therapeutic Approaches

Pathological processes, such as inflammation and oxidative stress, are involved in the genesis and ongoing development of numerous disorders. Transcriptional factor for erythroid 2-related factor 2 (Nrf2), a nuclear factor, can control expression of different antioxidants, cytoprotective proteins, and enzymes; therefore, it involves a major participation in protecting against oxidative injury [122]. Activated Nrf2 has been suggested as a strategy for treating diabetics. In this condition, naringenin, a bioactive flavonoid present in the citrus fruit, protected the pancreatic beta cells against oxidative damage in vitro. Streptozotocin induced diabetes in the mice by triggering Nrf2 [123].

5.2.1 Nrf2 Role in Neurodegenerative Diseases

The declining brain commonly has oxidative neuroinflammation and mitochondrial dysfunction, with studies showing decreased Nrf2 expression and gene-related changes in aging individuals. Actually, aging is a major cause of both the onset and progression of neurodegenerative disorders. Therefore, Nrf2 is an attractive target in clinical intervention and in the sequence part of neurodegeneration, defining the involvement of Nrf2 in AD, PD, MS, and ALS [124].

Nrf2 Alzheimer's Disease

AD is a common neurodegenerative disease that affects 55 million people against dementia worldwide. In the brain of a person with AD, Nrf2 is mostly in the cytoplasm compared with age-related healthy matches for reducing nuclear functions. Therefore, a decrease in Nrf2 level can be associated with a high prediction of Aβ.

5.2.2 NADPH

NADPH oxidases are a set of enzymes that can lead to ROS (reactive oxygen species). The NOX (NADPH oxidase 1) and NOX2 oxidases are the main sources of ROS in the arterial wall under conditions such as hypertension, hypercholesterolemia, diabetes, and aging, and they are a major contributor to oxidative stress, endothelial dysfunction, and vascular inflammation, leading to arterial remodeling and atherogenesis [125].

Isoforms of NADPH Oxidase in Vascular Diseases

Since the first discovery in endothelial cells and VSMCs, it is the fastest-growing area with the largest experimental evidence in the body, also including studies in eliminating mice, exposing NADPH oxidase as a basal source of free radicals in different types of cardiovascular disorders, as well as in leukocytes and brain cells. NADPH oxidases are a group of multi-subunit enzyme complexes that function to generate superoxide and hydrogen peroxide.

5.2.3 Targeted Drug Delivery Systems to Reduce Oxidative Damage

The process of pathogenesis and development of severe obstructive pulmonary disorder (COPD) can involve oxidative stress, which can play a vital role in it. In animal models, the combination of medications and nutritional results was considered to be a beneficial therapeutic system [126]. One of the necessary organs is the mitochondria, which play a key role in the aging process. Increased mitochondrial size was observed in the senescent cells, while defective mitochondria can cause mitochondrial oxidative stress. Also, the mitochondrial oxidative stress is targeted by defective mitochondria. The interaction between broken mitochondria and mitochondrial oxidative stress can contribute to the progression and development of aging and age-related processes [127]. Oxidative stress may cause many disorders such as cancer, asthma, cardiovascular diseases, diabetes, and neurodegenerative diseases. Generally, the ROS are explored by somebody's antioxidant system, including molecules such as vitamins, tocopherols, thioredoxin, ascorbic acid, reduced glutathione, and transferrin, as well as enzymes like glutathione peroxidase (GPx), glutathione reductase (GR), superoxide dismutase (SOD), and catalase. A decrease in these protective systems can lead to oxidative stress-stimulated diseases under pathophysiological conditions [128].

5.3 Lifestyle Interventions

Oxygen is the most important source for life, and it acts as a powerful oxidizing agent. The oxidative event, such as the generation of free radicals, may cause damage within the cell. These free radicals were entered into our body not only internally but also by external sources like environmental pollution, cigarette smoke, pesticides, and toxic metals, which also adds damage to our body [129].

5.3.1 Diet and Exercise Intervention

Hypertension, also known as high blood pressure, can affect 50 million people in the United States, and it is a major cause of renal insufficiency, congestive heart failure, stroke, and coronary artery disease [130]. Recently, oxidative stress has been the main factor because it is a causative factor of hypertension. Scientific proof shows that feeding animals a diet high in fat or concentrated carbohydrates can cause oxidative stress, reducing the availability of nitric oxide and causing abnormalities in the endothelial cells. Adapting of a lower-fat diet consisting of vegetables, fruits, and low-fat dairy in humans has been shown to help reduce BP in both normotensive and hypertensive individuals. Therefore, more exercise can help decrease hypertensive BP for individuals. Nowadays, most recommendations focus on lifestyle modifications, including weight loss and reducing salt and alcohol intake, to lower blood pressure [131].

5.3.2 Environmental Stress and Factors

Animals, plants, and humans are affected by metabolic changes because of environmental stresses; this may affect the reduction of antioxidant production or extend the production of both reactive oxygen and nitrogen species (RONs).

Pollution

The exogenous pollutant may contribute to the generation of free radicals in daily life, and it is difficult to avoid them. Various environmental factors and pollutants, which are secreted from chemical toxins, may induce the entry of ROS. In the environment, the most dangerous type of pollution is air pollution [132]. Also, we all know that air pollution is the biggest issue where human health is affected by respiratory morbidity, mortality, and cardiovascular diseases [133].

6 Future Directions and Research Gaps

6.1 Biomarker Discovery

The present review examines recent advancements to identify biomarkers associated with several stresses in humans and animals, and the potential prognosis and therapeutic value in stress-related disorders, and also deliberates on future perspectives [134].

6.1.1 Types of Biomarkers

The biomarkers were differentiated on different bases such as characteristics, application, and genetics and molecular biology methods. *Characteristics*: Based on characteristics, they can be classified into two types: imaging and non-imaging biomarkers. Imaging biomarkers are used to identify or visualize the lesion or disease in positron emission tomography, computed tomography, and magnetic resonance imaging. Non-imaging biomarkers are biochemical biomarkers considered as molecular biomarkers with biophysical properties; they will be measured by biological samples. They include cellular structure or biophysical substances, which are nucleic acid biomarkers such as gene mutation or polymorphism, peptides, quantitative gene expression analysis, proteins, metabolites, lipids, and other small molecules. *Application*: On an application basis, biomarkers can be divided into prognostic, diagnostic, and therapeutic biomarkers. Diagnostic biomarkers help in diagnosing or determining a disease. Prognostic biomarkers are helpful in predicting the disease outcome. Therapeutic biomarkers help to monitor the developmental treatment of a disease [135]. *Genetics and Molecular Biology Methods*: In this method, biomarkers are classified into three types: (1) type 0, (2) type 1, and (3) type 2. Type 0 biomarkers are natural history biomarkers used for measuring progression in a disease over time and are called clinical indicators. Type 1 biomarkers are drug-related action biomarkers. They include efficacy biomarkers (showing the therapeutic effects of a drug), mechanism biomarkers (providing information about the function of a drug), and toxicity biomarkers (indicating harmful effects of a drug). Type 2 biomarkers act as surrogate markers and provide alternative clinical outcomes of a disease. Furthermore, they help to predict the effects of therapeutic interventions. Therefore, biomarkers are classified into various types, and each of them has its related stress mechanism, disease, system, or organ involved. This classification details how each biomarker helps clarify the diagnostic, pathophysiological, and clinical importance.

6.2 Personalized Medicine

The genetic risk prediction can utilize the genetic information to personalize the prediction of the result and the causes of harm. Many examples were discussed for toxicogenetics, such as Mendelian traits being highly reflective before the Human Genome Project started in 1990 [136].

6.2.1 Genetics Before the Human Genome Project

From the 1860s until the late 1980s, Mendel generally identified the genes of each one as having a few minor alleles and a wild-type allele, which sometimes caused disorders. Their gene is considered to follow the pattern of simple Mendelian inheritance of a distinct monogenic trait.

6.2.2 Environment–Gene Interactions

Gene–environment interactions are about analyzing the complex connection between genetic and environmental influences. It assesses the disease risk and involves exploring gene–environment interaction related to susceptibility to illness. The scientists aim to understand genetic tendencies in collaboration with factors such as diet, lifestyle choices, and pollutants. It is likely to develop diseases by integrating genomics with its environmental elements. This research facilitates the prevention strategies and personalized treatments personalized to individuals based on genetic makeup and environmental influences [137]. Genetic susceptibility to aplastic anemia can be affected by environmental factors, such as exposure to toxins or infections. However, some genetic variants can increase the individual's sensitivity to the environment and make the cause of aplastic anemia worse. One of the examples is a person with genetic factors who can attack the immune response or DNA repair processes and be more susceptible to the impact of environmental exposure, leading to the growth of aplastic anemia [138]. Genetic screening played an important role in finding a person at risk of developing aplastic anemia and providing personalized medicine approaches. The specific genetic variants that are associated with aplastic anemia and personalized medicine strategies were identified by analyzing the genetic profiles and healthcare providers' activity. Personalized medicine strategies, like targeted treatments and preventive measures based on genetic risk factors, promise to improve patients' outcomes and maintain aplastic anemia successfully.

6.2.3 Impacts for Personalized Medicine

The interchange of susceptibility and HIV-induced inflammation on the pathogenesis of aplastic anemia (AA) highlights the importance of individualized medicine in maintaining and preventing this condition. If changing genetic information is the understanding of inflammatory mechanisms, personalized medicine can supplement the diagnosis, improve outcomes, and personalize interventions to individual patient needs. This can help to give a variety of key impacts for maintaining aplastic anemia in HIV-positive patients.

6.3 Emerging Therapies

Some of the future treatments and therapies for oxidative damage were identified and associated with specific mechanisms and current research trends.

6.3.1 Cellular Therapies

In cellular therapies, the major treatment is stem cell therapy to reduce oxidative damage. Mainly, the mesenchymal stem cells indirectly helped to reduce oxidative stress by secreting growth factors and cytokines; these factors will promote healing and reduce inflammation. In the future, researchers will focus on strengthening the techniques functions to reduce oxidative damage more successfully.

6.3.2 Nanotechnology

Nanotechnology is a process of delivering an antioxidant to decrease or reduce the cause of oxidative stress. These nanoparticles are drawn to release their antioxidants directly to areas that oxidative damage can affect, which may enhance treatment effectiveness. In particular, gold nanoparticles are being studied for their antioxidant capabilities.

6.3.3 Dietary Treatment

Diets abundant in antioxidants from sources like vegetables, fruits, nuts, and whole grains will help to decrease oxidative stress. Some of these examples are berries rich in polyphenols; they can be found to increase the activity of antioxidant enzymes. Clinical trials have recently calculated that the impact of diets rich in omega-3 fatty acids and low in refined sugar on oxidative stress markers [139].

6.3.4 Gene Therapy

Here are some of the gene therapy techniques.

(a) *Enhance Antioxidant Enzyme*: In the future, gene therapies will focus on introducing genes that encode important antioxidant enzymes like glutathione peroxidase and SOD (superoxide dismutase). It helps to increase expression in targeted tissues; their therapies can help to reduce oxidative stress [140].
(b) *CRISPR/Cas9 Technology*: CRISPR/Cas9 is a technology that can be used to modify genes involved in oxidative stress response. The targeting gene can have the ability to control ROS (reactive oxygen species) production, and scientists have their goal in strengthening the cellular response to oxidative damage.
(c) *Combining Stem Cell Therapy with Gene Therapy*: Combining gene therapy with stem cell therapy will enhance the tissue repair capability. Stem cells engineered to express antioxidant enzymes might have protective effects against oxidative damage in diseases such as neurodegenerative diseases.

7 Conclusion

In summary, oxidative stress plays a significant role in the development and advancement of various chronic illnesses, including cancer, diabetes, cardiovascular disease, neurodegenerative conditions, and inflammatory disorders. ROS and the body's antioxidant defenses cause inflammation and tissue dysfunction, all of which contribute to disease onset and worsen its course, ultimately impacting patients quality of life. Exploring how oxidative stress contributes to chronic diseases reveals potential for targeted therapeutic approaches. Antioxidant treatments, lifestyle changes, and dietary interventions aimed at boosting the body's own antioxidant defenses could help reduce disease impact. However, the complexity of oxidative pathways and individual variability presents challenges in achieving clinical effectiveness. Future research should delve into the specific role of oxidative stress within different disease contexts, aiming to identify novel biomarkers for early diagnosis and develop personalized strategies to combat oxidative damage. Oxidative stress is a pathological process; it can cause various chronic diseases. Mostly, it was involved in numerous disorders; here, some of the transcription factors were used to reduce this oxidative stress. Nuclear factor erythroid 2-related factor 2 (Nrf2) is one of them, and another one is NADPH. These factors can help to reduce the effect of oxidative damage. Nrf2 regulates the expression of various antioxidants, and it is vital for different diseases like neurodegenerative disease, Alzheimer's disease, PD, and ALS. Nrf2 is the most clinically relevant intervention for the sequence section. NADPH contains a group of enzymes that lead to the production of reactive oxygen species, which is also a main contribution to oxidative stress. It can also have a basal effect on oxidative stress for various types of cardiovascular diseases, as well as on leukocytes and brain cells. The enzymes of superoxide dismutase and hydrogen peroxide can generate the necessary function for reducing oxidative stress. Taking a comprehensive approach to managing oxidative stress—encompassing prevention, therapeutic interventions, and lifestyle modifications—has the potential to significantly benefit public health by reducing both the prevalence and severity of chronic diseases. This chapter underscores the critical need for continued research into oxidative mechanisms and the development of innovative strategies to more effectively address chronic disease management in today's society.

References

1. Sharifi-Rad, M., Anil Kumar, N. V., Zucca, P., Varoni, E. M., Dini, L., Panzarini, E., Rajkovic, J., Tsouh Fokou, P. V., Azzini, E., Peluso, I., Prakash Mishra, A., Nigam, M., El Rayess, Y., Beyrouthy, M. El, Polito, L., Iriti, M., Martins, N., Martorell, M., Docea, A. O., Sharifi-Rad, J. (2020). Lifestyle, oxidative stress, and antioxidants: back and forth in the pathophysiology of chronic diseases. *Frontiers in Physiology, 11*. Frontiers Media S.A. https://doi.org/10.3389/fphys.2020.00694.
2. Li, H., Horke, S., & Förstermann, U. (2014). Vascular oxidative stress, nitric oxide and atherosclerosis. *Atherosclerosis, 237*(1), 208–219. Elsevier Ireland Ltd. https://doi.org/10.1016/j.atherosclerosis.2014.09.001.
3. Taverne, Y. J. H. J., Bogers, A. J. J. C., Duncker, D. J., & Merkus, D. (2013). Reactive oxygen species and the cardiovascular system. *Oxidative Medicine and Cellular Longevity*. https://doi.org/10.1155/2013/862423
4. Gandhi, S., & Abramov, A. Y. (2012). Mechanism of oxidative stress in neurodegeneration. In *Oxidative medicine and cellular longevity*. Hindawi Limited. https://doi.org/10.1155/2012/428010.
5. Salehi, B., Shetty, M. S., Anil Kumar, N. V., Živković, J., Calina, D., Docea, A. O., Emamzadeh-Yazdi, S., Kılıç, C. S., Goloshvili, T., Nicola, S., Pignata, G., Sharopov, F., Del Mar Contreras, M., Cho, W. C., Martins, N., & Sharifi-Rad, J. (2019). Veronica plants—drifting from farm to traditional healing, food application, and phytopharmacology. *Molecules, 24*(13). MDPI AG. https://doi.org/10.3390/molecules24132454.
6. Aroor, A. R., Mandavia, C., Ren, J., Sowers, J. R., & Pulakat, L. (2012). Mitochondria and Oxidative Stress in the Cardiorenal Metabolic Syndrome. *Cardiorenal Medicine, 2*(2), 87–109. https://doi.org/10.1159/000335675
7. Karam, B. S., Chavez-Moreno, A., Koh, W., Akar, J. G., & Akar, F. G. (2017). Oxidative stress and inflammation as central mediators of atrial fibrillation in obesity and diabetes. *Cardiovascular Diabetology, 16*(1). BioMed Central Ltd. https://doi.org/10.1186/s12933-017-0604-9.
8. Zhou, Y., Li, Y., Zheng, J., Liu, K., & Zhang, H. (2015). Detecting of gastric cancer by Bcl-2 and Ki67. *International journal of clinical and experimental pathology, 8*(6). www.ijcep.com/.
9. Mishra, A. P., Salehi, B., Sharifi-Rad, M., Pezzani, R., Kobarfard, F., Sharifi-Rad, J., & Nigam, M. (2018). Programmed cell death, from a cancer perspective: an overview. *Molecular Diagnosis and Therapy, 22*(3), 281–295. Springer International Publishing. https://doi.org/10.1007/s40291-018-0329-9.

10. Jaramillo, M. C., & Zhang, D. D. (2013). The emerging role of the Nrf2-Keap1 signaling pathway in cancer. *Genes and Development, 27*(20), 2179–2191. https://doi.org/10.1101/gad.225680.113.
11. Prata, C., Angeloni, C., & Maraldi, T. (2024). Strategies to counteract oxidative stress and inflammation in chronic-degenerative diseases 2.0. *International Journal of Molecular Sciences, 25*(9). Multidisciplinary Digital Publishing Institute (MDPI). https://doi.org/10.3390/ijms25095026.
12. Tomioka, H., Iwamoto, E., Itakura, H., & Hirai, K. (2001). Generation and characterization of a fairly stable triplet carbene. www.nature.com.
13. ɰuračková, Z. (2014). Free radicals and antioxidants for non-experts. In *Systems biology of free radicals and antioxidants* (vol. 9783642300189, pp. 3–38). Springer-Verlag Berlin Heidelberg. https://doi.org/10.1007/978-3-642-30018-9_2.
14. Phaniendra, A., Jestadi, D. B., & Periyasamy, L. (2015). Free radicals: properties, sources, targets, and their implication in various diseases. *Indian Journal of Clinical Biochemistry, 30*(1), 11–26. Springer India. https://doi.org/10.1007/s12291-014-0446-0.
15. Kuppusamy, P., & Zweiers, J. L. (1989). The journal of biological chemistry characterization of free radical generation by xanthine oxidase evidence for hydroxyl radical generation*, *264*(17).
16. Sen, S., Chakraborty, R., Sridhar, C., Reddy, Y. S. R., & De, B. (2010). Free radicals, antioxidants, diseases and phytomedicines: current status and future prospect. *International Journal of Pharmaceutical Sciences Review and Research, 3*(1). www.globalresearchonline.net.
17. J H Festos, B. H. (1894). Oxidation of tartaric acid in presexce of iron. 899 LXXZIl.-Oxictutioiz of Tcxriaric Acid iia prcseizce of Jrolz.
18. Bedwell, S., Deant, R. T., & Jessup, W. (1989). The action of defined oxygen-centred free radicals on human low-density lipoprotein. *Biochemical Journal, 262.*
19. Haber, F., Weiss, J., Seph, J. O., & Eiss, W. (1934). The catalytic decom position o f hydrogen peroxide by iron salts* By Fritz Haber. https://royalsocietypublishing.org/.
20. Kanofsky, J.R. (1989). Singlet oxygen production by biological systems. *Chemico-Biological Interactions, 70.*
21. Sies, H., & Menck, C. F. M. (1992). Singlet oxygen induced DNA damage. *Mutation Research, 275.*
22. Esterbauer, H., Gebicki, J., Puhl, H., & Jirgens, G. (1992). The role of lipid peroxidation and antioxidants in oxidative modification of ldl. *Free Radical Biology & Medicine, 13.*
23. Hansford, R. G., Hogue, B. A., & Mildaziene2, V. (1997). Dependence of H_2O_2 formation by rat heart mitochondria on substrate availability and donor age.
24. Newsholme, P., Cruzat, V. F., Keane, K. N., Carlessi, R., & De Bittencourt, P. I. H. (2016). Molecular mechanisms of ROS production and oxidative stress in diabetes. *Biochemical Journal 473*(24), 4527–4550. Portland Press Ltd. https://doi.org/10.1042/BCJ20160503C
25. Schieber, M., & Chandel, N. S. (2014). ROS function in redox signaling and oxidative stress. *Current Biology, 24*(10). Cell Press. https://doi.org/10.1016/j.cub.2014.03.034.
26. Pizzino, G., Irrera, N., Cucinotta, M., Pallio, G., Mannino, F., Arcoraci, V., Squadrito, F., Altavilla, D., & Bitto, A. (2017). Oxidative stress: harms and benefits for human health. *Oxidative Medicine and Cellular Longevity, 2017*. Hindawi Limited. https://doi.org/10.1155/2017/8416763.
27. Bhatti, J. S., Bhatti, G. K., & Reddy, P. H. (2017a). Mitochondrial dysfunction and oxidative stress in metabolic disorders—a step towards mitochondria based therapeutic strategies. *Biochimica et Biophysica Acta—Molecular Basis of Disease, 1863*(5), 1066–1077. Elsevier B.V. https://doi.org/10.1016/j.bbadis.2016.11.010.
28. Bhandary, B., Marahatta, A., Kim, H. R., & Chae, H. J. (2013). An involvement of oxidative stress in endoplasmic reticulum stress and its associated diseases. *International Journal of Molecular Sciences, 14*(1), 434–456. https://doi.org/10.3390/ijms14010434.
29. Chaudhari, N., Talwar, P., Parimisetty, A., d'Hellencourt, C. L., & Ravanan, P. (2014). A molecular web: endoplasmic reticulum stress, inflammation, and oxidative stress. *Frontiers in Cellular Neuroscience, 8*(JULY). Frontiers Research Foundation. https://doi.org/10.3389/fncel.2014.00213.

30. Shamu1, C. E., & Walter, P. (1996). Oligomerization and phosphorylation of the Irelp kinase during intracellular signaling from the endoplasmic reticulum to the nucleus. *The EMBO Journal, 15*(12). https://www.embopress.org.
31. Ong, G., & Logue, S. E. (2023). Unfolding the interactions between endoplasmic reticulum stress and oxidative stress. *Antioxidants, 12*(5). MDPI. https://doi.org/10.3390/antiox12050981.
32. Walter, P., & Ron, D. (2011). The unfolded protein response: from stress pathway to homeostatic regulation. *Science, 334*(6059), 1081–1086. American Association for the Advancement of Science. https://doi.org/10.1126/science.1209038.
33. Zhu, C., Johansen, F.-E., & Prywes, R. (1997). Interaction of ATF6 and serum response factor. *Molecular and Cellular Biology, 17*(9).
34. Aranda-Rivera, A. K., Cruz-Gregorio, A., Arancibia-Hernández, Y. L., Hernández-Cruz, E. Y., & Pedraza-Chaverri, J. (2022). RONS and Oxidative Stress: An Overview of Basic Concepts. *Oxygen, 2*(4), 437–478. https://doi.org/10.3390/oxygen2040030
35. Birben, E., Sahiner, U. M., Sackesen, C., Erzurum, S., & Kalayci, O. (2012).Oxidative stress and antioxidant defense.
36. De Jager, T. L., Cockrell, A. E., & Du Plessis, S. S. (2017). Ultraviolet lightinduced generation of reactive oxygen species. *Advances in Experimental Medicine and Biology, 996*, 15–23. SpringerNew York LLC. https://doi.org/10.1007/978-3-319-56017-5_2.
37. Kimeswenger, S., Schwarz, A., Födinger, D., Müller, S., Pehamberger, H., Schwarz, T., & Jantschitsch, C. (2016). Infrared A radiation promotes survival of human melanocytes carrying ultraviolet radiation-induced DNA damage. *Experimental Dermatology, 25*(6), 447–452. https://doi.org/10.1111/exd.12968
38. Deliconstmtinos, G., Villiotou, V., & Stawides, J. C. (1996). Alterations of nitric oxide synthase and xanthine oxidase activities of human keratinocytes by ultraviolet B radiation potential role for peroxynitrite in skin inflammation. *Biochemical Pharmacology, 51.*
39. Alizadeh, E., Orlando, T. M., & Sanche, L. (2015). Biomolecular damage induced by ionizing radiation: The direct and indirect effects of low-energy electrons on DNA. *Annual Review of Physical Chemistry, 66*, 379–398. https://doi.org/10.1146/annurev-physchem-040513-103605
40. Rai, Y., Anita, Kumari, N., Singh, S., Kalra, N., Soni, R., & Bhatt, A. N. (2021). Mild mitochondrial uncoupling protects from ionizing radiation induced cell death by attenuating oxidative stress and mitochondrial damage. *Biochimica et Biophysica Acta—Bioenergetics, 1862*(1). https://doi.org/10.1016/j.bbabio.2020.148325.
41. Szumiel, I. (2015). Ionizing radiation-induced oxidative stress, epigenetic changes and genomic instability: the pivotal role of mitochondria. *International Journal of Radiation Biology, 91*(1), 1–12. Informa Healthcare. https://doi.org/10.3109/09553002.2014.934929.
42. Paithankar, J. G., Saini, S., Dwivedi, S., Sharma, A., & Chowdhuri, D. K. (2021). Heavy metal associated health hazards: an interplay of oxidative stress and signal transduction. *Chemosphere, 262.* https://doi.org/10.1016/j.chemosphere.2020.128350.
43. Ghosh, M., & Singh, S. P. (2005). A review on phytoremediation of heavy metals and utilization of its byproducts. *Applied Ecology and Environmental Research, 3*(1), 1–18. Corvinus University of Budapest. https://doi.org/10.15666/aeer/0301_001018.
44. Zhang, P.-Y., Xu, X., Li, X.-C., & Zhang, P. (2014). Cardiovascular diseases: oxidative damage and antioxidant protection.
45. Halliwell, B. (2006). Reactive species and antioxidants. Redox biology is a fundamental theme of aerobic life. *Plant Physiology, 141*(2), 312–322. American Society of Plant Biologists. https://doi.org/10.1104/pp.106.077073.
46. Unsal, V., Dalkiran, T., Çiçek, M., & Kölükçü, E. (2020). The role of natural antioxidants against reactive oxygen species produced by cadmium toxicity: a review. *Advanced Pharmaceutical Bulletin, 10*(2), 184–202. Tabriz University of Medical Sciences. https://doi.org/10.34172/apb.2020.023.
47. Sepasi Tehrani, H., & Moosavi-Movahedi, A. A. (2018). Catalase and its mysteries. *Progress in Biophysics and Molecular Biology, 140*, 5–12. Elsevier Ltd. https://doi.org/10.1016/j.pbiomolbio.2018.03.001.

48. Arthur, J. R. (2000). The glutathione peroxidases. CMLS *Cellular and Molecular Life Sciences, 57*.
49. Margis, R., Dunand, C., Teixeira, F. K., & Margis-Pinheiro, M. (2008). Glutathione peroxidase family - An evolutionary overview. *FEBS Journal, 275*(15), 3959–3970. https://doi.org/10.1111/j.1742-4658.2008.06542.x
50. Knapen, M. F. C. M., Zusterzeel, P. L. M., Peters, W. H. M., & Steegers, E. A. P. (1999). Glutathione and glutathione-related enzymes in reproduction A review a a. *European Journal of Obstetrics & Gynecology and Reproductive Biology, 82*.
51. Xu, Y., Zhou, X., Shi, C., Wang, J., & Wu, Z. (2015). α-Lipoic acid protects against the oxidative stress and cytotoxicity induced by cadmium in HepG2 cells through regenerating glutathione regulated by glutamate-cysteine ligase. *Toxicology Mechanisms and Methods, 25*(8), 596–603. https://doi.org/10.3109/15376516.2015.1044150
52. Kundu, S., Sengupta, S., Chatterjee, S., Mitra, S., & Bhattacharyya, A. (2009). Cadmium induces lung inflammation independent of lung cell proliferation: a molecular approach. *Journal of Inflammation, 6*. https://doi.org/10.1186/1476-9255-6-19.
53. Beyer, C. E., Steketee, J. D., & Saphier, D. (1998). Antioxidant Properties of Melatonin-An Emerging Mystery. *Biochemical pharmacology, 56*.
54. Cho, E. H., & Koh, P. O. (2010). Proteomic identification of proteins differentially expressed by melatonin in hepatic ischemia-reperfusion injury. *Journal of Pineal Research, 49*(4), 349–355. https://doi.org/10.1111/j.1600-079X.2010.00799.x
55. Niki, E., & Traber, M. G. (2012). A history of vitamin e. *Annals of Nutrition and Metabolism, 61*(3), 207–212. https://doi.org/10.1159/000343106.
56. Akano, M. N., Nodera, A. O., Aito, E. S., Anabe, M. T., Ajima, K. Y., Akahashi, J. T., Van, N., & Huyen, C. (2008). Effect of Astaxanthin in Combination with-Tocopherol or Ascorbic Acid against Oxidative Damage in Diabetic ODS Rats. *Journal of Nutritional Science and Vitaminology, 54*.
57. Kocot, J., Luchowska-Kocot, D., Kiełczykowska, M., Musik, I., & Kurzepa, J. (2017). Does vitamin c influence neurodegenerative diseases and psychiatric disorders? *Nutrients, 9*(7). MDPI AG. https://doi.org/10.3390/nu9070659.
58. May, J. M., & Harrison, F. E. (2013). Role of vitamin C in the function of the vascular endothelium. *Antioxidants and Redox Signaling, 19*(17), 2068–2083. https://doi.org/10.1089/ars.2013.5205b.
59. Münzel, T., Gori, T., Bruno, R. M., & Taddei, S. (2010). Is oxidative stress a therapeutic target in cardiovascular disease? *European Heart Journal, 31*(22), 2741–2749. https://doi.org/10.1093/eurheartj/ehq396.
60. Panche, A. N., Diwan, A. D., & Chandra, S. R. (2016). Flavonoids: an overview. *Journal of Nutritional Science, 5*. Cambridge University Press. https://doi.org/10.1017/jns.2016.41.
61. Murphy, M. P. (2009). How mitochondria produce reactive oxygen species. *Biochemical Journal, 417*(1), 1–13. https://doi.org/10.1042/BJ20081386.
62. Chen, S. Da, Wang, X. Y., Nisar, M. F., Lin, M., & Zhong, J. L. (2019). Heme oxygenases: cellular multifunctional and protective molecules against UV-induced oxidative stress. In oxidative medicine and cellular longevity (vol, 2019). Hindawi Limited. https://doi.org/10.1155/2019/5416728.
63. Dash, U. C., Bhol, N. K., Swain, S. K., Samal, R. R., Nayak, P. K., Raina, V., Panda, S. K., Kerry, R. G., Duttaroy, A. K., & Jena, A. B. (2024). Oxidative stress and inflammation in the pathogenesis of neurological disorders: Mechanisms and implications. *Acta Pharmaceutica Sinica B*. https://doi.org/10.1016/j.apsb.2024.10.004
64. Finkel, T. & Holbrook, N. J. (2000). Oxidants, oxidative stress and the biology of ageing.
65. Hsueh, Y. M., Lin, P., Chen, H. W., Shiue, H. S., Chung, C. J., Tsai, C. T., Huang, Y. K., Chiou, H. Y., & Chen, C. J. (2005). Genetic polymorphisms of oxidative and antioxidant enzymes and arsenic-related hypertension. *Journal of Toxicology and Environmental Health—Part A, 68*(17–18), 1471–1484. https://doi.org/10.1080/15287390590967414
66. Valgimigli, L. (2023). Lipid Peroxidation and Antioxidant Protection. *Biomolecules, 13*(9). Multidisciplinary Digital Publishing Institute (MDPI). https://doi.org/10.3390/biom13091291.

67. Shome, A., Sarkhel, R., Apoorva, S., Nair, S. S., Chauhan, T. K. S., Bhure, S. K., & Mahawar, M. (2020). Role of protein repair enzymes in oxidative stress survival and virulence of Salmonella. *Annals of Microbiology, 70*(1). BioMed Central Ltd. https://doi.org/10.1186/s13213-020-01597-2.
68. Schumacher, B., Pothof, J., Vijg, J., & Hoeijmakers, J. H. J. (2021). The central role of DNA damage in the ageing process. *Nature, 592*(7856), 695–703. Nature Research. https://doi.org/10.1038/s41586-021-03307-7.
69. pacifici1991 (1). (n.d.).
70. Radak, Z., Zhao, Z., Goto, S., & Koltai, E. (2011). Age-associated neurodegeneration and oxidative damage to lipids, proteins and DNA. *Molecular Aspects of Medicine, 32*(4–6), 305–315. https://doi.org/10.1016/j.mam.2011.10.010
71. Bhat, A. H., Dar, K. B., Anees, S., Zargar, M. A., Masood, A., Sofi, M. A., & Ganie, S. A. (2015). Oxidative stress, mitochondrial dysfunction and neurodegenerative diseases; a mechanistic insight. *Biomedicine and Pharmacotherapy, 74*, 101–110. Elsevier Masson s.r.l. https://doi.org/10.1016/j.biopha.2015.07.025.
72. Chatterjee, S. (2016). Oxidative stress, inflammation, and disease. In *Oxidative stress and biomaterials* (pp. 35–58). Elsevier. https://doi.org/10.1016/B978-0-12-803269-5.00002-4.
73. Moynagh, P. N. (2005). The NF-κB pathway. *Journal of Cell Science, 118*(20), 4589–4592. https://doi.org/10.1242/jcs.02579
74. He, F., Ru, X., & Wen, T. (2020). NRF2, a transcription factor for stress response and beyond. *International Journal of Molecular Sciences, 21*(13), 1–23. MDPI AG. https://doi.org/10.3390/ijms21134777.
75. Egbujor, M. C., Saha, S., Buttari, B., Profumo, E., & Saso, L. (2021). Activation of Nrf2 signaling pathway by natural and synthetic chalcones: a therapeutic road map for oxidative stress. *Expert Review of Clinical Pharmacology, 14*(4), 465–480. Taylor and Francis Ltd. https://doi.org/10.1080/17512433.2021.1901578.
76. Rezatabar, S., Karimian, A., Rameshknia, V., Parsian, H., Majidinia, M., Kopi, T. A., Bishayee, A., Sadeghinia, A., Yousefi, M., Monirialamdari, M., & Yousefi, B. (2019). RAS/MAPK signaling functions in oxidative stress, DNA damage response and cancer progression. *Journal of Cellular Physiology, 234*(9), 14951–14965. Wiley-Liss Inc. https://doi.org/10.1002/jcp.28334.
77. Dubois-deruy, E., Peugnet, V., Turkieh, A., & Pinet, F. (2020). Oxidative stress in cardiovascular diseases. *Antioxidants, 9*(9), 1–15. https://doi.org/10.3390/antiox9090864
78. Madamanchi, N. R., Vendrov, A., & Runge, M. S. (2005). Oxidative stress and vascular disease. *Arteriosclerosis, Thrombosis, and Vascular Biology, 25*(1), 29–38. https://doi.org/10.1161/01.ATV.0000150649.39934.13.
79. Diaconu, C. (2019). Oxidative stress and heart failure. *Archives of the Balkan Medical Union, 54*(2), 219–221. Balkan Medical Union. https://doi.org/10.31688/ABMU.2019.54.2.219.
80. Münzel, T., Gori, T., Keaney, J. F., Maack, C., & Daiber, A. (2015). Pathophysiological role of oxidative stress in systolic and diastolic heart failure and its therapeutic implications. *European Heart Journal, 36*(38), 2555–2564. Oxford University Press. https://doi.org/10.1093/eurheartj/ehv305.
81. Hung, C. W., Chen, Y. C., Hsieh, W. L., Chiou, S. H., & Kao, C. L. (2010a). Ageing and neurodegenerative diseases. *Ageing Research Reviews, 9*(SUPPL). https://doi.org/10.1016/j.arr.2010.08.006.
82. Emerit, J., Edeas, M., & Bricaire, F. (2004). Neurodegenerative diseases and oxidative stress. *Biomedicine and Pharmacotherapy, 58*(1), 39–46. Elsevier Masson SAS. https://doi.org/10.1016/j.biopha.2003.11.004.
83. Inflammation and neurodegenerative diseases 1–4. (2006). https://academic.oup.com/ajcn/article-abstract/83/2/470S/4650314b.
84. Liu, Z., Zhou, T., Ziegler, A. C., Dimitrion, P., & Zuo, L. (2017). Oxidative stress in neurodegenerative diseases: from molecular mechanisms to clinical applications. *Oxidative Medicine and Cellular Longevity, 2017*. Hindawi Limited. https://doi.org/10.1155/2017/2525967.

85. Hung, C. W., Chen, Y. C., Hsieh, W. L., Chiou, S. H., & Kao, C. L. (2010b). Ageing and neurodegenerative diseases. *Ageing Research Reviews, 9*(SUPPL). https://doi.org/10.1016/j.arr.2010.08.006.
86. Lassmann, H., & van Horssen, J. (2016). Oxidative stress and its impact on neurons and glia in multiple sclerosis lesions. *Biochimica et Biophysica Acta—Molecular Basis of Disease, 1862*(3), 506–510. https://doi.org/10.1016/j.bbadis.2015.09.018
87. Quincozes-Santos, A., Santos, C. L., de Souza Almeida, R. R., da Silva, A., Thomaz, N. K., Costa, N. L. F., Weber, F. B., Schmitz, I., Medeiros, L. S., Medeiros, L., Dotto, B. S., Dias, F. R. P., Sovrani, V., & Bobermin, L. D. (2021). Gliotoxicity and glioprotection: the dual role of glial cells. *Molecular Neurobiology, 58*(12), 6577–6592. Springer. https://doi.org/10.1007/s12035-021-02574-9.
88. Klaunig, J. E., & Wang, Z. (2018). Oxidative stress in carcinogenesis. *Current Opinion in Toxicology, 7*, 116–121. Elsevier B.V. https://doi.org/10.1016/j.cotox.2017.11.014.
89. Mena, S., Ortega, A., & Estrela, J. M. (2009). Oxidative stress in environmental-induced carcinogenesis. *Mutation Research—Genetic Toxicology and Environmental Mutagenesis*, 674(1–2), 36–44. https://doi.org/10.1016/j.mrgentox.2008.09.017.
90. Cheung, E. C., & Vousden, K. H. (2022). The role of ROS in tumour development and progression. *Nature Reviews Cancer, 22*(5), 280–297. Nature Research. https://doi.org/10.1038/s41568-021-00435-0.
91. Aggarwal, V., Tuli, H. S., Varol, A., Thakral, F., Yerer, M. B., Sak, K., Varol, M.,Jain, A., Khan, M. A., & Sethi, G. (2019). Role of reactive oxygen species in cancer progression: Molecular mechanisms and recent advancements. *Biomolecules, 9*(11). MDPI AG. https://doi.org/10.3390/biom9110735.
92. Ling, X. C., & Kuo, K. L. (2018). Oxidative stress in chronic kidney disease. *Renal Replacement Therapy, 4*(1). BioMed Central Ltd. https://doi.org/10.1186/s41100-018-0195-2.
93. Kinnula, V. L., Fattman, C. L., Tan, R. J., & Oury, T. D. (2005). Oxidative stress in pulmonary fibrosis: A possible role for redox modulatory therapy. *American Journal of Respiratory and Critical Care Medicine, 172*(4), 417–422. https://doi.org/10.1164/rccm.200501-017PP.
94. Barnes, P. J. (2022). Oxidative stress in chronic obstructive pulmonary disease. *Antioxidants, 11*(5). MDPI. https://doi.org/10.3390/antiox11050965
95. Simon, J. A. (1992). Vitamin C and cardiovascular disease: a review. *Journal of the American College of Nutrition, 11*(2), 107–125. https://doi.org/10.1080/07315724.1992.12098232
96. Münzel, T., Gori, T., Bruno, R. M., & Taddei, S. (2010b). Is oxidative stress a therapeutic target in cardiovascular disease? *European Heart Journal, 31*(22), 2741–2749. https://doi.org/10.1093/eurheartj/ehq396.
97. Gross, M. (2004). Flavonoids and cardiovascular disease. *Pharmaceutical Biology, 42*(SUPPL), 21–35. https://doi.org/10.1080/13880200490893483.
98. Natural antioxidants in oxidative stress. (n.d.).
99. Pietta, P. G. (2000). Flavonoids as antioxidants. *Journal of Natural Products, 63*(7), 1035–1042. https://doi.org/10.1021/np9904509.
100. Rao, A. V., & Balachandran, B. (2002). Role of oxidative stress and antioxidants in neurodegenerative diseases. *Nutritional Neuroscience, 5*(5), 291–309. https://doi.org/10.1080/1028415021000033767.
101. Olufunmilayo, E. O., Gerke-Duncan, M. B., & Holsinger, R. M. D. (2023). Oxidative stress and antioxidants in neurodegenerative disorders. *Antioxidants, 12*(2). MDPI. https://doi.org/10.3390/antiox12020517.
102. Choi, J. J., Choi, J. W., Kang, C. D., Chen, X., Wu, C. F., Ko, K. H., & Kim, W. K. (2007). Hydrogen peroxide induces the death astrocytes through the down-regulation of the constitutive nuclear factor-kappaB activity. *Free Radical Research, 41*(5), 555–562. https://doi.org/10.1080/10715760601173010
103. Navarro-Yepes, J., Zavala-Flores, L., Anandhan, A., Wang, F., Skotak, M., Chandra, N., Li, M., Pappa, A., Martinez-Fong, D., Del Razo, L. M., Quintanilla-Vega, B., & Franco, R. (2014). Antioxidant gene therapy against neuronal cell death. *Pharmacology and Therapeutics, 142*(2), 206–230. Elsevier Inc. https://doi.org/10.1016/j.pharmthera.2013.12.007.

104. Perry, T. L., Wee Yong, N., Hansen, S., Jones, K., Bergeron, C., Foulks I, J. G., & Wright, J. M. (1987). 7-Tocopherol and//-carotene do not protect marmosets against the dopaminergic neurotoxicity of N-methyl-4-phenyl-1,2,3,6-tetrahydropyridine. *Journal of the Neurological Sciences, 81.*
105. Wee Yong, V., Perry, T. L., & Krisman, A. A. (1986). Depletion ofglutathione in brainstem of mice caused by n-Methyl-4-Phenyl-I,2,3,6-Tetrahydropyridine is prevented by antioxidant pretreatment.
106. Zhao 2005. (n.d.).
107. Nanjo, F., Goto, K., Seto, R., Suzuki, M., Sakai, M., & Hara, Y. (1996). Brief Communication scavenging effects of tea catechins and their derivatives on l,l-diphenyl-2-picrylhydrazyl radical. *Free Radical Biology & Medicine, 21*(6).
108. Nie, G., Wei, T., Shen, S., & Zhao, B. (2001). Polyphenol protection of DNA against damage.
109. Guo, Q., Rimbach, G., Moini, H., Weber, S., & Packer, L. (2002). ESR and cell culture studies on free radical-scavenging and antioxidant activities of isoflavonoids. *Toxicology, 179.* www.elsevier.com/locate/toxicol.
110. Ohigashi, T., Ueno, M., Nonaka, S., Nakanoma, T., Furukawa, Y., Deguchi, N., & Murai, M. (2000). Tyrosine kinase inhibitors reduce BCL-2 expression and induce apoptosis in androgen-dependent cells. http://www.ajpcell.org.
111. Zeng, H., Chen, Q., & Zhao, B. (2004). Genistein ameliorates β-amyloid peptide (25–35)-induced hippocampal neuronal apoptosis. *Free Radical Biology and Medicine, 36*(2), 180–188. https://doi.org/10.1016/j.freeradbiomed.2003.10.018
112. Rahimi, R., Nikfar, S., Larijani, B., & Abdollahi, M. (2005). A review on the role of antioxidants in the management of diabetes and its complications. *Biomedicine and Pharmacotherapy, 59*(7), 365–373. https://doi.org/10.1016/j.biopha.2005.07.002
113. Maritim, A. C., Moore, B. H., Sanders, R. A., & Watkins Iii, J. B. (n.d.). Effect of melatonin on phase I and II biotransformation enzymes in streptozotocin-induced diabetic rats.
114. Smith 2004. (n.d.).
115. Ruhnau, K.-J., Meissner, H. P., Finn, J.-R., Reljanovic, M., Lobisch, M., Schu È Tte, K., Nehrdich, D., Tritschler, H. J., Mehnert, H., & Ziegler, D. (n.d.). 1040±1043 (1999) ã 1999 British Diabetic Association. *Diabetic Medicine, 16.*
116. Thakur, P., Kumar, A., & Kumar, A. (2018). Targeting oxidative stress through antioxidants in diabetes mellitus. *Journal of Drug Targeting, 26*(9), 766–776. Taylor and Francis Ltd. https://doi.org/10.1080/1061186X.2017.1419478.
117. Calabrese, V., Cornelius, C., Leso, V., Trovato-Salinaro, A., Ventimiglia, B., Cavallaro, M., Scuto, M., Rizza, S., Zanoli, L., Neri, S., & Castellino, P. (2012). Oxidative stress, glutathione status, sirtuin and cellular stress response in type 2 diabetes. *Biochimica et Biophysica Acta—Molecular Basis of Disease, 1822*(5), 729–736. https://doi.org/10.1016/j.bbadis.2011.12.003
118. Heller, R., Unbehaun, A., Schellenberg, B., Mayer, B., Werner-Felmayer, G., & Werner, E. R. (2001). L-ascorbic acid potentiates endothelial nitric oxide synthesis via a chemical stabilization of tetrahydrobiopterin. *Journal of Biological Chemistry, 276*(1), 40–47. https://doi.org/10.1074/jbc.M004392200
119. Sánchez-Valle, V., Chávez-Tapia, N. C., Uribe, M., & Méndez-Sánchez, N. (2012). Send orders of reprints at reprints@benthamscience.org role of oxidative stress and molecular changes in liver fibrosis: a review. *Current Medicinal Chemistry, 19.*
120. Li, S., Tan, H. Y., Wang, N., Zhang, Z. J., Lao, L., Wong, C. W., & Feng, Y. (2015). The role of oxidative stress and antioxidants in liver diseases. *International Journal of Molecular Sciences, 16*(11), 26087–26124. MDPI AG. https://doi.org/10.3390/ijms161125942.
121. Li, G., Ye, Y., Kang, J., Yao, X., Zhang, Y., Jiang, W., Gao, M., Dai, Y., Xin, Y., Wang, Q., Yin, Z., & Luo, L. (2012). L-Theanine prevents alcoholic liver injury through enhancing the antioxidant capability of hepatocytes. *Food and Chemical Toxicology, 50*(2), 363–372. https://doi.org/10.1016/j.fct.2011.10.036
122. Egea, J., González-Rodríguez, Á., Gómez-Guerrero, C., & Moreno, J. A. (2019). Editorial: role of Nrf2 in disease: novel molecular mechanisms and therapeutic approaches. *Frontiers in Pharmacology, 10.* Frontiers Media S.A. https://doi.org/10.3389/fphar.2019.01149.

123. Rajappa, R., Sireesh, D., Salai, M. B., Ramkumar, K. M., Sarvajayakesavulu, S., & Madhunapantula, S. R. V. (2019). Treatment with naringenin elevates the activity of transcription factor Nrf2 to protect pancreatic β-cells from streptozotocin-induced diabetes in vitro and in vivo. *Frontiers in Pharmacology, 9*(JAN). https://doi.org/10.3389/fphar.2018.01562.
124. Mayer, C., Riera-Ponsati, L., Kauppinen, S., Klitgaard, H., Erler, J. T., & Hansen, S. N. (2024). Targeting the NRF2 pathway for disease modification in neurodegenerative diseases: mechanisms and therapeutic implications. *Frontiers in Pharmacology, 15*. Frontiers Media SA. https://doi.org/10.3389/fphar.2024.1437939.
125. Drummond, G. R., Selemidis, S., Griendling, K. K., & Sobey, C. G. (2011). Combating oxidative stress in vascular disease: NADPH oxidases as therapeutic targets. *Nature Reviews Drug Discovery, 10*(6), 453–471. https://doi.org/10.1038/nrd3403.
126. Xu, Y., Liu, H., & Song, L. (2020). Novel drug delivery systems targeting oxidative stress in chronic obstructive pulmonary disease: a review. *Journal of Nanobiotechnology, 18*(1). BioMed Central Ltd. https://doi.org/10.1186/s12951-020-00703-5.
127. Lee, Y. H., Kuk, M. U., So, M. K., Song, E. S., Lee, H., Ahn, S. K., Kwon, H. W., Park, J. T., & Park, S. C. (2023). Targeting mitochondrial oxidative stress as a strategy to treat aging and age-related diseases. *Antioxidants, 12*(4). MDPI. https://doi.org/10.3390/antiox12040934.
128. Prasad, N., Ramteke, P., Dholia, N., & Yadav, U. C. S. (2017). Therapeutic interventions to block oxidative stress-associated pathologies. In *Immunity and inflammation in health and disease: emerging roles of nutraceuticals and functional foods in immune support* (pp. 341–362). Elsevier. https://doi.org/10.1016/B978-0-12-805417-8.00027-5.
129. Aseervatham, G. S. B., Sivasudha, T., Jeyadevi, R., & Arul Ananth, D. (2013). Environmental factors and unhealthy lifestyle influence oxidative stress in humans—An overview. *Environmental Science and Pollution Research International, 20*(7), 4356–4369. https://doi.org/10.1007/s11356-013-1748-0.
130. Ho H, & H HO. (n.d.). H i _ H H ECG 4-a.
131. Roberts, C. K., Vaziri, N. D., & Barnard, R. J. (2002). Effect of diet and exercise intervention on blood pressure, insulin, oxidative stress, and nitric oxide availability. *Circulation, 106*(20), 2530–2532. https://doi.org/10.1161/01.CIR.0000040584.91836.0D
132. Global Air Quality and Pollution Hajime Akimoto. (n.d.). http://science.sciencemag.org/
133. Simkhovich, B. Z., Kleinman, M. T., & Kloner, R. A. (2008). Air pollution and cardiovascular injury. Epidemiology, toxicology, and mechanisms. *Journal of the American College of Cardiology 52*(9), 719–726. https://doi.org/10.1016/j.jacc.2008.05.029
134. Dhama, K., Latheef, S. K., Dadar, M., Samad, H. A., Munjal, A., Khandia, R., Karthik, K., Tiwari, R., Yatoo, M. I., Bhatt, P., Chakraborty, S., Singh, K. P., Iqbal, H. M. N., Chaicumpa, W., & Joshi, S. K. (2019). Biomarkers in stress related diseases/disorders: diagnostic, prognostic, and therapeutic values. *Frontiers in Molecular Biosciences, 6*. Frontiers Media S.A. https://doi.org/10.3389/fmolb.2019.00091.
135. Drucker, E., & Krapfenbauer, K. (2013). Pitfalls and limitations in translation from biomarker discovery to clinical utility in predictive and personalised medicine. http://www.epmajournal.com/content/4/1/7.
136. Nebert, D. W., Zhang, G., & Vesell, E. S. (2013). Genetic risk prediction: individualized variability in susceptibility to toxicants. *Annual Review of Pharmacology and Toxicology, 53*, 355–375. https://doi.org/10.1146/annurev-pharmtox-011112-140241.
137. Arora, M. (2024). Investigating genetic variation in disease susceptibility. *Universal Research Reports, 11*(2), 79–92. https://doi.org/10.36676/urr.v11.i2.10.
138. Obeagu, E. I., & Kanu, S. N. (2024). Understanding the genetic susceptibility to aplastic anemia in HIV. *Elite Journal of Laboratory Medicine, 2*(9). https://epjournals.com/journals/EJLM.
139. Halliwell, B. (2012). Free radicals and antioxidants: Updating a personal view. *Nutrition Reviews, 70*(5), 257–265. https://doi.org/10.1111/j.1753-4887.2012.00476.x
140. Chu, F. F., Doroshow, J. H., & Esworthy, R. S. (1993). Expression, characterization, and tissue distribution of a new cellular selenium-dependent glutathione peroxidase. *GSHPx-GI. Journal of Biological Chemistry, 268*(4), 2571–2576. https://doi.org/10.1016/s0021-9258(18)53812-6

Elucidating the Impact of Reactive Molecules on Cell Signaling Networks to Illuminate Their Contribution to the Etiology and Progression of Human Diseases

Harsath Ram Govindan Radha Krishnan, Pandian Chithamparam, and Azhaguchamy Muthukumaran

Abstract Reactive species, including reactive oxygen species (ROS), reactive nitrogen species (RNS), and reactive sulfur species (RSS), are highly reactive molecules essential for cellular metabolism. Reactive species are essential for normal cellular function, but when their levels become disrupted, it can result in oxidative stress, inflammation, and cellular damage. In this review, we examine the intricate functions of reactive species in cellular metabolism, emphasizing their role in signaling pathways and the progression of diseases. This chapter examines the processes that regulate the formation and elimination of reactive molecules, along with their interplay with essential cellular components such as proteins, lipids, and genetic material. Additionally, we examine the effects of reactive species on various significant health issues, such as cancers, neurological conditions, and cardiovascular disease. We also explore the therapeutic possibilities of focusing on reactive species for the treatment and prevention of diseases.

Keywords Reactive species · Oxidative stress · Free radicals · Antioxidants · Reactive oxygen species

1 Introduction

Oxygen, nitrogen, sulfur, and other high-reactivity compounds are indispensable for numerous biological activities and significantly contribute to metabolic processes [1]. Reactive oxygen compounds, such as superoxides, hydrogen peroxide, and hydroxyl radicals, are crucial for immune responses, cellular communication, and redox regulation in typical cellular metabolism [2]. Reactive oxygen species lead to oxidative

H. R. G. R. Krishnan · P. Chithamparam · A. Muthukumaran (✉)
Department of Biotechnology, Kalasalingam Academy of Research and Education, Krishnankoil, Tamil Nadu, India
e-mail: a.muthukumaran@klu.ac.in

K. Kasinathan et al. (eds.), *Recent Advances in Oxidative Stress Associated Chronic Diseases Volume 1*, https://doi.org/10.1007/978-981-96-9670-3_11

imbalances, resulting in cellular damage and interfering with vital biomolecules, which may contribute to diseases such as cancer, aging, coronary artery disease, and dementia [3]. Cellular imbalance arises when oxygen-reactive compounds surpass the protective systems of the body, resulting in damage. The regulation of reactive oxygen species is essential for the maintenance of health. The cellular defense system of the body employs enzymes such as catalase and superoxide dismutase to transform ROS into nontoxic molecules, whereas non-enzymatic antioxidants such as vitamins C and E mitigate ROS through chemical interactions [4]. Nitric oxide synthase is an essential cellular enzyme that generates reactive nitrogen species (RNS), comprising nitric oxide, nitrite, and peroxynitrite, which play a crucial role in numerous physiological functions [5]. Nitric oxide, a principal reactive nitrogen species (RNS), serves as a powerful vasodilator that modulates blood flow and blood pressure, while also engaging in immune response and neurotransmission. Nitrogen-containing reactive compounds have the potential to disturb cellular equilibrium, induce oxidative stress, and damage tissues, thereby elevating the risk of chronic illnesses such as cardiovascular diseases, cancer, and neurological disorders [6]. RNS can also react with ROS, forming reactive nitrogen–oxygen species and amplifying oxidative damage. Antioxidant defenses, such as glutathione and vitamin C, regulate RNS levels, mitigating potential harm [7]. Highly reactive sulfur-containing compounds are produced through specific biochemical pathways, and they significantly influence various cellular processes, including intercellular communication, protection against oxidative stress, and the regulation of inflammatory responses. RSS include hydrogen sulfide (H_2S), sulfide (S_2^-), and thiol radicals (RS), which are produced by various cellular enzymes, such as cystathionine gamma-lyase and cystathionine beta-synthase [8]. Hydrogen sulfide, a primary RSS, acts as a potent antioxidant, protecting cells from oxidative stress and damage while regulating inflammation, apoptosis, and cellular metabolism. RSS play a significant role in several vital bodily functions, such as heart health, cognitive impairment, and host defense [9]. Additionally, RSS have therapeutic potential in diseases such as tumor growth, atherosclerosis, and dementia, where oxidative stress and inflammation play critical roles.

2 Reactive Oxygen Species (ROS)

ROS are chemically reactive species, including free radicals that originate from molecular oxygen, which possess instability and high reactivity due to the presence of unpaired electrons [10]. Oxygen and nitrogen are the primary elements in biological systems that produce highly reactive molecules, but other elements can also influence their production [11]. Under standard conditions, the main oxygen-free radical is the superoxide anion (O_2^-), and mitochondria are the main source. The inefficiency associated with electron transfer in enzymes of the respiratory chain results in electron leakage to molecular oxygen, particularly from complexes I and III, thereby contributing to the production of O_2^- [12]. The rate of production for complex IV rises with elevated levels of oxygen and glucose, similar to diabetes, along with hypoxic situations caused by the accumulation of electrons due to restricted oxygen availability. Roughly 2% of the O_2 utilized by mitochondria is transformed into O_2^- rather than H_2O [13]. Owing to its charge, O_2^-, is unable to traverse the membrane and is retained within the mitochondrial matrix, whereas superoxide may be generated by electron leaks originating from the ER. The endoplasmic reticulum (ER) generates 25% of oxygen (O_2) within cells via oxidative processes, which encompass protein folding and ER stress, during which several efforts are undertaken to refold proteins that are misfolded [14]. Superoxide production is enabled by enzymatic mechanisms such as cytochrome P450 and NADPH oxidase, with significant elevations noted during gestational periods. The generation of reactive oxygen species can be initiated by external sources, including growth hormones, pharmaceutical agents, and detrimental compounds [15]. The xanthine dehydrogenase enzyme plays a vital role in pathological conditions, transforming hypoxanthine, xanthine, and purines into uric acid while utilizing NAD^+ as the electron acceptor [16]. During the hypoxic phase, the breakdown of ATP results in an elevation of hypoxanthine levels, which subsequently enhances enzyme activity during the reperfusion phase of ischemia–reperfusion injury. Superoxide dismutase enzymes diminish the reactivity of superoxide by transforming it into the less potent hydrogen peroxide (H_2O_2). It belongs to the group of reactive oxygen species (ROS) because it is significant in both the production and elimination of free radicals. Its non-polarity nature makes it easier for it to pass across cell membranes and function as a mediator molecule in different signaling cascades, which improves signal transmission inside cells [17]. Glutathione peroxidase and catalase are essential enzymes that work together to efficiently convert H_2O_2 into H_2O through a regulated detoxification process. The coordination of antioxidant enzymes is essential in preventing the production of toxic hydroxyl ions resulting from an imbalance in O_2^- and H_2O_2 levels, which interact with biological molecules via diffusion [17]. Excessive superoxide accumulation can lead to a reaction with nitric oxide (NO^-) to create peroxynitrite ($ONOO^-$).

2.1 *Sources of ROS*

Oxygen-containing reactive molecules known as reactive oxygen species (ROS) are formed through an electron transfer reaction involving oxygen, leading to the creation of highly reactive and unstable free radical species, such as superoxide ($O_2^{\cdot-}$), hydroxyl radical ($OH^{\cdot}$), and hydrogen peroxide (H_2O_2) [1]. ROS are produced from internal and external sources (Fig. 1). Endogenous sources include the mitochondrial electron transport chain, NADPH oxidases, xanthine oxidase, and other enzymes, which produce ROS as byproducts of normal cellular metabolism [3]. Additionally, immune cells, such as neutrophils and macrophages, produce ROS through the activation of NADPH oxidase during respiratory bursts. The generation of reactive oxygen species (ROS) can be triggered by various external influences, including exposure to ionizing radiation, ultraviolet light, and environmental toxins such as air pollution, cigarette smoke, and certain chemicals like pesticides and heavy

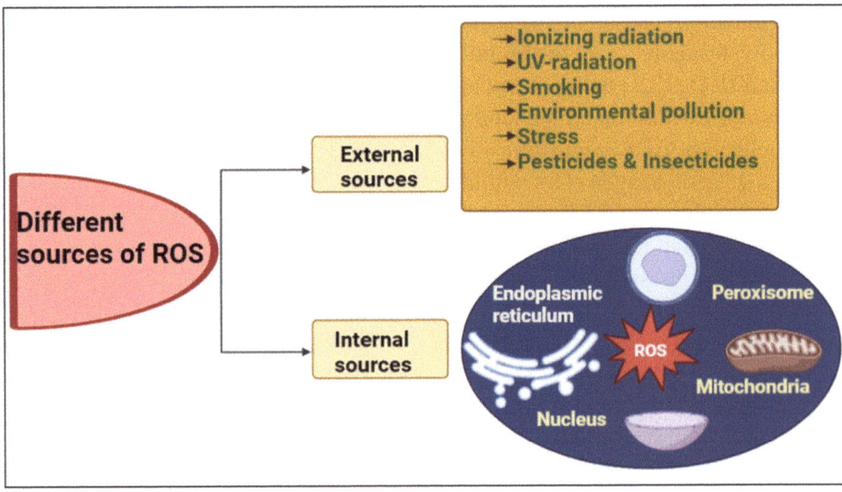

Fig. 1 Various sources of reactive oxygen species

metals [18]. Physiological sources of ROS include exercise-induced oxidative stress, ischemia–reperfusion injury, and inflammation. Furthermore, ROS can also be generated through cellular signaling pathways, such as cytokine signaling and growth factor signaling. Pathological sources of ROS include cancer cells, neurological disorders, cardiovascular diseases, insulin deficiency, and obesity [4].

2.2 Types of ROS

ROS are a broad category that includes a wide variety of highly reactive oxygen-based molecules, encompassing superoxide ($O_2^{\cdot-}$), hydrogen peroxide (H_2O_2), hydroxyl radical ($OH^{\cdot}$), singlet oxygen ($_1O^2$), peroxyl radical ($ROO^{\cdot}$), alkoxyl radical ($RO^{\cdot}$), nitrogen dioxide ($NO_2^{\cdot}$), and hypochlorous acid (HOCl) [3] (Fig. 2). Superoxide, which is generated by the mitochondrial electron transport chain and NADPH oxidases, is a relatively weak oxidizing agent that readily transforms into hydrogen peroxide, a stable and membrane-permeable signaling molecule. The Fenton reaction produces hydroxyl radical, a harmful reactive species, while singlet oxygen oxidizes biomolecules and aids in photodynamic therapy. Peroxyl and alkoxyl radicals undermine the integrity of cellular membranes [19]. Myeloperoxidase generates hypochlorous acid, an effective antibiotic, while nitric oxide synthase and NADPH oxidases yield nitrogen dioxide, which oxidizes biomolecules and leads to inflammation and cancer.

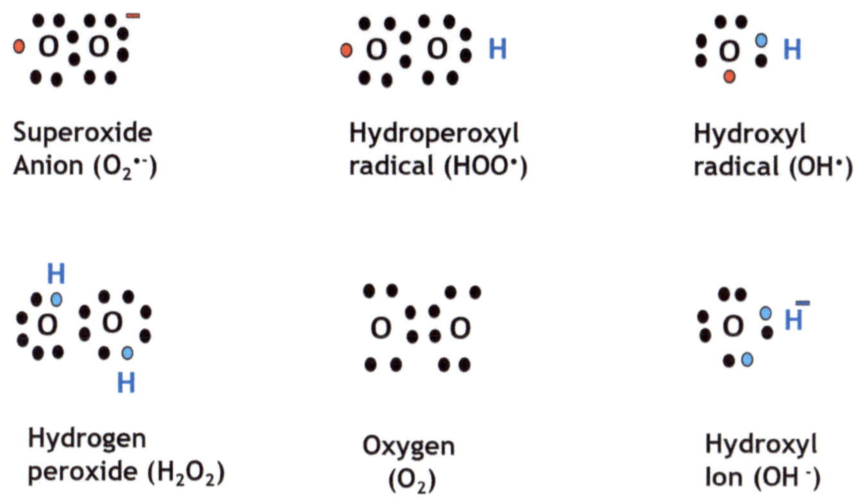

Fig. 2 Types of reactive oxygen species

2.3 Production of ROS

The body's normal metabolic processes give rise to free radicals and other ROS formation. Within cells, these reactive molecules are generated relentlessly through various chemical reactions, including those facilitated by enzymes and those that occur spontaneously [20]. Enzymatic reactions in the respiratory chain, phagocytosis, prostaglandin synthesis, and the cytochrome P-450 system, as well as non-enzymatic reactions and ionizing events, all contribute to the generation of free radicals [21]. Mc Cord and Fridovich identified superoxide dismutase (SOD) in 1969, highlighting its significance in physiological processes and enzymatic roles [22]. According to the study, hydroxyl radicals facilitate the production of cGMP and the stimulation of guanylate cyclase, indicating that organisms have evolved to efficiently utilize free radicals in biological functions [23]. Several endogenous free radicals and ROS sources include arachidonate pathways, inflammation, ischemia/reperfusion injury, mitochondria, phagocytosis, peroxisomes, and xanthine oxidase. The mitochondria, the primary source of ATP in mammalian cells, are crucial for human survival and also produce a significant amount of superoxide [24]. Energy conversion produces superoxide oxygen-free radicals, which are associated with disease advancement, occurring close to the location of in vivo generation of OH [25]. Peroxisomes play an essential role in the utilization of oxygen and in metabolic processes that depend on oxygen, generating hydrogen peroxide via the use of oxygen, which is accountable for the oxidation of different molecules [26] (Fig. 3). Catalase, an organelle enzyme, breaks down H_2O_2 and prevents accumulation [27]. Peroxisomes control enzyme levels, but dysfunction can cause oxidative stress in the cytosol [28]. Phagocytic cells, such as neutrophils, are capable of sensing foreign particles and triggering a series of respiratory burst reactions upon receiving a stimulus [29]. The enzyme

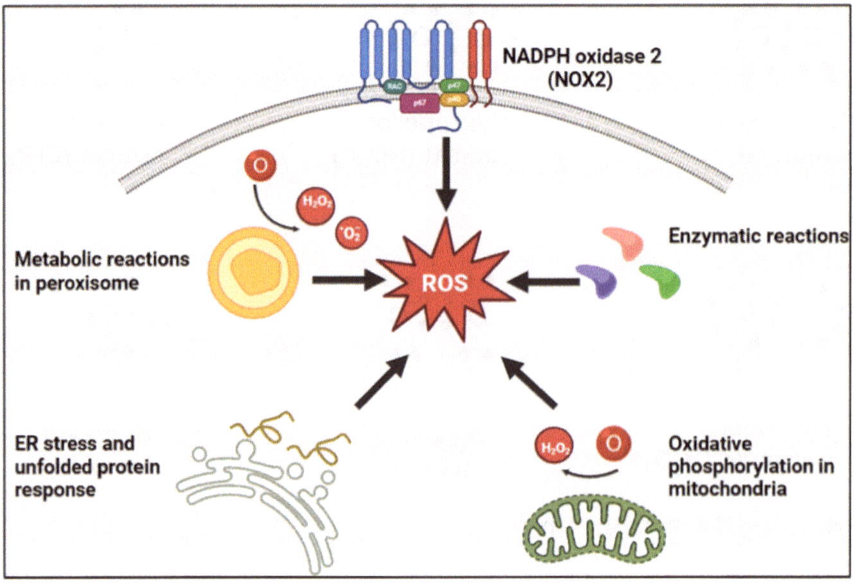

Fig. 3 Production of reactive oxygen species

NADPH oxidase in neutrophils is crucial for the generation of superoxide radicals (O_2^-), which initiates the respiratory burst, an essential step in the eradication of bacterial infections. Lipid peroxidation, a free radical-based process, can cause significant biochemical damage by oxidizing polyunsaturated fatty acids (PUFAs) in cell membranes [24]. The hydroxyl radical produces free radicals through the removal of hydrogen atoms, thereby facilitating the formation of lipid radicals. The addition of oxygen results in the creation of reactive peroxyl radicals, which subsequently attack fatty acids, leading to the formation of lipid hydroperoxide (LOOH) along with an additional radical [30]. Lipid peroxidation persists, leading to the formation of compounds such as isopropanes, malonaldehyde, and alkanes, which act as indicators and have been validated in neurodegenerative diseases [26].

2.4 *Targets of ROS*

2.4.1 Atomic Targets

ROS can establish chemical bonds with particular atoms in complex biomolecules, and these interactions are reversible in some cases, with a restricted range of reactivity within the molecule. Thus, ROS responses are characterized by a lack of specificity attributable to complementarity-based molecular interactions, such as the binding of insulin to its receptor, which are optimal for transmitting signals through a defined pathway [31]. On the other hand, based on the source and quantity, ROS interactions with specific macromolecular atoms could transmit signals to multiple pathways, influencing various subcellular areas or an extensive cellular region [32]. These variations can lead to fluctuations in ROS levels from stimulatory to inhibitory effects, which can change the metabolic profile of the cell via various signaling pathways. Sulfur is a unique biomolecule that is highly reactive and crucial to cellular signaling pathways [33]. However, ROS primarily interact with a specific group of sulphur atoms found in the side chains of methionine or cysteine residues in peptides or proteins rather than with all sulfur atoms. Most ROS-reactive cysteines in proteins are found in environments that support thiolate's involvement in active site reactions [34]. The activity of specific cysteine thiol groups is subtly altered by the presence of nearby amino acid side chains with low pKa values, although the full complexity of factors governing their behavior remains to be elucidated.

2.4.2 Molecular Targets

Proteins

Some proteins regulated by ROS include ataxia–telangiectasia mutated (ATM) kinase, calmodulin-dependent kinase II, caspases, epidermal growth factor receptor, forkhead box O (FOXO) transcription factors, GTP cyclohydrolase, glucose uptake regulator, guanylyl cyclase, heat-shock proteins, metalloproteinases, phosphatase, histone deacetylases, phosphatase and tensin homolog (PTEN), protein kinase B, protein kinase G-Iα, prolyl hydroxylase, zinc-finger proteins, α1-antitrypsin, α2-macroglobulin, secretory leukocyte protease inhibitor, and the ryanodine receptor [35]. ROS oxidation of the cysteine sulfhydryl group in phosphatases facilitates signal transmission, activating kinases and facilitating ligand binding. Elevated ROS or temporary inactivation may initiate this process [36]. The whole proteome analysis indicates that ROS regulate phosphatases, leading to phosphorylation. Protein cysteine residues possess the capability to detect reactive oxygen species (ROS) and regulate their concentrations through feedback mechanisms [37]. For instance, the intermolecular oxidation of the transcription factor FOXO4 leads to the protein associating with p300/CBP acetylase and the acetylation of lysine residues. In the same way, the oxidation of cysteine residues by ATM (ataxia–telangiectasia mutated) kinase enhances glucose flow through the pentose phosphate pathway, consequently raising NADPH levels [38]. Enzymes involved in ROS metabolism, such as glutathione reductase, methionine sulfoxide reductases, and peroxiredoxins, depend on NADPH for the reduction process and facilitate the oxidation of glutathione and thioredoxin [39]. The precise control of ROS levels is essential for the effective function of ATM in both hematopoiesis and angiogenesis. ROS promote transcription activation through various mechanisms (Fig. 4). For example, the interaction between superoxide anion and H_2O_2 with iron–sulfur redox centers facilitates the activation of the bacterial transcription factors OxyR and SoxR, resulting in the

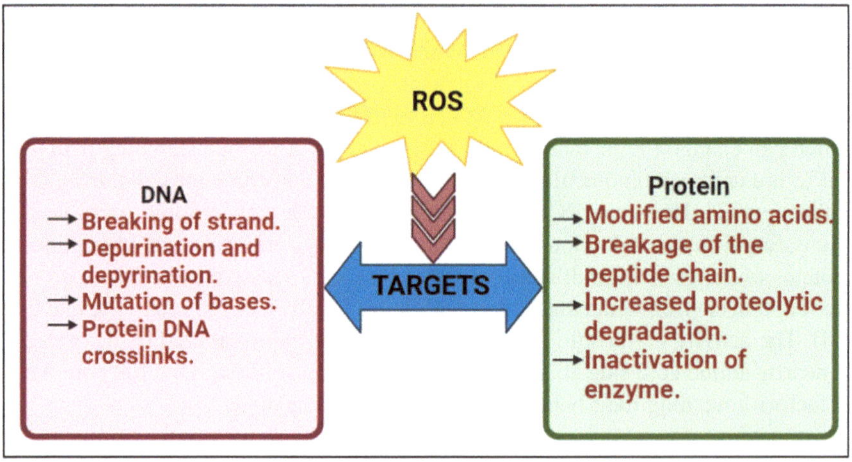

Fig. 4 Molecular targets of ROS

production of antioxidant enzymes [40]. Furthermore, ROS-induced protein phosphorylation activates the Jun or NF-κB transcription factors, thereby facilitating their translocation into the nucleus through ubiquitylation and degradation [14].

DNA

Instead of depending on DNA-binding proteins, mitochondrial ROS have the ability to directly oxidize DNA, which initiates transcription [41]. ROS from hypoxic mitochondria oxidize the VEGF promoter, leading to mitochondria migration along microtubules in pulmonary artery endothelial cells and their aggregation near the nucleus [42]. ROS produced by mitochondria induce oxidative modifications in the VEGF promoter area, facilitating the recruitment of HIF1α. Thus, ROS are highlighted as a significant mediator within the physiological signaling cascade that occurs between the mitochondria and the nucleus. Another instance is that DNA oxidation regulates genes through activating receptors such as the androgen receptor, activator protein 1, estrogen receptor, thyroxin receptor, and retinoic acid receptor, which necessitates site-specific DNA breaks in target gene-regulatory regions mediated by topoisomerase IIb [43]. H_2O_2 is released by the lysine-specific histone demethylase 1 when the estrogen receptor triggers it. The resultant H_2O_2 forms 8-oxoguanine by locally oxidizing bases in gene-regulatory areas. The enzyme 8-oxoguanine activates 8-oxoguanine DNA glycosylase 1 (OGG1) to induce single-stranded DNA breaks [44]. The DNA breaks activate topoisomerase IIb, which twists the DNA to repair damage and allows the transcription initiation complex to access specific promoters. This way, ROS from various sources, such as cytosol, endosomal membranes, mitochondria, NOXs, and plasma, can influence transcription factors.

In contrast, ROS from nuclear flavoenzymes or perinuclear mitochondria can target DNA.

2.5 *Effect of ROS on Cellular Signaling*

ROS play a wide range of roles in signaling as a secondary messenger at normal concentrations, significantly impacting physiological responses. The increase in ROS has been correlated with various growth factors. Various external factors, such as cytokines, epidermal growth factor (EGF), platelet-derived growth factor (PDGF), and tumor necrosis factor (TNF), promote the production of ROS [45]. ROS signaling is primarily triggered by the oxidation of the thiol group in cysteine residues, an amino acid with a low pKa. ROS have been linked to elevated activity of mitogen-activated protein kinase (MAPK), either by oxidizing cysteine residues or activating tyrosine kinases [46]. The mitogen-activated protein kinase (MAPK) pathway, made up of three kinases, regulates biological functions including cell growth, differentiation, survival, and apoptosis through extracellular signals. Extracellular signal-regulated protein kinases 1 and 2 (ERK1/2), Jun N-terminal kinase (JNK; also known as stress-activated protein kinase, SAPK), and p38 kinase are the three main MAPK pathways that regulate different cellular functions [47]. These enzymes that phosphorylate serine/threonine residues are triggered by external signals (as mentioned before) or by environmental pressure. ROS impact other protein tyrosine phosphatases (PTPs) and protein kinases (PKs) that are responsive to changes in redox levels [48]. AKT, mammalian target of rapamycin (mTOR), phosphoinositide 3-kinase (PI3K), and phosphatase and tensin homolog on chromosome 10 (PTEN) are the key proteins involved in various biological processes [49]. Growth factors like platelet-derived growth factor (PDGF), epidermal growth factor (EGF), and ROS produced by NOX and mitochondria activate PI3K; whereas, ROS inhibit PI3K's primary inhibitor, PTEN [50]. Advances in proteomics have revealed more than 300 redox-responsive proteins, highlighting the substantial influence of ROS on cellular activity. ROS play a crucial role in regulating transcription factors linked to growth and metabolism, including hypoxia-inducible factor 1α (HIF-1α), nuclear factor kappa B (NF-κB), heat shock factor 1 (HSF1), nuclear factor erythroid 2-related factor 2 (Nrf2), and p53, all of which respond to changes in redox signaling. Within the cytosol, nuclear factor erythroid 2-related factor 2 (Nrf2) promotes the ubiquitination and proteasomal degradation of Keap1 [51]. Moreover, ROS oxidize key reactive cysteines on Keap1, causing Keap1-Nrf2 dissociation and subsequent Nrf2 nuclear translocation [52]. ROS are essential in the process of signal transduction and greatly influence cellular pathways by altering cell physiology, metabolism, and proliferation at low to moderate levels.

2.6 Role of ROS in Diseases

ROS significantly contribute to the onset and advancement of a multitude of disorders. In cancer, ROS induce genetic instability, promote tumor growth, and enhance metastasis. At the same time, in neurodegenerative diseases like Alzheimer's, Parkinson's, and Huntington's, ROS damage neuronal cells, contributing to cognitive decline and neuroinflammation [53] (Fig. 5). Additionally, ROS promote atherosclerosis, hypertension, and cardiac remodeling in cardiovascular diseases and impairs insulin signaling, contributing to pancreatic beta-cell damage and complications in diabetes [9]. Obesity-related ROS induce inflammation, insulin resistance, and metabolic dysfunction, whereas in inflammatory diseases, such as arthritis, asthma, and inflammatory bowel disease, ROS exacerbate inflammation and tissue damage [54]. Furthermore, ROS accumulate with age, contributing to cellular senescence and organismal decline. ROS are generated during an immune response to eliminate pathogens but potentially damage host cells in infectious diseases [55]. Autoimmune diseases, like rheumatoid arthritis and lupus, involve ROS-mediated tissue damage and autoimmune responses. On the other hand, ROS-induced lung tissue destruction is a hallmark of respiratory disorders such as pulmonary fibrosis and chronic obstructive pulmonary disease (COPD) [56]. Moreover, ROS-mediated renal tissue injury intensifies kidney disorders such as chronic kidney disease and kidney failure, impairing gut barrier functionality and resulting in gastrointestinal disorders such as irritable bowel syndrome (IBS). Finally, ROS-induced hepatic injury and retinal tissue injury are associated with numerous ailments, including non-alcoholic fatty liver disease (NAFLD) and age-related macular degeneration (AMD), and play a role in age-related macular degeneration and diabetic retinopathy [57].

2.7 Therapeutic Applications

ROS are promising anticancer agents due to their ability to induce apoptosis, suppress cell division, and enhance the sensitivity of cancer cells to radiation and chemotherapy [58]. ROS-driven antimicrobial treatments employ ROS to disrupt bacterial membranes and DNA, creating a novel method for addressing antibiotic-resistant infections. Additionally, ROS encourage tissue regeneration and repair by stimulating growth factors and cytokines, which assist in wound healing [55]. ROS also have anti-inflammatory effects, reducing tissue damage and promoting the resolution of inflammation. Moreover, ROS-based therapies are being investigated in the context of neurodegenerative diseases, such as Alzheimer's and Parkinson's, where oxidative stress contributes to disease progression [53]. Other potential applications include cardiovascular disease treatment, immune system modulation, and gene therapy.

Selenium nanoparticles (SeNPs) provide therapeutic advantages in cardiovascular diseases owing to their multifaceted actions, including the reduction of oxidative

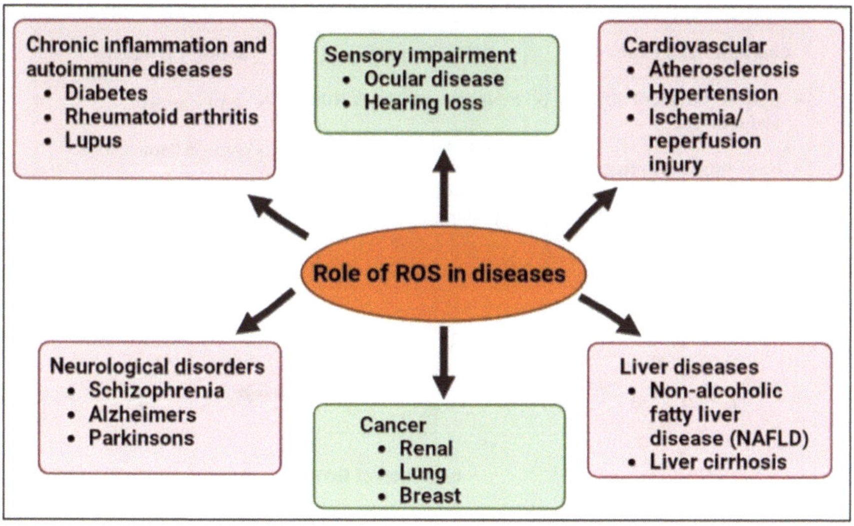

Fig. 5 Role of ROS in diseases

stress, enhancement of antioxidant enzyme activity, modulation of pro-inflammatory mediators, and vasodilatory effects [59]. SeNPs have antioxidant properties that prevent severe heart failure, control high blood pressure, and reduce ischemia–reperfusion injury, while also reducing ROS-mediated lipid peroxidation, fibrosis, and vascular constriction. Moreover, SeNPs' biocompatibility, low toxicity, and ability to cross-cell membranes enhance their therapeutic potential [60] (Fig. 6). Researchers suggest SeNPs as a novel adjunctive therapy for cardiovascular diseases, including atherosclerosis, hypertension, and cardiac failure, bringing renewed optimism to patients who have exhausted conventional therapeutic avenues [61].

3 Reactive Nitrogen Species

When nitrogen molecules accept a proton, they transform into free radicals or reactive chemical entities, known as RNS [62]. Nitric oxide (NO), a transparent gas with unpaired electrons, easily penetrates cellular membranes and interacts with various components due to its ability to dissolve in water and organic solvents. The biochemical production of NO is mediated by the enzyme nitric oxide synthase, which orchestrates the molecular union of L-arginine and O_2 to yield NO [63]. In addition to molecular oxygen, cofactors such as flavin adenine dinucleotide (FAD), flavin mononucleotide (FMN), tetrahydrobiopterin (BH_4), and nicotinamide adenine dinucleotide (NADPH) synergistically enable the formation of NO [7]. The synthesis reaction involves an oxidation of five electrons, with three coming from NADPH and two from O_2. The expression of nitric oxide synthase (NOS) is governed by three

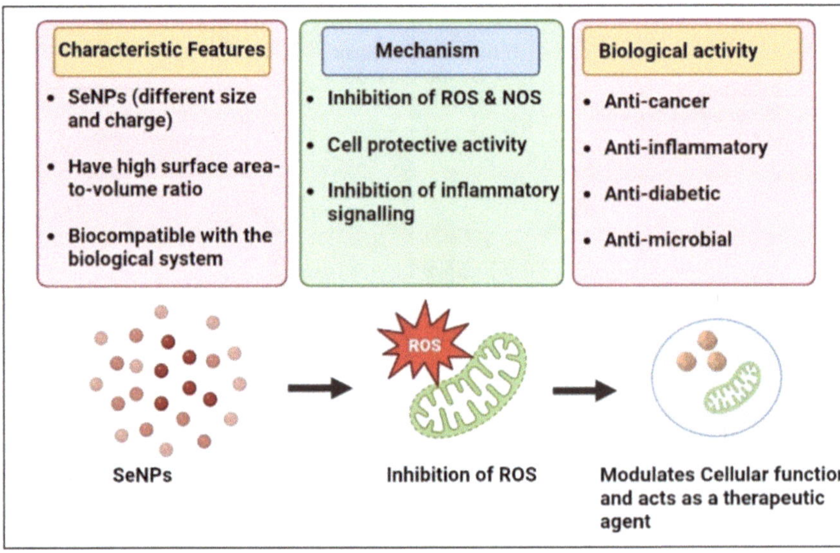

Fig. 6 Therapeutic application of SeNPs in ROS-induced diseases

genes, resulting in constitutive and inducible variants that possess distinct kinetic and biochemical characteristics. The genes produce three isoforms of the NOS enzyme: NOS_1, NOS_2, and NOS_3, with the latter being endothelial (eNOS), neuronal (nNOS), and inducible (iNOS). Calcium ions are crucial in activating the eNOS and nNOS enzyme variants, which are why they are called constitutive enzymes [64]. These enzymes produce minimal levels of NO that act as a communication channel between nervous system cells and smooth muscles. As a result, it activates cytoplasmic guanylate cyclase, causing an increase in levels of 3′,5′-cyclic guanosine monophosphate (cGMP) inside cells [5]. Various enzymes are regulated by cGMP to maintain the intracellular calcium concentration. The enzyme becomes deactivated as intracellular ionized calcium concentration decreases, producing low nNOS and eNOS isoforms. Endotoxin and/or other cytokines activate macrophages and other cell types, particularly phagocytic leukocytes, producing the iNOS isoform [65]. Despite calcium levels, iNOS is still active when arginine is present. Consequently, it can help produce nitric oxide in significant quantities over a long time. Nitric oxide is essential to manage blood pressure, improve renal function, and modulate vascular tone. NO is a vital vasodilator in endothelial cells within the blood vessels [66]. NO activation triggers the synthesis of cGMP in smooth muscle cells by activating soluble guanylate cyclase (sGC). Increasing cell cGMP levels trigger protein kinase activation, causing blood vessel dilation and smooth muscle relaxation [65]. Nitric oxide (NO) undergoes a reaction with hydrogen peroxide (H_2O_2) and hypochlorous acid (HClO) to generate dinitrogen trioxide (N_2O_3), nitrite (NO_2^-), and nitrate (NO_3^-).

$$O_2^{-\cdot} + {}^{\cdot}NO \rightarrow ONOO^-$$

Peroxynitrite formation is critical to the biochemical reaction between superoxide and nitric oxide radicals in the body's natural environment. The interaction between superoxide radicals and nitric oxide is indispensable for controlling ROS levels and regulating redox processes [36]. The peroxynitrite anion is formed when the superoxide radical reacts with NO in water.

$$ONOO^- + H+ \rightarrow ONOOH \rightarrow {}^{\cdot}OH + NO_2$$

Peroxynitrite is a powerful oxidant that enables it to initiate membrane lipid and protein oxidation, ultimately producing harmful nitro-derivatives like nitrotyrosine [67]. It may also oxidize and deactivate proteins, lipids, enzymes, DNA, and thiol groups, which increases its toxicity.

$$ONOOH + RS^- \rightarrow RSOH + NO_2^-$$

Nitric oxide undergoes a reaction with lipophilic peroxyl radicals, forming alkyl peroxynitrate (LOONO) as part of the lipid peroxidation chain reaction [36]. NO scavenges peroxyl radicals, preventing lipid peroxidation and allowing for the safe degradation of LOONO products without producing harmful free radicals.

3.1 Sources of RNS

RNS are generated from various endogenous and exogenous sources. Endogenously, RNS are produced through enzymatic reactions involving nitric oxide synthase (NOS), which catalyzes the conversion of L-arginine to citrulline and nitric oxide (NO) [68] (Fig. 7). The enzyme nitric oxide synthase (NOS) exists in three primary forms: neuronal NOS (nNOS), endothelial NOS (eNOS), and inducible NOS (iNOS), each producing NO in different cellular contexts [69]. Other enzymes, such as xanthine oxidoreductase, aldehyde dehydrogenase, and cytochrome P450, contribute to RNS production. Exogenously, RNS can be generated from extrinsic factors like air pollution, cigarette smoke, and dietary nitrate-rich foods. Furthermore, certain medical conditions, including inflammation, infection, and cancer, can increase RNS production. Other sources of RNS include lifestyle factors, such as physical activity and UV exposure, which can potentially induce NO production in various cell types [70]. Overall, the diverse origins of RNS highlight thc complex interplay between cellular metabolism, environmental factors, and disease states in regulating RNS production and biological activity.

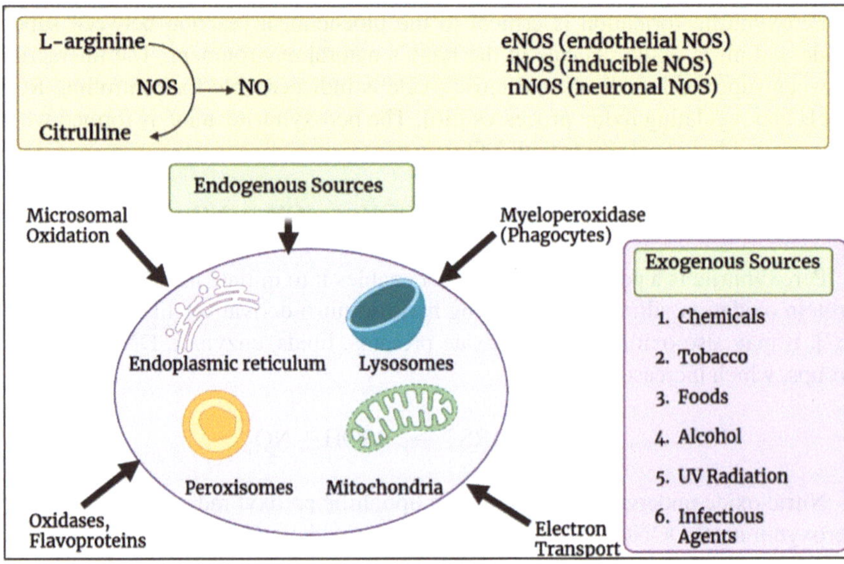

Fig. 7 Various sources of reactive nitrogen species

3.2 Types of RNS

RNS comprise a diverse group of nitrogen-containing molecules, including nitric oxide (NO), nitrite (NO_2^-), peroxynitrite ($ONOO^-$), nitrogen dioxide (NO_2), nitroso peroxocarbonate ($ONOOCO_2$), and S-nitroso thiols (RSNOs) (Table 1). NO is a primary RNS, acting as a signaling molecule and vasodilator, while nitrite (NO_2^-) and nitrogen dioxide (NO_2) are reactive intermediates formed through NO oxidation [4]. Peroxynitrite ($ONOO^-$), a potent oxidant, is generated from the reaction between NO and superoxide (O_2^-). While S-nitroso thiols (RSNOs), including GSNO, serve as reservoirs for nitric oxide and play a role in the S-nitrosation of proteins, nitroso peroxocarbonate ($ONOOCO_2$) is a strong oxidizing agent formed when nitric oxide reacts with carbonate radicals [71]. Moreover, nitroxyl (HNO), nitrosyl cations (NO^+), and nitronium ions (NO^{2+}) are other RNS that possess distinct chemical properties and biological effects [72]. They engage in complex functions in physiological mechanisms and disease-related processes such as oxidative stress, cell signaling, and immune response.

Table 1 Diverse types of RNS and their production site

Reactive nitrogen species		
S. no	Types of RNS	Site of production
1	Nitric oxide (NO)	Endothelial cells, neurons, macrophages
2	Nitrite (NO_2^-)	Mitochondria, endothelial cells, macrophages
3	Peroxynitrite ($ONOO^-$)	Macrophages, neutrophils, endothelial cells
4	Nitrogen dioxide (NO_2)	Activated macrophages, neutrophils
5	Nitroso peroxycarbonate ($ONOOCO_2$)	Activated macrophages

3.3 Cellular Signaling and Regulations of Reactive Nitrogen Species

Nitric oxide synthase (NOS) governs cellular signaling and the production of reactive nitrogen species (RNS) by transforming L-arginine into citrulline and nitric oxide (NO), producing reactive radicals when engaging with oxygen or superoxide [73]. Cellular signaling pathways, including MAPK, NF-κB, and PI3K/Akt, regulate the interaction between nitric oxide (NO) and reactive radicals to generate reactive nitrogen species (RNS). Redox-sensitive proteins, including glutathione and thioredoxin, regulate the RNS activity, influencing cellular metabolism, gene expression, protein function, and signaling pathways [74]. Moreover, mitochondrial function and energy utilization influence RNS synthesis and signaling, while transcription factors and RNS-sensing proteins maintain redox equilibrium. Dysregulation of RNS signaling contributes to various diseases, including cancer, cardiovascular disease, and neurodegeneration [73].

3.4 Role in Diseases

RNS are effective in various diseases, contributing to oxidative stress, inflammation, and tissue damage. In cancer, RNS induce DNA damage, mutagenesis, and carcinogenesis, promoting tumor development and progression [56]. Cardiovascular disease is exacerbated by RNS-mediated oxidative stress, damaging vascular tissues and leading to atherosclerosis, hypertension, and cardiac dysfunction. Alzheimer's, Parkinson's, and Huntington's are some of the neurodegenerative disorders caused by RNS-induced oxidative damage, neuroinflammation, and protein misfolding [74]. Mitochondrial dysfunction and insulin signaling impairment from RNS are associated with metabolic issues such as obesity and diabetes, whereas inflammation and oxidative stress exacerbate respiratory conditions like asthma and COPD. Additionally, RNS have antimicrobial properties but can contribute to tissue damage and inflammation during infectious diseases. Multiple sclerosis, lupus, rheumatoid arthritis, and various other autoimmune diseases can lead to tissue injury and inflammation as a result of an RNS-modulated immune response. Hepatitis and liver cirrhosis result from inflammation and fibrosis, while kidney disorders such as kidney damage and fibrosis are induced by RNS-mediated oxidative stress [71]. Finally, cellular senescence, inflammation, and oxidative stress associated with aging are all associated with RNS accumulation.

3.5 Therapeutic Applications

RNS-mediated therapies are employed to address infectious diseases, cancer, and inflammatory disorders due to their antimicrobial, anti-inflammatory, and prooxidant characteristics [75]. The investigation into the treatment of heart disease, high blood pressure, and erectile dysfunction is focusing on S-nitroso thiols and nitric oxide donors, while RNS-modulating compounds show promise for cancer therapy [74]. Antimicrobial materials and coatings were developed to address RNS-mediated antibiotic-resistant infections and investigate RNS-based therapies for neurodegenerative diseases such as Parkinson's and Alzheimer's [76]. Polyphenols and thiols serve as examples of RNS-scavenging antioxidants utilized to counteract oxidative damage in several diseases [77]. Researchers are investing in nanoparticles that target RNS for the purposes of drug delivery and imaging applications, in addition to RNS-modulated stem cell therapies aimed at improving tissue repair and regeneration [78]. Research on RNS is therefore enhancing its therapeutic potential, with effective modulation being essential for the development of new treatments and improved patient outcomes in the context of complex diseases.

4 Reactive Sulfur Species

Sulfur is found in its most reduced oxidation state as reactive sulfur species, such as thiols, hydrogen sulfide (H_2S), and carbonyl sulfide (COS). The trans-sulfuration pathway transforms L-cysteine into H_2S and includes 3-mercaptopyruvate sulfur transferase (3-MST), cystathionine-γ-lyase (CSE), and cystathionine-β-synthase (CBS) [79]. It can bind to metal centers in cytochrome c oxidase and hemoglobin, act as an antioxidant, or participate in downstream signaling via S-sulfhydrating proteins. Sulfide quinone oxidoreductase (SQR) is a mitochondrial flavoprotein that can oxidize H_2S to create a momentary persulfide intermediate that can modify proteins or be used as a precursor for generating thiosulfate by the enzyme sulfur transferase rhodanese [80]. Sulfite is produced by the metabolism of L-cysteine by aspartate aminotransferase (AAT) and cysteine dioxygenase (CDO), which are in equilibrium with SO_2 [81]. It is also possible to produce it from thiosulfate with the help of thiosulfate sulfur transferase (TST). A prevalent technique for oxidizing sulfite is S-sulfonation, resulting in sulfate through interactions with proteins, and is either eliminated in urine or taken up by cells through sulfate transporters [82]. ATP sulfurylase is an enzyme that initiates the process of sulfate assimilation in all living organisms, transforming adenosine-5′-phosphosulfate into 3′-phosphoadenosine-5′-phosphosulfate, thereby promoting sulfonation and the detoxification of xenobiotics [79]. Hydrogen sulfide (H_2S) is an essential reactive sulfur species that has greatly impacted redox biology in the last two decades. Hydrogen sulfide, also known as H_2S, has been recognized for a long time as a significant gaseous form of sulfur due to its toxicity and lethal characteristics [83]. For the past 15 years, it has been recognized that sulfur dioxide, a harmful byproduct of hydrogen sulfide oxidation, serves as a crucial signaling molecule that leads to acid rain and allergic responses in some individuals. Hydrated sulphites are employed in the food sector as preservatives and antioxidants and show vasodilatory effects [82].

4.1 Sources of RSS

Various internal and external sources, such as different biological, ecological, and lifestyle elements, play a role in the production of RSS. RSS are generated endogenously as byproducts of cellular metabolism, involving the enzymatic actions of cystathionine γ-lyase, cystathionine β-synthase, and mercaptopyruvate sulfurtransferase, which facilitate the transformation of sulfur-containing amino acids into RSS [84]. Additionally, RSS are formed through the decomposition of hydrogen sulfide (H_2S), which is produced by the enzymatic activity of H_2S-producing enzymes, such as cystathionine γ-lyase and cystathionine β-synthase, in various tissues, including the brain, liver, and kidney. External source of RSS include sulfur-rich foods like garlic and cruciferous vegetables, which contain sulfur compounds that the body transforms into RSS, whereas environmental pollutants including sulfur dioxide (SO_2) and sulfur trioxide (SO_3) found in air pollution and industrial emissions can also produce RSS [85] (Fig. 8). Additionally, the generation of RSS is influenced by sulfur compounds present in cosmetics, gut bacteria, medical therapies, mitochondria, the microbiome, and pharmaceuticals, owing to oxidative stress and the degradation of amino acids [19]. Certain health conditions, like liver and kidney diseases, can elevate RSS production, emphasizing the complex connection between RSS and disease progression [86]. In addition to other factors, behavioral patterns, including meal plan, exercise, and smoking, also influence RSS production, with a diet rich in sulfur-containing foods and regular exercise contributing to increased RSS production, while smoking and other forms of oxidative stress can lead to decreased RSS production.

4.2 Types of RSS

RSS encompass a diverse range of sulfur-containing molecules, including hydrogen sulfide (H_2S), methyl mercaptan (CH_3SH), dimethyl sulfide ($(CH_3)_2S$), sulfur

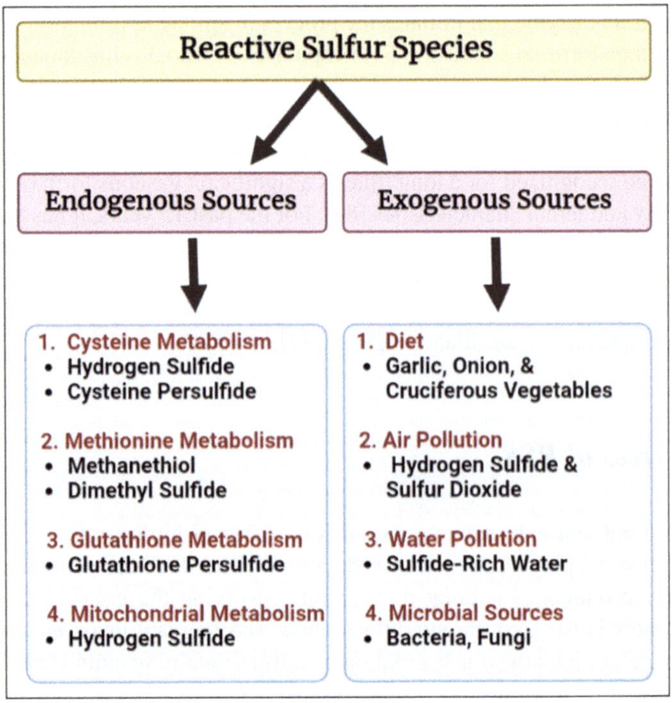

Fig. 8 Different sources of reactive sulfur species

dioxide (SO_2), sulfur trioxide (SO_3), thiol radicals ($RS^{\cdot}$), sulfur-centered radicals ($R\text{-}S^{\cdot}$), and peroxysulfur radicals ($RSOO^{\cdot}$) [87] (Fig. 9). H_2S, a potent vasodilator and neuroprotectant, is produced endogenously through enzymatic activity and is key in cellular signaling and homeostasis. The metabolic processes of gut microbes generate volatile organic sulfur compounds such as dimethyl sulfide and methyl mercaptan, which are associated with health issues like trimethylaminuria and halitosis [19]. Environmental pollutants like sulfur dioxide and sulfur trioxide, produced from fossil fuel combustion and industrial operations, can cause cardiovascular and respiratory illnesses, with oxidative stress contributing to cellular harm [88]. Peroxysulfur radicals (RSS), generated from thiols interacting with ROS, influence cellular redox equilibrium and reduce oxidative stress; additionally, other RSS, including sulfenic acid (R-SOH), sulfenic amide ($R\text{-}S(O)NH_2$), and sulfonamide ($R\text{-}S(O_2)NH_2$), have also been recognized.

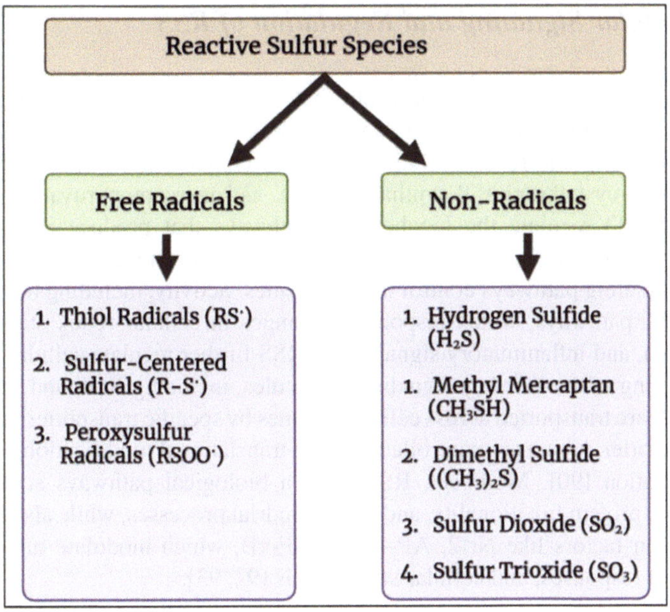

Fig. 9 Various types of reactive sulfur species

4.3 Biological Effects of RSS

Hydrogen sulfide (H_2S) and other reactive sulfur species serve as important signaling molecules that regulate various cellular processes such as vasodilation, angiogenesis, and neuroprotection. They are essential for preserving cellular homeostasis and reducing oxidative stress [89]. H_2S also has anti-inflammatory properties, which help to reduce inflammation by preventing the synthesis of pro-inflammatory cytokines. The antibacterial effects of RSS can obstruct the spread of infections and impact cellular metabolism, thereby inducing modifications in lipid and carbohydrate metabolism. In contrast, the excessive production of reactive species can lead to harmful outcomes, resulting in cellular damage, increased oxidative stress, and inflammation. These factors are implicated in the emergence of a range of diseases, such as heart disease, cancer, neurodegenerative conditions, and respiratory diseases [90]. Reactive sulfur species (RSS) have the potential to modify the activity of enzymes, receptors, and ion channels which in turn impacts cellular signaling pathways and may facilitate disease progression. In addition, reactive sulfur species (RSS) engage with other reactive entities, such as reactive oxygen species (ROS) and reactive nitrogen species (RNS), thereby intensifying oxidative stress and contributing to cellular damage [91]. The effects of reactive species on biological systems are modulated by their engagement with other biomolecules, including DNA, proteins, and lipids, which can lead to post-translational modifications, lipid peroxidation, and genetic mutations.

4.4 Cellular Signaling and Regulation of RSS

H_2S is an essential signaling molecule that affects numerous cellular functions, including inflammation, apoptosis, and metabolism, via complex mechanisms in cellular signaling and RSS regulation [92]. The enzymes such as cystathionine γ-lyase (CSE), cystathionine β-synthase (CBS), and mercaptopyruvate sulfurtransferase (MPST) regulate the biochemical pathways that produce reactive sulfur species by converting sulfur-containing amino acids into signaling molecules [93]. Various signaling pathways control these enzymes' activity, including the PI3K/Akt and MAPK pathways, which respond to changes in cellular redox status, energy metabolism, and inflammatory signals [89]. RSS further regulate cellular processes by interacting with other key signaling molecules, including ROS and nitric oxide (NO). RSS are transported across cell membranes by specific transporters such as the H_2S transporter, which are controlled by post-translational modifications, including S-sulfhydration [90]. Moreover, RSS govern biological pathways such as gene expression, protein functionality, and mitochondrial processes, while also impacting transcription factors like Nrf2, AP-1, and NF-κB, which modulate inflammation, antioxidant responses, and cellular stress levels [92, 93].

4.5 Role of RSS in Diseases

RSS have been implicated in a wide range of pathological conditions, with notable associations observed in the development and progression of malignancies, cardiovascular afflictions, neurodegenerative syndromes, and respiratory disorders. RSS can act as pro-tumorigenic or anti-tumorigenic agents in cancer, depending on their concentration and cellular context [94]. As an instance, hydrogen sulfide (H_2S) is recognized for its ability to promote angiogenesis and tumor progression, induce apoptosis, and inhibit cellular proliferation [56]. Neurodegenerative diseases like Alzheimer's and Parkinson's are associated with RSS, which can lead to oxidative stress, dysfunction of mitochondria, and neuronal injury [19]. Exposure to RSS could increase the chances of developing respiratory illnesses like bronchial asthma and emphysema by intensifying oxidative damage and inflammatory responses. Besides respiratory issues, RSS have been linked to various other health concerns, including metabolic disorders, kidney disease, and liver disease, emphasizing a complex and intricate relationship between RSS and the progression of diseases [94].

4.6 Therapeutic Applications

The diverse therapeutic applications of RSS, including its signaling and antioxidant properties, establish it as a promising alternative to conventional antibiotics in the management of ailments such as cancer [95]. Hydrogen sulfide (H_2S), a potent reactive species, has shown promise in inhibiting tumor growth and angiogenesis, making it an appealing alternative to conventional antibiotics [90]. RSS are beneficial in reducing oxidative stress and inflammation in neurodegenerative diseases like Alzheimer's and Parkinson's, as well as in cardiovascular diseases like atherosclerosis and hypertension [86]. Reactive sulfur species have potential therapeutic applications in the realm of tissue regeneration, thereby augmenting collagen production and facilitating the formation of blood vessels, which in turn expedites the healing process [91]. RSS are employed in numerous therapeutic strategies, encompassing diabetes management, enhancing insulin sensitivity and glucose metabolism, treating liver diseases, and mitigating oxidative stress and inflammation [95]. Moreover, RSS-based treatments may relieve respiratory conditions like COPD and asthma, as they appear to decrease inflammation and free radical damage.

5 Conclusion

Reactive species, encompassing ROS, RNS, and RSS, play a multifaceted role in modulating biological processes, including cellular signaling, antioxidant defenses, and disease pathogenesis. Maintaining a delicate balance between production and eliminating reactive species is essential for preserving cellular homeostasis, while disruptions lead to oxidative stress, inflammation, and disease progression. ROS, generated through mitochondrial respiration, enzymatic reactions, or environmental stressors, can damage biomolecules, activate signaling pathways, and regulate cellular processes. RNS, acquired through nitric oxide synthase activity or ecological exposure, participate in antimicrobial defense, neurotransmission, and vascular regulation. RSS, generated through enzymatic or non-enzymatic reactions, modulate cellular redox states and signaling pathways. Numerous health issues, such as cardiovascular diseases, neurological disorders, cancer, and infectious diseases, are linked to the dysregulation of reactive species. It is crucial to understand the intricate relationships between biological systems and reactive species to develop effective treatment strategies that provide new opportunities for disease prevention and health promotion.

References

1. Konno, T., Melo, E. P., Chambers, J. E., & Avezov, E. (2021). Intracellular sources of ROS/ H_2O_2 in health and neurodegeneration: Spotlight on endoplasmic reticulum. *Cells, 10*(2), 1–26.
2. Di Meo, S., Reed, T. T., Venditti, P., & Victor, V. M. (2016). Role of ROS and RNS sources in physiological and pathological conditions. *Oxidative Medicine and Cell Longevity, 2016*.
3. Ray, P. D., Huang, B. W., & Tsuji, Y. (2012). Reactive oxygen species (ROS) homeostasis and redox regulation in cellular signaling. *Cell Signal, 24*(5), 981–990. Available from: https://doi.org/10.1016/j.cellsig.2012.01.008

4. Odyniec, M. L., Sedgwick, A. C., Swan, A. H., Weber, M., Tang, T. M. S., Gardiner, J. E., et al. (2018). 'AND'-based fluorescence scaffold for the detection of ROS/RNS and a second analyte. *Chemical Communications, 54*(61), 8466–8469.
5. Moreau, M., Lindermayr, C., Durner, J., & Klessig, D. F. (2010). NO synthesis and signaling in plants - where do we stand? *Physiologia Plantarum, 138*(4), 372–383.
6. Pacher, P., Beckman, J. S., & Liaudet, L. (2007). Nitric oxide and peroxynitrite in health and disease. *Physiological Reviews, 87*(1), 315–424.
7. Del Río, L. A. (2015). ROS and RNS in plant physiology: An overview. *Journal of Experimental Botany, 66*(10), 2827–2837.
8. Zhu, C. Q., Zhang, J. H., Sun, L. M., Zhu, L. F., Abliz, B., Hu, W. J., et al. (2018). Hydrogen sulfide alleviates aluminum toxicity via decreasing apoplast and symplast al contents in rice. *Frontiers in Plant Science, 9*, 1–14.
9. Benchoam, D., Cuevasanta, E., Möller, M. N., & Alvarez, B. (2019). Hydrogen sulfide and persulfides oxidation by biologically relevant oxidizing species. *Antioxidants, 8*(2), 1–23.
10. Touyz, R. M., & Delles, C. (2019). *Textbook of vascular medicine* (pp. 1–503).
11. Das, K., & Roychoudhury, A. (2014). Reactive oxygen species (ROS) and response of antioxidants as ROS-scavengers during environmental stress in plants. *Frontiers in Environmental Science, 2*(DEC), 1–13.
12. Kreslavski, V. D., Los, D. A., Allakhverdiev, S. I., & Kuznetsov, V. V. (2012). Signaling role of reactive oxygen species in plants under stress. *Russian Journal of Plant Physiology, 59*(2), 141–154.
13. Murphy, M. P. (2009). How mitochondria produce reactive oxygen species. *The Biochemical Journal, 417*(1), 1–13.
14. Finkel, T. (2011). Signal transduction by reactive oxygen species. *Journal of Cell Biology, 194*(1), 7–15.
15. Bulut, N., Kocyigit, U. M., Gecibesler, I. H., Dastan, T., Karci, H., Taslimi, P., et al. (2018). Synthesis of some novel pyridine compounds containing bis-1,2,4-triazole/thiosemicarbazide moiety and investigation of their antioxidant properties, carbonic anhydrase, and acetylcholinesterase enzymes inhibition profiles. *Journal of Biochemical and Molecular Toxicology, 32*(1), 2–11.
16. Gülçin, I. (2012). Antioxidant activity of food constituents: An overview. *Archives of Toxicology, 86*(3), 345–391.
17. Rezai, M., Bayrak, Ç., Taslimi, P., Gülçin, I., & Menzek, A. (2018). The first synthesis and antioxidant and anticholinergic activities of 1-(4,5-dihydroxybenzyl)pyrrolidin-2-one derivative bromophenols including natural products. *Turkish Journal of Chemistry, 42*(3), 808–825.
18. Villamena, F. A. (2013). Chemistry of reactive species. *Molecular Basis of Oxidative Stress: Chemistry, Mechanisms and Disease Pathogenesis*, 1–48.
19. Olson, K. R. (2020). Reactive oxygen species or reactive sulfur species: Why we should consider the latter. *Journal of Experimental Biology, 223*(4).
20. Ingle, M. E., Watkins, D., Rosario, Z., VélezVega, C. M., Calafat, A. M., Ospina, M., et al. (2020). An exploratory analysis of urinary organophosphate ester metabolites and oxidative stress among pregnant women in Puerto Rico. *Science of the Total Environment, 703*, 1161–1208.
21. Jomova, K., Raptova, R., Alomar, S. Y., Alwasel, S. H., Nepovimova, E., Kuca, K., et al. (2023) Reactive oxygen species, toxicity, oxidative stress, and antioxidants: Chronic diseases and aging [Internet]. *Archives of Toxicology, 97*, 2499–2574. Available from: https://doi.org/10.1007/s00204-023-03562-9
22. Ishigami, M. (1998). Superoxide dismutase. *Nippon rinsho Japanese Journal of Clinical Medicine, 56*(22), 160–168.
23. Mittal, C. K., & Murad, F. (1977). Activation of guanylate cyclase by superoxide dismutase and hydroxyl radical: A physiological regulator of guanosine 3',5'-monophosphate formation. *Proceedings of the National Academy of Sciences, U S A, 74*(10), 4360–4364.

24. Kovacic, P., Pozos, R. S., Somanathan, R., Shangari, N., & O'Brien, P. J. (2005). Mechanism of mitochondrial uncouplers, inhibitors, and toxins: focus on electron transfer, free radicals, and structure-activity relationships. *Current Medicinal Chemistry, 12*(22), 2601–2623.
25. Leonard, S. S., Harris, G. K., & Shi, X. (2004). Metal-induced oxidative stress and signal transduction. *Free Radical Biology & Medicine, 37*(12), 1921–1942.
26. DeCoursey, T. E., & Ligeti, E. (2005). Regulation and termination of NADPH oxidase activity. *Cellular and Molecular Life Sciences, 62*(19–20), 2173–2193.
27. Valko, M., Rhodes, C. J., Moncol, J., Izakovic, M., & Mazur, M. (2006). Free radicals, metals and antioxidants in oxidative stress-induced cancer. *Chemico-Biological Interactions, 160*(1), 1–40.
28. Borges, F., Fernandes, E., & Roleira, F. (2012). Progress towards the discovery of xanthine oxidase inhibitors. *Current Medicinal Chemistry, 9*(2), 195–217.
29. Ridnour, L. A., Isenberg, J. S., Espey, M. G., Thomas, D. D., Roberts, D. D., & Wink, D. A. (2005). Nitric oxide regulates angiogenesis through a functional switch involving thrombospondin-1. *Proceedings of the National Academy of Sciences U S A, 102*(37), 13147–13152.
30. Preiser, J. C. (2012). Oxidative stress. *Journal of Parenteral and Enteral Nutrition, 36*(2), 147–154.
31. Thorpe, G. W., Fong, C. S., Alic, N., Higgins, V. J., & Dawes, I. W. (2004). Cells have distinct mechanisms to maintain protection against different reactive oxygen species: Oxidative-stress-response genes. *Proceedings of the National Academy of Sciences U S A., 101*(17), 6564–6569.
32. Nathan, C. (2003). Specificity of a third kind: Reactive oxygen and nitrogen intermediates in cell signaling. *The Journal of Clinical Investigation, 111*(6), 769–778.
33. Ferrer-Sueta, G., Manta, B., Botti, H., Radi, R., Trujillo, M., & Denicola, A. (2011). Factors affecting protein thiol reactivity and specificity in peroxide reduction. *Chemical Research in Toxicology, 24*(4), 434–450.
34. Winterbourn, C. C., & Hampton, M. B. (2008). Thiol chemistry and specificity in redox signaling. *Free Radical Biology & Medicine, 45*(5), 549–561.
35. Kuiper, J. W. P., Sunm, C., Magalha, M. A. O., & Glogauer, M. (2016) Rac regulates PtdInsP 3 signaling and the chemotactic compass through a redox-mediated feedback loop. *Blood, The Journal of the American Society of Hematology, 118*(23), 6164–6172.
36. Paulsen, C. E., Truong, T. H., Garcia, F. J., Homann, A., Gupta, V., Leonard, S. E., et al. (2012). Peroxide-dependent sulfenylation of the EGFR catalytic site enhances kinase activity. *Nature Chemical Biology, 8*(1), 57–64. Available from: https://doi.org/10.1038/nchembio.736
37. Erickson, J. R., Joiner, M., Ling, A., Guan, X., Kutschke, W., Yang, J., Oddis, C. V., et al. (2008) A dynamic pathway for calcium-independent activation of CaMKII by methionine oxidation. *Cell, 133*(3), 462–474.
38. Okuno, Y., Nakamura-Ishizu, A., Otsu, K., Suda, T., & Kubota, Y. (2012). Pathological neoangiogenesis depends on oxidative stress regulation by ATM. *Nature Medicine, 18*(8), 1208–1216. Available from: https://doi.org/10.1038/nm.2846
39. Guo, Z., Kozlov, S., Lavin, M. F., Person, M. D., & Paull, T. T. (2010). ATM activation by oxidative stress. *Science, 330*(6003), 517–521.
40. Ito, K., Hirao, A., Arai, F., Takubo, K., Matsuoka, S., Miyamoto, K., et al. (2006). Reactive oxygen species act through p38 MAPK to limit the lifespan of hematopoietic stem cells. *Nature Medicine, 12*(4), 446–451.
41. Ruchko, M. V., Gorodnya, O. M., Pastukh, V. M., Swiger, B. M., Middleton, N. S., Wilson, G. L., et al. (2009). Hypoxia-induced oxidative base modifications in the VEGF hypoxia-response element are associated with transcriptionally active nucleosomes. *Free Radical Biology and Medicine, 46*(3), 352–359. Available from: https://doi.org/10.1016/j.freeradbiomed.2008.09.038
42. Al-Mehdi, A. B., Pastukh, V. M., Swiger, B. M., Reed, D. J., Patel, M. R., Bardwell, G. C., et al. (2012). Perinuclear mitochondrial clustering creates an oxidant-rich nuclear domain required for hypoxia-induced transcription. *Science Signaling, 5*(231).

43. Perillo, B., Ombra, M. N., Bertoni, A., Cuozzo, C., Sacchetti, S., Sasso, A., et al. (2008). DNA oxidation as triggered by H3K9me2 demethylation drives estrogen-induced gene expression. *Science, 319*(5860), 202–226.
44. Amente, S., Lania, L., Avvedimento, E. V., & Majello, B. (2010). DNA oxidation drives Myc mediated transcription. *Cell Cycle, 9*(15), 3074–3076.
45. Groitl, B., & Jakob, U. (2014). Thiol-based redox switches. *Biochimica et Biophysica Acta - Proteins and Proteomics, 1844*(8), 1335–1343. Available from: https://doi.org/10.1016/j.bbapap.2014.03.007
46. Brown, M. D., & Sacks, D. B. (2009). Protein scaffolds in MAP kinase signalling. *Cellular Signaling, 21*(4), 462–469. Available from: https://doi.org/10.1016/j.cellsig.2008.11.013
47. Winter-Vann, A. M., & Johnson, G. L. (2007). Integrated activation of MAP3Ks balances cell fate in response to stress. *Journal of Cellular Biochemistry, 102*(4), 848–858.
48. Junttila, M. R., Li, S., & Westermarck, J. (2008). Phosphatase-mediated crosstalk between MAPK signaling pathways in the regulation of cell survival. *The FASEB Journal, 22*(4), 954–965.
49. Ramos, J. W. (2008). The regulation of extracellular signal-regulated kinase (ERK) in mammalian cells. *International Journal of Biochemistry & Cell Biology, 40*(12), 2707–2719.
50. Bogoyevitch, M. A., Ngoei, K. R. W., Zhao, T. T., Yeap, Y. Y. C., & Ng, D. C. H. (2010). c-Jun N-terminal kinase (JNK) signaling: Recent advances and challenges. *Biochimica et Biophysica Acta - Proteins and Proteomics, 1804*(3), 463–475. Available from: https://doi.org/10.1016/j.bbapap.2009.11.002
51. Wang, Y., Branicky, R., Noë, A., & Hekimi, S. (2018). Superoxide dismutases: Dual roles in controlling ROS damage and regulating ROS signaling. *Journal of Cell Biology, 217*(6), 1915–1928.
52. Paul, S., Ghosh, S., Mandal, S., Sau, S., & Pal, M. (2018). NRF2 transcriptionally activates the heat shock factor 1 promoter under oxidative stress and affects survival and migration potential of MCF7 cells. *Journal of Biological Chemistry, 293*(50), 19303–19316. Available from: https://doi.org/10.1074/jbc.RA118.003376
53. Sahiner, U. M., Birben, E., Erzurum, S., Sackesen, C., & Kalayci, Ö. (2018). Oxidative stress in asthma: Part of the puzzle. *Pediatric Allergy and Immunology, 29*(8), 789–800.
54. Li, X., Fang, P., Mai, J., Choi, E. T., Wang, H., & Yang, X. F. (2013). Targeting mitochondrial reactive oxygen species as novel therapy for inflammatory diseases and cancers. *Journal of Hematology & Oncology, 6*(1), 1–19.
55. Manna, P., & Jain, S. K. (2015). Obesity, oxidative stress, adipose tissue dysfunction, and the associated health risks: Causes and therapeutic strategies. *Metabolic Syndrome and Related Disorders, 13*(10), 423–444.
56. Winyard, P. G., Ryan, B., Eggleton, P., Nissim, A., Taylor, E., Lo Faro, M. L., et al. (2011). Measurement and meaning of markers of reactive species of oxygen, nitrogen and sulfur in healthy human subjects and patients with inflammatory joint disease. *Biochemical Society Transactions, 39*(5), 1226–1232.
57. Chattopadhyay, M., Khemka, V. K., Chatterjee, G., Ganguly, A., Mukhopadhyay, S., & Chakrabarti, S. (2015). Enhanced ROS production and oxidative damage in subcutaneous white adipose tissue mitochondria in obese and type 2 diabetes subjects. *Molecular and Cellular Biochemistry, 399*(1–2), 95–103.
58. Pang, C. N. I., & Wilkins, M. R. (2018). Protein post-translational modification prediction. *Encyclopedia of Bioinformatics and Computational Biology ABC of Bioinformatics, 1–3*, 15–27.
59. Chandramohan, S., Naveenkumar, S., Kaviyarasu, K., Lavakumar, V., Sowmya, C., Santhanakumar, M., et al. (2021). Bio-distribution of selenium nanoparticles (SeNPs) to the Wistar rats and its breastfed offspring. *Journal of Drug Delivery Science and Technology, 61*, 102299. Available from: https://doi.org/10.1016/j.jddst.2020.102299
60. Zhou, Y., Li, H., & Xia, N. (2021). The interplay between adipose tissue and vasculature: Role of oxidative stress in obesity. *Frontiers in Cardiovascular Medicine, 8*, 1–14.

61. Kalishwaralal, K., Jeyabharathi, S., Sundar, K., Selvamani, S., Prasanna, M., & Muthukumaran, A. (2018). A novel biocompatible chitosan–Selenium nanoparticles (SeNPs) film with electrical conductivity for cardiac tissue engineering application. *Materials of Science and Engineering: C, 92*, 151–160. Available from: https://doi.org/10.1016/j.msec.2018.06.036
62. Del Río, L. A., Sandalio, L. M., Corpas, F. J., Palma, J. M., & Barroso, J. B. (2006). Reactive oxygen species and reactive nitrogen species in peroxisomes. Production, scavenging, and role in cell signaling. *Plant Physiology, 141*(2), 330–335.
63. Demidchik, V. (2015). Mechanisms of oxidative stress in plants: From classical chemistry to cell biology. *Environmental and Experimental Botany, 109*, 212–228. Available from: https://doi.org/10.1016/j.envexpbot.2014.06.021
64. Silberschatz, G. (2018). Chapter 16 : Security. 1694, 155–161.
65. Corpas, F. J., Chaki, M., Fernández-Ocaña, A., Valderrama, R., Palma, J. M., Carreras, A., et al. (2008). Metabolism of reactive nitrogen species in pea plants under abiotic stress conditions. *Plant and Cell Physiology, 49*(11), 1711–1722.
66. Smith, S. R., Gillard, J. T. F., Kustka, A. B., McCrow, J. P., Badger, J. H., Zheng, H., et al. (2016). Transcriptional orchestration of the global cellular response of a model pennate diatom to diel light cycling under iron limitation. *PLoS Genetics, 12*(12), 1–39.
67. Sikora, A., Zielonka, J., Lopez, M., Joseph, J., & Kalyanaraman, B. (2009). Direct oxidation of boronates by peroxynitrite: Mechanism and implications in fluorescence imaging of peroxynitrite. *Free Radical Biology and Medicine, 47*(10), 1401–1407. Available from: https://doi.org/10.1016/j.freeradbiomed.2009.08.006
68. Xu, J., Hou, Q.M., Khare, T., Verma, S. K., & Kumar, V. (2019). Exploring miRNAs for developing climate-resilient crops: A perspective review. *Science of the Total Environment, 653*, 91–104. Available from: https://doi.org/10.1016/j.scitotenv.2018.10.340
69. Garcia, V., & Sessa, W. C. (2019). Endothelial NOS: Perspective and recent developments. *British Journal of Pharmacology, 176*(2), 189–196.
70. Zhang, A., Jiang, M., Zhang, J., Ding, H., Xu, S., Hu, X., et al. (2007). Nitric oxide induced by hydrogen peroxide mediates abscisic acid-induced activation of the mitogen-activated protein kinase cascade involved in antioxidant defense in maize leaves. *New Phytologist, 175*(1), 36–50.
71. Weidinger, A., & Kozlov, A. V. (2015). Biological activities of reactive oxygen and nitrogen species: Oxidative stress versus signal transduction. *Biomolecules, 5*(2), 472–484.
72. Pourova, J., Kottova, M., Voprsalova, M., & Pour, M. (2010). Reactive oxygen and nitrogen species in normal physiological processes. *Acta Physiologica, 198*(1), 15–35.
73. Adams, L., Franco, M. C., & Estevez, A. G. (2015). Reactive nitrogen species in cellular signaling. *Experimental Biology and Medicine, 240*(6), 711–717.
74. Martínez, M. C., & Andriantsitohaina, R. (2009). Reactive nitrogen species: Molecular mechanisms and potential significance in health and disease. *Antioxidants Redox Signal, 11*(3), 669–702.
75. Shnayder, N. A., Petrova, M. M., Popova, T. E., Davidova, T. K., Bobrova, O. P., Trefilova, V. V., et al. (2021). Prospects for the personalized multimodal therapy approach to pain management via action on no and nos. *Molecules, 26*(9).
76. Jackson, M. J., Papa, S., Bola, J., Bruckdorfer, R., Carlsen, H., Elliott, R. M., et al. (2002). Antioxidants, reactive oxygen and nitrogen species. *Molecular Aspects of Medicine, 23*(1–3), 209–285.
77. Ramalingam, M., & Kim, S. J. (2012). Reactive oxygen/nitrogen species and their functional correlations in neurodegenerative diseases. *Journal of Neural Transmission, 119*(8), 891–910.
78. Fang, F. C. (2004). Antimicrobial reactive oxygen and nitrogen species: Concepts and controversies. *Nature Reviews Microbiology, 2*(10), 820–832.
79. Gruhlke, M. C. (2019). Reactive sulfur species. *Reactive Oxygen, Nitrogen and Sulfur Species in Plants, 2*, 715–728.
80. Gruhlke, M. C. H., & Slusarenko, A. J. (2012). The biology of reactive sulfur species (RSS). *Plant Physiology and Biochemistry, 59*, 98–107. Available from: https://doi.org/10.1016/j.plaphy.2012.03.016

81. Fukudome, M., Shimada, H., Uchi, N., Osuki, K. I., Ishizaki, H., Murakami, E. I., et al. (2020). Reactive sulfur species interact with other signal molecules in root nodule symbiosis in lotus japonicus. *Antioxidants, 9*(2), 1–13.
82. Giles, G. I., Nasim, M. J., Ali, W., & Jacob, C. (2017). The reactive sulfur species concept: 15 years on. *Antioxidants, 6*(2), 1–29.
83. Kaur, P., Handa, N., Verma, V., Bakshi, P., Kalia, R., Sareen, S., et al. (2019). Cross talk among reactive oxygen, nitrogen and sulfur during abiotic stress in plants. *Reactive Oxygen, Nitrogen and Sulfur Species in Plants: Production, Metabolism, Signaling and Defense Mechanisms, 2*, 857–871.
84. Gebicka, L., & Krych-Madej, J. (2019). The role of catalases in the prevention/promotion of oxidative stress. *Journal of Inorganic Biochemistry, 197*, 110699. Available from: https://doi.org/10.1016/j.jinorgbio.2019.110699
85. Giles, G. I., Tasker, K. M., & Jacob, C. (2001). Hypothesis: The role of reactive sulfur species in oxidative stress. *Free Radical Biology & Medicine, 31*(10), 1279–1283.
86. Kolluru, G. K., Shen, X., & Kevil, C. G. (2020). Reactive sulfur species: A new redox player in cardiovascular pathophysiology. *Arteriosclerosis, Thrombosis, and Vascular Biology, 40*(4), 874–884.
87. Wu, L., Sedgwick, A. C., Sun, X., Bull, S. D., He, X. P., & James, T. D. (2019). Reaction-based fluorescent probes for the detection and imaging of reactive oxygen, nitrogen, and sulfur species. *Accounts of Chemical Research, 52*(9), 2582–2597.
88. Finkel, T. (2000). Redox-dependent signal transduction. *FEBS Letters, 476*, 52–54.
89. Yuan, S., Shen, X., & Kevil, C. G. (2017). Beyond a gasotransmitter: Hydrogen sulfide and polysulfide in cardiovascular health and immune response. *Antioxidants Redox Signaling, 27*(10), 634–653.
90. Fukuto, J. M., Ignarro, L. J., Nagy, P., Wink, D. A., Kevil, C. G., Feelisch, M., et al. (2018). Biological hydropersulfides and related polysulfides—a new concept and perspective in redox biology. *FEBS Letters, 592*(12), 2140–2152.
91. Mani, S., Li, H., Untereiner, A., Wu, L., Yang, G., Austin, R. C., et al. (2013). Decreased endogenous production of hydrogen sulfide accelerates atherosclerosis. *Circulation, 127*(25), 2523–2534.
92. Mishanina, T. V., Libiad, M., & Banerjee, R. (2015). Biogenesis of reactive sulfur species for signaling by hydrogen sulfide oxidation pathways. *Nature Chemical Biology, 11*(7), 457–464. Available from: https://doi.org/10.1038/nchembio.1834
93. Iciek, M., Kowalczyk-Pachel, D., Bilska-Wilkosz, A., Kwiecién, I., Górny, M., & Wøodek, L. (2016). S-sulfhydration as a cellular redox regulation. *Bioscience Reports, 36*(2), 1–17.
94. Valko, M., Leibfritz, D., Moncol, J., Cronin, M. T. D., Mazur, M., & Telser, J. (2007). Free radicals and antioxidants in normal physiological functions and human disease. *International Journal of Biochemistry & Cell Biology, 39*(1), 44–84.
95. Yan, K. C., Sedgwick, A. C., Zang, Y., Chen, G. R., He, X. P., Li, J., et al. (2019). Sensors, imaging agents, and theranostics to help understand and treat reactive oxygen species related diseases. *Small Methods, 3*(7).

Antioxidants in Action: A New Era of Innovation in Health, Advancing Disease Prevention and Nutritional Wellness

Manikandan Vani Raju, Meenakshi Kaniyur Chandrasekaran, Mithra Jayan, Gopalakrishnan Velliyur Kanniappan, Vijaya Chitra Arumugam, and Rathi Muthaiyan Ahalliya

Abstract Antioxidants are essential molecules that respond to oxidative damage caused by free radicals, which arise from metabolic processes and external factors such as pollution, radiation, and unhealthy diets. Oxidative stress, resulting from an imbalance between free radical production and antioxidant defense, is linked to chronic diseases, including cancer, cardiovascular disorders, neurodegenerative conditions, and premature aging. While the human body possesses enzymatic antioxidants such as superoxide dismutase (SOD), catalase, and glutathione peroxidase, dietary antioxidants like polyphenols, flavonoids, carotenoids, and vitamins C and E play a crucial role in reinforcing these defenses. A diet rich in antioxidant-containing foods, such as fruits, vegetables, nuts, and whole grains, is vital for preventing oxidative damage and supporting immune and metabolic functions. Antioxidants are classified into natural and synthetic types. Natural antioxidants, present in plant-based meals and extracts, help to reduce oxidative stress and improve cellular health. Synthetic antioxidants, such as butylated hydroxytoluene (BHT) and butylated hydroxyanisole (BHA), are commonly used in food preservation to prevent spoilage and lipid oxidation, although concerns exist regarding their long-term safety. The modern lifestyle, characterized by high consumption of processed foods, environmental pollutants, and chronic stress, exacerbates oxidative stress, increasing the

---

M. Vani Raju · M. Kaniyur Chandrasekaran · R. Muthaiyan Ahalliya (✉)
Department of Biochemistry, Karpagam Academy of Higher Education, Coimbatore, Tamil Nadu, India
e-mail: rathi.muthaiyan@kahedu.edu.in

M. Jayan · R. Muthaiyan Ahalliya
Department of Biotechnology, Karpagam Academy of Higher Education, Coimbatore, Tamil Nadu, India

G. Velliyur Kanniappan
Department of Physiology, Saveetha Medical College and Hospital, Saveetha Institute of Medical and Technical Sciences (SIMATS), Chennai, Tamil Nadu, India

V. C. Arumugam
Department of Microbiology, Sri Ramakrishnma College of Arts & Science for Women, Coimbatore, Tamil Nadu, India

K. Kasinathan et al. (eds.), *Recent Advances in Oxidative Stress Associated Chronic Diseases Volume 1*, https://doi.org/10.1007/978-981-96-9670-3_12

risk of metabolic and inflammatory diseases. Consequently, promoting antioxidant-rich diets, reducing dependency on synthetic additives, and incorporating lifestyle modifications such as regular exercise and stress management are essential strategies for mitigating oxidative stress-related diseases. Advances in nutraceuticals, functional foods, and personalized nutrition are increasing antioxidant absorption and efficacy. Future research should focus on improving the stability and bioavailability of natural antioxidants while developing safer synthetic replacements. Raising public awareness and integrating evidence-based dietary recommendations will be crucial in leveraging antioxidants for disease prevention and overall health maintenance.

1 Introduction

Antioxidants are molecules that inhibit or prevent oxidative damage triggered by free radicals, which are unstable molecules produced by metabolic processes as well as external factors such as pollution, radiation, and poor diets. Oxidative stress, caused by an imbalance between free radical generation and antioxidant defense, has been related to a variety of long-term diseases, including cancer, cardiovascular disorders, neurological diseases, and premature aging [1]. The human body possesses innate antioxidant defense mechanisms, which include enzymatic antioxidants such as superoxide dismutase (SOD), catalase, and glutathione peroxidase, which help to neutralize reactive oxygen species (ROS). However, external sources of antioxidants from diet, particularly plant-based chemicals like polyphenols, flavonoids, carotenoids, and vitamins C and E, are needed to strengthen these natural defenses. These dietary antioxidants contribute to immunological response, inflammatory regulation, and general metabolic balance in addition to preventing cellular damage. A diet high in foods that contain antioxidants, such as fruits, vegetables, nuts, and whole grains, is crucial since inadequate antioxidant intake can raise the risk of diseases linked to oxidative stress [2]. Antioxidants are broadly divided into natural and synthetic forms. Natural antioxidants are found in foods such as fruits, vegetables, and plant extracts. Polyphenols, flavonoids, carotenoids, and vitamins all contribute to lowering oxidative damage and improving cellular health [3]. Synthetic antioxidants such as butylated hydroxytoluene (BHT) and butylated hydroxyanisole (BHA) are widely employed in food processing to avoid spoilage and lipid oxidation. While synthetic antioxidants enhance shelf life and stability, concerns have been raised about their possible toxicity and long-term health implications [4]. Dietary imbalances, a high intake of processed and fast meals, exposure to pollutants, and chronic stress are all characteristics of modern lifestyles that are contributing to oxidative damage and the rise in non-communicable diseases. The addiction to highly processed foods, which frequently lack natural antioxidants and vital minerals, has resulted in widespread shortages and weakened the body's defenses against oxidative damage [5]. Furthermore, exterior factors such as pollution, UV radiation, and smoking increase free radical generation, which accelerates cellular aging and increases the risk of inflammatory disorders [6]. Sedentary work situations, combined with poor lifestyle choices, enhance oxidative stress, increasing the risk of metabolic disorders, impaired immune function, and chronic illnesses like obesity, diabetes, and hypertension. To offset these consequences, there is a rising emphasis on the need for antioxidant-rich diets and lifestyles. Public health advice encourages the use of real foods high in antioxidants, therefore decreasing the usage of synthetic food additives and lifestyle changes such as regular physical activity and stress management practices like meditation and mindfulness. Furthermore, advances in functional foods, nutraceuticals, and bioactive chemicals provide new prospects for improving antioxidant absorption and efficacy in reducing oxidative stress-related illnesses. Personalized nutrition approaches, led by genetic and metabolic analysis, are also gaining popularity for adjusting antioxidant intake to match individual health requirements [7]. This chapter aims to give an in-depth examination of the significance of antioxidants in human health, classification, and relevance in the context of contemporary lifestyle concerns. It tries to understand the processes by which antioxidants combat oxidative stress, as well as their role in chronic disease prevention. The discussion will cover a thorough comparison of natural and synthetic antioxidants, including their origins, benefits, limitations, and potential health hazards. This chapter intends to raise awareness of the critical role of antioxidants in improving general well-being and longevity in the face of modern health issues by integrating scientific evidence and current research trends.

2 Natural Antioxidants

Natural antioxidants are compounds found in nature, such as fruits and vegetables, that help our bodies neutralize free radicals. They protect and delay cell damage. Flavonoids, carotenoids, polyphenols, and vitamins are crucial cellular protective antioxidant molecules that aid in the inhibition of cell and tissue deterioration. According to studies, people are opting for natural antioxidants instead of synthetic antioxidants due to their well-being and possible health values. Natural antioxidants are an important part of a healthy diet due to their presence in many plant-based foods [8]. Natural antioxidants primarily function against ROS, which are responsible for a variety of diseases such as cardiovascular, neurodegenerative, and cancer. Research has shown that vitamin C and polyphenols prevent atherosclerosis and some cancers, while flavonoids are good for cardiac health [9]. Natural antioxidants scavenge free radicals, lower inflammation, and strengthen the body's own endogenous antioxidant systems. Foods high in antioxidants, like green leafy vegetables, berries, and nuts, are helpful in general health and disease prevention [10]. Natural antioxidants have useful purposes in the food and pharmaceutical sectors beyond their health advantages. They are used in cosmetics and medicinal formulations that protect against oxidative stress-related damage and food preservatives used to stop oxidation-related deterioration. The trend toward functional compounds and natural food preservatives has encouraged more study on their effectiveness and extraction techniques [11]. The possibility of natural antioxidants in disease prevention and health maintenance is widening with continuous scientific developments, thereby supporting the growing practice of incorporating them into regular diets and medical treatments.

2.1 Sources of Natural Antioxidants

2.1.1 Plant-Based Antioxidants

Plant-based antioxidants are often found in a variety of fruits, vegetables, grains, and medicinal plants. These chemicals are essential for eliminating free radicals, reducing oxidative damage, and lowering the chances of chronic illnesses. Flavonoids, also known as polyphenols, are found in citrus fruits, berries, tea, apples, and onions. These antioxidants assist cells in resisting oxidative damage and promoting cardiovascular health. Research reveals that flavonoids contribute to anti-inflammatory and anti-carcinogenic characteristics by neutralizing free radicals. Carrots, tomatoes, spinach, and sweet potatoes have high levels of carotenoids such as beta-carotene, lutein, and lycopene. They are powerful antioxidants that promote eye health and lower the incidence of some malignancies. Studies show that carotenoids can protect biological membranes and tissues from oxidative damage. Coffee, red wine, dark chocolate, and olive oil have high levels of polyphenols, including phenolic acids. These compounds have considerable antioxidant activity because they neutralize free radicals and modulate signaling pathways related to inflammation and cellular aging [12].

2.1.2 Animal-Based Antioxidants

While plant-based antioxidants are well studied, numerous natural antioxidants originating from animals serve critical roles in shielding biological systems from oxidative damage. Glutathione is a tripeptide antioxidant produced in the body and present in meat, fish, eggs, and dairy products. It serves as an important detoxifier, scavenging ROS and promoting cellular health. Studies demonstrate that dietary intake of glutathione can improve the body's ability to fight oxidative stress and promote immunological function [13]. CoQ10 is a fat-soluble antioxidant found in organ meat, fatty fish, and chicken. It is essential for mitochondrial energy production and serves as a strong antioxidant, protecting cells from oxidative stress, especially in cardiovascular regions. Research emphasizes the beneficial effects of CoQ10 in avoiding atherosclerosis and neurological disorders [14]. Dietary antioxidants, as presented in Table 1, are key to minimizing oxidative stress, enhancing immunity, and lowering the risk of cardiovascular disease and cancer.

Table 1 The significance and functions of plant and animal-based antioxidants

S. no	Category	Antioxidant	Sources	Functions
1	Plant-based antioxidants	Flavonoids	Citrus fruits, tea, berries, apples, onions	Scavenge free radicals, enhance cardiovascular health, and have anti-inflammatory and anti-carcinogenic properties
		Carotenoids	Carrots, tomatoes, spinach, sweet potatoes	Protect biological membranes, support eye health, and reduce cancer risk
		Polyphenols	Coffee, red wine, dark chocolate, olive oil	Neutralize free radicals and modulate inflammation and aging pathways
2	Animal-based antioxidants	Glutathione	Meat, fish, eggs, dairy products	Detoxifies reactive oxygen species and enhances immune function
		Coenzyme Q10	Organ meat, fatty fish, chicken	Supports mitochondrial energy production and protects cardiovascular and neurological health

2.2 *Mechanisms of Action*

Natural antioxidants defend the body through a variety of processes, the most important of which are free radical neutralization and the enhancement of enzymatic defense systems. These systems serve to reduce oxidative stress. Antioxidants function through two basic mechanisms: free radical scavenging and enzymatic antioxidant pathways.

2.2.1 Free Radical Scavenging

The primary mechanism of antioxidants is free radical scavenging, which reduces ROS and reactive nitrogen species (RNS), remnants of cellular metabolism. ROS, including superoxide anions, hydrogen peroxide, and hydroxyl radicals, may lead to lipid peroxidation, protein oxidation, and DNA damage if not managed. Natural antioxidants protect against this damage via numerous molecular mechanisms. Antioxidants provide hydrogen atoms to free radicals, transforming them into stable, non-reactive molecules. This mechanism is essential for phenolic antioxidants, which include flavonoids and polyphenols found in green tea, berries, and red wine [15]. Antioxidants contribute an electron to neutralize free radicals, hence lowering their reactivity. Carotenoids, such as beta-carotene and lycopene, have significant single-electron transfer action, reducing lipid peroxidation in cell membranes [16]. Sequential proton loss electron transfer involves a two-step process in which antioxidants lose a proton, resulting in a charged species that contributes an electron to neutralize free radicals. Research has proved that polyphenols, such as gallic acid and epigallocatechin gallate (EGCG), exploit this pathway to protect cells from oxidative damage [17]. Metal chelation involves transition metals like iron (Fe^{2+}) and copper (Cu^{+}), which can stimulate the production of hydroxyl radicals via Fenton reactions. Antioxidants such as flavonoids and tannins bind to these metals, limiting their potential to produce free radicals [18]. Some antioxidants react directly with free radicals, forming stable, non-reactive complexes that further inhibit oxidative chain reactions. Research on vitamin C has revealed its potential to engage in radical adduct formation, stabilizing ROS before they may trigger damage [19].

2.2.2 Enzymatic Antioxidant Pathways

In addition to their direct antioxidant capabilities, natural substances help the body's enzymatic defense mechanisms, which are constantly working to eliminate ROS. SOD prevents oxidative damage by converting superoxide anions to hydrogen peroxide. Natural polyphenols such as resveratrol and quercetin have been shown to increase SOD activity, which improves cell protection against oxidative stress. Catalase (CAT) changes hydrogen peroxide into oxygen and water, reducing the accumulation of harmful peroxides [20]. Glutathione peroxidase (GPx) detoxifies the lipid hydroperoxides and hydrogen peroxide by altering them to nontoxic molecules. This enzyme is dependent on glutathione (GSH), a natural antioxidant found in animal-based diets and generated in the body. Research reveals that plant-derived antioxidants, such as curcumin and EGCG, improve GPx activity and enhance its protective benefits [21]. Natural antioxidants influence cellular antioxidant defenses via signaling mechanisms through the Nrf2-Keap1 pathway activation and NF-κB inhibition, as shown in Fig. 1. The transcriptional factor Nrf2 controls the expression of antioxidant enzymes. Certain polyphenols, such as sulforaphane from cruciferous vegetables, activate the Nrf2 pathway, which increases cellular adaptability to oxidative stress [22]. The NF-κB pathway influences inflammation and oxidative stress. Natural antioxidants, such as flavonoids and omega-3 fatty acids, prevent NF-κB activation and reduce inflammation-related oxidative stress [23].

2.3 Health Benefits of Natural Antioxidants

Natural antioxidants provide significant health benefits because they lower oxidative damage, a major part in the occurrence of chronic illnesses and aging. Their primary advantages include the avoidance of oxidative stress illnesses, as well as anti-aging and immune function enhancement.

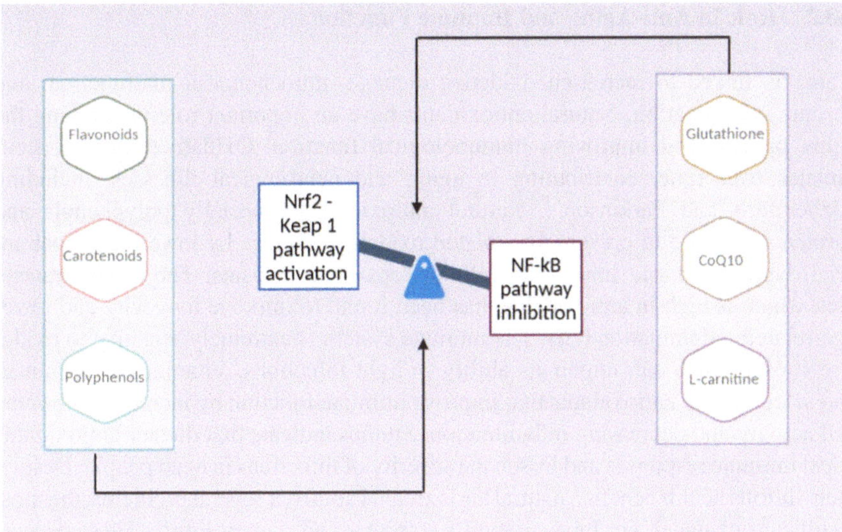

Fig. 1 The image presents a diagram illustrating the balance between Nrf2-Keap1 pathway activation and NF-kB pathway inhibition

2.3.1 Prevention of Oxidative Stress-Related Diseases

Oxidative stress is described as an imbalance between ROS and the body's antioxidant defense system. This imbalance is related to a variety of chronic conditions, including heart disease, cancer, and diabetes. Oxidative stress hastens the progression of cardiovascular diseases by promoting lipid peroxidation, endothelial dysfunction, and inflammation. Natural antioxidants, including flavonoids, carotenoids, and vitamins C and E, protect blood vessels from oxidative damage, improve endothelial function, and reduce the risk of atherosclerosis. Studies demonstrate that antioxidants found in fruits and vegetables, such as resveratrol and polyphenols, protect against cardiovascular damage. ROS can cause DNA mutations, resulting in uncontrolled cell growth and tumor formation. Natural antioxidants such as polyphenols, curcumin, and astaxanthin have been found to inhibit cancer cell growth, improve apoptosis, and prevent tumor formation. According to studies, astaxanthin, a potent antioxidant derived from marine sources, has high anti-cancer properties and can protect against oxidative DNA damage. Oxidative stress induces resistance to insulin and pancreatic beta-cell dysfunction, both of which contribute to diabetes development. Natural antioxidants, including quercetin, catechins, and anthocyanins, have been shown to improve the response to insulin and reduce inflammation in diabetics. Polyphenolic antioxidants regulate glucose metabolism and alleviate diabetes complications induced by oxidative stress [24].

2.3.2 Role in Anti-Aging and Immune Function

Aging is linked to increased oxidative damage, mitochondrial malfunction, and chronic inflammation. Natural antioxidants have an important role in delaying the aging process and improving immunological function. Oxidative damage accumulates over time, contributing to aging and neurological illnesses, including Alzheimer's and Parkinson's. Natural antioxidants, especially polyphenols and carotenoids, serve to prevent age-related oxidative damage by lowering mitochondrial dysfunction and improving cellular repair mechanisms. The Mediterranean diet, which is high in antioxidants, has been found to improve longevity and lower age-related inflammation [25]. The immune system is extremely sensitive to oxidative stress, which can impair its ability to fight infections. Vitamin C, vitamin E, and selenium are antioxidants that improve immune function by increasing immune cell activity and decreasing inflammation. Studies indicate that dietary antioxidants boost immune responses and lessen the severity of infections in aged people. Despite their various health benefits, natural antioxidants confront several problems, the most significant of which are bioavailability, stability, and dietary limits. These aspects reduce their efficacy and limit their utility as medicinal and functional food additives.

2.3.3 Bioavailability and Stability

The efficiency of natural antioxidants is largely determined by their bioavailability, which is the proportion of a substance that enters the bloodstream and exerts biological effects. Many natural antioxidants, including polyphenols, carotenoids, and flavonoids, have low bioavailability due to their low solubility, quick metabolism, and poor absorption in the gastrointestinal tract. For example, despite its significant antioxidant qualities, resveratrol and curcumin have low absorption and are quickly removed from the body, decreasing their therapeutic efficacy [26]. Stability is another significant difficulty. Natural antioxidants can deteriorate in a variety of situations, including heat, light, and oxygen exposure. Polyphenols, for example, are oxidized and structurally altered during food processing and storage, lowering their effectiveness. Encapsulation approaches, such as nano formulations and micelle-based delivery systems, have been proven in studies to greatly boost the stability and bioavailability of antioxidants like astaxanthin and quercetin, improving their usefulness in therapeutic applications.

2.3.4 Dietary Limitations

Natural antioxidant intake through diet is frequently uneven due to factors such as food processing, seasonal availability, and personal dietary preferences. While fruits, vegetables, and whole grains are high in antioxidants, modern diets frequently lack adequate amounts of these nutrients, resulting in poor antioxidant consumption. Also, food procedures like frying and extended heating can diminish the antioxidant content of foods, decreasing their efficiency [27]. Furthermore, the bioefficacy of dietary antioxidants differs by person due to genetic differences, gut microbiota makeup, and metabolic variables. Some people may digest and use particular antioxidants more efficiently than others, altering their overall health. According to research, specific nutrition techniques, such as targeted supplementation and functional food formulations, can assist in overcoming these constraints and maximize antioxidant consumption for diverse groups [28].

3 Synthetic Antioxidants

Antioxidants are chemicals that prevent other substances from oxidizing. Food items that are resistant to this type of chemical change require antioxidants to prevent food degradation caused by oxidation. Although routinely used synthetic antioxidants (such as BHT, TBHQ, and BHA) are highly effective, customer demand for processed foods containing natural ingredients is increasing. Fruits, vegetables, tea, herbs, and spice plants are examples of plants that have long been used for their health benefits. These raw materials contain carotenoids, phenolic compounds, and antioxidant additions, which reduce lipid oxidation. Food phenolics induce antioxidant activity primarily through their roles as reducing agents, hydrogen donors, and singlet oxygen quenchers. Moreover, certain phenolics can chelate metal ions, which catalyze oxidation reactions [29]. Chemical compounds are mixed in laboratories to produce synthetic antioxidants, which are mostly used in food preservation. Their outstanding properties and availability make them excellent food additives for preventing rancidification. Since they are not found in nature, they are chemically manufactured substances that are added to food as additives to support avoiding lipid oxidation [30]. BHA, BHT, TBHQ, and propyl octyl and dodecyl gallates are examples of synthetic antioxidants that are primarily phenolic [31]. The initial use of BHA and BHT was to prevent petroleum from oxidative gumming. The maximum quantity of BHT that can be added to food is 100 mg/kg of fat, although twice as much TBHQ and BHA can be added. Using synthetic antioxidants excessively or improperly can cause cancer [32]. Synthetic phenolic antioxidants (SPAs) are highly effective lipophilic antioxidants present in a wide range of diets. When compared to naturally occurring antioxidants like tocopherols, which are produced by plants and animals, SPAs have the advantages of a stable structure, high thermal stability, and significant antioxidant activity. They can also be manufactured in huge numbers to fulfill demand. However, due to their widespread use, SPAs can be found in a wide range of biological samples (e.g., human serum, urine, and marine mollusks) as well as environmental matrices. SPAs are also used in stable settings and are resistant to temperature-induced deterioration. As a result, they have been commonly found in mollusks, surface and groundwater, indoor and outdoor dust, municipal sewage sludge, aged landfill leachate, human nails, urine, and serum. The amount of BHT in environmental matrices has received a lot of attention, making it one of the most used SPAs. A concentration of BHT ranging from one trillion parts (ppm) has been found by certain researchers in soil, surface water, ground water, indoor dust, and household sewage [33].

3.1 Food Preservation

Antioxidants can increase food stability, prevent or stop food oxidation, and increase the time that food can be stored. Various sources classify antioxidants into two main groups: synthetic and natural. Food can contain synthetic antioxidants, of which phenolic antioxidants are more prevalent, which include BHT, BHA, tert-butylhydroquinone (TBHQ), and propyl gallate (PG) [34]. Living organisms, enzymes, and atmospheric oxygen cause fat-rich food to change color and smell, lowering its nutritional value and perhaps affecting human health. To inhibit lipid oxidation, antioxidants are commonly added to lipids and high-fat diets. The phenolic antioxidants TBHQ, BHT, PG, and BHA are widely used because of their low cost and high chemical stability. TBHQ, BHA, and BHT can be used in meats and edible oils. In dried pork slices, antioxidants were utilized in a complicated combination of 100 mg/kg TBHQ + 100 mg/kg BHA + 100 mg/kg citric acid. Furthermore, the combination of BHT and BHA preserves frozen bacon slices, pork chops, and carp fowl.

3.2 Mechanisms of Action

Natural antioxidants, such as polyphenols and terpenoids, act by scavenging reactive oxygen and nitrogen species, including superoxide, hydroxyl, and peroxyl radicals. These compounds are effective in inhibiting lipid oxidation in food matrices, which is crucial for maintaining the quality and shelf life of high-fat foods like meat and fish. Essential oils, another category of natural antioxidants, exhibit their effects through similar mechanisms, providing both antioxidant and antimicrobial properties that help in preserving food products [35, 36].

3.3 Health Benefits

Natural antioxidants derived from fruits, vegetables, and other plant sources are associated with numerous health benefits. They are known to improve health status and reduce the probability of rising diseases due to their antioxidant and antimicrobial activities. These compounds, such as those found in fruits like grapes and olives, interrelate with different radicals and show various mechanisms of action that contribute to their health-promoting effects [37].

3.4 Cost-Effective Alternatives to Natural Antioxidants

Natural antioxidants are often sourced from agricultural waste and by-products, making them a cost-effective alternative to synthetic antioxidants. For instance, extracts from fruit skins, seeds, and pulp are valuable due to their high content of bioactive compounds like catechin and gallic acid [38]. The use of plant-based antioxidants not only provides economic benefits but also aligns with consumer preferences for natural and sustainable food additives. Natural antioxidants play a crucial role in the preservation of processed foods by extending their shelf life and maintaining quality. They are used in various forms, such as free or encapsulated extracts, and are combined into films and coatings to prevent microbial growth and slow oxidative reactions [39]. Essential oils have been shown to enhance the oxidative and microbial stability of foods, making them effective natural preservatives [36].

3.5 Toxicity and Side Effects

Health Risks: SPAs such as BHA, BHT, and TBHQ have been linked to various health issues, including hepatic toxicity, endocrine disruption, and potential carcinogenicity. These compounds can cause DNA damage and promote tumor development, particularly through their transformation products like BHT-Q, which can be more toxic than the parent compounds [40]. Developmental Toxicity: Studies on zebrafish have shown that SPAs can cause developmental issues such as decreased heart rates, body lengths, and various morphological deformities. These effects are linked to interference with critical biological pathways [41]. Endocrine Disruption: SPAs have been shown to disturb steroidogenesis, affecting hormone levels and potentially causing estrogenic effects, which raises concerns about their impact on reproductive health.

3.6 Regulatory Concerns

Environmental Contamination: SPAs are found in numerous environmental matrices, including indoor dust and water bodies, leading to widespread human exposure through food, dust ingestion, and personal care products [42, 40]. *Regulatory Challenges*: There is a need for stricter regulatory frameworks to assess the long-term health impacts of SPAs. Current regulations may not adequately address the possible risks associated with their widespread use in food and consumer products [43].

3.7 *Role of Antioxidants in Plants*

3.7.1 Defense Mechanisms Against Environmental Stress

Plants experience a variety of environmental challenges, such as drought, salinity, severe temperatures, and disease infections, which cause oxidative stress. Antioxidants are crucial in defending plants from ROS, which can cause damage to cells. Plants' antioxidant defense system is made up of both enzymatic and non-enzymatic components that contribute to redox balance and stress resilience. Plants use a variety of antioxidant enzymes, including SOD, CAT, peroxidases (POX), ascorbate peroxidase (APX), glutathione reductase (GR), and glutathione peroxidase. These enzymes act cooperatively to neutralize ROS and avoid oxidative stress [44]. In addition to enzymatic antioxidants, plants produce non-enzymatic compounds such as ascorbic acid (vitamin C), tocopherols (vitamin E), carotenoids, glutathione, and phenolic compounds. These molecules scavenge ROS and help regulate cellular redox balance [45]. Antioxidant systems not only scavenge ROS but also play a role in signaling pathways that trigger stress responses. Plants adjust their antioxidant levels based on environmental cues to balance ROS production and detoxification [46]. Environmental circumstances such as salinity, drought, and temperature extremes induce oxidative stress, which requires strong antioxidant agents. Both enzymatic and non-enzymatic antioxidants work to defend cellular structures, sustain metabolic function, and enhance plant tolerance [47]. ROS act as signaling molecules that modulate the expression of antioxidant genes. For example, hydrogen peroxide (H_2O_2) has been shown to upregulate antioxidant enzyme genes, such as SOD, CAT, and ascorbate peroxidase (APX), in transgenic tobacco plants, enhancing their stress tolerance [48]. Epigenetic alterations such as DNA methylation, histone acetylation, and chromatin remodeling all influence antioxidant gene expression. Redox regulation affects gene expression by altering transcription factors such as NF-κB and AP-1, which respond to oxidative stress signals. Studies in rice and wheat have shown that antioxidant defense transcription is dynamically controlled in response to environmental stress. A transcriptome investigation found that genes encoding SOD, CAT, and APX have distinct patterns of expression under cold and drought stress, revealing species-specific regulatory mechanisms [49]. Overexpression of multiple antioxidant enzymes in plants has been shown to improve stress tolerance. For example, transgenic tobacco plants expressing CuZnSOD, APX, and dehydroascorbate reductase (DHAR) exhibited higher resistance to oxidative stress and increased antioxidant enzyme activity [50]. Plant hormones like abscisic acid (ABA), ethylene, and cytokinin also affect the expression of antioxidant defense genes. Overexpression of cytokinin oxidase/dehydrogenase in tobacco plants increased antioxidant gene expression, resulting in improved stress tolerance [51]. Plant antioxidant enzyme expression is regulated through a complex mechanism that includes ROS signaling, epigenetic changes, transcriptional reprogramming, genetic engineering, and hormone regulation. Understanding these pathways helps to improve plant resilience to environmental stress through targeted breeding and biotechnological interventions.

3.7.2 Applications of Antioxidants in Crop Improvement

The role of antioxidants in crop improvement is pivotal in enhancing stress tolerance, ensuring sustainable agriculture, and increasing crop productivity. By mitigating oxidative stress, plants can withstand abiotic and biotic challenges such as drought, salinity, extreme temperatures, and pathogen attacks. Advances in genetic engineering, agronomic practices, and exogenous applications of antioxidants have contributed significantly to improving crop resilience.

Genetic Engineering for Antioxidant Enhancement

Overexpression of antioxidant genes in crops has shown promising results in improving stress tolerance. For example, transgenic tobacco plants expressing multiple antioxidant enzymes (APX, DHAR, and CuZnSOD) exhibited improved resistance to oxidative damage, salinity, and drought. In Arabidopsis, seed-specific overexpression of Mn-SOD, CAT, and tocopherol biosynthesis genes improved growth and primary seedling growth under oxidative stress [52]. Transgenic sweet potato and potato plants engineered with both CuZnSOD and APX in chloroplasts exhibited improved tolerance to multiple environmental stresses, including chilling, heat, and drought.

Exogenous Application of Antioxidant Compounds

Field studies have demonstrated that applying antioxidant agents like ascorbate and glutathione improves plant resilience to oxidative stress. This approach, known as "*antioxidant agriculture*," integrates genetic engineering, agronomic practices, and microbial inoculation to reinforce plant antioxidant defenses [53]. The application of 24-epibrassinolide and spermine in maize significantly upregulated antioxidant defense genes, reducing drought-induced oxidative damage [54].

Manipulation of Redox Signaling for Stress Adaptation

Antioxidants play a vital role in modulating gene expression through redox signaling. For instance, research on ROS-mediated gene regulation has identified key transcription factors that control stress response pathways in plants [55]. In wheat, overexpression of the TaPRX-2A gene improved drought tolerance by enhancing antioxidant enzyme activity and reducing ROS levels [56].

Improving Crop Yield and Productivity

Genetic and biochemical interventions targeting antioxidant pathways can significantly enhance crop productivity. Studies have demonstrated that ROS-regulated genes influence not only stress tolerance but also growth and yield [57]. The application of 5-aminolevulinic acid (ALA) has been shown to regulate photosynthesis, nutrient uptake, and antioxidant metabolism, thereby promoting growth under stress conditions [58].

Crop-Specific Antioxidant Strategies

A relative analysis of antioxidant gene families in finger rice, sorghum, millet, and foxtail discovered key targets for improving drought tolerance and oxidative stress resilience. Rice cultivars engineered with OsHBP1b, a nuclear-localized transcription factor, exhibited enhanced salinity and drought tolerance by improving antioxidant enzyme activity.

3.8 Antioxidants in Animal Health

3.8.1 Livestock Productivity and Health

Antioxidants play a vital role in maintaining livestock health and enhancing productivity by mitigating oxidative stress, improving immune function, and ensuring high-quality animal products such as milk and meat.

Impact on Ruminant Health and Productivity

Oxidative stress in livestock is a significant concern, especially due to environmental challenges like heat stress, poor nutrition, and metabolic imbalances. Antioxidants such as vitamin E and selenium help maintain redox balance, improve immune responses, and enhance meat and milk quality. Supplementing livestock feed with natural antioxidants, including plant-based polyphenols and flavonoids, has been shown to enhance growth performance, decrease oxidative damage, and improve overall health. Oxidative stress negatively impacts milk yield and quality. Research has shown that antioxidant supplementation, particularly vitamin E and selenium, improves lactation efficiency and milk composition. The use of antioxidant-enriched feed can help maintain redox homeostasis in dairy cows, reducing metabolic disorders and improving reproductive health. Antioxidants influence meat quality by preventing lipid peroxidation and maintaining freshness. Farm animal diets supplemented with antioxidants like selenium and vitamin E lead to improved meat quality, texture, and shelf life [59].

3.8.2 Prevention of Age-Related Disorders

Oxidative stress contributes to aging and age-related disorders in livestock, affecting their longevity, reproductive health, and productivity. Aging in livestock leads to increased oxidative stress, which impairs cellular function. Supplementation with antioxidants like vitamin E, flavonoids, and selenium reduces oxidative damage and improves overall health [60]. Antioxidants help maintain reproductive efficiency in aging animals by preventing oxidative damage to sperm and egg cells, thus supporting fertility and reducing reproductive disorders.

3.9 Antioxidants in Human Health

3.9.1 Nutrition and Dietary Sources

Antioxidants are essential in human nutrition, as they aid in neutralizing damaging free radicals, lowering the risk of chronic diseases. Plant-based foods such as fruits (berries, citrus), vegetables (spinach, kale), seeds, nuts, and complete grains all contain antioxidants. These foods contain some of the most valuable antioxidants, including vitamins, carotenoids, flavonoids, and polyphenols. Studies highlight that a diet rich in natural antioxidants can significantly lower the risk of cardiac diseases, neurodegenerative disorders, and some cancers. Functional foods enriched with antioxidants, such as green tea, dark chocolate, and omega-3-rich fish, have been shown to provide additional health benefits. Dietary supplementation with vitamin C, vitamin E, and polyphenols is widely used for boosting immune function and combating oxidative stress [61].

3.9.2 Therapeutic Applications

Antioxidants are used in medical treatments and disease prevention strategies due to their ability to counteract oxidative damage. Antioxidants have a vital role in reducing inflammation and preventing cardiac diseases, neurodegenerative disorders like Alzheimer's, and diabetes [55]. Antioxidant compounds are incorporated into pharmaceutical formulations to treat oxidative stress-related diseases. Polyphenols, flavonoids, and vitamins are extensively researched for their role in anti-aging, immune support, and cancer prevention [62].

4 Impact of Modern Lifestyles on Antioxidant Efficacy

Modern lifestyles, characterized by dietary habits, physical activity levels, and environmental exposures, significantly influence the antioxidant efficacy of the body. Antioxidants play a crucial part in neutralizing oxidative stress, which is implicated in the cause of various diseases such as diabetes, cardiac diseases, and cancer. Several studies have highlighted the effect of lifestyle choices on antioxidant capacity. For instance, a study assessing the antioxidant capacity in healthy individuals found that nonsmoking, regular physical activity, and vitamin or mineral supplementation were associated with increased antioxidant capacity. Conversely, smoking, psychological stress, alcohol consumption, and poor intake of vegetables and fruits remained linked to decreased antioxidant potential [63]. Similarly, the Mediterranean diet, rich in vegetables, fruits, and olive oil, has been shown to enhance antioxidant properties and decrease the risk of metabolic and cardiovascular illnesses [25, 64]. Exercise has a complex relationship with oxidative stress; while regular moderate exercise can enhance antioxidant defenses, acute intense exercise may increase oxidative stress. This stress, however, can stimulate the body's endogenous antioxidant mechanisms, a concept known as hormesis. Dietary strategies, including adequate intake of vitamins and minerals, can support these defenses, although excessive supplementation may have adverse effects [65]. The role of antioxidant supplementation in mitigating oxidative stress-related diseases is debated. While some studies suggest benefits in reducing disease risk, others have not demonstrated significant clinical benefits. The effectiveness of supplementation may depend on issues like the type of antioxidant, dosage, and individual genetic variability.

4.1 Lifestyle Factors Affecting Antioxidant Needs

Exposure to pollutants and toxins is a significant lifestyle factor that increases the body's antioxidant needs. Modern lifestyles often involve contact with a wide range of substances, which can induce oxidative stress by increasing the stages of prooxidant factors. This exposure can lead to physical and functional alterations at the cellular level, contributing to diseases such as cardiac diseases and cancer [66]. Additionally, environmental influences like xenobiotic pollution and diet-borne toxins are known to exacerbate oxidative stress, further increasing the demand for antioxidants to counteract these effects [67]. The consumption of processed foods is another lifestyle factor that affects antioxidant needs. Diets high in processed foods often lack essential nutrients and antioxidants, which are crucial for maintaining oxidative balance. The Western dietary pattern, categorized by high intakes of processed foods and poor consumption of vegetables and fruits, is related to increased oxidative stress. This dietary pattern fails to provide adequate antioxidants, which are necessary to combat oxidative damage and support immune function, especially during times of increased stress such as the COVID-19 pandemic [68]. Moreover, excessive consumption of fast foods contributes to obesity and oxidative stress, highlighting the role of a balanced diet rich in natural antioxidants [69]. Sedentary lifestyles significantly contribute to oxidative stress, particularly as individuals age. Lack of physical activity is a prooxidant factor that increases oxidative stress markers, leading to higher antioxidant needs. Studies show that sedentary lifestyles are related to improved oxidative stress, especially in older populations, suggesting that regular physical activity is crucial for maintaining antioxidant defenses [70]. Additionally, psychological stress, often linked with sedentary lifestyles, negatively impacts antioxidant capacity, further emphasizing the need for lifestyle modifications to enhance antioxidant defenses [63]. Oxidative stress is an important factor in the development of various diseases, including cardiac diseases, neurodegenerative diseases, diabetes, and cancer. Modern lifestyle issues such as a low diet, lack of exercise, and exposure to pollution exacerbate oxidative damage, contributing to disease prevalence.

4.2 Links Between Lifestyle and Chronic Diseases

Oxidative Damage and Chronic Conditions: Oxidative damage is implicated in the cause of numerous diseases, including cardiac diseases, diabetes, neurodegenerative disorders, and cancer. It results from an imbalance among prooxidant and antioxidant factors, leading to cellular damage and disease progression [71–73]. (i) *Lifestyle Factors*: Modern lifestyle choices, such as consumption of processed foods, exposure to chemicals, and physical inactivity, significantly contribute to oxidative damage. These factors are related to the growth of chronic diseases through increased oxidative damage and inflammation [74, 73]. (ii) *Antioxidant Interventions*: Antioxidant-rich diets and lifestyle changes can mitigate oxidative stress. Studies show that interventions focusing on nutrition, physical activity, and stress management can improve oxidative stress parameters and reduce disease risk [75, 74]. (iii) *Inflammation and Aging*: Aging and chronic inflammation, or "inflammaging," are closely linked with oxidative stress, further increasing the risk of age-related diseases like cardiovascular diseases [76, 71]. (iv) *Oxidative Balance Score (OBS)*: The OBS is a tool used to assess the balance of oxidative damage-related exposures from food and lifestyle. Higher OBS is related to reduced prevalence of diseases such as kidney disease and obstructive pulmonary disease, emphasizing the importance of antioxidant-rich foods and lifestyles [75, 77, 78].

4.3 Strategies for Improving Antioxidant Intake

4.3.1 Integrative Approaches Combining Natural and Synthetic Sources

Integrating natural and synthetic antioxidants can be a strategic approach to enhance antioxidant intake. Natural antioxidants, such as those found in edible mushrooms, have been shown to effectively improve antioxidant defenses because of their bioactive compounds, such as polyphenols and vitamins [79]. However, synthetic antioxidants are also generally used in the food industry, despite concerns about their potential toxicity and health risks [80]. The mixture of both sources can possibly offer a stable approach, leveraging the efficacy of synthetic antioxidants while minimizing their adverse effects by incorporating natural alternatives [81].

4.3.2 Personalized Nutrition

Personalized nutrition is an emerging strategy that tailors antioxidant supplementation to an individual's specific needs. This approach considers the unique antioxidant status of each person, optimizing supplementation to improve health outcomes and exercise performance [82]. By focusing on personalized interventions, it is possible to address specific antioxidant deficiencies, leading to consistent improvements in redox profiles and overall health [82]. This method highlights the importance of data-driven, individualized treatments in enhancing antioxidant intake and managing oxidative stress [83].

5 Future Directions and Research Gaps

Innovations in antioxidant delivery systems, particularly through nanotechnology and bioengineering, are advancing rapidly. These innovations aim to enhance the bioavailability, stability, and targeted distribution of antioxidants, addressing the limitations of traditional methods.

5.1 Innovations in Antioxidant Delivery Systems

Nanoparticles are used to increase the stability, controlled release, and site-specific delivery of antioxidants. This approach enhances the bioavailability and efficacy of antioxidants by overcoming issues like poor absorption and degradation during delivery [84, 26, 85]. Biopolymer-based nano-drug delivery methods offer biocompatibility, biodegradability, and less immunogenicity, making them promising for targeted and controlled release of antioxidants [86]. Techniques such as nanoemulsion-based encapsulation are being explored to improve oxidative stability and solubility of natural antioxidants, particularly in food applications.

5.2 Development of Safer Synthetic Alternatives

Advances in the synthesis of sustainable antioxidants from natural monomers, such as lignin and quercetin, are being explored for their free radical scavenging abilities and possible applications in food packaging and drug delivery. The usage of natural antioxidants from plants is being emphasized as a safer alternative to synthetic antioxidants, particularly in food and pharmaceutical applications [38, 86].

5.3 Bridging the Gap Between Research and Practical Applications

Despite promising preclinical results, the clinical translation of antioxidant therapies faces challenges such as low bioavailability and stability. Nanoparticle-based delivery systems are being investigated to address these issues and enhance clinical applicability [84, 87]. The integration of natural antioxidants into practical applications, such as food preservation and pharmaceuticals, is being pursued to meet consumer demands for natural and sustainable products [38, 88].

6 Conclusion

Antioxidants play a major role in mitigating oxidative damage-related diseases like cardiovascular disorders, diabetes, and neurodegenerative problems in today's lifestyle, which is characterized by excessive stress, processed food intake, and environmental contaminants. Natural antioxidants produced from plants and animals, such as flavonoids, carotenoids, and coenzyme Q10, provide significant health advantages. Synthetic antioxidants, on the other hand, offer cost-effective and stable alternatives while raising concerns about potential toxicity and regulatory constraints. A balance of natural and synthetic antioxidants is necessary for supporting general health in plants, animals, and humans. Furthermore, the increased prevalence of chronic diseases emphasizes the importance of smart dietary and lifestyle changes to optimize antioxidant consumption. Antioxidants must be integrated into modern lifestyles in a holistic manner. This includes promoting antioxidant-rich foods, including fruits, vegetables, nuts, and whole grains, while avoiding processed and artificially preserved foods. Personalized nutrition, advances in antioxidant delivery technologies, and lifestyle changes such as regular exercise and stress management can all help to boost their effectiveness. Future research should focus on creating safer synthetic alternatives as well as enhancing the stability and bioavailability of natural antioxidants. In future decades, increasing public awareness and implementing evidence-based dietary guidelines will be critical to utilize antioxidants in the inhibition and therapy of lifestyle-related diseases.

References

1. Paital, B. (2020). Antioxidants for human health. In *Bulletin of medical and clinical research* (pp. 22–26). IOR International Press. https://doi.org/10.34256/br2012
2. Chib, A., Gupta, N., Bhat, A., Anjum, N., & Yadav, G. (2020). Role of antioxidants in food. *International Journal of Chemical Studies*, *8*(1), 2354–2361. https://doi.org/10.22271/chemi.2020.v8.i1aj.8621
3. Swaroopa, G., Srinath, D., Swaroopa, G., & Srinath, D. (2017). Antioxidants and their roll in human health: A review. *International Journal of Agricultural Science and Research, 7*(5), 9–18. https://doi.org/10.24247/ijasroct20172
4. Singh, P., Paul, V., & Dubey, R. (2022). Antioxidant foods. *International Journal of Advances in Agricultural Science and Technology*, *9*(4), 1–7. https://doi.org/10.47856/ijaast.2022.v09i04.001
5. Salim, F., Adnan, N., Shuib, N. S., & Mohd Yusof, R. (2022). Antioxidants for Health Management. *Jurnal Intelek*, *17*(1), 55. https://doi.org/10.24191/ji.v17i1.15851
6. Yang, C. S., Ho, C.-T., Zhang, J., Wan, X., Zhang, K., & Lim, J. (2018). Antioxidants: Differing meanings in food science and health science. *Journal of Agricultural and Food Chemistry, 66*(12), 3063–3068. https://doi.org/10.1021/acs.jafc.7b05830
7. Ashraf, M. V., Khan, S., Misri, S., Gaira, K. S., Rawat, S., Rawat, B., Khan, M. A. H., Shah, A. A., Asgher, M., & Ahmad, S. (2024). High-altitude medicinal plants as promising source of phytochemical antioxidants to combat lifestyle-associated oxidative stress-induced disorders. *Pharmaceuticals, 17*(8), 975. https://doi.org/10.3390/ph17080975
8. Usoro, E. (2024). Comparative analysis of natural vs. synthetic antioxidants in Nigeria. *Journal of Chemistry*, *3*(3), 27–37. https://doi.org/10.47672/jchem.2515
9. Nemzer, B. V., Yashin, A. Y., Vedenin, A. N., Yashin, Y. I., Yashunsky, D. V., Nifantiev, N. E., & Kalita, D. (2019). Selected powerful natural antioxidants: Structure, food sources, antioxidant activities, and important health benefits. *Journal of Food Research, 8*(1), 60. https://doi.org/10.5539/jfr.v8n1p60
10. Khan, N., Bagade, M. V., Giri, M. R., & Bhoyar, L. A. (2023). The study of natural antioxidants in human health and diseases. *International Journal of Advanced Research in Science, Communication and Technology*, 266–267. https://doi.org/10.48175/IJARSCT-14233
11. Harun, N. F., & Ab Hamid, F. H. (2021). An overview of the extraction methods of plant-based natural antioxidant compounds. *Malaysian Journal of Chemical Engineering and Technology*. https://doi.org/10.24191/mjcet.v4i2.13055
12. Xu, D.-P., Li, Y., Meng, X., Zhou, T., Zhou, Y., Zheng, J., Zhang, J.-J., & Li, H.-B. (2017). Natural antioxidants in foods and medicinal plants: Extraction, assessment and resources. *International Journal of Molecular Sciences, 18*(1), 96. https://doi.org/10.3390/ijms18010096
13. Torres, R. C., Manalo, C. O., & Manongsong, E. B. (2019). Preparation of natural antioxidant health supplements from philippine-grown medicinal plants. *Open Journal of Pharmaceutical Science and Research*, 134–140. https://doi.org/10.36811/ojpsr.2019.110007
14. Castro, C. S., Cotrim, C. F. C., Santos, I. R. dos, Santos, K. J. G. dos, Paula, J. A. M. de, Caramori, S. S., Almeida, L. M. de, Bailão, E. F. L. C., & Borges, L. L. (2022). Application of

natural antioxidants in animal reproduction. *Ciência Animal Brasileira, 23*. https://doi.org/10.1590/1809-6891v23e-73601e
15. Zeb, A. (2020). Concept, mechanism, and applications of phenolic antioxidants in foods. *Journal of Food Biochemistry, 44*(9). https://doi.org/10.1111/jfbc.13394
16. Heymann, T., Schmitz, L. M., Lange, J., & Glomb, M. A. (2020). Influence of β-carotene and lycopene on oxidation of ethyl linoleate in one- and disperse-phased model systems. *Journal of Agricultural and Food Chemistry, 68*(9), 2747–2756. https://doi.org/10.1021/acs.jafc.9b07862
17. Markovic, Z. (2016). Study of the mechanisms of antioxidative action of different antioxidants. *Journal of Serbian Society for Computational Mechanics, 10*(1), 135–150. https://doi.org/10.5937/jsscm1601135M
18. Ivanova, A., Gerasimova, E., & Gazizullina, E. (2020). Study of antioxidant properties of agents from the perspective of their action mechanisms. *Molecules, 25*(18), 4251. https://doi.org/10.3390/molecules25184251
19. Robert, G., & Wagner, J. R. (2024). Scavenging of alkylperoxyl radicals by addition to ascorbate: An alternative mechanism to electron transfer. *Antioxidants, 13*(10), 1194. https://doi.org/10.3390/antiox13101194
20. Marković, Z. S., & Milenković, D. A. (2020). Different theoretical approaches in the study of antioxidative mechanisms. In *Computational modeling in bioengineering and bioinformatics* (pp. 211–256). Elsevier. https://doi.org/10.1016/B978-0-12-819583-3.00007-2
21. Kövesi, B., Worlanyo, A. P., Kulcsár, S., Ancsin, Z., Erdélyi, M., Zándoki, E., Mézes, M., & Balogh, K. (2024). Curcumin mitigates ochratoxin A-induced oxidative stress and alters gene expression in broiler chicken liver and kidney. *Acta Veterinaria Hungarica, 72*(1), 41–50. https://doi.org/10.1556/004.2024.01016
22. Nimse, S. B., & Pal, D. (2015). Free radicals, natural antioxidants, and their reaction mechanisms. *RSC Advances, 5*(35), 27986–28006. https://doi.org/10.1039/C4RA13315C
23. Aydin, C., & Seyidoglu, N. (2021). Natural antioxidants to the rescue? In *Antioxidants—benefits, sources, mechanisms of action*. IntechOpen. https://doi.org/10.5772/intechopen.96132
24. García-Aguilar, A., & Guillén, C. (2022). Targeting pancreatic beta cell death in type 2 diabetes by polyphenols. *Frontiers in Endocrinology*. https://doi.org/10.3389/fendo.2022.1052317
25. Chatzianagnostou, K., Del Turco, S., Pingitore, A., Sabatino, L., & Vassalle, C. (2015). The mediterranean lifestyle as a non-pharmacological and natural antioxidant for healthy aging. *Antioxidants, 4*(4), 719–736. https://doi.org/10.3390/antiox4040719
26. Khalil, I., Yehye, W. A., Etxeberria, A. E., Alhadi, A. A., Dezfooli, S. M., Julkapli, N. B. M., Basirun, W. J., & Seyfoddin, A. (2019). Nanoantioxidants: Recent trends in antioxidant delivery applications. *Antioxidants, 9*(1), 24. https://doi.org/10.3390/antiox9010024
27. Rodríguez-Cano, A. M., González-Ludlow, I., Suárez-Rico, B. V., Montoya-Estrada, A., Piña-Ramírez, O., Parra-Hernández, S. B., Reyes-Muñoz, E., Estrada-Gutierrez, G., Calzada-Mendoza, C. C., & Perichart-Perera, O. (2022). Ultra-processed food consumption during pregnancy and its association with maternal oxidative stress markers. *Antioxidants, 11*(7), 1415. https://doi.org/10.3390/antiox11071415
28. Vaiserman, A., Koliada, A., Zayachkivska, A., & Lushchak, O. (2020). Nanodelivery of natural antioxidants: An anti-aging perspective. *Frontiers in Bioengineering and Biotechnology*. https://doi.org/10.3389/fbioe.2019.00447
29. Hęś, M., Dziedzic, K., Górecka, D., Jędrusek-Golińska, A., & Gujska, E. (2019). Aloe vera (L.) Webb.: Natural sources of antioxidants—a review. *Plant Foods for Human Nutrition (Dordrecht, Netherlands), 74*(3), 255–265. https://doi.org/10.1007/s11130-019-00747-5
30. Amarachukwu Uzombah, T. (2022). The implications of replacing synthetic antioxidants with natural ones in the food systems. In *Natural food additives*. IntechOpen. https://doi.org/10.5772/intechopen.103810
31. Nanditha, B., & Prabhasankar, P. (2008). Antioxidants in bakery products: A review. *Critical Reviews in Food Science and Nutrition, 49*(1), 1–27. https://doi.org/10.1080/10408390701764104
32. Mika, M., Antończyk, A., & Wikiera, A. (2023). Influence of synthetic antioxidants used in food technology on the bioavailability and metabolism of lipids—in vitro studies. *Polish Journal of Food and Nutrition Sciences*, 95–107. https://doi.org/10.31883/pjfns/161366

33. Wang, W., Xiong, P., Zhang, H., Zhu, Q., Liao, C., & Jiang, G. (2021). Analysis, occurrence, toxicity and environmental health risks of synthetic phenolic antioxidants: A review. *Environmental Research, 201*, Article 111531. https://doi.org/10.1016/j.envres.2021.111531
34. Xu, X., Liu, A., Hu, S., Ares, I., Martínez-Larrañaga, M.-R., Wang, X., Martínez, M., Anadón, A., & Martínez, M.-A. (2021). Synthetic phenolic antioxidants: Metabolism, hazards and mechanism of action. *Food Chemistry, 353*, Article 129488. https://doi.org/10.1016/j.foodchem.2021.129488
35. Hassoun, A., & Emir Çoban, Ö. (2017). Essential oils for antimicrobial and antioxidant applications in fish and other seafood products. *Trends in Food Science & Technology, 68*, 26–36. https://doi.org/10.1016/j.tifs.2017.07.016
36. Pateiro, M., Barba, F. J., Domínguez, R., Sant'Ana, A. S., Mousavi Khaneghah, A., Gavahian, M., Gómez, B., & Lorenzo, J. M. (2018). Essential oils as natural additives to prevent oxidation reactions in meat and meat products: A review. *Food Research International*, *113*, 156–166. https://doi.org/10.1016/j.foodres.2018.07.014
37. Petcu, C. D., Mihai, O. D., Tăpăloagă, D., Gheorghe-Irimia, R.-A., Pogurschi, E. N., Militaru, M., Borda, C., & Ghimpețeanu, O.-M. (2023). Effects of plant-based antioxidants in animal diets and meat products: A review. *Foods, 12*(6), 1334. https://doi.org/10.3390/foods12061334
38. Fadda, A., Sanna, D., Sakar, E. H., Gharby, S., Mulas, M., Medda, S., Yesilcubuk, N. S., Karaca, A. C., Gozukirmizi, C. K., Lucarini, M., Lombardi-Boccia, G., Diaconeasa, Z., & Durazzo, A. (2022). Innovative and sustainable technologies to enhance the oxidative stability of vegetable oils. *Sustainability, 14*(2), 849. https://doi.org/10.3390/su14020849
39. Bensid, A., El Abed, N., Houicher, A., Regenstein, J. M., & Özogul, F. (2022). Antioxidant and antimicrobial preservatives: Properties, mechanism of action and applications in food—a review. *Critical Reviews in Food Science and Nutrition, 62*(11), 2985–3001. https://doi.org/10.1080/10408398.2020.1862046
40. Liu, R., & Mabury, S. A. (2020). Synthetic phenolic antioxidants: a review of environmental occurrence, fate, human exposure, and toxicity. *Environmental Science & Technology, 54*(19), 11706–11719. https://doi.org/10.1021/acs.est.0c05077
41. Yang, X., Song, W., Liu, N., Sun, Z., Liu, R., Liu, Q. S., Zhou, Q., & Jiang, G. (2018). Synthetic phenolic antioxidants cause perturbation in steroidogenesis in vitro and in vivo. *Environmental Science & Technology, 52*(2), 850–858. https://doi.org/10.1021/acs.est.7b05057
42. Ji, X., Liang, J., Wang, Y., Liu, X., Li, Y., Liu, Q., & Liu, R. (2023). Synthetic antioxidants as contaminants of emerging concern in indoor environments: Knowns and unknowns. *Environmental Science & Technology, 57*(51), 21550–21557. https://doi.org/10.1021/acs.est.3c06487
43. Stoia, M., & Oancea, S. (2022). Low-molecular-weight synthetic antioxidants: Classification, pharmacological profile, effectiveness and trends. *Antioxidants, 11*(4), 638. https://doi.org/10.3390/antiox11040638
44. Rajput, V. D., Harish, Singh, R. K., Verma, K. K., Sharma, L., Quiroz-Figueroa, F. R., Meena, M., Gour, V. S., Minkina, T., Sushkova, S., & Mandzhieva, S. (2021). Recent developments in enzymatic antioxidant defence mechanism in plants with special reference to abiotic stress. *Biology*, *10*(4), 267. https://doi.org/10.3390/biology10040267
45. Abdulfatah, H. (2022). Non-enzymatic antioxidants in stressed plants: A review. *Journal of University of Anbar for Pure Science*, *16*(2), 25–37. https://doi.org/10.37652/juaps.2022.176435
46. Bhattacharjee, S. (2019). ROS and antioxidants: Relationship in green cells. In *Reactive oxygen species in plant biology* (pp. 33–63). Springer India. https://doi.org/10.1007/978-81-322-3941-3_2
47. Mishra, N., Jiang, C., Chen, L., Paul, A., Chatterjee, A., & Shen, G. (2023). Achieving abiotic stress tolerance in plants through antioxidative defense mechanisms. *Frontiers in Plant Science*. https://doi.org/10.3389/fpls.2023.1110622
48. Wan, X., Tan, J., Lu, S., Lin, C., Hu, Y., & Guo, Z. (2009). Increased tolerance to oxidative stress in transgenic tobacco expressing a wheat oxalate oxidase gene via induction of antioxidant enzymes is mediated by H_2O_2. *Physiologia Plantarum, 136*(1), 30–44. https://doi.org/10.1111/j.1399-3054.2009.01210.x

49. Ermakov, A., Bobrovskikh, A., Zubairova, U., Konstantinov, D., & Doroshkov, A. (2019). Stress-induced changes in the expression of antioxidant system genes for rice (*Oryza sativa* L.) and bread wheat (*Triticum aestivum* L.). *PeerJ*, *7*, e7791. https://doi.org/10.7717/peerj.7791
50. Lee, Y.-P., Kim, S.-H., Bang, J.-W., Lee, H.-S., Kwak, S.-S., & Kwon, S.-Y. (2007). Enhanced tolerance to oxidative stress in transgenic tobacco plants expressing three antioxidant enzymes in chloroplasts. *Plant Cell Reports, 26*(5), 591–598. https://doi.org/10.1007/s00299-006-0253-z
51. Lubovská, Z., Dobrá, J., Štorchová, H., Wilhelmová, N., & Vanková, R. (2014). Cytokinin oxidase/dehydrogenase overexpression modifies antioxidant defense against heat, drought and their combination in *Nicotiana tabacum* plants. *Journal of Plant Physiology, 171*(17), 1625–1633. https://doi.org/10.1016/j.jplph.2014.06.021
52. Xi, D.-M., Liu, W.-S., Yang, G.-D., Wu, C.-A., & Zheng, C.-C. (2010). Seed-specific over-expression of antioxidant genes in Arabidopsis enhances oxidative stress tolerance during germination and early seedling growth. *Plant Biotechnology Journal, 8*(7), 796–806. https://doi.org/10.1111/j.1467-7652.2010.00509.x
53. Sun, Y., Xie, X., & Jiang, C.-J. (2024). Antioxidant agriculture for stress-resilient crop production: Field practice. *Antioxidants, 13*(2), 164. https://doi.org/10.3390/antiox13020164
54. Talaat, N. B., Ibrahim, A. S., & Shawky, B. T. (2022). Enhancement of the expression of ZmBZR1 and ZmBES1 regulatory genes and antioxidant defense genes triggers water stress mitigation in maize (*Zea mays* L.) plants treated with 24-epibrassinolide in combination with spermine. *Agronomy*, *12*(10), 2517. https://doi.org/10.3390/agronomy12102517
55. Mishra, S., Ganapathi, T. R., Pandey, G. K., Foyer, C. H., & Srivastava, A. K. (2024). Meta-analysis of antioxidant mutants reveals common alarm signals for shaping abiotic stress-induced transcriptome in plants. *Antioxidants & Redox Signaling, 41*(1–3), 42–55. https://doi.org/10.1089/ars.2023.0361
56. Su, P., Sui, C., Niu, Y., Li, J., Wang, S., Sun, F., Yan, J., & Guo, S. (2023). Comparative transcriptomic analysis and functional characterization reveals that the class III peroxidase gene TaPRX-2A regulates drought stress tolerance in transgenic wheat. *Frontiers in Plant Science*. https://doi.org/10.3389/fpls.2023.1119162
57. Srivastava, A. K., & Suprasanna, P. (2015). Redox-regulated mechanisms: Implications for enhancing plant stress tolerance and crop yield. In *Elucidation of abiotic stress signaling in plants* (pp. 191–205). Springer New York. https://doi.org/10.1007/978-1-4939-2211-6_7
58. Akram, N. A., & Ashraf, M. (2013). Regulation in plant stress tolerance by a potential plant growth regulator, 5-aminolevulinic acid. *Journal of Plant Growth Regulation, 32*(3), 663–679. https://doi.org/10.1007/s00344-013-9325-9
59. Skaperda, Z., Veskoukis, A., & Kouretas, D. (2019). Farm animal welfare, productivity and meat quality: Interrelation with redox status regulation and antioxidant supplementation as a nutritional intervention (Review). *World Academy of Sciences Journal*. https://doi.org/10.3892/wasj.2019.19
60. Gurusaran, P. L., Naresh, M., Hariharan, M., & Varun, A. (2024). Antioxidant a feed additive: Unravelling the impact of antioxidants supplementation on productivity and wellness in animals. *Chronicle of Aquatic Science*, *10*(01), 94–101. https://doi.org/10.61851/coas.v1i10.09
61. Salami, S. A., Guinguina, A., Agboola, J. O., Omede, A. A., Agbonlahor, E. M., & Tayyab, U. (2016). Review: In vivo and postmortem effects of feed antioxidants in livestock: A review of the implications on authorization of antioxidant feed additives. *Animal, 10*(8), 1375–1390. https://doi.org/10.1017/S1751731115002967
62. Das, P., Lakra, N., Nutan, K. K., Singla-Pareek, S. L., & Pareek, A. (2019). A unique bZIP transcription factor imparting multiple stress tolerance in rice. *Rice, 12*(1), 58. https://doi.org/10.1186/s12284-019-0316-8
63. Lesgards, J.-F., Durand, P., Lassarre, M., Stocker, P., Lesgards, G., Lanteaume, A., Prost, M., & Lehucher-Michel, M.-P. (2002). Assessment of lifestyle effects on the overall antioxidant capacity of healthy subjects. *Environmental Health Perspectives, 110*(5), 479–486. https://doi.org/10.1289/ehp.02110479

64. Gantenbein, K. V., & Kanaka-Gantenbein, C. (2021). Mediterranean diet as an antioxidant: The impact on metabolic health and overall wellbeing. *Nutrients, 13*(6), 1951. https://doi.org/10.3390/nu13061951
65. Pingitore, A., Lima, G. P. P., Mastorci, F., Quinones, A., Iervasi, G., & Vassalle, C. (2015). Exercise and oxidative stress: Potential effects of antioxidant dietary strategies in sports. *Nutrition, 31*(7–8), 916–922. https://doi.org/10.1016/j.nut.2015.02.005
66. Dato, S., Crocco, P., D'Aquila, P., De Rango, F., Bellizzi, D., Rose, G., & Passarino, G. (2013). Exploring the role of genetic variability and lifestyle in oxidative stress response for healthy aging and longevity. *International Journal of Molecular Sciences, 14*(8), 16443–16472. https://doi.org/10.3390/ijms140816443
67. Bjørklund, G., Shanaida, M., Lysiuk, R., Butnariu, M., Peana, M., Sarac, I., Strus, O., Smetanina, K., & Chirumbolo, S. (2022). Natural compounds and products from an anti-aging perspective. *Molecules, 27*(20), 7084. https://doi.org/10.3390/molecules27207084
68. Trujillo-Mayol, I., Guerra-Valle, M., Casas-Forero, N., Sobral, M. M. C., Viegas, O., Alarcón-Enos, J., Ferreira, I. M., & Pinho, O. (2021). Western dietary pattern antioxidant intakes and oxidative stress: Importance during the SARS-CoV-2/COVID-19 pandemic. *Advances in Nutrition, 12*(3), 670–681. https://doi.org/10.1093/advances/nmaa171
69. Taherkhani, S., Suzuki, K., & Ruhee, R. T. (2021). A brief overview of oxidative stress in adipose tissue with a therapeutic approach to taking antioxidant supplements. *Antioxidants, 10*(4), 594. https://doi.org/10.3390/antiox10040594
70. Sánchez-Rodríguez, M. A., Zacarías-Flores, M., Correa-Muñoz, E., Arronte-Rosales, A., & Mendoza-Núñez, V. M. (2021). Oxidative stress risk is increased with a sedentary lifestyle during aging in Mexican women. *Oxidative Medicine and Cellular Longevity, 2021*(1). https://doi.org/10.1155/2021/9971765
71. Prata, C., Angeloni, C., & Maraldi, T. (2024). Strategies to counteract oxidative stress and inflammation in chronic-degenerative diseases 2.0. *International Journal of Molecular Sciences, 25*(9), 5026. https://doi.org/10.3390/ijms25095026
72. Reddy, V. P. (2023). Oxidative stress in health and disease. *Biomedicines, 11*(11), 2925. https://doi.org/10.3390/biomedicines11112925
73. Sharifi-Rad, M., Anil Kumar, N. V., Zucca, P., Varoni, E. M., Dini, L., Panzarini, E., Rajkovic, J., Tsouh Fokou, P. V., Azzini, E., Peluso, I., Prakash Mishra, A., Nigam, M., El Rayess, Y., Beyrouthy, M. E., Polito, L., Iriti, M., Martins, N., Martorell, M., Docea, A. O., & Sharifi-Rad, J. (2020). Lifestyle, oxidative stress, and antioxidants: Back and forth in the pathophysiology of chronic diseases. *Frontiers in Physiology*. https://doi.org/10.3389/fphys.2020.00694
74. Husain, S., & Englert, H. (2021). Power of lifestyle medicine: Improving oxidative stress profile of high risk lifestyle disease patients in rural germany. *Current Developments in Nutrition, 5*, 23. https://doi.org/10.1093/cdn/nzab033_023
75. Chen, W., & Zhang, W. (2024). Association between oxidative balance score and chronic obstructive pulmonary disease: A cross-sectional study. *Medicine, 103*(40), Article e39883. https://doi.org/10.1097/MD.0000000000039883
76. De Almeida, A. J. P. O., de Almeida Rezende, M. S., Dantas, S. H., de Lima Silva, S., de Oliveira, J. C. P. L., de Lourdes Assunção Araújo de Azevedo, F., Alves, R. M. F. R., de Menezes, G. M. S., dos Santos, P. F., Gonçalves, T. A. F., Schini-Kerth, V. B., & de Medeiros, I. A. (2020). Unveiling the role of inflammation and oxidative stress on age-related cardiovascular diseases. *Oxidative Medicine and Cellular Longevity, 2020*, 1–20. https://doi.org/10.1155/2020/1954398
77. Xu, Z., Xue, Y., Wen, H., & Chen, C. (2023). Association of oxidative balance score and lung health from the National Health and Nutrition Examination Survey 2007–2012. *Frontiers in Nutrition*. https://doi.org/10.3389/fnut.2022.961950
78. Zhu, L., Ruan, X., Wang, J., Yan, Y., Tang, C., & Xu, Y. (2024). Higher oxidative balance score is linearly associated with reduced prevalence of chronic kidney disease in individuals with metabolic syndrome: Evidence from NHANES 1999–2018. *Frontiers in Nutrition*. https://doi.org/10.3389/fnut.2024.1442274
79. Kozarski, M., Klaus, A., Jakovljevic, D., Todorovic, N., Vunduk, J., Petrović, P., Niksic, M., Vrvic, M., & Van Griensven, L. (2015). Antioxidants of edible mushrooms. *Molecules, 20*(10), 19489–19525. https://doi.org/10.3390/molecules201019489

80. De Oliveira, V. S., Ferreira, F. S., Cople, M. C. R., da Labre, T. S., Augusta, I. M., Gamallo, O. D., & Saldanha, T. (2018). Use of natural antioxidants in the inhibition of cholesterol oxidation: A review. *Comprehensive Reviews in Food Science and Food Safety, 17*(6), 1465–1483. https://doi.org/10.1111/1541-4337.12386
81. Lourenço, S. C., Moldão-Martins, M., & Alves, V. D. (2019). Antioxidants of natural plant origins: From sources to food industry applications. *Molecules, 24*(22), 4132. https://doi.org/10.3390/molecules24224132
82. Margaritelis, N. V., Paschalis, V., Theodorou, A. A., Kyparos, A., & Nikolaidis, M. G. (2018). Antioxidants in personalized nutrition and exercise. *Advances in Nutrition, 9*(6), 813–823. https://doi.org/10.1093/advances/nmy052
83. Poljsak, B. (2011). Strategies for reducing or preventing the generation of oxidative stress. *Oxidative Medicine and Cellular Longevity, 2011*, 1–15. https://doi.org/10.1155/2011/194586
84. Ashok, A., Andrabi, S. S., Mansoor, S., Kuang, Y., Kwon, B. K., & Labhasetwar, V. (2022). Antioxidant therapy in oxidative stress-induced neurodegenerative diseases: Role of nanoparticle-based drug delivery systems in clinical translation. *Antioxidants, 11*(2), 408. https://doi.org/10.3390/antiox11020408
85. Kumar, H., Bhardwaj, K., Nepovimova, E., Kuča, K., Singh Dhanjal, D., Bhardwaj, S., Bhatia, S. K., Verma, R., & Kumar, D. (2020). Antioxidant functionalized nanoparticles: A combat against oxidative stress. *Nanomaterials, 10*(7), 1334. https://doi.org/10.3390/nano10071334
86. Stevanović, M., & Filipović, N. (2024). A review of recent developments in biopolymer nano-based drug delivery systems with antioxidative properties: Insights into the last five years. *Pharmaceutics, 16*(5), 670. https://doi.org/10.3390/pharmaceutics16050670
87. Sezgin-Bayindir, Z., Losada-Barreiro, S., Fernández-Bravo, S., & Bravo-Díaz, C. (2023). Innovative delivery and release systems for antioxidants and other active substances in the treatment of cancer. *Pharmaceuticals, 16*(7), 1038. https://doi.org/10.3390/ph16071038
88. Losada-Barreiro, S., Sezgin-Bayindir, Z., Paiva-Martins, F., & Bravo-Díaz, C. (2022). Biochemistry of antioxidants: Mechanisms and pharmaceutical applications. *Biomedicines, 10*(12), 3051. https://doi.org/10.3390/biomedicines10123051

Recombinant Enzyme Therapies: A Pioneering Approach to Combat Oxidative Stress-Related Diseases

Mithra Jayan, Manikandan Vani Raju, Meenakshi Kaniyur Chandrasekaran, Gopalakrishnan Velliyur Kanniappan, and Nirmaladevi Panneerselvam

Abstract Oxidative stress is induced by an imbalance in reactive oxygen species (ROS) production and the body's antioxidant defenses. While ROS serve essential roles in cellular signaling and immunology, their excessive buildup damages DNA, lipids, and proteins, contributing to serious diseases like cancer, neurological disorders, and cardiovascular disease. Oxidative DNA damage, such as strand breakage and 8-hydroxyguanine production, causes genomic instability and increases cancer risk. Lipid peroxidation produces damaging byproducts such as malondialdehyde (MDA) and 4-hydroxynonenal (HNE), which impair membrane integrity and worsen neurological and cardiovascular disorders. Similarly, protein oxidation reduces enzymatic activity, resulting in neurodegeneration in disorders like Alzheimer's and Parkinson's disease. ROS production plays a part in cardiovascular diseases by causing endothelial dysfunction, low-density lipoprotein (LDL) oxidation, and nitric oxide (NO) dysregulation, which promotes atherosclerosis, hypertension, and heart failure. Recombinant enzyme treatments, which enhance antioxidant enzyme activity, are a promising technique for combating oxidative stress-related damage. Advances in protein engineering, gene therapy, and nanotechnology have increased enzyme stability, specificity, and targeted delivery. These therapies have potential for treating oxidative stress-related disorders by minimizing cellular damage and restoring equilibrium. However, immunogenicity, high production costs, and regulatory barriers impede their clinical adoption. Future advances in gene-editing technologies such as CRISPR, precision medicine, and bioprocessing developments could improve the efficacy and accessibility of recombinant enzyme

M. Jayan · N. Panneerselvam (✉)
Department of Biotechnology, Karpagam Academy of Higher Education, Coimbatore, Tamil Nadu, India
e-mail: nirmaladevi.panneerselvam@kahedu.edu.in

M. Vani Raju · M. Kaniyur Chandrasekaran · N. Panneerselvam
Department of Biochemistry, Karpagam Academy of Higher Education, Coimbatore, Tamil Nadu, India

G. Velliyur Kanniappan
Department of Physiology, Saveetha Medical College and Hospital, Saveetha Institute of Medical and Technical Sciences (SIMATS), Tamil Nadu, Chennai, India

K. Kasinathan et al. (eds.), *Recent Advances in Oxidative Stress Associated Chronic Diseases Volume 1*, https://doi.org/10.1007/978-981-96-9670-3_13

therapy. As research advances, these medicines have the potential to transform treatment approaches for oxidative stress-related disorders by providing focused and individualized interventions that improve patient outcomes.

1 Introduction

Oxidative stress is caused by an imbalance in the body's ability to produce ROS and neutralize them with antioxidants. While ROS have a significant part in normal cellular signaling and immune defense, excessive buildup causes severe damage to critical macromolecules such as DNA, lipids, and proteins. The oxidative alteration of these biological components has an important role in the etiology of many diseases, including cancer, neurological disorders, and cardiovascular diseases. One of the most concerning results of oxidative stress is DNA damage, which can lead to mutations, chromosomal instability, and, eventually, cancer. ROS may trigger strand breaks, base alterations, and the production of DNA adducts, which disrupt normal genetic functioning. 8-hydroxyguanine, a major indicator of oxidative DNA damage, has been related to an elevated risk of cancer and other chronic disorders [34]. When oxidative stress overwhelms the repair mechanisms of the cell, it leads to genomic instability, increasing the likelihood of malignant transformation [54]. Furthermore, excessive ROS generation contributes to lipid peroxidation, in which free radicals attack polyunsaturated fatty acids in cell membranes, releasing harmful byproducts such as MDA and HNE. This process undermines membrane integrity, disrupts cellular signaling pathways, and contributes to the development of a variety of chronic diseases, including neurodegenerative disorders and cardiovascular ailments. Elevated levels of lipid peroxidation products have been detected in patients with AD and atherosclerosis, further highlighting their role in disease progression [41]. As well as lipids and DNA, protein oxidation caused by ROS significantly affects cellular function. Oxidation leads to structural modifications, including carbonylation, nitration, and cross-linking, which impair enzymatic activities and protein–protein interactions. These modifications are especially damaging in neurodegenerative illnesses like Alzheimer's and Parkinson's, where oxidative stress-induced protein degradation affects normal brain function [63]. Aggregated, oxidized proteins accumulate in affected neurons, exacerbating disease progression [23]. Similarly, in cardiovascular diseases, oxidative modifications of key proteins involved in endothelial function and inflammation contribute to vascular dysfunction, hypertension, and heart failure. The role of oxidative stress in cancer is well-documented, as chronic oxidative stress to DNA and lipids promotes carcinogenesis. ROS-induced mutations in tumor suppressor genes and oncogenes contribute to uncontrolled cell proliferation and tumor progression. Inflammatory processes driven by oxidative stress further exacerbate tumorigenesis by creating a pro-tumorigenic microenvironment that fosters angiogenesis and metastasis [65]. In addition to cancer, oxidative stress plays an important role in neurodegenerative disorders, where it accelerates neuronal damage through mitochondrial dysfunction and the aggregation of misfolded proteins. In AD, oxidative damage is involved in amyloid plaque deposition, tau hyperphosphorylation, and synaptic loss, leading to progressive cognitive decline [11]. Likewise, in Parkinson's disease, ROS contribute to the disintegration of dopaminergic neurons in the substantia nigra, motor dysfunction, and disease progression [81]. Cardiac diseases, including atherosclerosis, hypertension, and heart failure, are also strongly related to oxidative stress. ROS play an important role in endothelial dysfunction, which is a precursor to atherosclerosis and other cardiovascular conditions. Oxidative stress encourages the oxidation of low-density LDL, a vital stage in plaque development within artery walls, resulting in atherosclerosis and an increased risk of heart attacks and strokes. Additionally, increased ROS levels impair NO signaling, a key regulator of vascular tone, contributing to hypertension and vascular stiffening [54]. Clinical studies have shown that patients with heart attack exhibit significantly high oxidative stress markers, suggesting that ROS play a direct role in disease progression and cardiac remodeling [56].

2 Basics of Recombinant Enzyme Technology

2.1 Overview of Recombinant Enzyme Technology

Recombinant enzyme technology involves the use of genetic engineering to produce enzymes with improved properties for industrial, pharmaceutical, and medical applications. This technology enables large-scale production of enzymes with enhanced stability, specificity, and catalytic efficiency. By manipulating genetic material, scientists can design enzymes that meet specific functional requirements, optimizing their performance in various applications.

2.2 Genetic Engineering and Recombinant DNA Techniques

Genetic engineering plays a fundamental role in recombinant enzyme production. The process begins with the isolation and identification of a gene encoding the desired enzyme, followed by its insertion into an expression vector, typically a plasmid. This vector serves as a carrier to introduce the gene into a host organism such as bacteria, yeast, or mammalian cells. Once inside the host, the recombinant DNA directs the production of the target enzyme using the host's cellular machinery. Recombinant DNA technology has been widely applied in producing therapeutic enzymes such as erythropoietin, coagulation modulators, and thrombolytic agents [47]. Recent advances in site-specific recombinase technology have improved the efficiency of gene modifications, allowing precise control over gene expression and protein production [35]. Additionally, homologous recombination techniques have been used to enhance enzyme stability and catalytic activity, making them more suitable for industrial applications.

2.3 Expression Systems for Recombinant Enzyme Production

The choice of an expression system is crucial in determining the yield, activity, and stability of recombinant enzymes. Various prokaryotic and eukaryotic systems are available, each offering unique advantages and limitations.

2.3.1 Bacterial Expression Systems

Escherichia coli is the most often used bacterial expression system because of its quick growth, ease of genetic manipulation, and cost-effectiveness. However, it lacks the ability to perform post-translational modifications required for certain eukaryotic proteins. Advances in bacterial expression technology have improved recombinant protein yield through metabolic engineering, strong promoters, and secretion signals [39]. Additionally, novel bacterial strains and vector elements have been developed to enhance the stability and solubility of complex proteins [76].

2.3.2 Yeast and Fungal Expression Systems

Yeast systems, particularly *Saccharomyces cerevisiae* and *Pichia pastoris*, offer the advantage of eukaryotic post-translational modifications while maintaining high protein yield and ease of culture. *P. pastoris* has become a preferred system for pharmaceutical applications due to its capability to express large amounts of recombinant proteins with proper folding and glycosylation [64]. Similarly, fungal expression systems such as *Hypocrea jecorina* have been used for cellulase production, with recombination-based engineering enhancing thermostability and enzymatic efficiency.

2.3.3 Mammalian Expression Systems

Mammalian cells, including Chinese hamster ovary (CHO) and HEK293 cells, are commonly used for creating complex beneficial proteins that need human-like glycosylation patterns. Although these systems are more expensive and grow more slowly compared to microbial hosts, they ensure proper protein folding and functional activity. Mammalian expression systems are widely used for producing biologics such as monoclonal antibodies and hormone therapies [83].

2.3.4 Plant-Based and Algal Expression Systems

Plants and algae provide cost-effective and scalable platforms for recombinant enzyme production. *Chlorella spp.* has been explored for large-scale protein expression because of its eukaryotic cellular machinery, environmental sustainability, and low production costs [22]. Additionally, transient expression in plants allows for rapid production of recombinant proteins, significantly reducing development time compared to microbial and mammalian systems [68].

2.4 Types of Recombinant Enzymes

Antioxidant enzymes protect cells from the oxidative damage caused by ROS. These enzymes eliminate damaging free radicals, thereby reducing oxidative stress-related diseases like cancer, neurological disorders, and cardiovascular disease. Superoxide dismutase (SOD), glutathione peroxidase (GPX), and catalase (CAT) are three of the most significant antioxidant enzymes, which have multiple applications that are given in Table 1.

2.4.1 Superoxide Dismutase (SOD)

Superoxide dismutase is an important antioxidant enzyme that promotes the conversion of superoxide radicals into H_2O_2 and molecular oxygen, reducing oxidative damage in cells. Research has shown that SOD has a protective role in numerous biological systems, which include erythrocytes, the brain, and tissues prone to oxidative stress [13]. Its therapeutic applications include treatments for neurodegenerative diseases, inflammation, and ischemia–reperfusion injury. SOD-based therapies, such as recombinant human SOD (rhSOD), have been explored for their capability to reduce oxidative damage in Fanconi anemia and other oxidative stress-related conditions [60].

2.4.2 Catalase (CAT)

Catalase is responsible for decomposing hydrogen peroxide, a consequence of SOD action, into oxygen and water, thus avoiding cellular damage. It is widely distributed in tissues and plays a significant role in shielding erythrocytes from oxidative haemolysis. CAT has been investigated for its part in reducing oxidative stress in conditions such as diabetes, where its activity is altered in the brain tissues of diabetic patients [72].

Table 1 Applications of recombinant enzymes in preventing oxidative stress

Recombinant enzyme	Source/host	Function	Application in oxidative stress
Superoxide dismutase (SOD)	*Escherichia coli*, yeast, mammalian cells	Catalyzes the conversion of superoxide radicals (O_2^-) into H_2O_2 and oxygen (O_2)	Reduces oxidative damage by scavenging superoxide radicals, commonly used in inflammatory and neurodegenerative diseases like Alzheimer's and Parkinson's
Catalase (CAT)	*Escherichia coli*, yeast, mammalian cells	Catalyzes the breakdown of H_2O_2 into water (H_2O) and oxygen (O_2)	Helps protect cells from hydrogen peroxide-induced oxidative damage; used in wound healing and in preventing oxidative damage in cancer therapy
Glutathione peroxidase (GPx)	*Escherichia coli*, yeast, mammalian cells	Reduces H_2O_2 and organic peroxides to water and alcohol, respectively, using glutathione as a cofactor	Reduces cellular oxidative damage, is used in therapies for liver diseases and neurodegenerative disorders, and is an antioxidant supplement in various diseases
Peroxiredoxins	*Escherichia coli*, yeast, mammalian cells	Reduces H_2O_2 and organic peroxides, protecting cells from oxidative damage	Used to mitigate oxidative stress in diseases like cardiovascular diseases, cancer, and diabetes
Thioredoxin	*Escherichia coli*, yeast, mammalian cells	Acts as an electron donor to reduce disulfide bonds in proteins, playing a role in maintaining cellular redox balance	Used in gene therapy and the treatment of oxidative stress-related diseases, such as neurodegenerative conditions and cancer
NADPH oxidase (NOX)	Yeast, mammalian cells	Generates ROS as part of immune responses but can also cause oxidative stress if dysregulated	Recombinant NOX enzymes can be used to study the role of ROS in immune system activation and inflammation

(continued)

Table 1 (continued)

Recombinant enzyme	Source/host	Function	Application in oxidative stress
Glutathione reductase	*Escherichia coli*, yeast, mammalian cells	Regenerates reduced glutathione (GSH) from its oxidized form (GSSG), critical for maintaining cellular antioxidant defenses	Used in therapies for conditions involving chronic oxidative stress, such as cystic fibrosis, liver disease, and aging
Methionine sulfoxide reductase (Msr)	*Escherichia coli*, yeast, mammalian cells	Reduces oxidized methionine residues in proteins, reversing oxidative modifications	Important for protein repair in cells, used in neurodegenerative disease research and as a potential therapeutic tool in aging

2.4.3 Glutathione Peroxidase (GPX)

Glutathione peroxidase plays a crucial role in detoxifying H_2O_2 and lipid peroxides by utilizing glutathione as a reducing agent. It is especially vital to protect red blood cells and other tissues from oxidative stress [3]. GPX activity was studied in various biological contexts, including its part in fetal development and oxidative stress regulation in maternal and cord blood erythrocytes [59].

2.4.4 Proteolytic Enzymes

Proteolytic enzymes such as trypsin, chymotrypsin, and bromelain are widely used in therapeutic applications for their ability to facilitate wound healing, reduce inflammation, and support digestive health. These enzymes help with tissue regeneration and repair by breaking down proteins into smaller peptides and amino acids. Clinical studies have demonstrated that the combination of bromelain and rutoside with trypsin significantly enhances anti-inflammatory effects, reducing pain and swelling associated with injuries and post-surgical recovery [43]. Additionally, trypsin–chymotrypsin therapy has been shown to accelerate wound healing by improving erythema reduction, minimizing edema, and promoting faster recovery in patients undergoing surgery [20]. Furthermore, proteolytic enzymes derived from natural sources such as crab collagenase have been investigated for their potential in tissue engineering and wound healing applications [2].

2.4.5 Thrombolytic Enzymes

Thrombolytic enzymes, including tissue plasminogen activator (tPA) and streptokinase, are critical in dissolving blood clots and restoring blood flow in conditions such as stroke and myocardial infarction. These enzymes convert plasminogen into plasmin, which breaks down fibrin clots, preventing severe cardiovascular complications. Recombinant tPA has become a gold-standard therapy for ischemic stroke, significantly improving patient outcomes when administered within the therapeutic window. Advances in thrombolytic therapy have led to the development of modified enzyme formulations with enhanced stability and reduced risk of hemorrhagic complications. Research continues to explore safer and more effective thrombolytic agents for broader clinical use.

2.4.6 Digestive Enzymes

Digestive enzymes, including amylase, lipase, and proteases, play an important role in the breakdown and absorption of nutrients. They are commonly used in the treatment of pancreatic insufficiency, lactose intolerance, and other gastrointestinal disorders. Enzyme replacement therapy (ERT) with digestive enzymes was shown to improve digestive function in patients with exocrine pancreatic insufficiency, allowing for better nutrient absorption and reducing symptoms such as bloating and malabsorption. Oral enzyme supplements are widely used for their efficacy in enhancing digestion and supporting gut health.

2.4.7 Detoxifying Enzymes

Detoxifying enzymes like alcohol dehydrogenase and cytochrome P450 have a vital role in metabolizing toxins, drugs, and harmful substances in the liver. Cytochrome P450 enzymes are involved in the metabolism of pharmaceuticals and endogenous compounds, influencing drug interactions and therapeutic efficacy. Research has demonstrated that targeting cytochrome P450 pathways can aid in managing conditions such as diabetic wound healing, where oxidative stress and impaired detoxification mechanisms contribute to delayed recovery [100]. Moreover, these enzymes are being studied for their role in reducing drug-induced toxicity and improving personalized medicine strategies.

2.5 Enzyme Production and Challenges

Selecting an appropriate expression system is crucial for producing recombinant enzymes efficiently. Bacterial systems, particularly *Escherichia coli*, are widely used due to their rapid growth, high yield, and ease of genetic manipulation. However, they lack post-translational modifications, which can affect enzyme functionality, particularly for therapeutic applications [78]. Yeast expression systems, such as *Pichia pastoris* and *Saccharomyces cerevisiae*, offer an alternative for producing eukaryotic proteins with post-translational modifications. Yeasts have been successfully used to express therapeutic enzymes and industrial biocatalysts with improved stability and activity [64]. Research has also demonstrated that yeast systems can efficiently produce active, native-form enzymes, making them suitable for pharmaceutical applications [15]. Cell expression systems are preferred for more complex proteins that need human-like post-translational modifications. These systems ensure suitable folding and glycosylation, which are critical for therapeutic enzyme activity. However, they come with higher costs, slower growth rates, and the need for specialized infrastructure, making them less scalable for industrial enzyme production [10].

2.5.1 Scalability and Cost Considerations

Scaling up enzyme production from lab-scale to industrial levels presents economic and technical challenges. Factors such as fermentation optimization, nutrient costs, and yield efficiency influence production feasibility. Bacterial and yeast-based systems offer advantages in scalability due to their rapid growth and ease of culture, making them more cost-effective for large-scale enzyme production [52]. In contrast, mammalian cell-based production systems require expensive culture media and controlled bioreactors, significantly increasing production costs. Strategies such as optimizing gene expression, using high-density cell cultures, and engineering more efficient promoters have been explored to enhance enzyme yields while reducing costs [28]. Additionally, transgenic plant-based systems are being explored as a cost-effective alternative, but challenges with protein yield and regulatory concerns remain [9].

2.5.2 Purification and Formulation Issues

The purification of recombinant enzymes is one of the costliest and technically challenging steps in production. Affinity chromatography, ion-exchange methods, and ultrafiltration techniques are commonly used to isolate high-purity enzymes. However, enzyme aggregation, improper folding, and host-cell protein contamination can hinder efficiency [94]. Recent advancements in protein engineering and immobilization techniques have improved enzyme stability and formulation. Strategies such as PEGylation and glycoengineering have been employed to enhance enzyme half-life and therapeutic efficacy, particularly in pharmaceuticals [89]. Furthermore, innovative expression systems such as *Pichia pastoris* have been optimized for large-scale enzyme secretion, reducing the need for complex purification steps [96].

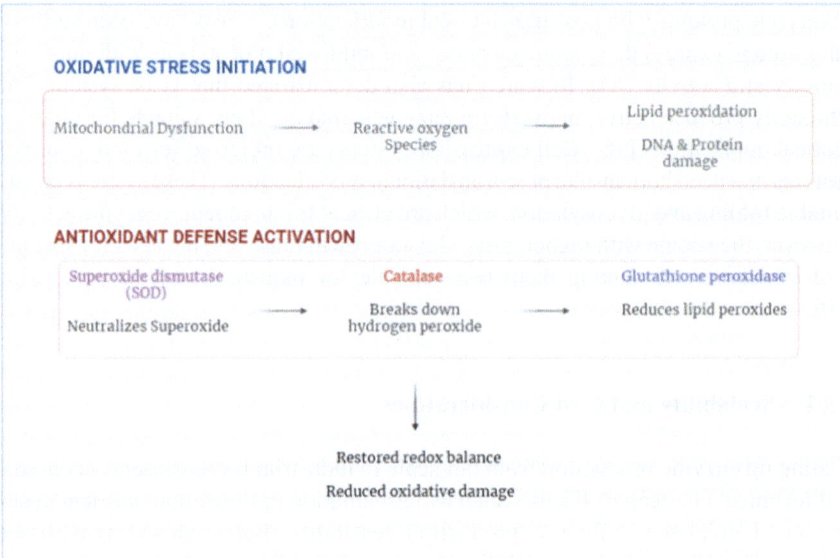

Fig. 1 The progression of oxidative stress initiation and the activation of antioxidant defense mechanisms to restore cellular balance

3 Mechanisms of Oxidative Stress and Antioxidant Defense

3.1 Generation of Oxidative Stress

Oxidative stress occurs when the body's antioxidant defense mechanisms are unable to counteract the creation of ROS and reactive nitrogen species (RNS). ROS and RNS are naturally created during cellular metabolism and play critical roles in cell signaling, immune response, and death. However, excessive ROS and RNS production can induce oxidative damage, which can contribute to the pathophysiology of a range of diseases, including neurological disorders, cardiovascular diseases, and cancer. Figure 1 demonstrates the process of initiating oxidative stress and then activating antioxidant defense systems to restore cellular redox balance.

3.2 ROS and RNS (Reactive Nitrogen Species) Generation

ROS refers to free radicals like superoxide anion ($O_2\bullet^-$), hydroxyl radical (•OH), and non-radical compounds such as hydrogen peroxide (H_2O_2). These species are byproducts of oxygen metabolism; while they play important functions in cellular signaling, excessive buildup causes oxidative stress. Excessive levels of RNS, such as NO and peroxynitrite ($ONOO^-$), can cause cellular damage. Mitochondria are significant sources of both ROS and RNS. The electron transport chain (ETC) in mitochondria promotes ATP generation but also produces ROS as a byproduct. ETC complexes I and III are highly susceptible to electron leakage, which results in the formation of superoxide radicals. Nitric oxide synthases (NOSs) in cells also contribute to RNS synthesis, which can result in $ONOO^-$ accumulation and oxidative damage.

3.3 Sources of ROS in Cells

3.3.1 Mitochondria

Because of their role in oxidative phosphorylation, mitochondria remain the principal sources of ROS. During ATP synthesis, electrons can break out from the ETC and combine with molecular oxygen to create superoxide radicals. This ROS production is normally controlled by mitochondrial antioxidant mechanisms, like SOD and glutathione peroxidase, which assist in neutralizing excess ROS. However, malfunction in these protective processes may contribute to oxidative stress, contributing to neurodegenerative illnesses as well as Parkinson's and Alzheimer's [80].

3.3.2 Enzymatic Sources

Several enzyme systems contribute to ROS production in cells. NADPH oxidases (NOX enzymes), which generate ROS, are components of the immune response against infections. While beneficial in controlled amounts, chronic overactivation of NOX enzymes has been linked to inflammatory diseases and cardiovascular disorders [79]. Xanthine oxidase catalyzes the oxidation of hypoxanthine to xanthine, producing ROS in the process. This enzyme is particularly relevant in ischemia–reperfusion injury, where a sudden repair of oxygen supply leads to a burst of ROS production [1]. Myeloperoxidase (MPO), found in neutrophils, generates hypochlorous acid (HOCl), a potent oxidant that contributes to tissue damage in inflammatory diseases [82].

3.3.3 Environmental Factors

Exogenous sources of ROS comprise radiation, pollutants, and toxins, which can significantly contribute to oxidative stress. Ultraviolet (UV) radiation exposure triggers ROS production in skin cells, leading to oxidative stress that accelerates skin aging and increases the risk of skin cancer. Air pollution, particularly particulate matter and heavy metals, can stimulate NOX enzymes and mitochondrial dysfunction, further exacerbating oxidative stress [6]. Additionally, chemicals and drugs such as certain pharmaceuticals and environmental toxins promote ROS formation by interfering with mitochondrial function or inducing lipid peroxidation, further increasing cellular damage [91].

3.4 Consequences of Oxidative Stress on Cell Function and Survival

Oxidative damage has a significant impact on cell fate by altering cellular activities such as signaling, metabolism, and gene expression. Cells respond to oxidative stress through a delicate balance between survival and death pathways. While moderate oxidative stress can activate protective mechanisms, excessive stress leads to cellular dysfunction and apoptosis.

3.4.1 Impact on Cellular Survival and Death Pathways

Oxidative stress influences both pro-survival and pro-death pathways. At low-to-moderate levels, ROS can activate protective mechanisms such as antioxidant enzyme expression and heat shock proteins, which help maintain cellular homeostasis. However, when oxidative damage exceeds the cell's repair capacity, programmed cell death pathways such as apoptosis, necrosis, pyroptosis, and autophagy are activated. The severity and duration of oxidative stress determine whether certain pathways are activated. Studies have revealed that oxidative stress can cause cell death through mitochondrial malfunction, DNA damage, and the activation of death receptors [14].

3.4.2 Mitochondrial Dysfunction and Apoptosis

Mitochondria have an important role in oxidative stress-induced apoptosis. When exposed to excessive ROS, mitochondria release cytochrome c, which activates the caspase cascade, leading to programmed cell death. This mechanism is particularly relevant in neurodegenerative diseases such as Parkinson's and Alzheimer's, where mitochondrial dysfunction and oxidative stress contribute to neuronal loss [19]. Additionally, oxidative stress disrupts mitochondrial membrane potential, leading to energy depletion and increased cell vulnerability [24].

3.4.3 DNA Damage and Mutagenesis

Oxidative stress-induced DNA damage adds to genomic instability, which is an important element in cancer development. ROS can cause DNA strand breakage, base alterations, and chromosomal aberrations, which increase the chance of mutation and tumor formation. If the harm is extensive, cells either undergo programmed cell death or accumulate mutations that promote carcinogenesis. The regulation of survival and death signals by oxidative stress is particularly relevant in tumor cells, where manipulating redox signaling can influence therapy outcomes [48].

3.4.4 Redox Signaling and Autophagy

Autophagy, a process that removes damaged organelles and misfolded proteins, is also regulated by oxidative stress. Moderate ROS levels can stimulate autophagy as a survival mechanism, helping cells eliminate damaged components. However, excessive oxidative stress can impair autophagic flux, leading to cellular dysfunction and increased susceptibility to diseases such as cancer and neurodegeneration [66].

3.4.5 Inflammation and Immune Response

Chronic inflammation is primarily driven by oxidative damage. ROS can activate NF-κB and other pro-inflammatory signaling pathways, resulting in cytokines that worsen tissue damage. This inflammatory response is important in cardiovascular and autoimmune diseases, as well as metabolic problems. Research demonstrates that oxidative stress-induced inflammation leads to the advancement of illnesses like atherosclerosis and rheumatoid arthritis [57].

3.4.6 Aging and Oxidative Damage

Oxidative stress is also connected to aging and age-related disorders. The deposition of oxidative stress over time affects cellular function and reduces the ability of cells to cope with stress. This process contributes to age-related conditions like cognitive decline, cardiac diseases, and reduced tissue repair capacity. However, a concept known as "*positive oxidative stress*" suggests that moderate ROS levels may induce adaptive responses that enhance longevity and stress resistance [97].

4 Recombinant Enzymes in the Management of Oxidative Stress

4.1 *The Concept of Enzyme Replacement Therapy (ERT)*

4.1.1 Mechanism and Rationale for Recombinant Enzyme Therapy

Enzyme replacement therapy involves the administration of recombinant enzymes to compensate for deficient or absent enzymes in patients with specific metabolic disorders. The rationale behind ERT is to restore the normal function of metabolic pathways by supplementing the missing enzyme, thereby reducing the accumulation of substrates that cause disease symptoms. For instance, in Pompe disease, recombinant human alpha-glucosidase (rhGAA) is used to address the deficiency of acid rhGAA, which is crucial for breaking down glycogen in the lysosome [90]. Similarly, vestronidase alfa, a recombinant human β-glucuronidase, is used for Mucopolysaccharidosis VII (MPS VII) to enhance enzyme uptake and activity in target tissues [17]. The mechanism often involves receptor-mediated endocytosis, where the recombinant enzyme is taken up by cells and directed to lysosomes to perform its function [29].

4.1.2 Benefits and Limitations in Oxidative Stress Modulation

ERT has shown potential benefits in modulating oxidative stress, which is often a secondary complication in metabolic disorders. For example, in Pompe disease, oxidative stress levels inversely correlate with the efficacy of rhGAA, and the use of antioxidants alongside ERT has been shown to improve enzyme activity and processing [90]. This suggests that managing oxidative stress can enhance the therapeutic outcomes of ERT. However, there are limitations, such as the potential for hypersensitivity reactions to the recombinant enzymes, which can necessitate desensitization protocols to continue treatment [84]. Additionally, the blood–brain barrier poses an important challenge for ERT in treating central nervous system manifestations of lysosomal storage diseases, although intrathecal administration has been demonstrated as a potential solution [30].

4.2 *Key Recombinant Antioxidant Enzymes*

Antioxidant enzymes shield cells from oxidative stress by counteracting ROS. Among these, SOD, GPx, and CAT are essential in sustaining cellular homeostasis and stopping oxidative damage, as shown in Fig. 2.

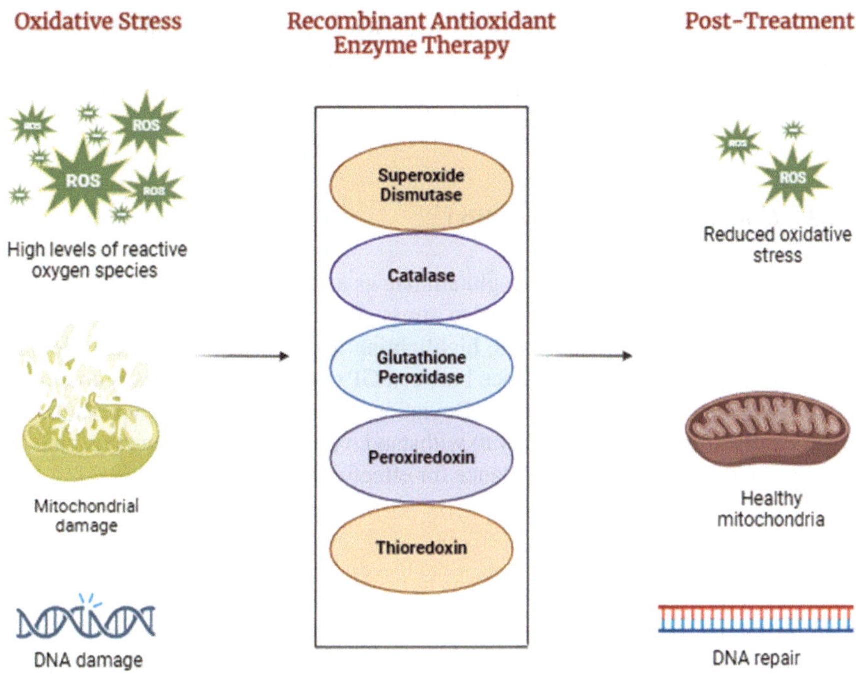

Fig. 2 The process of recombinant antioxidant enzyme therapy in mitigating oxidative stress

4.2.1 Superoxide Dismutase (SOD)

SOD catalyzes the transformation of ($O_2{\bullet}^-$) to oxygen and hydrogen peroxide, which reduces oxidative stress. There are three forms of SOD: Cu/Zn-SOD (SOD1), Mn-SOD (SOD2), and EC-SOD (SOD3). Cu/Zn-SOD is found in the cytoplasm, Mn-SOD in the mitochondria, and EC-SOD in the extracellular spaces. Each type has a unique role and compartmentalization within the cell, contributing to its specific functions in oxidative stress management [37, 38, 98]. SOD has been explored for its therapeutic potential in treating ischemia–reperfusion injuries. However, high concentrations of SOD can lead to increased production of hydroxyl radicals, which can be toxic. Conjugating SOD with CAT has shown promise in mitigating this toxicity by reducing H_2O_2 levels, thus enhancing its therapeutic efficacy [55].

4.2.2 Catalase

CAT is essential for converting hydrogen peroxide, a byproduct of SOD activity, into oxygen and water, thereby avoiding cellular damage from H_2O_2 accumulation. This detoxification process is crucial for sustaining cellular health and preventing oxidative stress-related harm [42, 62]. CAT, often in conjunction with other antioxidant enzymes like SOD and GPx, is vital in protecting cells from oxidative damage. Its role in detoxifying H_2O_2 makes it a potential therapeutic agent in conditions characterized by oxidative stress [55].

4.2.3 Glutathione Peroxidase (GPx)

GPx reduces H_2O_2 to water using glutathione as a substrate, thus playing a serious role in protecting cells from oxidative stress. It works synergistically with SOD to ensure efficient removal of ROS, highlighting its importance in cellular defense mechanisms [7, 38, 62]. The balance between GPx and SOD activities is crucial for cellular resistance to oxidative stress. Deviations from the optimal activity ratios can significantly affect the cell's ability to withstand oxidative damage, emphasizing the importance of maintaining this balance for effective antioxidant defense [7].

4.2.4 Other Enzymes

Peroxiredoxins and thioredoxin are other antioxidant enzymes that contribute to cellular defense against oxidative stress. They work alongside SOD, CAT, and GPx to provide a comprehensive antioxidant defense system. These enzymes have potential therapeutic applications in managing diseases related to oxidative stress, though further research is desired to fully know their roles and therapeutic potential [42, 62].

5 Role of Recombinant Enzymes in Various Disease Models

5.1 Cancer

Recombinant enzymes play a significant role in cancer treatment and research. For instance, enzymes like l-asparaginase are used in the treatment of leukemia by depleting asparagine, which is vital for the survival of cancer cells [12]. Additionally, proteases, including cysteine cathepsins, are involved in tumor progression and metastasis by degrading the extracellular matrix, thus facilitating cancer cell invasion [8, 87]. The BRCA tumor suppressor network, which involves recombinase enzymes, is crucial for repairing chromosome damage through homologous recombination, thereby preventing tumor development [101].

5.2 Cardiovascular Diseases

In cardiovascular diseases, recombinant enzymes such as alteplase, reteplase, and Tenecteplase are used to treat myocardial infarction by dissolving blood clots [12]. Paraoxonase enzymes, which have antioxidant properties, play a protective role against cardiovascular diseases by preventing oxidative stress-induced cellular injury [46]. Proteases are also involved in remodeling the extracellular matrix, which affects arterial structure and function, contributing to cardiovascular disease pathology [8].

5.3 Neurodegenerative Disorders (E.G., Alzheimer's, Parkinson's)

Recombinant enzymes have therapeutic potential in neurodegenerative disorders such as Alzheimer's and Parkinson's diseases. APE1, a multifunctional enzyme, is highly expressed in the central nervous system and has a role in maintaining tissue homeostasis, with implications in neurodegenerative and neuroinflammatory diseases [69]. Cysteine cathepsins are implicated in neurodegeneration by contributing to protein misfolding and aggregation, which are seals of diseases like Alzheimer's and Parkinson's [8, 87]. Additionally, paraoxonase enzymes are linked to the progression of neurological disorders due to their role in modulating oxidative stress and inflammation [46].

5.4 Clinical Applications of Recombinant Enzymes

5.4.1 Neurological Disorders

Recombinant enzymes have shown immense potential as therapeutic agents for the treatment of neurological conditions like AD and Parkinson's disease (PD). Because of their capacity to regulate neuroinflammation and promote the breakdown of abnormal protein aggregates, they may have neuroprotective effects.

5.4.2 Parkinson's Disease and Alzheimer's Disease

In PD, the accumulation of misfolded α-synuclein proteins leads to neuronal damage. Recombinant enzymes that target lysosomal pathways have shown promise in enhancing the degradation of these aggregates. For instance, studies have highlighted the role of lysosomal enzymes in mitigating neurodegeneration by promoting the clearance of toxic protein aggregates [103]. Furthermore, artificial enzymes, or "*cluster zymes,*" have been engineered to exhibit high catalytic activity and selectivity. These enzymes can modulate neuroinflammation by reducing peroxide levels in the brain, thereby offering potential therapeutic benefits for PD patients. AD is characterized by the accumulation of amyloid-beta plaques and tau tangles, leading to cognitive decline. Recombinant enzymes capable of degrading amyloid-beta peptides have been identified, offering potential therapeutic avenues. Additionally, enzymes involved in post-translational modifications, such as transglutaminases, have been implicated in the aggregation of proteins like amyloid-beta and tau. Modulating the activity of these enzymes may influence disease progression [85]. Furthermore, developments in molecular treatments, including gene editing and gene silencing, have the potential to target the root causes of AD. Such methods aim to either inhibit the expression of genes that trigger disease pathogenesis or fix genetic mutations [45].

5.4.3 Cardiovascular Diseases

Mechanisms of Oxidative Stress in Heart Disease

Oxidative stress plays a key role in the pathophysiology of heart disease by disrupting cellular function, promoting inflammation, and triggering apoptosis in cardiac cells. ROS are produced in the heart through mitochondrial dysfunction, NADPH oxidase activation, and ischemia–reperfusion injury, leading to damage to lipids, proteins, and DNA. The accumulation of ROS contributes to endothelial dysfunction, atherosclerosis, and myocardial infarction by impairing NO bioavailability and promoting inflammatory signaling pathways. Studies have demonstrated that oxidative stress-induced cardiac remodeling results in fibrosis, hypertrophy, and decreased cardiac contractility, which are hallmarks of heart failure [71].

Use of Recombinant Enzymes in Cardio Protection

Recombinant antioxidant enzymes have emerged as a promising therapeutic approach for mitigating oxidative stress-induced heart damage. Peroxiredoxin-6 and chimeric proteins with SOD and peroxidase activity have shown cardioprotective properties by maintaining myocardial contractility and reducing lipid peroxidation during oxidative stress [44]. Additionally, recombinant glutathione peroxidase 4 (GPx4) has demonstrated efficacy in protecting against myocardial ischemia by reducing infarct size, improving mitochondrial function, and preventing cardiomyocyte apoptosis via ROS scavenging mechanisms [51]. Recent advances have also explored gene treatment approaches using recombinant adeno-associated virus (AAV) vectors to deliver antioxidant genes. In a study utilizing AAV9 to deliver an anti-NF-κB p65 ribozyme, researchers observed reduced apoptosis and improved cell viability in oxidative stress-exposed cardiac cells, highlighting the potential of targeted gene therapy in cardio protection. These findings suggest that recombinant enzymes, either as direct therapeutic agents or via gene delivery, hold great promise in mitigating oxidative stress-related cardiac diseases.

5.4.4 Cancer Therapy

Targeting Oxidative Stress to Sensitize Cancer Cells

Oxidative stress is a hallmark of cancer cells, as they often exhibit raised levels of ROS because of metabolic dysregulation and genetic mutations. While moderate ROS levels promote tumor growth and survival, excessive oxidative stress can push cancer cells beyond their antioxidant defense capacity, leading to cell death. This vulnerability is being exploited as a therapeutic strategy to selectively sensitize cancer cells to treatment while sparing normal cells.

Mechanisms of Oxidative Stress in Cancer Therapy

Cancer cells rely on redox homeostasis to maintain their survival despite increased oxidative stress. However, disrupting this balance by enhancing ROS production or inhibiting antioxidant defenses can trigger oxidative damage and apoptosis. Research has shown that histone deacetylase (HDAC) inhibitors have a key role in adaptable oxidative stress pathways in cancer, making them a promising target for therapy. Studies on non-small cell lung cancer (NSCLC) and hepatocellular carcinoma (HCC) suggest that targeting HDACs can enhance oxidative stress and sensitize tumors to treatment [50]. Additionally, combinatorial therapies that simultaneously deplete antioxidants like GSH while increasing ROS generation have demonstrated strong anticancer effects. For example, the combination of redox-active compounds such as azines and copper (II) complexes has been found to synergistically disrupt redox homeostasis, effectively eliminating cancer cells with minimal toxicity to normal tissues [4].

Targeted Therapies Leveraging Oxidative Stress

Targeted therapies that promote oxidative stress have shown significant promise in hematologic malignancies. Several novel agents, including proteasome inhibitors and Bcl-2-targeted compounds, have been found to elevate intracellular ROS levels, contributing to cancer cell cytotoxicity. Understanding the redox-altering effects of these therapies may help design more effective multi-regimen strategies to overcome drug resistance in cancers such as leukemia and lymphoma [21]. In addition, new research highlights that cancer cells heavily depend on antioxidant enzyme systems to counteract oxidative damage. By targeting these enzymatic pathways, researchers aim to push cancer cells past their oxidative stress tolerance threshold, inducing selective tumor cell death. This approach has been suggested as a novel therapeutic avenue to impair cancer cell metabolism and progression [40].

5.5 *Clinical Implications and Future Directions*

The use of oxidative stress as a therapeutic strategy continues to gain traction, with increasing evidence supporting the efficacy of redox-modulating agents in sensitizing tumors to chemotherapy and radiation therapy. While challenges remain, such as ensuring selective targeting of cancer cells while minimizing harm to normal tissues, the development of ROS-enhancing drugs and antioxidant inhibitors holds important potential for improving cancer therapy outcomes. Future research will likely focus on optimizing combination therapies that exploit oxidative stress pathways while reducing side effects in clinical settings.

5.5.1 Recombinant Enzymes as Adjuncts in Cancer Treatment

Recombinant enzymes have emerged as promising adjunct therapies in cancer treatment by enhancing drug efficacy, targeting metabolic vulnerabilities, and reducing chemotherapy-related side effects. These engineered enzymes can be used in combination with existing cancer treatments to increase tumor sensitivity, improve drug delivery, and disrupt cancer cell metabolism.

5.5.2 Recombinant Enzymes in Gene Therapy

Gene therapy utilizing recombinant enzymes has shown significant potential in enhancing cancer treatment. A recombinant adenovirus expressing *Escherichia coli* cytosine deaminase was developed for breast cancer therapy, enabling the change of a nontoxic prodrug (5-fluorocytosine) into its cytotoxic form, 5-fluorouracil, directly within tumor cells. This enzyme/prodrug system effectively sensitized breast cancer cells and inhibited tumor growth in preclinical models, demonstrating the feasibility of enzyme-directed gene therapy in cancer treatment.

5.5.3 Recombinant Enzymes in Tumor-Targeting Strategies

Genetically engineered bacteria have been explored as tumor-targeting vectors capable of delivering recombinant enzymes directly to cancer cells. These bacterial systems can infiltrate tumor microenvironments and express therapeutic enzymes, improving localized treatment efficacy. The use of bacteria to carry recombinant enzyme-producing genes to tumors represents a novel approach to enhancing cancer therapies, particularly for difficult-to-treat solid tumors.

5.5.4 Amino Acid-Depleting Enzymes in Cancer Therapy

Certain cancers depend on specific amino acids for survival, making amino acid-depleting recombinant enzymes an effective therapeutic strategy. Recombinant arginine deiminase (rADI) has been investigated for its ability to deplete arginine, an essential amino acid for various tumors, including HCC and melanoma. However, resistance to rADI treatment has been observed in some cell lines due to the activity of endogenous arginosuccinate synthetase, highlighting the need for combination strategies to enhance efficacy.

5.5.5 Recombinant Enzymes in Cancer-Associated Anemia

Recombinant human erythropoietin (r-HuEPO) has been used to treat anemia in cancer patients undergoing chemotherapy. By stimulating red blood cell production, r-HuEPO helps mitigate fatigue and improve the quality of life in cancer-affected people. Clinical studies have shown that r-HuEPO reduces transfusion requirements and improves hematologic parameters in chemotherapy-treated patients, making it a valuable supportive therapy in oncology.

5.6 Future Directions in Recombinant Enzyme Therapy

As research continues, recombinant enzymes are being integrated into personalized cancer treatment strategies. Advances in enzyme engineering, gene therapy, and tumor-targeting delivery systems will further enhance their effectiveness in cancer therapy. The development of immune-evasive and tumor-specific enzyme variants may provide additional treatment options, reducing off-target effects while maximizing therapeutic potential.

5.6.1 Respiratory Diseases

Potential of Recombinant Enzymes in Lung Disease Management

Recombinant enzyme therapy has arisen as an auspicious approach for managing various lung diseases, including cystic fibrosis, chronic obstructive pulmonary disease (COPD), and genetic disorders such as alpha-1 antitrypsin (AAT) deficiency. By targeting underlying molecular dysfunctions, recombinant enzymes can help improve lung function, reduce disease progression, and enhance patients' quality of life.

Recombinant Human DNase (rhDNase) for Cystic Fibrosis

Cystic fibrosis (CF) is characterized by thick, sticky mucus in the lungs due to an accumulation of extracellular DNA from disintegrated inflammatory cells. Recombinant human DNase (rhDNase), also known as dornase alfa, enzymatically degrades extracellular DNA, reducing mucus viscosity and improving mucociliary clearance. Clinical studies have demonstrated that daily nebulization of rhDNase enhances lung function and decreases the frequency of pulmonary exacerbations in CF patients, making it a standard therapy for managing the disease [40].

Alpha-1 Antitrypsin Therapy for Lung Protection

Alpha-1 antitrypsin (AAT) deficiency is a genetic disorder that leads to unregulated neutrophil elastase activity, causing lung tissue destruction and predisposing individuals to emphysema. Recombinant human alpha-1 antitrypsin (rAAT) has been developed to augment lung defenses against neutrophil elastase. Studies in animal models have shown that rAAT administration significantly increases lung antielastase capacity, suggesting potential benefits for patients with AAT deficiency-related lung disease [18].

Recombinant Acid Sphingomyelinase for Niemann-Pick Disease

Niemann-Pick disease type B is a lysosomal storage disorder that results in sphingomyelin accumulation in the lungs, leading to respiratory complications. Recombinant acid sphingomyelinase (rhASM) therapy has demonstrated efficacy in clearing lysosomal sphingomyelin storage, reducing lung inflammation, and improving pulmonary function. Intranasal administration of rhASM in animal models has shown promising results, supporting its potential as a targeted therapy for lung involvement in Niemann–Pick disease [102].

Gene Therapy Approaches with Recombinant Enzymes

Gene therapy strategies utilizing recombinant enzymes have been explored for inherited lung diseases such as CF. Recombinant adenoviruses have been used to deliver therapeutic genes to lung epithelial cells, achieving high transgene expression and prolonged therapeutic effects. While early challenges included immune responses to viral vectors, recent advancements in recombinant adenovirus design have improved gene delivery efficiency and minimized inflammation, making this a viable approach for lung disease treatment [33].

5.7 *Other Potential Applications*

Recombinant antioxidant enzymes have shown significant potential in managing various diseases by reducing oxidative stress, which plays a critical role in autoimmune disorders, diabetes, and aging-related conditions. These enzymes help mitigate the harmful effects of ROS, regulate inflammatory pathways, and support cellular homeostasis.

5.7.1 Recombinant Antioxidant Enzymes in Autoimmune Diseases

Oxidative stress is a major contributor to autoimmune diseases such as rheumatoid arthritis, multiple sclerosis, and lupus. It promotes inflammation by triggering immune responses and damaging tissues. Enzymatic free radical scavengers, such as recombinant SOD and CAT, have been studied for their ability to reduce ROS accumulation and inflammatory cytokine release. Research suggests that enhancing the efficiency of these antioxidant enzymes using colloidal carriers, such as liposomes and nanoparticles, can improve their therapeutic efficacy in autoimmune disorders [86]. Additionally, recombinant Galectin-1, a glycan-binding protein with immunoregulatory properties, has been found to restore immune tolerance and suppress autoimmune inflammation in aging-related salivary gland autoimmunity. Treatment with recombinant Galectin-1 significantly reduced inflammation in an animal model of Sjögren's syndrome, highlighting its potential as a therapeutic intervention for age-related autoimmune diseases [58].

5.7.2 Recombinant Antioxidant Enzymes in Diabetes

Diabetes is characterized by increased oxidative stress due to chronic hyperglycaemia, leading to pancreatic beta-cell dysfunction and vascular complications. The antioxidative enzyme CAT plays a crucial role in diminishing the production of hydrogen peroxide, a highly toxic ROS that damages pancreatic cells. Studies indicate that recombinant antioxidant therapy could help reduce oxidative stress and improve insulin secretion in diabetic patients [67]. Gene therapy approaches using recombinant enzymes have also been explored in diabetes management. A study on type 1 diabetes (T1D) demonstrated that adeno-associated virus-mediated IL-10 gene therapy helped prevent autoimmune recurrence in islet transplantation by inducing the expression of antioxidant enzymes, such as manganese superoxide dismutase (Mn-SOD) and heme oxygenase-1 (HO-1). These findings suggest that recombinant enzyme-based gene therapies may improve graft survival and diabetes management [99].

5.7.3 Recombinant Antioxidant Enzymes in Aging-Related Conditions

Aging is associated with increased oxidative stress and a decline in the body's natural antioxidant defense mechanisms. This imbalance contributes to neurodegenerative diseases, cardiovascular disorders, and metabolic syndromes. Research on peroxiredoxin-6 (Prdx6), a multitask antioxidant enzyme with both peroxidase and phospholipase activities, suggests its involvement in protecting against age-related diseases. Knockout studies in mice have shown that the absence of Prdx6 accelerates oxidative damage and metabolic dysfunctions, underscoring its potential as a therapeutic target for aging-related conditions [70]. Furthermore, recombinant SOD fused with cell-penetrating peptides has been developed to enhance its permeability and therapeutic potential. This novel approach has demonstrated the ability to reduce oxidative damage in various tissues, including the brain, heart, and skin, making it a promising candidate for treating oxidative stress-related diseases, such as neurodegeneration, cardiovascular disorders, and premature aging [95].

6 Challenges and Limitations of Recombinant Enzyme Therapy

There are many challenges in the efficient administration of recombinant enzymes for therapeutic applications, especially with regard to pharmacokinetics and bioavailability.

6.1 Pharmacokinetics and Bioavailability

6.1.1 Enzyme Stability, Half-Life, and Tissue Penetration

Many enzymes in the physiological environment are inherently unstable, which is a major problem. This leads to a short half-life and rapid degradation, requiring regular administration of high doses. Achieving enough tissue penetration, particularly for target organs such as the brain, remains a significant challenge. Many therapeutic enzymes are significantly restricted from passing through the blood–brain barrier [27].

6.2 Challenges in Systemic Delivery

Intravenous administration, the most common delivery method, often leads to limited enzyme distribution to target tissues. Cellular uptake of enzymes can be inefficient, requiring high doses to reach therapeutic levels. The immune system's potential to recognize recombinant enzymes as foreign can trigger antibody production, neutralizing enzyme activity and causing adverse reactions.

6.3 Immune Response and Toxicity

Recombinant enzymes can trigger immune responses, leading to adverse effects that reduce efficacy and increase toxicity. Studies highlight the following concerns: Immunogenicity of recombinant enzymes remains a major challenge, as immune responses can lead to neutralizing antibodies, reducing therapeutic effectiveness [77]. Toxicity issues arise from enzyme instability, improper cellular uptake, and off-target effects [25].

6.4 Immunogenicity of Recombinant Enzymes

Recombinant enzymes used for enzyme replacement therapies (ERTs) often exhibit high immunogenicity, necessitating immunosuppressive treatments to maintain efficacy [77]. Structural modifications, such as PEGylation, can help reduce immunogenicity and improve half-life [25].

6.5 Cost-Effectiveness and Scalability

Recombinant human trypsin is significantly more expensive than its animal-derived counterpart, limiting its application [73]. Enzyme replacement therapies remain expensive, making them inaccessible for many patients, particularly in low-income regions [74]. Advances in molecular pharming (plant-based production) may provide a scalable and affordable solution for enzyme production [36]. Some recombinant enzyme therapies, such as recombinant Factor VIII for hemophilia A, have been found cost-effective under certain healthcare settings [16].

6.6 Regulatory Hurdles

The approval of recombinant enzyme therapies is complex, involving strict regulations: Regulatory agencies like the FDA and EMA impose extensive requirements for safety, efficacy, and manufacturing quality, slowing down approvals [26]. Gene and enzyme therapies face regulatory obstacles due to evolving guidelines and difficulties in defining clinical endpoints [75]. Differences in regulatory frameworks across regions add further challenges, leading to delays in commercialization [31].

7 Future Directions in Recombinant Enzyme Therapies

7.1 Advancements in Enzyme Engineering

The future of recombinant enzyme therapies lies in the continuous improvement of enzyme stability, specificity, and efficacy. Genetic modifications are being utilized to enhance enzyme performance under physiological and industrial conditions. Advanced protein engineering techniques such as rational design, directed evolution, and computational modeling have been employed to create enzymes with optimized catalytic activity and resistance to degradation [32]. Additionally, novel recombinant enzymes are being developed for diverse applications, ranging from drug metabolism enhancement to bioremediation. These engineered enzymes not only exhibit improved functionality but also contribute to the cost-effective and scalable production of enzyme-based therapies [53].

7.2 Gene Therapy Approaches

Recombinant enzyme therapy is being revolutionized by gene therapy strategies, allowing for the sustained, long-term production of therapeutic enzymes within the body. Targeted gene delivery systems, including viral and non-viral vectors, are being developed to enable precise gene insertion, leading to endogenous production of antioxidant enzymes to counteract oxidative stress [49]. Additionally, CRISPR and other gene-editing technologies hold promise for optimizing enzyme expression and reducing immune responses, ultimately improving treatment outcomes [61]. These advancements offer the potential for one-time, curative treatments, particularly for inherited metabolic disorders.

7.3 Nanotechnology in Enzyme Delivery

Nanotechnology is playing a crucial role in enhancing the delivery and stability of recombinant enzymes. Nano formulations, including liposomes, polymeric nanoparticles, and metal–organic frameworks, are being engineered to provide controlled and targeted enzyme release [93]. These nanocarriers improve enzyme bioavailability, reduce degradation, and enhance penetration into specific tissues. Furthermore, advancements in nanomedicine are enabling the precise localization of therapeutic enzymes, minimizing off-target effects and improving treatment efficiency [5].

7.4 Personalized Medicine

The future of recombinant enzyme therapies is increasingly shifting toward personalized medicine, where treatments are tailored to individual oxidative stress profiles. Advances in biomarker discovery are enabling patient stratification based on genetic and metabolic factors [88]. By integrating high-throughput screening and omics technologies, clinicians can determine the most effective enzyme therapy for each patient. This precision approach not only enhances therapeutic efficacy but also minimizes adverse reactions, making enzyme therapies safer and more efficient [92].

8 Conclusion

Recombinant enzyme therapies have emerged as a promising approach for managing oxidative stress-related disorders by supplementing or enhancing enzymatic activity within the body. Advances in protein engineering, gene therapy, and nanotechnology have significantly improved the stability, specificity, and delivery of recombinant enzymes. These therapies have been applied across various fields, including metabolic disorders, neurodegenerative diseases, and inflammatory conditions, providing a targeted approach to mitigate oxidative damage. Despite these advancements, challenges such as immunogenicity, high production costs, and regulatory complexities remain barriers to widespread clinical adoption. The integration of recombinant enzymes into modern medicine has the potential to revolutionize treatment paradigms for oxidative stress-related diseases. These enzymes can be utilized both as standalone treatments and in combination with other therapies to improve therapeutic efficiency. The growing use of gene-editing tools like CRISPR and targeted delivery mechanisms offers hope for long-term, patient-specific treatments. As enzyme therapies become more refined and personalized, their role in precision medicine is expected to expand, offering tailored treatments that align with individual patient profiles. While recombinant enzyme therapies hold immense potential, several challenges must still be addressed to enable their widespread adoption. Key areas for improvement include reducing immunogenicity, enhancing cost-effectiveness, and streamlining regulatory pathways to facilitate approvals. Advances in bioprocessing, synthetic biology, and computational modeling will play critical roles in overcoming these hurdles. The future of recombinant enzyme therapy is likely to be deeply intertwined with personalized medicine, where patient-specific biomarkers guide treatment choices for optimal outcomes. By addressing these challenges, recombinant enzyme therapies can become a cornerstone of modern medicine, offering safe, effective, and tailored treatments for oxidative stress-related diseases.

References

1. Adam-Vizi, V. (2005). Production of reactive oxygen species in brain mitochondria: Contribution by electron transport chain and non-electron transport chain sources. *Antioxidants & Redox Signaling, 7*(9–10), 1140–1149. https://doi.org/10.1089/ars.2005.7.1140
2. Adamyan, A. A., Glyantsev, S. P., Sakharov, I. Y., & Savvina, T. V. (1992). Experimental morphologic study of the action of collagenase of the crab Paralithodes camtschatica on wound healing. *Bulletin of Experimental Biology and Medicine, 114*(6), 1898–1901. https://doi.org/10.1007/BF00840491
3. Agostoni, A., Gerli, G. C., Beretta, L., Bianchi, M., Vignali, M., & Bombelli, F. (1980). Superoxide dismutase, catalase and glutathione peroxidase activities in maternal and cord blood erythrocytes. *Clinical Chemistry and Laboratory Medicine, 18*(11). https://doi.org/10.1515/cclm.1980.18.11.771
4. Akladios, F. N., Andrew, S. D., & Parkinson, C. J. (2015). Selective induction of oxidative stress in cancer cells via synergistic combinations of agents targeting redox homeostasis. *Bioorganic & Medicinal Chemistry, 23*(13), 3097–3104. https://doi.org/10.1016/j.bmc.2015.05.006
5. Ali, M., Ishqi, H. M., & Husain, Q. (2020). Enzyme engineering: Reshaping the biocatalytic functions. *Biotechnology and Bioengineering, 117*(6), 1877–1894. https://doi.org/10.1002/bit.27329
6. Aliev, G., Li, Y., H. Palacios, H., & E. Obrenovich, M. (2011). Oxidative stress induced mitochondrial DNA deletion as a hallmark for the drug development in the context of the cerebrovascular diseases. *Recent Patents on Cardiovascular Drug Discovery, 6*(3), 222–241. https://doi.org/10.2174/157489011797376942
7. Amstad, P., Moret, R., & Cerutti, P. (1994). Glutathione peroxidase compensates for the hypersensitivity of Cu,Zn-superoxide dismutase overproducers to oxidant stress. *Journal of Biological Chemistry, 269*(3), 1606–1609. https://doi.org/10.1016/S0021-9258(17)42068-0
8. Apoorva, O. S., Shukla, K., Khurana, A., & Chaudhary, N. (2024). Proteases: Role in various human diseases. *Current Pharmaceutical Biotechnology, 26*. https://doi.org/10.2174/0113892010316162240910103659
9. Arya, R., Bhattacharya, A., & Saini, K. S. (2008). *Dictyostelium discoideum*—a promising expression system for the production of eukaryotic proteins. *The FASEB Journal, 22*(12), 4055–4066. https://doi.org/10.1096/fj.08-110544
10. Aydin, H., Azimi, F. C., Cook, J. D., & Lee, J. E. (2012). A convenient and general expression platform for the production of secreted proteins from human cells. *Journal of Visualized Experiments, 65*. https://doi.org/10.3791/4041

11. Bagchi, M., Zafra-Stone, S., Bagchi, D., & Patel, S. (2007). Oxidative stress and neurodegeneration. In *Veterinary toxicology* (pp. 313–334). Elsevier. https://doi.org/10.1016/B978-012370467-2/50117-6
12. Baldo, B. A. (2015). Enzymes approved for human therapy: Indications mechanisms and adverse effects. *BioDrugs, 29*(1), 31–55. https://doi.org/10.1007/s40259-015-0116-7
13. Bannister, W. H., & Bannister, J. V. (1987). Factor analysis of the activities of superoxide dismutase, catalase and glutathione peroxidase in normal tissues and neoplastic cell lines. *Free Radical Research Communications, 4*(1), 1–13. https://doi.org/10.3109/10715768709088082
14. Battistelli, M., Malatesta, M., & Meschini, S. (2016). Oxidative stress to promote cell death or survival. *Oxidative Medicine and Cellular Longevity, 2016*(1). https://doi.org/10.1155/2016/2054650
15. Borzooee, F., & Larijani, M. (2019). Pichia pastoris as a host for production and isolation of mutagenic AID/APOBEC enzymes involved in cancer and immunity. *New Biotechnology, 51*, 67–79. https://doi.org/10.1016/j.nbt.2019.02.006
16. Bullement, A., McMordie, S. T., Hatswell, A. J., Li, N., & Wilson, K. (2020). Cost-effectiveness analysis of recombinant factor VIII Fc-Fusion protein (rFVIIIFc) for the treatment of severe Hemophilia A in Italy incorporating real-world dosing and joint health data. *PharmacoEconomics - Open, 4*(1), 133–142. https://doi.org/10.1007/s41669-019-0158-8
17. Cadaoas, J., Boyle, G., Jungles, S., Cullen, S., Vellard, M., Grubb, J. H., Jurecka, A., Sly, W., & Kakkis, E. (2020). Vestronidase alfa: Recombinant human β-glucuronidase as an enzyme replacement therapy for MPS VII. *Molecular Genetics and Metabolism, 130*(1), 65–76. https://doi.org/10.1016/j.ymgme.2020.02.009
18. Casolaro, M. A., Fells, G., Wewers, M., Pierce, J. E., Ogushi, F., Hubbard, R., Sellers, S., Forstrom, J., Lyons, D., Kawasaki, G., et al. (1987). Augmentation of lung antineutrophil elastase capacity with recombinant human alpha-1-antitrypsin. *Journal of Applied Physiology, 63*(5), 2015–2023. https://doi.org/10.1152/jappl.1987.63.5.2015
19. Chan, P. H. (2005). Mitochondrial dysfunction and oxidative stress as determinants of cell death/survival in stroke. *Annals of the New York Academy of Sciences, 1042*(1), 203–209. https://doi.org/10.1196/annals.1338.022
20. Chandanwale, A., Langade, D., Sonawane, D., & Gavai, P. (2017). A randomized, clinical trial to evaluate efficacy and tolerability of trypsin: Chymotrypsin as compared to serratiopeptidase and trypsin: Bromelain: Rutoside in wound management. *Advances in Therapy, 34*(1), 180–198. https://doi.org/10.1007/s12325-016-0444-0
21. Chandra, J. (2009). Oxidative stress by targeted agents promotes cytotoxicity in hematologic malignancies. *Antioxidants & Redox Signaling, 11*(5), 1123–1137. https://doi.org/10.1089/ars.2008.2302
22. Chen, C., & Ward, V. C. A. (2024). Recombinant protein expression and its biotechnological applications in Chlorella spp. *SynBio, 2*(2), 223–239. https://doi.org/10.3390/synbio2020013
23. Chen, Z., & Zhong, C. (2014). Oxidative stress in Alzheimer's disease. *Neuroscience Bulletin, 30*(2), 271–281. https://doi.org/10.1007/s12264-013-1423-y
24. Choi, K., Kim, J., Kim, G., & Choi, C. (2009). Oxidative stress-induced necrotic cell death via mitochondira-dependent burst of reactive oxygen species. *Current Neurovascular Research, 6*(4), 213–222. https://doi.org/10.2174/156720209789630375
25. Cioni, P., Gabellieri, E., Campanini, B., Bettati, S., & Raboni, S. (2022). Use of exogenous enzymes in human therapy: Approved drugs and potential applications. *Current Medicinal Chemistry, 29*(3), 411–452. https://doi.org/10.2174/0929867328666210713094722
26. Conlon, T. J., & Mavilio, F. (2018). The pharmacology of gene and cell therapy. *Molecular Therapy - Methods & Clinical Development, 8*, 181–182. https://doi.org/10.1016/j.omtm.2017.12.007
27. de la Fuente, M., Lombardero, L., Gómez-González, A., Solari, C., Angulo-Barturen, I., Acera, A., Vecino, E., Astigarraga, E., & Barreda-Gómez, G. (2021). Enzyme therapy: Current challenges and future perspectives. *International Journal of Molecular Sciences, 22*(17), 9181. https://doi.org/10.3390/ijms22179181

28. Desai, P. N., Shrivastava, N., & Padh, H. (2010). Production of heterologous proteins in plants: Strategies for optimal expression. *Biotechnology Advances, 28*(4), 427–435. https://doi.org/10.1016/j.biotechadv.2010.01.005
29. Desnick, R. J., & Schuchman, E. H. (2012). Enzyme replacement therapy for lysosomal diseases: Lessons from 20 years of experience and remaining challenges. *Annual Review of Genomics and Human Genetics, 13*(1), 307–335. https://doi.org/10.1146/annurev-genom-090711-163739
30. Dickson, P., McEntee, M., Vogler, C., Le, S., Levy, B., Peinovich, M., Hanson, S., Passage, M., & Kakkis, E. (2007). Intrathecal enzyme replacement therapy: Successful treatment of brain disease via the cerebrospinal fluid. *Molecular Genetics and Metabolism, 91*(1), 61–68. https://doi.org/10.1016/j.ymgme.2006.12.012
31. Drago, D., Foss-Campbell, B., Wonnacott, K., Barrett, D., & Ndu, A. (2021). Global regulatory progress in delivering on the promise of gene therapies for unmet medical needs. *Molecular Therapy - Methods & Clinical Development, 21*, 524–529. https://doi.org/10.1016/j.omtm.2021.04.001
32. Dubey, A., & Verma, A. K. (2019). Enzyme engineering for enzyme activity improvement. In *Enzymes in Food Biotechnology* (pp. 675–689). Elsevier. https://doi.org/10.1016/B978-0-12-813280-7.00039-6
33. Engelhardt, J. F., Litzky, L., & Wilson, J. M. (1994). Prolonged transgene expression in cotton rat lung with recombinant adenoviruses defective in E2a. *Human Gene Therapy, 5*(10), 1217–1229. https://doi.org/10.1089/hum.1994.5.10-1217
34. Fogarasi, E., Croitoru, M. D., Fülöp, I., & Muntean, D.-L. (2016). Is the oxidative stress really a disease? *Acta Medica Marisiensis, 62*(1), 112–120. https://doi.org/10.1515/amma-2015-0070
35. Gaj, T., Sirk, S. J., & Barbas, C. F. (2014). Expanding the scope of site-specific recombinases for genetic and metabolic engineering. *Biotechnology and Bioengineering, 111*(1), 1–15. https://doi.org/10.1002/bit.25096
36. Gomes, C., Oliveira, F., Isabel Vieira, S., & Sofia Duque, A. (2020). Prospects for the production of recombinant therapeutic proteins and peptides in plants: Special focus on Angiotensin I-Converting Enzyme Inhibitory (ACEI) peptides. In *Genetic Engineering—A Glimpse of Techniques and Applications*. IntechOpen. https://doi.org/10.5772/intechopen.84419
37. Grankvist, K., Marklund, S. L., & Täljedal, I. B. (1981). CuZn-superoxide dismutase, Mn-superoxide dismutase, catalase and glutathione peroxidase in pancreatic islets and other tissues in the mouse. *Biochemical Journal, 199*(2), 393–398. https://doi.org/10.1042/bj1990393
38. Guan, T., Song, J., Wang, Y., Guo, L., Yuan, L., Zhao, Y., Gao, Y., Lin, L., Wang, Y., & Wei, J. (2017). Expression and characterization of recombinant bifunctional enzymes with glutathione peroxidase and superoxide dismutase activities. *Free Radical Biology and Medicine, 110*, 188–195. https://doi.org/10.1016/j.freeradbiomed.2017.06.005
39. Gupta, S. K., & Shukla, P. (2016). Advanced technologies for improved expression of recombinant proteins in bacteria: Perspectives and applications. *Critical Reviews in Biotechnology, 36*(6), 1089–1098. https://doi.org/10.3109/07388551.2015.1084264
40. Hatem, E., & Azzi, S. (2016). Oxidative stress in carcinogenesis and therapy. *Journal of Cell Signaling, 01*(01). https://doi.org/10.4172/2576-1471.1000102
41. Heslop, C. L., Hill, J. S., & Frohlich, J. J. (2006). 251 Biomarkers of oxidative stress: Relative predictive value for cardiovascular mortality. *Journal of Investigative Medicine, 54*(1), S123–S123. https://doi.org/10.2310/6650.2005.X0004.250
42. Ighodaro, O. M., & Akinloye, O. A. (2018). First line defence antioxidants-superoxide dismutase (SOD), catalase (CAT) and glutathione peroxidase (GPX): Their fundamental role in the entire antioxidant defence grid. *Alexandria Journal of Medicine, 54*(4), 287–293. https://doi.org/10.1016/j.ajme.2017.09.001
43. John, J., Khemnar, B. M., Divekar, G. H., & Patil, N. (2022). The bromelain and rutoside advantage in systemic enzyme therapy: pharmacological basis of combination with trypsin. *International Journal of Basic & Clinical Pharmacology, 11*(6), 658. https://doi.org/10.18203/2319-2003.ijbcp20222608

44. Karaduleva, E. V., Mubarakshina, E. K., Sharapov, M. G., Volkova, A. E., Pimenov, O. Y., Ravin, V. K., Kokoz, Yu. M., & Novoselov, V. I. (2016). Cardioprotective effect of modified peroxiredoxins in retrograde perfusion of isolated rat heart under conditions of oxidative stress. *Bulletin of Experimental Biology and Medicine, 160*(5), 639–642. https://doi.org/10.1007/s10517-016-3237-1
45. Katta, M., Mathew, B. A., Chaturvedi, P., Ludhiadch, A., & Munshi, A. (2023). Advanced molecular therapies for neurological diseases: Focus on stroke, alzheimer's disease, and parkinson's disease. *Neurological Sciences, 44*(1), 19–36. https://doi.org/10.1007/s10072-022-06356-6
46. Khalaf, F. K., Connolly, J., Khatib-Shahidi, B., Albehadili, A., Tassavvor, I., Ranabothu, M., Eid, N., Dube, P., Khouri, S. J., Malhotra, D., Haller, S. T., & Kennedy, D. J. (2023). Paraoxonases at the heart of neurological disorders. *International Journal of Molecular Sciences, 24*(8), 6881. https://doi.org/10.3390/ijms24086881
47. Kishore, K., & Krishan, P. (2009). Pharmacology of recombinant or genetically engineered drugs. *Journal of Young Pharmacists, 1*(2), 141. https://doi.org/10.4103/0975-1483.55747
48. Kuwabara, M., Asanuma, T., Niwa, K., & Inanami, O. (2008). Regulation of cell survival and death signals induced by oxidative stress. *Journal of Clinical Biochemistry and Nutrition, 43*(2), 51–57. https://doi.org/10.3164/jcbn.2008045
49. Labrou, N. (Ed.). (2019). *Therapeutic enzymes: Function and clinical implications* (Vol. 1148). Springer Singapore. https://doi.org/10.1007/978-981-13-7709-9
50. Lawless, M. W., O'Byrne, K. J., & Gray, S. G. (2010). Targeting oxidative stress in cancer. *Expert Opinion on Therapeutic Targets, 14*(11), 1225–1245. https://doi.org/10.1517/14728222.2010.526933
51. Liu, C., Li, B., Yan, Q., Niu, S., Zhao, Y., Xiong, C., Zhang, T., & Wei, J. (2021). Protective effects and mechanisms of recombinant human glutathione peroxidase 4 on isoproterenol-induced myocardial ischemia injury. *Oxidative Medicine and Cellular Longevity, 2021*(1). https://doi.org/10.1155/2021/6632813
52. Liu, L., Yang, H., Shin, H., Chen, R. R., Li, J., Du, G., & Chen, J. (2013). How to achieve high-level expression of microbial enzymes. *Bioengineered, 4*(4), 212–223. https://doi.org/10.4161/bioe.24761
53. Liu, Q., Xun, G., & Feng, Y. (2019). The state-of-the-art strategies of protein engineering for enzyme stabilization. *Biotechnology Advances, 37*(4), 530–537. https://doi.org/10.1016/j.biotechadv.2018.10.011
54. Loft, S., & Møller, P. (2011). *Chapter 11. Biomarkers of exposure: Oxidative stress to DNA and lipids—Relation to air pollution* (pp. 160–173). https://doi.org/10.1039/9781849733540-00160
55. Mao, G. D., Thomas, P. D., Lopaschuk, G. D., & Poznansky, M. J. (1993a). Superoxide dismutase (SOD)-catalase conjugates. Role of hydrogen peroxide and the Fenton reaction in SOD toxicity. *Journal of Biological Chemistry, 268*(1), 416–420. https://doi.org/10.1016/S0021-9258(18)54167-3
56. Markesbery, W. R. (1999). The role of oxidative stress in Alzheimer disease. *Archives of Neurology, 56*(12), 1449. https://doi.org/10.1001/archneur.56.12.1449
57. Martindale, J. L., & Holbrook, N. J. (2002). Cellular response to oxidative stress: Signaling for suicide and survival*. *Journal of Cellular Physiology, 192*(1), 1–15. https://doi.org/10.1002/jcp.10119
58. Martínez Allo, V. C., Hauk, V., Sarbia, N., Pinto, N. A., Croci, D. O., Dalotto-Moreno, T., Morales, R. M., Gatto, S. G., Manselle Cocco, M. N., Stupirski, J. C., Deladoey, Á., Maronna, E., Marcaida, P., Durigan, V., Secco, A., Mamani, M., Dos Santos, A., Catalán Pellet, A., Pérez Leiros, C., & Toscano, M. A. (2020). Suppression of age-related salivary gland autoimmunity by glycosylation-dependent galectin-1-driven immune inhibitory circuits. *Proceedings of the National Academy of Sciences, 117*(12), 6630–6639. https://doi.org/10.1073/pnas.1922778117
59. Mavelli, I., Autuori, F., Dini, L., Spinedi, A., Ciriolo, M. R., & Rotilio, G. (1981). Correlation between superoxide dismutase, glutathione peroxidase and catalase in isolated rat hepatocytes

during fetal development. *Biochemical and Biophysical Research Communications, 102*(3), 911–916. https://doi.org/10.1016/0006-291X(81)91624-7
60. Mavelli, I., Ciriolo, M. R., Rotilio, G., De Sole, P., Castorino, M., & Stabile, A. (1982). Superoxide dismutase, glutathione peroxidase and catalase in oxidative hemolysis. A study of Fanconi's anemia erythrocytes. *Biochemical and Biophysical Research Communications, 106*(2), 286–290. https://doi.org/10.1016/0006-291X(82)91107-X
61. Mehrzad, R., & Liu, P. Y. (2018). Developing a gene therapy clinical trial: Tips and tricks. In *Gene therapy in reconstructive and regenerative surgery* (pp. 103–110). Springer International Publishing. https://doi.org/10.1007/978-3-319-78957-6_8
62. Michiels, C., Raes, M., Toussaint, O., & Remacle, J. (1994). Importance of SE-glutathione peroxidase, catalase, and CU/ZN-SOD for cell survival against oxidative stress. *Free Radical Biology and Medicine, 17*(3), 235–248. https://doi.org/10.1016/0891-5849(94)90079-5
63. Milatovic, D., Zaja-Milatovic, S., & Gupta, R. C. (2016). Oxidative stress and excitotoxicity. In *Nutraceuticals* (pp. 401–413). Elsevier. https://doi.org/10.1016/B978-0-12-802147-7.00029-2
64. Mohammadzadeh, R., Karbalaei, M., Soleimanpour, S., Mosavat, A., Rezaee, S. A., Ghazvini, K., & Farsiani, H. (2021a). Practical methods for expression of recombinant protein in the Pichia pastoris system. *Current Protocols, 1*(6). https://doi.org/10.1002/cpz1.155
65. Moreira, P. I., Smith, M. A., Zhu, X., Nunomura, A., Castellani, R. J., & Perry, G. (2005). Oxidative stress and neurodegeneration. *Annals of the New York Academy of Sciences, 1043*(1), 545–552. https://doi.org/10.1196/annals.1333.062
66. Navarro-Yepes, J., Burns, M., Anandhan, A., Khalimonchuk, O., del Razo, L. M., Quintanilla-Vega, B., Pappa, A., Panayiotidis, M. I., & Franco, R. (2014). Oxidative stress, redox signaling, and autophagy: Cell death versus survival. *Antioxidants & Redox Signaling, 21*(1), 66–85. https://doi.org/10.1089/ars.2014.5837
67. Nikolic, D., Stanimirovic, J., Bjelogrlic, P., & Isenovic, E. (2014). Oxidative stress and the role of antioxidative treatment in diabetes mellitus. *Oxidants and Antioxidants in Medical Science, 3*(1), 9. https://doi.org/10.5455/oams.070314.rv.014
68. Nosaki, S., Hoshikawa, K., Ezura, H., & Miura, K. (2021). Transient protein expression systems in plants and their applications. *Plant Biotechnology, 38*(3), 21.0610a. https://doi.org/10.5511/plantbiotechnology.21.0610a
69. Oliveira, T. T., Coutinho, L. G., de Oliveira, L. O. A., Timoteo, A. R., Farias, G. C., & Agnez-Lima, L. F. (2022). APE1/Ref-1 role in inflammation and immune response. *Frontiers in Immunology, 13*. https://doi.org/10.3389/fimmu.2022.793096
70. Pacifici, F., Della Morte, D., Capuani, B., Pastore, D., Bellia, A., Sbraccia, P., Di Daniele, N., Lauro, R., & Lauro, D. (2019). Peroxiredoxin6, a multitask antioxidant enzyme involved in the pathophysiology of chronic noncommunicable diseases. *Antioxidants & Redox Signaling, 30*(3), 399–414. https://doi.org/10.1089/ars.2017.7427
71. Park, E.-S., Kang, D.-H., Kang, J. C., Jang, Y. C., Lee, M.-J., Chung, H.-J., Yi, K. Y., Kim, D.-E., Kim, B., & Shin, H.-S. (2017). Cardioprotective effect of KR-33889, a novel PARP inhibitor, against oxidative stress-induced apoptosis in H9c2 cells and isolated rat hearts. *Archives of Pharmacal Research, 40*(5), 640–654. https://doi.org/10.1007/s12272-017-0912-3
72. Pigeolet, E., Corbisier, P., Houbion, A., Lambert, D., Michiels, C., Raes, M., Zachary, M.-D., & Remacle, J. (1990). Glutathione peroxidase, superoxide dismutase, and catalase inactivation by peroxides and oxygen derived free radicals. *Mechanisms of Ageing and Development, 51*(3), 283–297. https://doi.org/10.1016/0047-6374(90)90078-T
73. Ponomarenko, S. V. (2018). Economic barriers restraining the production of therapeutic recombinant human trypsin. Use of innovative technologies to overcome them. *PHARMACOECONOMICS. Modern Pharmacoeconomics and Pharmacoepidemiology, 11*(3), 58–66. https://doi.org/10.17749/2070-4909.2018.11.3-058-066
74. Puetz, J., & Wurm, F. M. (2019). Recombinant proteins for industrial versus pharmaceutical purposes: A review of process and pricing. *Processes, 7*(8), 476. https://doi.org/10.3390/pr7080476

75. Reape, K. Z., & High, K. A. (2023). Trial by "Firsts": Clinical trial design and regulatory considerations in the development and approval of the first AAV gene therapy product in the United States. *Cold Spring Harbor Perspectives in Medicine, 13*(5), Article a041312. https://doi.org/10.1101/cshperspect.a041312
76. Rosano, G. L., & Ceccarelli, E. A. (2014). Recombinant protein expression in microbial systems. *Frontiers in Microbiology, 5*. https://doi.org/10.3389/fmicb.2014.00341
77. Safary, A., Akbarzadeh Khiavi, M., Mousavi, R., Barar, J., & Rafi, M. A. (2018a). Enzyme replacement therapies: what is the best option? *BioImpacts, 8*(3), 153–157. https://doi.org/10.15171/bi.2018.17
78. Sayut, D. J., Kambam, P. K. R., Herrick, W. G., & Sun, L. (2014). Enzyme production in *Escherichia coli*. In *Manual of industrial microbiology and biotechnology* (pp. 539–548). ASM Press. https://doi.org/10.1128/9781555816827.ch37
79. Sevilla, F., Jiménez, A., & Lázaro, J. J. (2015). What do the plant mitochondrial antioxidant and redox systems have to say under salinity, drought, and extreme temperature? In *Reactive oxygen species and oxidative damage in plants under stress* (pp. 23–55). Springer International Publishing. https://doi.org/10.1007/978-3-319-20421-5_2
80. Sieprath, T., Corne, T. D. J., Willems, P. H. G. M., Koopman, W. J. H., & De Vos, W. H. (2016). *Integrated high-content quantification of intracellular ROS levels and mitochondrial morphofunction* (pp. 149–177). https://doi.org/10.1007/978-3-319-28549-8_6
81. Simonian, N. A., & Coyle, J. T. (1996). Oxidative stress in neurodegenerative diseases. *Annual Review of Pharmacology and Toxicology, 36*(1), 83–106. https://doi.org/10.1146/annurev.pa.36.040196.000503
82. Smallwood, M. J., Nissim, A., Knight, A. R., Whiteman, M., Haigh, R., & Winyard, P. G. (2018). Oxidative stress in autoimmune rheumatic diseases. *Free Radical Biology and Medicine, 125*, 3–14. https://doi.org/10.1016/j.freeradbiomed.2018.05.086
83. Sodoyer, R. (2004). Expression systems for the production of recombinant pharmaceuticals. *BioDrugs, 18*(1), 51–62. https://doi.org/10.2165/00063030-200418010-00005
84. Spataro, F., Carlucci, P., Loverre, T., Macchia, L., & Di Bona, D. (2023). Hypersensitivity reaction during enzyme replacement therapy in lysosomal storage disorders. A systematic review of desensitization strategies. *Pediatric Allergy and Immunology, 34*(6). https://doi.org/10.1111/pai.13981
85. Spencer, B., Verma, I., Desplats, P., Morvinski, D., Rockenstein, E., Adame, A., & Masliah, E. (2014). A neuroprotective brain-penetrating endopeptidase fusion protein ameliorates alzheimer disease pathology and restores neurogenesis. *Journal of Biological Chemistry, 289*(25), 17917–17931. https://doi.org/10.1074/jbc.M114.557439
86. Srivastava, S., Singh, D., Patel, S., & Singh, M. R. (2017). Role of enzymatic free radical scavengers in management of oxidative stress in autoimmune disorders. *International Journal of Biological Macromolecules, 101*, 502–517. https://doi.org/10.1016/j.ijbiomac.2017.03.100
87. Stoka, V., Vasiljeva, O., Nakanishi, H., & Turk, V. (2023). The role of cysteine protease cathepsins B, H, C, and X/Z in neurodegenerative diseases and cancer. *International Journal of Molecular Sciences, 24*(21), 15613. https://doi.org/10.3390/ijms242115613
88. Taipa, M. Â., Fernandes, P., & de Carvalho, C. C. C. R. (2019). *Production and purification of therapeutic enzymes* (pp. 1–24). https://doi.org/10.1007/978-981-13-7709-9_1
89. Tandon, S., Sharma, A., Singh, S., Sharma, S., & Sarma, S. J. (2021). Therapeutic enzymes: Discoveries, production and applications. *Journal of Drug Delivery Science and Technology, 63*, Article 102455. https://doi.org/10.1016/j.jddst.2021.102455
90. Tarallo, A., Damiano, C., Strollo, S., Minopoli, N., Indrieri, A., Polishchuk, E., Zappa, F., Nusco, E., Fecarotta, S., Porto, C., Coletta, M., Iacono, R., Moracci, M., Polishchuk, R., Medina, D. L., Imbimbo, P., Monti, D. M., De Matteis, M. A., & Parenti, G. (2021). Correction of oxidative stress enhances enzyme replacement therapy in Pompe disease. *EMBO Molecular Medicine, 13*(11). https://doi.org/10.15252/emmm.202114434
91. Tsutsui, H., Kinugawa, S., & Matsushima, S. (2011). Oxidative stress and heart failure. *American Journal of Physiology-Heart and Circulatory Physiology, 301*(6), H2181–H2190. https://doi.org/10.1152/ajpheart.00554.2011

92. Vanella, R., Kovacevic, G., Doffini, V., Fernández de Santaella, J., & Nash, M. A. (2022). High-throughput screening, next generation sequencing and machine learning: Advanced methods in enzyme engineering. *Chemical Communications, 58*(15), 2455–2467. https://doi.org/10.1039/D1CC04635G
93. da Silva, V., Amatto, I., Gonsales da Rosa-Garzon, N., de Oliveira, A., Simões, F., Santiago, F., da Silva, P., Leite, N., Raspante Martins, J., & Cabral, H. (2022). Enzyme engineering and its industrial applications. *Biotechnology and Applied Biochemistry, 69*(2), 389–409. https://doi.org/10.1002/bab.2117
94. Wan, W., Wang, D., Gao, X., & Hong, J. (2011). Expression of family 3 cellulose-binding module (CBM3) as an affinity tag for recombinant proteins in yeast. *Applied Microbiology and Biotechnology, 91*(3), 789–798. https://doi.org/10.1007/s00253-011-3373-5
95. Wang, X.-L., & Jiang, R.-W. (2022). Therapeutic potential of superoxide dismutase fused with cell-penetrating peptides in oxidative stress-related diseases. *Mini-Reviews in Medicinal Chemistry, 22*(17), 2287–2298. https://doi.org/10.2174/1389557522666220228150127
96. Xie, Y., Han, X., & Miao, Y. (2018). An effective recombinant protein expression and purification system in *Saccharomyces cerevisiae. Current Protocols in Molecular Biology, 123*(1). https://doi.org/10.1002/cpmb.62
97. Yan, L.-J. (2014). Positive oxidative stress in aging and aging-related disease tolerance. *Redox Biology, 2*, 165–169. https://doi.org/10.1016/j.redox.2014.01.002
98. Zelko, I. N., Mariani, T. J., & Folz, R. J. (2002). Superoxide dismutase multigene family: A comparison of the CuZn-SOD (SOD1), Mn-SOD (SOD2), and EC-SOD (SOD3) gene structures, evolution, and expression. *Free Radical Biology and Medicine, 33*(3), 337–349. https://doi.org/10.1016/S0891-5849(02)00905-X
99. Zhang, Y. C., Pileggi, A., Agarwal, A., Molano, R. D., Powers, M., Brusko, T., Wasserfall, C., Goudy, K., Zahr, E., Poggioli, R., Scott-Jorgensen, M., Campbell-Thompson, M., Crawford, J. M., Nick, H., Flotte, T., Ellis, T. M., Ricordi, C., Inverardi, L., & Atkinson, M. A. (2003). Adeno-associated virus-mediated IL-10 gene therapy inhibits diabetes recurrence in syngeneic islet cell transplantation of NOD mice. *Diabetes, 52*(3), 708–716. https://doi.org/10.2337/diabetes.52.3.708
100. Zhao, H., Chen, J., Chai, J., Zhang, Y., Yu, C., Pan, Z., Gao, P., Zong, C., Guan, Q., Fu, Y., & Liu, Y. (2017). Cytochrome P450 (CYP) epoxygenases as potential targets in the management of impaired diabetic wound healing. *Laboratory Investigation, 97*(7), 782–791. https://doi.org/10.1038/labinvest.2017.21
101. Zhao, W., Wiese, C., Kwon, Y., Hromas, R., & Sung, P. (2019). The BRCA tumor suppressor network in chromosome damage repair by homologous recombination. *Annual Review of Biochemistry, 88*(1), 221–245. https://doi.org/10.1146/annurev-biochem-013118-111058
102. Ziegler, R. J., Brown, C., Barbon, C. M., D'Angona, A. M., Schuchman, E. H., Andrews, L., Thurberg, B. L., McPherson, J. M., Karey, K. P., & Cheng, S. H. (2009). Pulmonary delivery of recombinant acid sphingomyelinase improves clearance of lysosomal sphingomyelin from the lungs of a murine model of Niemann-Pick disease. *Molecular Genetics and Metabolism, 97*(1), 35–42. https://doi.org/10.1016/j.ymgme.2009.01.008
103. Zunke, F. (2020). The function of lysosomes and their role in Parkinson's disease. *Neuroforum, 26*(1), 43–51. https://doi.org/10.1515/nf-2019-0035

MIX
Papier aus verantwortungsvollen Quellen
Paper from responsible sources
FSC® C105338
FSC
www.fsc.org

If you have any concerns about our products,
you can contact us on
ProductSafety@springernature.com

In case Publisher is established outside the EU,
the EU authorized representative is:
Springer Nature Customer Service Center GmbH
Europaplatz 3, 69115 Heidelberg, Germany

Printed by Libri Plureos GmbH
in Hamburg, Germany